A HISTORY OF SCIENCE

HELLENISTIC SCIENCE AND CULTURE IN THE LAST THREE CENTURIES B. C.

BY

GEORGE SARTON

The Norton Library
W · W · NORTON & COMPANY · INC ·
NEW YORK

FIRST PUBLISHED IN THE NORTON LIBRARY 1970
BY ARRANGEMENT WITH HARVARD UNIVERSITY PRESS

The companion to this volume, *A History of Science: Ancient Science Through the Golden Age of Greece,* is also available in a Norton Library edition (N525).

To Charles and Dorothea Waley Singer
in grateful remembrance
of forty years of friendship
and labor in the same
vineyard

PUBLISHED SIMULTANEOUSLY IN CANADA
BY GEORGE J. MCLEOD LIMITED, TORONTO

Books That Live
The Norton imprint on a book means that in the publisher's estimation it is a book not for a single season but for the years.
W. W. Norton & Company, Inc.

SBN 393 00526 7

PRINTED IN THE UNITED STATES OF AMERICA

2 3 4 5 6 7 8 9 0

PREFACE

This volume is entitled *Hellenistic science and culture*, in spite of the fact that it deals with Roman culture and Latin letters as well as with Greek letters and the culture of Eastern Europe, Egypt, and Western Asia. This is legitimate, because Hellenistic ideals were dominant everywhere – in science, arts, and letters. Even the Latin literature of that time was deriving its main sustenance and best inspiration from Greek models.

The gigantic personalities of Alexander the Great and Aristotle stood at the threshold of a new age. It was an age of unrest, wars, and rebellions, but it was also an age of scientific and artistic creation. Jealous fortune did not allow the first to live long; Alexander died in 323 at the age of thirty-three; the older man, Aristotle, was permitted to live longer and died in the following year at sixty-two. This was fortunate, for more time was needed to do his work than to conquer the world.

The book which I now offer to the reader is devoted to the three centuries that followed their deaths and preceded the birth of Christianity. It was a period of renaissance out of the ashes of pure Hellenism. During the first of those three centuries the main center was Alexandria; during the last two the leadership was shared by Alexandria with Pergamon, Rhodos, Antioch, and other Greek cities, and more and more as the years went by with Rome.

The Hellenistic world was international to a degree, polyglot and inspired by many religious faiths. The outstanding language was Greek, but the importance of Latin increased gradually with the success of Roman arms. Under Greek tutelage a fantastic mixture was allowed to brew, involving, in the first place, Hellenic and Roman materials, but also Egyptian, Jewish, Persian, Syrian, Anatolian ingredients, and dashes of many other kinds – Asiatic, Indian, African. During those three centuries, geometry, astronomy, anatomy, and grammar were established for ever; technology and medicine blossomed out. The creative work might be done in many places of Western Asia, Northern Africa, or Europe, but the leaven was always Greek.

Looking at it from a different angle, this age of renaissance and transition witnessed two gigantic struggles: first, the rivalry between Greek ideals, on the one side, and Asiatic or Egyptian ones, on the other; second, the rude impact of Rome upon both sides. Everything was in the crucible, including religion itself. In this context Greek ideals were pagan and the Hellenistic age witnessed their death struggle against Asiatic and Egyptian mysteries, on the one side, and against Judaism, on the other.

Let us move higher still and contemplate Hellenistic science in a deeper perspective. Modern science was set up during those three centuries upon so strong a basis because the Greek genius was being grafted upon the Roman body.[1] Inbreeding is always dangerous and often sterile; there was no physical inbreeding in those days and the creations of a genius that might be — and often was — unhealthy were preserved and protected by Roman vigor. The Augustan age was a political climax, a period of peace during which the physical and intellectual conquests of three centuries could be integrated and secured.

When one speaks of the culture of a nation, what does one mean? The creators of art and science are very few; they may be encouraged or discouraged by the public around them, or they may be left almost completely alone. If one wished to be more correct, the intensity of a national culture should be represented by two factors, the first symbolizing the general educational level, and the second the exceptional merit of a small elite of pioneers. The first of these factors would be a measurable quantity; [2] the second a potential very difficult to estimate. In ancient times, there was no public education, except that of the forum, the theater, and the street, and the general degree of illiteracy was very high. In arts and letters these shortcomings might be minimized by natural gifts. The number of people who could appreciate the beauty of a statue or enjoy a play was certainly far greater than the number who could take any interest in a geometric proposition, in a planetary theory, or even in a medical system. In short, the early men of science were left very much to themselves and such a phrase as "the scientific culture of Alexandria in the third century B.C." does not cover any reality. There were men of science, but we can hardly speak of a scientific culture. In a sense, this is still true today; the real pioneers are so far ahead of the crowd (even a very literate crowd) that they remain almost alone; yet they receive encouragements from academies and scientific societies, even as their ancient predecessors might have received them more capriciously from kings or from powerful individuals.

Nevertheless, one cannot help speaking of the scientific or artistic culture of this or that nation at a given time, and when I do so I beg the reader to remember that this is a convenient form of speech which must not be taken too literally.

Even if ancient men of science were few and lonely, we should bear in

[1] It is illuminating to compare this graft with a later one. The development of Arabic science in the ninth century was insured by the grafting of the Persian genius upon the Arabic body. G. Sarton, "Islamic science," in T. Cuyler Young, ed., *Near Eastern culture and society* (Princeton: Princeton University Press, 1951), p. 87. Such periodic grafts seem necessary to start human progress in new directions.

[2] In modern democracies, at least, it could be measured or estimated by the degree of literacy of the population, by the proportion of graduates from primary and secondary schools or from institutions of higher learning, and by other objective tests.

mind that the Hellenic people produced a relatively large share of them. Its scientific potential, we might say, was exceptionally high.

From my days as a student at the University of Ghent in Flanders, my life has been dominated by two passions – the love of science, or call it the love of rationality, and the love of humanities. It occurred to me very early that one could not live reasonably without science nor gracefully without arts and letters. All that I have done, this book included, has been done in order to satisfy those two passions, without which my life would have become meaningless in my own eyes. I hope to communicate them to the reader and make him feel as I do that Euclid, Hērophilos, and Archimēdēs were as heroic and as necessary to our happiness as Theocritos and Virgil.

Humanities are inseparable from human creations, whether these be philosophic, scientific, technical, or artistic and literary. They exist in everything to which men have imparted their virtues or vices, their joys or sufferings. There are blood and tears in geometry as well as in art, blood and tears but also innumerable joys, the purest that men can experience themselves or share with others. The sharing continues to this very day; the main purpose of this book is to extend it to my own friends.

It would be very foolish to claim that a good poem or a beautiful statue is more humanistic or more inspiring than a scientific discovery; it all depends upon the relation obtaining between them and you. Some people will be more deeply moved by poetry than by astronomy; it all depends upon their own experience, mind, and sensibility.

I shall necessarily devote much more space to ancient science than to ancient arts and letters but shall often refer to them, because we could never understand Hellenistic culture without their gracious presence.

When I began my *Introduction to the history of science* after the First World War, I fondly believed, in my innocence, that I would be able to carry it through to the beginning of our century. Therefore, I generally avoided references to the future of any event I was dealing with; it seemed to me that it sufficed to explain its causes, that is, its past, and to deal with its fruits, that is, its future, only when I came to it. In this book my policy must be different; I shall try to evaluate the greatness of each achievement and that cannot be done except by giving some account, however brief, of its tradition. "Ye shall know them by their fruits."

Only a small part of the past is known to us. Innumerable scientific manuscripts, as well as poems and works of art, have been created only to be lost. Many have been completely lost; others are known to us only indirectly or in fragments. Sometimes fate has been more generous and has permitted them to reach us in their integrity. The extant books and monuments are not necessarily better than the lost ones, but they are the only ones that we are able to appreciate and the only ones that belong to our inheritance. The *Iliad*, the *Elements* of Euclid, the Parthenōn have never ceased to influence good men and to encourage the creation of new masterpieces; men have never ceased to own them in the measure of their own virtue.

It is very important to replace each deed in its temporal and spatial environment, but that is not enough. In this book it will be my duty and purpose to explain not only the ancient achievements but also their transmission. How were they bequeathed to our ancestors and to ourselves? What were their vicissitudes? What did our ancestors think of them? The outstanding event in the tradition of each ancient writing was its first publication in printed form, for its survival and integrity could not be assured until then. Therefore, in spite of the fact that I am not primarily a bibliophile, I shall always indicate the princeps edition of each important book; the princeps was like a rebirth, a rebirth to eternal life. Without trying to give a full bibliography of each item, I shall mention next to the first edition, the best one, the most convenient for reference, and the first and best translations into English.

Though I am dealing mainly with ancient science, the account of its tradition will entail short digressions on the science and scholarship of the Middle Ages, the Renaissance, and later times. Though I am dealing almost exclusively with Western science, it will be necessary sometimes to explain its eastern repercussions with special emphasis on the Arabic and Hebrew writings which were at times very closely enmeshed with our own.[3]

The whole past and the whole world are alive in my heart, and I shall do my best to communicate their presence to my readers. A deed happens in a definite place at a definite time, but if it be sufficiently great and pregnant, its virtue radiates everywhere in time and space. We ourselves are living here and now, but if we are generous enough, we can stretch our souls everywhere and everywhen else. If we succeed in doing so, we shall discover that our present embraces the past and the future and that the whole world is our province. All men are our brothers. As far as the discovery of truth is concerned, they are all working with the same purpose; they may be separated by the accidents of space and time and by the exigencies of race, religion, nationality, and other groupings; from the point of view of eternity they are working together.

The history of science, being the history of the discoveries and inventions that man has completed by the application of reason to nature, is necessarily, to a large extent, the history of rationalism. Rationalism, however, implies irrationalism; the search for the truth implies a fight against errors and superstitions. This was not always clear; errors and even superstitions are

[3] Medieval and Oriental digressions will be necessarily brief, but references to my *Introduction* will enable inquisitive readers to extend them easily as far as they may wish.

relative. The growth of science entailed the gradual purification of its methods and even of its spirit. Men of science have made abundant mistakes of every kind; their knowledge has improved only because of their gradual abandonment of ancient errors, poor approximations, and premature conclusions. It is thus necessary to speak not only of temporary errors but also of superstitions, which are nothing but persistent errors, foolish beliefs, and irrational fears. Superstitions are infinite in number and scope, however, and we cannot do more than refer to some of them occasionally. It would not do to ignore them altogether, if only because we should never forget the weakness and fragility of our minds.

The consciousness that superstitions are rife in our own society is a healthy shock to our self-conceit and a warning. If I had to explain the astounding scientific discoveries of our own time, I would feel it my duty to refer to the superstitious twilight that surrounds us, but it would be wrong to insist too much on it. That consciousness is helpful in another way; it leads us to judge ancient superstitions with more indulgence and with a sense of humor. We could not overlook them without falsifying the general picture nor judge them too severely without hypocrisy.

Where is my audience? Whom did I have in mind while I was studying and cogitating? I am writing for historians of science, or, more generally, for men of science who are anxious to know the origins of their knowledge and of the amenities and privileges of their social life. Some critics have accused me of being an adversary to the philologists and to the professional humanists. Such a reproach is unwarranted, but I said and repeat that my book is not addressed so much to philologists as to people whose training was (like my own) scientific. Hence, I must add bits of information that would be superfluous for philologists. Fortunately, such information can be given briefly and I give it with special pleasure. It is much easier to say in a few words what the Muses and the Parcae are, to justify the phrase Berenice's Hair, or to describe a palimpsest or a pentathlon, than to explain the solution of spherical triangles, the asymptotes or the evolutes of a conic, or the theory of epicycles. As far as scientific matters are concerned, I try to say enough to refresh the reader's memory but do not attempt to provide complete explanations, which would be equally unbearable to those who know and to those who don't.

As the whole of Hellenistic science and culture must be covered in a book of moderate length, which should enlighten the reader without overwhelming him, it is obvious that the author cannot deal with every part of the subject nor give every detail about each part. If the book were devoted exclusively to Apollōnios or to Lucretius it would be my duty to omit nothing concerning them, but I am obliged to deal with hundreds of men and to make them live without killing the reader.

The main difficulty of a synthesis lies in the choice of subjects. I have taken considerable pains to choose as well as possible the stories that I was to tell and the details of each of them. It is impossible to narrate the history of ancient science completely, but I have tried to be as comprehensive as my frame allowed and to offer the essential.

The division of the whole book into chapters treating separate fields was necessary for the sake of clearness, but it involves unavoidable repetitions because the men of the Hellenistic age were less specialized than those of our own century. The mathematicians might be also astronomers, mechanicians, or geographers. Hence, some great men reappear in many chapters. I have tried to tell the main story of each man in one chapter and to reintroduce him as briefly as possible whenever his encyclopedic tendencies made this necessary.

Some repetitions have been left; they are deliberate. There are fewer of them in this book than there were in my Harvard lectures; indeed, they are less necessary for the reader, who can refer at any time to any part of the book, than for the listener who had no such facilities (no table of contents and no index). Moreover, the lectures were spaced over half a year, while the reader is free to regulate his own speed.

The illustrations of this book have been carefully chosen to complete the text and to introduce the kind of intuitive precision that only graphic means make possible. The meaning, source, and genuineness of each illustration are explained in the legend attached to it; indeed, an illustration is worthless without such explanation. There are no portraits, for, as I have often repeated, ancient portraits are symbolic images without any immediate relation to the individuals represented; [4] they are not portraits as we understand them. The "Aristotle" heads of Vienna and Naples (very different but equally improbable), the "Epicuros" of New York, the "Menandros" of Boston, and others are not even "ideal" portraits from the sculptor's point of view but "ideal" attributions by scholars of the Roman age, of the Renaissance, or even later. The Naples "Aristotle" was first baptized "Solōn"; it was called "Solōn" by Schefold [5] as late as 1943; then it occurred to some bright archaeologist who had often seen Solōn and Aristotle in his dreams that that head looked more like the latter than the former! And there you are – at the moment a new "Aristotle" was born.

It is remarkable that philologists who are capable of carrying accuracy to

[4] With the possible exception of kings like Alexander who had sculptors or painters in attendance. G. Sarton, "Iconographic honesty," *Isis 30*, 222–235 (1939); "Portraits of ancient men of science," *Lychnos* (Uppsala, 1945), pp. 249–256, 1 fig.; *Horus* (42–43).

[5] Richard Delbrück, *Antike Porträts* (*Tabulae in usum scholarum*, ed. Johannes Lietzmann, 6; Bonn: Marcus and Weber, 1912); Anton Hekler, *Bildnisse berühmten Griechen* (Berlin, 1939); Karl Schefold, *Die Bildnisse der antiken Dichter, Redner und Denker* (Basel: Schwabe, 1943).

pedantic extremes in the case of words are as credulous as babies when it comes to "images," and yet an image is so full of information that ten thousand words would not add up to it. The prize specimen of such iconographic inconscience was given by Studniczka,[6] who clinched his argument to prove the genuineness of the Vienna "Aristotle" by remarking that Aristotle was an "Urgemaner" and did not the Vienna head have some resemblance to Melanchthon and Helmholtz? Ergo it must be Aristotle! [7]

The rank and file of philologists are sure that the Vienna head is a faithful portrait of Aristotle, for has not that been proved up to the hilt in Studniczka's memoir? They have not always read it, but they know of it and its very existence gives validity to the Vienna head, even as the gold stócked in Fort Knox bolsters up our bank notes.

The cause of such perversion is a weakness deeply rooted in human nature. Men want to have the likenesses of their greatest benefactors in order to be closer to them and to show their gratitude. Patricians of the Hellenistic age wanted to surround themselves with the busts of Homer, Sophoclēs, Plato, or Aristotle, even as the priests wanted statues of Apollōn and Aphroditē in their temples. Their wishes were satisfied. Similar wishes were made with the same intensity during the Renaissance and more statues were provided, some of which were Hellenistic or Roman, others brand-new. The whole iconography of ancient science is simply the fruit of wishful thinking.

To conclude, an "ancient portrait" of Euclid or Archimēdēs should be taken in the same spirit as we take a portrait of Isis, Asclēpios, or St. George.

Apart from explanatory diagrams, my illustrations represent ancient monuments and pages of old books, especially the title pages of the first Renaissance editions. There are no antiquities more impressive than the principes of the great classics. I would be grateful to the reader if he would examine them with attention and sympathy (almost every title page contains some curious information not given in my own text). These glorious pages will serve to illustrate not only antiquity but also the history of scholarship, the history of science during the Renaissance and later.

My sources are primarily the ancient writings and the ancient commentaries. Full use has been made of other histories, of many more of them than my references would suggest. In order to lighten my footnotes, I have generally avoided commonplace references, especially those that can be easily found in my *Introduction*. On the other hand, whenever I have drawn information from a newer publication, I have been careful to give its full title.

[6] Franz Studniczka (1860–1929), *Ein Bildnis des Aristoteles* (55 pp., 3 pls.; Leipzig: Edelmann, 1908). Not to be confused with another writing having almost the same title, *Das Bildnis des Aristoteles*, same author, same publisher, same year, but shorter — 35 pp., 3 pls.

[7] I simplify and exaggerate. Studniczka's astute suggestions were not offered by him as formal proofs, but were swallowed as such by credulous readers.

The reader is thus enabled to continue my investigations (and perhaps finally to reverse my judgment), if he is sufficiently interested to do so.

Irrespective of sources and texts, which can be named, some forty years of experience in my field as a scholar and a teacher have given me great confidence mixed with greater humility.

In many cases I have used previous writings of my own, and even used the same terms when I could not improve upon them, without bothering to quote myself explicitly. The chapter on Euclid was very largely derived, with kind permission, from one of my Montgomery Lectures at the University of Nebraska,[8] and the chapter on Hipparchos from my own article in the *Encyclopaedia Britannica.*[9]

My first teachers have been listed in the preface to Volume I[10] (p. xiv) and my gratitude to them grows as I myself grow older. I owe many thanks also to my friends of the History of Science Society and of the International Academy of the History of Science. It would take too long to enumerate them. It must suffice to name a few who died recently: in 1953, the physicist Henry Crew of Evanston, Illinois; in 1954, the mathematician Gino Loria of Genoa, the Semitist Solomon Gandz of Philadelphia, the historian Henri Berr and the mathematician Pierre Sergescu both of Paris; in 1955, the physician Max Neuburger of Vienna, the mathematician Raymond Clare Archibald of Providence, Rhode Island, the historian of science Adnan Adivar of Istanbul. They are all still alive in my heart.

My gratitude to the Harvard Library has been expressed many times and I must thank its officers once more, particularly Professor William Alexander Jackson, keeper of rare books. I owe very much to the late Professor Herbert Weir Smyth (1857–1937), thanks to whose generosity the Harvard Library is very rich in early Greek books. Help has been received from other libraries, notably the Boston Medical Library (Dr. Henry R. Viets), the Armed Forces Medical Library in Cleveland, Ohio (William Jerome Wilson, Dorothy M. Schullian), the New York Academy of Medicine (Janet Doe), the Yale Medical Library in New Haven, Connecticut (John F. Fulton, Madeline Stanton), the Pierpont Morgan Library in New York (Curt F. Bühler), the Henry E. Huntington Library in San Marino, California, the Library of Congress in Washington, D.C., the Princeton University Library in New Jersey, the Laurentian Library in Florence, the British Museum in London, the Bibliothèque Nationale in Paris, the John Rylands Library in Manchester, England, and the University Library in Cambridge, England.

I am indebted also to various museums, notably the William Hayes Fogg

[8] G. Sarton, *Ancient science and modern civilization* (Lincoln: University of Nebraska Press, 1954), pp. 3–36.

[9] *Encyclopaedia Britannica,* vol. 11, pp. 583–583B (1947).

[10] G. Sarton, *A history of science: Ancient science through the golden age of Greece* (Cambridge: Harvard University Press, 1952), hereafter referred to as Volume 1.

Art Museum of Harvard University, the Boston Museum of Fine Arts, the Metropolitan Museum of Art in New York, the National Gallery in Washington, D.C., the Vatican Museum in Rome, and the Museo Nazionale in Naples. I hope this list is complete; at any rate each indebtedness is acknowledged in its proper place.

Finally, I renew my thanks to the American Philosophical Society of Philadelphia for a grant-in-aid awarded to me on 13 October 1952.

George Sarton

Christmas 1955

NOTES ON THE USE OF THIS BOOK[11]

Chronology. Indications such as (III-1 B.C.) or (IV-1) after a name mean two things: first, that that person flourished in the first half of the third century B.C. or in the first half of the fourth century after Christ, and second, that a section was devoted to him in my *Introduction*, where information concerning him and bibliography may be found. When a person was not dealt with in my *Introduction*, his time is indicated in a different way, for example, Lysippos (fl. c. 328 B.C.), Terentius (c. 195–159). In the second case, it is not necessary to add B.C. Double dates are generally unambiguous; if we write X (175–125) and Y (125–175), it is clear enough that X flourished before Christ and Y after him. In Part I of this book, covering the third century, the letters B.C. are generally left out; in Part II, covering the second and first centuries, it was sometimes necessary to add them, and the closer one comes to the end of the pre-Christian age the more necessary it was. For example, Livy the historian was born in 59 and died in 17; it is essential to write (59 B.C.–A.D. 17); he might have died in 17 B.C., at the age of 42, instead of in A.D. 17 at the age of 75.

Geography. I am just as anxious to indicate where an event happened or a man lived as to indicate when. In the past even as now the same place names were often used in different regions. Many places were named Alexandria, Antiocheia, Berenicē, Neapolis (Newtown), Tripolis (Threetown). The reader is always told (as far as possible) which place was meant and what was its relation to better-known places in the neighborhood. For example, it does not satisfy me to say that Polybios hailed from Megalopolis and Strabōn from Amaseia, for does the reader know where those towns were located? Probably not. Therefore, I am careful to add that Megalopolis was in Arcadia in the center of the Peloponnēsos, and Amaseia south of the middle part of the Black Sea on the river Iris (Yeşil Irmak). If possible, I add also some details which will evoke the place more vividly and fix it in the reader's memory. I want him to visualize the place as well as to sense the time.

The names of regions, countries, cities, and physical features have changed

[11] Some of the notes printed in Volume 1, pp. xv–xvii, are not repeated.

repeatedly throughout the ages. In Western Asia, the same places may have names in Assyrian, Greek, Hebrew, Arabic, Syriac, Persian, Turkish, Latin (with possible variants in each language). I have often preferred for the reader's convenience to use a modern term, such as Dardanelles instead of Hellēspontos, or Red Sea instead of *Erythra thalassa*. I have also preferred to write Western Asia or a longer periphrase rather than a weasel expression like Near East (near to what?).

References. When mentioning a statement that occurs in a classical text, I generally do not refer to a definite edition (which may be out of the reader's grasp) but rather to book and chapter (say XII, 7) or to the ancient pagination reproduced in every scholarly edition. For example, the pagination of the Greek text of Plato by Henri Estienne (Paris, 1578), or Immanuel Bekker's pagination of the Greek Aristotle (Berlin, 1831), are canonized and available to every reader. Direct quotations of ancient texts have been restricted to the minimum and given in English;[12] scholars wishing to have the Greek (or Latin) original can find it easily enough.

Transcription of Greek type. As the cost of printing Greek type has become prohibitive, it is necessary not only to transliterate Greek words but also to transliterate them exactly. This vexed me at the beginning, but I am now reconciled to transliterations, because I can see their advantages. A word written in Greek type is more pleasant to the Hellenist than its transliteration, but it is enigmatic to non-Hellenists; the exact transliteration is equally clear to everybody. Greek words will be transliterated in the same way that we transcribe Sanskrit or Arabic ones; there is no loss.[13]

The only way to realize an exact transliteration is to transliterate each Greek letter by the same Roman letter (or the same combination of Roman letters). In other words, the transliteration must be adapted to the script, not to the pronunciation. The original spelling of each word is relatively stable (it has remained the same for over two thousand years), while its sound never ceases to vary from time to time and from place to place. To try to reproduce exactly the pronunciation of words is to follow a will-o'-the-wisp.

The Greek alphabet is transliterated as follows: *a*, *b* (not *v*), *g*, *d*, *e*, *z*, *ē*, *th*, *i*, *c*, *l*, *m*, *n*, *x*, *o*, *p*, *r* or *rh* (initial *rhō*), *s*, *t*, *y*, *ph*, *ch*, *ps*, *ō*.

The diphthongs ending in *i* (*ai*, *ei*, *oi*) are written as in Greek (not *ae*, *i*, *oe*, Latinwise). The *iota subscriptum* is left out. The diphthong *ou* is written *u*, for it has always been pronounced like *u* in English (as in *full* or *bull*) or in German (French *ou*). The other diphthongs ending in upsilon are kept as

[12] A few brief quotations of Latin verse or prose have been given in the original as well as in English.

[13] There is no important loss. I have not tried to reproduce the *iota subscriptum*, though it remained *adscriptum* till the thirteenth century, and have not indicated the Greek accents, which would have made printing too complex, especially when accents fell upon ē or ō. If one wishes to transliterate exactly Hebrew and Arabic words, greater difficulties occur; yet the English form is preferable, for it does not lock out the average reader.

they are, except when the upsilon occurs between two vowels; it is better then to consonantize it as in *evergetēs* (benefactor) *evagōgos* (docile) *evornis* (auspicious), *avos* (dry).

The letter *gamma* before another *gamma*, or before *c*, *ch*, *x*, is generally nasalized, and we transliterate it *n*. Thus, we shall write *angelos* (not *aggelos*, angel), *encephalos* (not *egcephalos*, brain), *enchelys* (not *egchelys*, anguilla, eel), *encyclos* (not *egcyclos*, circular).

The ending *-os* of many names has not been changed into *-us* as the Latin-speaking people did (Epicuros, not Epicurus).

Scholars of the Renaissance writing in Latin had some justification for Latinizing Greek words; we have none when writing in English. To write Greek words Latinwise is just as silly as to write Chinese in Japanese style. We are neither Romans nor Japanese; why should we imitate their mannerism in English spelling?

The English form Ptolemy is used for the astronomer of international fame, while we spell the royal names Ptolemaios. It is the more necessary to do so because the second names of kings are obviously Greek. It is better to avoid such bastard combinations as "Ptolemy Sōtēr," and write Ptolemaios Sōtēr or Philadelphos, Evergētes, Philopatōr, Philomētōr, Epiphanēs.

It is better to keep the final *n* in such names as Hērōn, Apollōn, Manethōn, but long usage makes it impossible to write Platōn instead of Plato. This suggests other inconsistencies which cannot be completely avoided without pedantry.[14]

No transcription is absolutely equal to the original script, but we have done our best within limits. The main point is to help the reader realize the difference between Greek and Latin and not to confuse the two languages. This is particularly useful when we wish to avoid confusion between Greek writers, such as Sallustios, Celsos, Phaidros, and Latin ones called Sallustius, Celsus, Phaedrus.

I have taken great pains to be consistent without pedantry and to guide the reader, as was my duty, without irritating him. I trust that he will meet me half-way.

PUBLISHER'S NOTE — The publishers acknowledge with thanks Miss May Sarton's generous cooperation during the preparation of the manuscript of this book for publication. They are also grateful to Duane H. D. Roller and C. Doris Hellman for observing and correcting a number of slips and oversights, to I. Bernard Cohen for assisting in the location of sources for checking references, and to Edward Grant for his careful reading of the proofs and making of the index.

[14] For example, the OCD writes Poseidon, and on the following page Posidonios! Many other inconsistencies occur in that carefully edited dictionary.

CONTENTS

CONTENTS

FOREWORD

George Sarton originally conceived his *History of Science* in eight or possibly nine volumes, to present the growth of science and scientific activity from its earliest beginnings to the present. At the time of his death, on 22 March 1956, he had completed this second volume, had checked and revised the typescript, and had chosen the illustrations.

The reader of the following pages will be impressed, as is always the case in Sarton's writings, with the breadth of Sarton's interests, the profundity of his understanding, and the ability to evoke the whole culture of a past age. Here is to be found the poet Virgil as well as the mathematician Archimedes, the subject of the fine arts along with the technical aspects of science. The text thus emphasizes Sarton's conception of the history of science as more than "simply an account of discoveries." Its purpose, he wrote, "is to explain the development of the scientific spirit, the history of man's reactions to truth, the history of the gradual revelation of truth, the history of the gradual liberation of our minds from darkness and prejudice."[1] He expressed this point of view in another way, as follows: "It is true that most men of letters and, I am sorry to add, not a few scientists, know science only by its material achievements, but ignore its spirit and see neither its internal beauty nor the beauty it extracts continually from the bosom of nature. Now I would say that to find in the works of science of the past, that which is not and cannot be superseded, is perhaps the most important part of our own quest. A true humanist must know the life of science as he knows the life of art and the life of religion." Sarton's wide range of interest was epitomized in the epigram from Terence which graces the first volume of the monumental *Introduction to the History of Science* (Carnegie Institution of Washington, 1927, 1931, 1947–48): "Homo sum, humani nihil a me alienum puto." And he held that the ultimate aim of the books in the present series was "to show the growth of the human spirit in its natural background."

It is precisely this set of qualities that makes for difficulty in planning a continuation of this series. Volume Three had never achieved the final state of planning so that the author's intentions for it can be known only in the most general way; the same is true for the later volumes. Harvard University Press is, however, making plans to continue this series, even though it is obvious that the work of others will be different in almost every way from what Dr. Sarton produced. The end result, it may be hoped, will nevertheless be what he wished: to illuminate "the progress of mankind"

[1] This and the following quotations are taken from passages reprinted in the George Sarton Memorial Issue of *Isis*, containing a bibliography of his writings, published by the History of Science Society (September 1957, vol. 48, pt. 3).

by focussing attention on the growth of "systematic positive knowledge, or what has been taken as such at different ages and in different places." On this score he further said: "I am not prepared to say that this development is more important than any other aspect of intellectual progress, for example, than the development of religion, of art, or of social justice. But it is equally important; and no history of civilization can be tolerably complete which does not give considerable space to the explanation of scientific progress."

The work that follows is presented to the public just as the author left it, save for the correction of some obvious typographical errors and slips of the pen. The proofs were read by Dr. Edward Grant, Instructor in the History of Science in Harvard University, who has also made the index.

I. Bernard Cohen

PART ONE

THE THIRD CENTURY

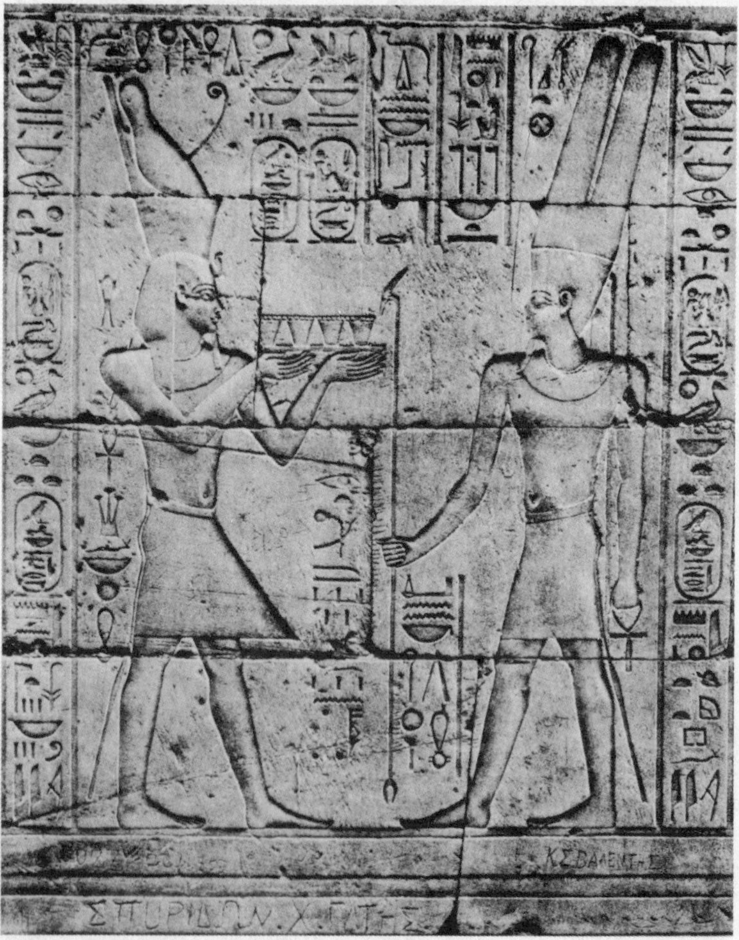

Fig. 1. Alexander the Great making offerings to the god Amon-Rē (Zeus-Ammōn). Alexander at the left is dressed like a Pharaoh and wears the double crown of Upper and Lower Egypt. He carries in both hands a tray holding four cups. The god on the right holds the *uas* scepter in his right hand and the symbol of life hangs from his left. This bas-relief is in the Temple of Luxor, of which Alexander had ordered the restoration; it probably dates from the end of the fourth century or the beginning of the third century B.C. The temple itself dates from the time of Amenhotep III (ruled 1411–1375). [Photograph borrowed from Friedrich Wilhelm von Bissing, *Denkmäler ägyptischer Sculptur* (Munich, 1914), pl. 114.]

I

THE ALEXANDRIAN RENAISSANCE

THE DISINTEGRATION OF ALEXANDER'S EMPIRE

The decadence and fall of Greece was completed by the Macedonian conquest; the battle of Chairōneia, won by Philip II in August 338, put an end to its independence. Two years later Philip was murdered and replaced by his son, Alexander III. This Alexander conquered a great part of the world within twelve years, from 334 to his death in 323, at the ripe age of thirty-three. The repercussions of this were very deep. The Alexandrian conquests put an end to the old Hellenism, but began a new period in history, the so-called Hellenistic age, which lasted three centuries, from, say, 330 B.C. to the establishment of the Roman Empire by Augustus in 30 B.C.

Alexander the Great closed an age and opened a new one; he created an empire which was universal and international, uniting under the Macedonian yoke peoples of many races, colors, languages, and religions, but the supreme culture and the supreme language were Greek. As Alexander's armies were Macedonian and Greek, he carried Greek culture into the very heart of Asia; it has been said that he Hellenized Western Asia,[1] but that statement must be qualified in many ways. For one thing, not only had Western Asia been Hellenized before him, but its western edge had been the very cradle of Greek science. Moreover, Alexander had not been dreaming only of a world empire but of a deeper kind of unity (*homonoia*, *concordia*). He was the first man, ahead of the Stoics and much ahead of the Christians, to think of the brotherhood of man.[2] For that reason he fully deserves to be immortalized under the name Alexander the Great. As he was not a pure Greek himself, but a Hellenized barbarian, it was easier for him than, say, for Plato to conceive such brotherhood and the fusion of races that it implied. He set the example, in 327, by marrying the Bactrian princess Roxana (Rhōxanē); [3] two years later, in Susa, he assigned to about

[1] Pierre Jouguet, *L'impérialisme macédonien et l'hellénisation de l'Orient* (Paris, 1926; English translation, London, 1928).

[2] Brotherhood, but with slavery! We should not judge him too severely, however, because that flagitious institution was still praised by good men in the United States a century ago and the Civil War (1861–1865) had to be fought in order to eradicate it.

[3] She fell into his hands when he captured a fortified place in Sogdianē, be-

eighty of his generals Asiatic wives whom he had richly dowered. He took a second wife, Barsinē, eldest daughter of Darios III, the last king of Persia, and perhaps a third, Parysatis, daughter of Artaxerxēs III Ōchos. Soon after Alexander's death, Barsinē was murdered by Roxana.

As to the Greek soldiers, camp followers, and settlers of every kind, they needed no persuasion to take native girls as wives or concubines. One should not exaggerate the importance of such blood mixtures because, however frequent they might be, they could affect only an infinitesimal part of the population.

There never were enough Greeks to Hellenize Egypt and Western Asia. Greece lost a large proportion of her most enterprising citizens and yet that Greek contingent was lost in the sea of Egyptian and Asiatic humanity. In spite of their cultural superiority, the Greeks could not help being submerged, and their orientalization was the unavoidable result. The influence of Asiatic wives and mothers was overwhelming in certain fields, such as folklore and religion. Therefore, one might claim that the Alexandrian empire helped to orientalize Eastern Europe. Instead of speaking of the Hellenization of Asia or the orientalization of Europe, it is safer to say that East and West were brought together, and in that region – southeastern Europe, northeastern Africa, western Asia, they have never ceased to be together, more or less.

Alexander died so young (at 33) that he left no heir (except a posthumous child), and no arrangements had been made to carry on the government. The empire that he had created was so heterogeneous and unwieldy that it is very doubtful whether Alexander himself could have preserved its integrity, but he was fortunate in that he died before its disintegration. When he was dying, he gave his signet ring to one of his generals, the Macedonian Perdiccas, son of Orontēs, but soon after his death the intense rivalries of other generals created a state of chaos. The end of the century and the beginning of the following (say 323–275) witnessed a succession of wars between them, the Wars of the Diadochoi (or successors), the description of which would be very complicated and does not concern our readers.

Leaving out the eastern satrapies, east of the Persian Gulf and southwest of the Oxus, the empire was broken into three main parts: Macedonia and Greece, ruled by the Antigonids; Western Asia, by the Seleucids; and Egypt, by the Ptolemies. After the constitution of those kingdoms (say by 275), they continued to be rivals, alternately allies and enemies. Any account of their jealousies, conflicts, and wars is made increasingly difficult by the consideration of internal secessions or revolts peculiar to each and by Roman intrigues which began in 212. The Romans used every discord as a pur-

yond the Oxus. Soon after his death, she gave birth to Alexander IV Aigos, who was acknowledged for a short time as a potential partner in government. Roxana and her son were protected by Alexander's mother, Olympias; they were put to death by Cassandros in 311. The little Alexander was only 12 at the time of his death.

chase for their own imperialism. For example, when the Attalid kings of Pergamon increased their power at the expense of the Seleucid kingdom, Rome was ready enough to help them (in 212 and later) and she contrived to be their heir in 130 B.C.

Each of these three or four kingdoms developed in its own way, according to its own geographic and anthropologic circumstances. We shall have opportunities of referring to one or another of them later. In this chapter, we must restrict ourselves chiefly to the Ptolemaic kingdom in Egypt.

When one speaks of Hellenistic times, however, one has in mind the Hellenized culture which developed all over the large territories that had constituted the Alexandrian empire as far west as Cyrenaica and as far east as the Indus. Those Hellenistic times may be said to continue roughly until the time of Christ; they were replaced gradually about the beginning of the Christian era by the Roman order. As far as the history of science was concerned, the Roman times were still to a large extent Greek, but they are no longer called Hellenistic; they are called Roman and later (after A.D. 325) Byzantine.

Indeed, the universality of the Greek language (as the vehicle of higher culture) was the outstanding characteristic of the Alexandrian world not only in Hellenistic but also during Roman times, at least in the eastern regions, which were by far the most cultured ones.

IRANIAN AND INDIAN INFLUENCES IN THE HELLENISTIC KINGDOMS

We shall devote most of our attention to the culture which flourished in Egypt, but before that it is worth while to insist upon the Oriental influences that were at play in all the Hellenistic kingdoms, because the reader is accustomed to the phrase "the Hellenization of the East" and not sufficiently aware of the Oriental reaction. Jewish influences which the reader would take more easily for granted will be left out at present.

We may take for granted also the local influences, Pharaonic in Egypt and Babylonian in the Seleucid kingdom. The ancient cultures were still alive, conspicuous, and impressive. It was an essential policy of the Ptolemies to pay full attention to the ancient Egyptian religion and of the Seleucids to respect and even to revive Babylonian knowledge and rites. The very great differences between the Ptolemaic and Seleucid kingdoms were due to natural conditions and to economic factors, but also and very markedly to their historical backgrounds, their religion and folklore.

Iranian influences were naturally considerable, because there had been exchanges of many kinds, pleasant and unpleasant, between the Greek colonists of Asia and the subjects of the Persian kings. Persian merchants must have been numerous in Milētos and the other cities of the Ionian confederacy. As far away in the West as Syracuse, King Gelōn (d. 478) received the visit of a *magos* (a mage)[4] who claimed to have sailed around

[4] The word "mage" is one of the most interesting in the English language. It is of

Africa, as had been done by Phoenicians under the orders of Necho and later of Darios the Great.[5] Ctēsias of Cnidos (end V B.C.) had explained Iranian culture in his *Persica*, and had not every educated Greek read the *Cyropaideia* of Xenophōn (IV–1 B.C.)? This was a political romance but no one could read it without being aware of Persia and realizing that there were good and noble Persians as well as evil ones.

Babylonia was a Persian satrapy from 538 and Egypt another from 525 to the Alexandrian conquest in 332, and during those two centuries many Persian institutions, usages, ideas, and words had taken root.

If we knew Iranian sources better than we do, it might be that many aspects of Greek culture could be traced back to them. It is possible, for example, that the theory of eléments originated in Persia and was diffused thence to the Greek world, to India, and to China.[6] That is mere speculation. About the reality of contacts between the Hellenistic kingdoms and Iran, however, and their multiplicity there can be no doubt.[7]

Greco-Indian relations are even more complex than the Greco-Iranian. They begin in the same way through the Ionian colonies, especially Milētos. Indian traders did not fail to reach those opulent markets, and Indian goods (including ideas) were brought also by middlemen. Other Indians visited Greece for the sake of obtaining wisdom or illustrating their own. The delightful story of Sōcratēs' interview with an Indian sage has already been told.[8] The earliest Greek accounts of India were given by Hērodotos (V B.C.), who recorded their cultivation and use of cotton, and by Ctēsias of Cnidos in his *Indica*.[9] Hippocratēs' contacts with Iranians are more dubious, though they would not have been difficult in the region of Cōs, or for that matter all over the Aegean Sea. The similarities between the Hippocratic *De flatibus* and Indian medicine are probably due to accidental convergence.[10]

Iranian origin but was promptly adopted in Greek (*magos*). It first meant a Zoroastrian priest, then a wise man, especially one who interpreted dreams. Its use in the New Testament (Matthew 11:1) popularized it in Christendom. The *magoi* became the Three Kings. The words magic, magician are derived ultimately from *magos*. Wisdom and magic were confused.

[5] A. J. Festugière, "Grecs et sages orientaux," *Revue de l'histoire des religions 130*, 29–41 (1945), p. 32. Necho (Necōs) was king of Egypt from 609 to 593, and Darios, of Persia from 521 to 485. For the two circumnavigations, see Volume 1, pp. 183, 299. On p. 299, the end of footnote 3, the reference should be to Necho instead of Sataspēs.

[6] Jean Przyluski, "La théorie des éléments et les origines de la science," *Scientia 54*, 1–9 (1933) [*Isis 21*, 434 (1934)]. See also his earlier paper, "L'influence iranienne en Grèce et dans l'Inde," *Revue de l'Université de Bruxelles* 37, 283–294 (1931–32) [*Isis 22*, 372 (1934–35)].

[7] For the interchange of religious ideas between Iran and Greece, see Joseph Bidez and Franz Cumont, *Les mages hellénisés. Zoroastre, Ostanès et Hystaspe d'après la tradition grecque* (2 vols.; Paris, 1938) [*Isis 31*, 458–462 (1939–40)]. Zōroastrēs (VII B.C.?) was the Zarathushtra of the Zendavesta; Ostanēs and Hystaspēs were later teachers of the same religion.

[8] Volume 1, p. 261.

[9] For details, see Volume 1, pp. 311, 327.

[10] Volume 1, pp. 372–373; for the meaning of convergence, see pp. 17–18. Jean Filliozat, "L'Inde et les échanges scientifiques dans l'humanité," *Cahiers d'histoire mondiale 1*, 353–367 (Paris, 1953).

All these Greco-Indian encounters were rare and limited in scope. When Alexander undertook his conquest of Asia, contacts occurred on a large scale. He reached the Indus and in the following centuries the northern part of India (to, say, lat. 22° N) was invaded by Greeks, who established kingdoms and settlements in various places.[11] The contact between Alexander and the Indian sages was the subject of a legendary cycle called "The colloquium of Alexander and the ten gymnosophistai," of which many forms appeared in ancient times.[12]

During the troubles that followed Alexander's death, an Indian adventurer named Chandragupta (Sandrocottos in Greek), who had met Alexander in his youth, managed to control a great part of northern India and created the Maurya empire, which lasted from his accession in 322 (or earlier) until A.D. 185. He established his capital in Pātaliputra.[13] The sophisticated Maurya culture was much influenced by the Persian one and hence Iranian influences might flow westward from northern India as well as from Iranian territories. Seleucos Nicatōr (king of Syria from 312 to 280) invaded Chandragupta's dominion in 305 but was obliged to withdraw. In the peace that followed, he ceded to Chandragupta the Panjāb and the mountains of the Hindu Kush, but received in return 500 war elephants. In 302, he sent Megasthenēs as his ambassador to the court of Pātaliputra. The results of Megasthenēs' experience were published by him under the title *Indica*. This work is unfortunately lost, and we have only fragments of it, but as far as we can judge from these it contained a large amount of information on Northern India. Many of his tales seemed incredible and, therefore, he was mistrusted by later historians, such as Polybios and Strabōn. He suffered the same fate as Hērodotos and Marco Polo, and it is possible that if the full text of his *Indica* were available he would be justified on many points, just as they have been.

At any rate, the Greek-reading people of the Hellenistic age were given the means of knowing much of that mysterious country; their knowledge was incomplete and sometimes faulty, but it was considerable.

Among the Indians who appeared in Egypt, some were merchants or travelers; others were Buddhist missionaries, especially during the time of the Maurya king Aśoka, who ruled over a great part of the peninsula (above 15°N) from 273 to 232. Aśoka was in relation with Ptolemaios Philadelphos of Egypt, as well as with Antiochos II of Syria and Antigonos of Macedonia. On the other hand, Ptolemaios Philadelphos sent an envoy to India in order

[11] Elaborate account by W. W. Tarn, *The Greeks in Bactria and India* (ed. 2, 591 pp., 2 pls., 3 maps; Cambridge: University Press, 1951; ed. 1, 1938).

[12] Summary by A. J. Festugière, "Trois rencontres entre la Grèce et l'Inde. I. Le colloque d'Alexandre et des dix gymnosophistes," *Revue de l'histoire des religions 125*, 33–40 (1942–43). The word *gymnosophistēs* means naked philosopher, the name given by Greeks to the Indian sages.

[13] Pātaliputra was built at the confluence of the Ganges and the Sọn rivers. Modern Patna, capital of Bihar province.

to obtain elephants and mahouts. The third century was an age of giant warships at sea and elephant warfare on land. Of course, the Seleucid kings being nearer to India were richer in elephants, but their Ptolemaic rivals did their best to obtain more elephants not only in India but also in Africa. Both species were used in battle; the first battle between Indian and African elephants was that of Rhapheia [14] in 217; the Africans were outnumbered and beaten. The trade in elephants implied other and easier forms of trade and cultural exchanges.

The most famous of the Yavana (= Greek) kings in India was Menandros. He is not well known to us and in the little that we know it is not easy to separate fact from fiction. He was king of Kābul and the Panjāb and finally ruled the whole of Greek India down to Kathiawar (western Gujarat, west coast, c. lat. 22° N) until his death, c. 150–143. He was so well known, however, to his Indian subjects under the name Milinda that he became the hero of a Buddhist treatise, the *Milindapañha* (the "Questions of Milinda"). It is not certain that he was a Buddhist himself, but, in the fashion of Hellenistic kings, he was friendly to the religion of the people around him. The *Milindapañha* is the only Indian book dealing with a Greek king; [15] it was probably written at the beginning of our era and is preserved in Pāli and Chinese versions (see note below).

Commercial and cultural relations between Egypt and India were subject to vicissitudes caused by the enmity of the Seleucid kingdom, but, even when the Syrian ways were closed, Egypt could reach India via the Red Sea and Arabia. The sea journey to India through the Bab el-Mandeb and the Arabian Sea could not be accomplished easily and safely before the discovery of the monsoon. It is possible that oriental sailors had been acquainted with it for a long time, but that knowledge did not become available to the Greeks until the time of Hippalos, c. 70 B.C.[16]

Greek domination in India had ceased completely before the Christian era, but trade continued in various ways. The best way of illustrating the importance of that trade at the end of the Hellenistic age is to recall Cleopatra's suggestion of abandoning the Mediterranean and ruling the Indian

[14] Rhapheia was a seaport at the Southwest end of Palestine, south of Gaza, on the edge of the desert.

[15] Tarn, *The Greeks in Bactria and India* (ed. 2), chap. 6, "Menander and his kingdom," pp. 225–269. My dating is derived from Tarn. In an excursus (pp. 414–436), Tarn compares the "Questions of Milinda" with the "Questions of Ptolemaios II" in the pseudo-Aristeas. We shall deal with the *Milindapañha* presently and with Aristeas later.

[16] The date is uncertain; it has been put as late as A.D. 50. I follow Rostovtzeff, *Isis 34*, 173 (1942–43). In his account of India, Megasthenēs (III–1 B.C.) called the monsoon Etēsian winds; later it was called *hippalos* after its discoverer. The name monsoon is considerably later still, because it is derived from the Arabic *mawsim* (season). See Henry Yule and A. C. Burnell, *Hobson-Jobson: A glossary of colloquial Anglo-Indian words and phrases, and of kindred terms, etymological, historical, geographical and discursive*, ed. William Crooke (London: Murray, 1903), p. 577.

seas instead! Tarn remarks about this, "She was not talking folly; she might have anticipated Albuquerque." [17] The only successors of Alexander who became legendary were Menandros and Cleopatra; they both deserved their extraordinary fame.

Milindapañha

The *Milindapañha* is a dialogue between King Milinda and the monk *Nāgasena*, the king asking many questions on various points of Buddhist doctrine. The whole of it as available in Pāli is very long, but the ancient kernel, consisting of a prologue (*pubbayoga*) and three books, is considerably shorter.[18] That ancient kernel was written during the first centuries of our era, certainly before the fifth century, because there are two versions of it in the Chinese Tripiṭaka,[19] versions made during the Eastern Chin dynasty (317–420). The Chinese translation was made not from the Pāli text that we have but from a Prākrit text that is probably more ancient.

The dialogue takes place at Sāgalā, Milinda's capital in the Panjāb, in the presence of a number of Yonakas (or Greeks). There is no doubt that Milinda is identical with Menandros; one may find in this work a few other Greek references (or words derived from the Greek)[20] and the beginning is perhaps a little more vivacious or less turgid than other Indian writings. Nevertheless, the *Milindapañha* is definitely Buddhist and Indian. It is not a part of the canon, yet it is considered an excellent piece of Buddhist literature. The reading of it is extremely instructive. It is as different as can be from Greek writings of the early centuries of our era. The comparison of Buddhist writings with Christian theological treatises of approximatively the same period, say some of the early patristic literature, would not be unfair; it would reveal abysmal differences.

The author of the *Milindapañha* had no knowledge of Greek or Greek literature, and his work remained entirely unknown in the West until recent times. On the other hand, its popularity in the Buddhist world was considerable; witness the Prākrit, Pāli, and Chinese texts already mentioned and versions into Singhalese, Burmese, Korean, and Annamese.

The Pāli text was edited by Vilhelm Trenckner (London, 1880); the Chinese versions by Paul Demiéville in the *Bulletin de l'Ecole française d'Extrême-Orient 24*, 1–258 (1924).

The English version of the Pāli text was published by T. W. Rhys Davids in *The sacred books of the East* (1890, 1894), vols. 35, 36. A French version of the ancient part was made from the Pāli by Louis Finot (Paris, 1923).

The *Milindapañha* is discussed in every history of Indian literature. See, for example, Moriz Winternitz, *Geschichte der Indischen Litteratur* (Leipzig, 1920), vol. 2, pp. 139–146; English translation (Calcutta, 1933), vol. 2.

SOME PRELIMINARY REMARKS ON THE EXCHANGE OF SCIENTIFIC IDEAS

The exchanges just dealt with concern literature and the reader may wonder whether none are relative to scientific ideas. We must bear in mind that

[17] W. W. Tarn and G. T. Griffith, *Hellenistic civilisation* (London: Arnold, ed. 3, 1952), p. 248. Albuquerque (Affonso o Grande, 1453–1515) conquered a part of India for Portugal in 1504.

[18] The long Pāli text covers 420 full pages in Trenckner's edition; the ancient part stops at p. 89, so is hardly more than one fifth of the whole.

[19] No. 1358 in the Catalogue of Bunyi Nanjio (Oxford, 1883; reprint, Tōkyō, 1930). On Chinese Tripiṭaka, see *Introduction*, vol. 3, pp. 466–468.

[20] For example, the word Alasanda in Book III is probably a corruption of Alexandria.

religious beliefs, literary conceits, or artistic motives are far more contagious than science, especially abstract science. There may be a popular hunger for knowledge, but that hunger is more easily satisfied with false knowledge than with truth. Superstitions, such as astrology, could travel far and wide but science did not. We shall notice some curious facts, however, in later chapters.

The best that Egypt and Babylonia had to offer had been assimilated by Greek minds in earlier times; they added little or nothing to that in the last pre-Christian centuries. The extraordinary astronomy developed during the Seleucid period in Mesopotamia contained many novelties, but these were not transmitted westward; their lunar and planetary theories remained so completely unknown in Europe that they were unable to affect astronomical progress there. Those astonishing discoveries were explained in cuneiform tablets which were not deciphered until recent days (1881 and later).[21] Some Babylonian observations were utilized by Hipparchos (II–2 B.C.), however, and these will be discussed when we speak of him.

As to the mathematical ideas of the Ancient East, those that had not yet been integrated into Greek science reached the West via Egypt, but that was in post-Christian days through two Alexandrians, Hērōn[22] and Diophantos (III–2).

What about the travel of scientific ideas in the opposite direction? There was very little of it. The Macedonian and Greek soldiers who conquered the East were more interested in warfare and administration, in political intrigues and economic exploitation, than in science. They certainly introduced improvements in what the Germans call "Kriegswissenschaft," and we, the "art of war"; they probably introduced technical refinements in the other arts and industries; and Greek physicians must have accompanied the soldiers and settlers. We shall come across some of them in other chapters. One remarkable exception was that of the astronomer Seleucos (II–1 B.C.), who explained Aristarchian astronomical views in Babylonia.

Illustrious men of science who continued the Greek tradition flourished in the East, but they belong mostly to post-Christian times, for the main waves of scientific thought were pushed eastward only by Christian intolerance. Greek astronomy did not appear in India until very late; it was late in starting, because it was chiefly posterior to Ptolemy (II–1), and it was not published in Sanskrit until the time of the Siddhānta treatises (V–1, or before).

[21] For an account of them, see Otto Neugebauer, *The exact sciences in antiquity* (Acta historica scientiarum naturalium et medicinalium, edidit Bibliotheca Universitatis Hauniensis, vol. IX; Copenhagen: Munksgaard, 1951; Princeton: Princeton University Press, 1952) [*Isis 43*, 69–73 (1952)], and Chapter XIX below. [A second edition of this work was published by Brown University Press in 1957.]

[22] That is, if we assume that Hērōn is not pre-Christian, as I first thought, and not placed in (I–1 B.C.) but rather in (I–2). He flourished probably after 62 and before 150; *Isis 32*, 263 (1947–49), *39*, 243 (1948).

Fig. 2. Amon-Rē, the sun-god. Part of granite relief of the time of Ptolemaios II Philadelphos (king 285–247). It comes probably from the temple of Isis at Bahbīt al-Ḥigāra in the central Delta and is now in the Boston Museum of Fine Arts. This reproduction does not show the two very tall feathers above the god's crown which help to identify him. His pectoral is a miniature shrine meaning "protection." The god is holding the symbol of life in his left hand, and was probably holding the scepter *uas*, meaning "lordship," in his right one. [Bernard V. Bothmer in *Bulletin of the Boston Museum of Fine Arts 51*, 1–6 (1953).]

In short, the Greek emigrants were too few [23] in pre-Christian times and too little interested in science and scholarship to affect and change Eastern minds and, on the other hand, the Asiatics did not feel the need of Greek thought (why should they have felt it?); they rejected it instinctively or assimilated only superficial manners and customs, never the substance and the informing spirit. Asiatic inertia was immense. As Tarn puts it: "In matters of the spirit, Asia knew that she could outstay the Greek, as she did." [24]

PTOLEMAIC EGYPT

Soon after Alexander's death, the Macedonian Ptolemaios,[25] son of Lagos, became the satrap of Egypt. He was a childhood friend of Alexander and, perhaps, his half brother; [26] he took part in all the campaigns of Asia and was one of Alexander's leading generals and best friends. This made it possible for him to write memoirs, now lost, which were the most valuable source of Arrianos' history. He extended the limits of his satrapy by the conquest of Palestine and Coilē-Syria, c. 320, and by the acquisition of the southwest coast of Anatolia and of the island of Cōs. He assumed the title

[23] Not in absolute numbers but in proportion to the Asiatic populace.

[24] Tarn, *Hellenistic civilisation*, p. 163.

[25] The name is Anglicized "Ptolemy," but when speaking of the kings of Egypt I shall always use the Greek form "Ptolemaios," reserving the form "Ptolemy" for the great astronomer of the second century after Christ. Ptolemy the astronomer is so great a man that he deserves an international name (or a separate name in each country); he belongs to the whole world, while the Lagid kings concern only Egypt and the Near East. The plural form "Ptolemies" may be used as well as "Ptolemaioi" to designate all the kings without ambiguity.

[26] His mother, Arsinoē, had been a concubine of Philip of Macedonia.

Fig. 3. Ptolemaios I Sōtēr (ruled from 323, king from 305 to 285) making an offering to Hathor, goddess of joy and love, identified by the Greeks with Aphroditē. The king's representation (to the right) is idealized. We know it is the king by the uraeus (or royal serpent) on his forehead as well as by the cartouche behind him which contains his prenomen, "The one whom Rē has chosen, the beloved of Amon," while the nomen, Ptolmis, is written in the left cartouche. The bas-relief was originally at Tarraneh, near Kafr Dā'ūd in the western Delta. It is now in the Boston Museum of Fine Arts. [Bernard V. Bothmer, *Bulletin of the Boston Museum of Fine Arts 50,* 49–56 (1952).]

of king in 306, the other diadochoi doing so at about the same time and for the same reason. He was the founder of the Ptolemaic or Lagid dynasty, the organizer of Ptolemaic Egypt. He was a good soldier and administrator, the creator of Egyptian prosperity and of the Alexandrian Renaissance. He ruled until 285 and was named Ptolemaios Sōtēr (the savior).

His last and most beloved wife, Berenicē, had given him a son, Ptolemaios Philadelphos (brother-loving), born in Cōs, who succeeded him in 285 and ruled until 247. He continued the efforts of his father with so much diligence and merit that when we describe the cultural renaissance it is hardly possible to separate the father from the son; what the first began or conceived, the latter completed and realized. He increased his patrimony and his power, exploring Upper Egypt and developing commercial relations with Ethiopia, the countries bordering the Red Sea, Arabia, and even India.

The third king was called Ptolemaios Evergetēs (the benefactor); he ruled from 247 to 222 and brought the Ptolemaic dynasty to its climax. He conquered Mesopotamia, Babylonia, and Susiana, and brought back to Egypt an immense booty, including the statues of Egyptian gods which had been taken away by Cambyses II (king of Iran 529–522). The decline began with his son, Ptolemaios IV Philopatōr, king from 222 to 205. We need not consider the others, but we may note that there were altogether fifteen royal Ptolemaioi. The last ruler is perhaps the best known of all, Queen Cleopatra, a woman of great beauty and of superior ability, an extraordinary polyglot.[27]

[27] Everybody knows Cleopatra and knows but one. Even as learned a man as F. Sherwood Taylor, in *The Alchemists* (New York: Schuman, 1949), p. 26, rejects the ascription of early alchemical Greek texts to Cleopatra, because the latter, he says, is an Egyptian queen! The name was fairly common in the Greek world and, *toutes proportions gardées,* there were probably as many Cleopatras in Ptolemaic Egypt as there were Victorias in Victorian England. Thirty-three of them are famous

The Romans paid her, unwillingly, the greatest possible tribute; they feared her, a woman, as they had feared no one since Hannibal.[28] She aimed to be empress of the Roman world, and might have succeeded, if her lover, Caesar, had been permitted to live. He was murdered in 44; she fell back on Antony, but the battle of Actium (31 B.C.) put an end to her dreams, and she committed suicide [29] in the following year, lest she be taken to Rome as a captive. The last Ptolemaios was Ptolemaios XIV Caesarion, the son of Caesar and Cleopatra, murdered by order of Octavian (Augustus) in 30 B.C. at the age of 17, a kind of Hellenistic *aiglon.* From that time on, Egypt was simply a Roman province. The golden age had lasted but a single century, the third, but, short as it was, it had been long enough for a few men of genius to create immortal achievements.

What kind of country was Egypt under the Lagid kings? I do not mean the physical country, which had not changed since Pharaonic days, a magnificent gift of the Nile. The geography and the physical climate were unchanged; but what about the political climate? One might claim that the latter had not changed very much either, except that the masters and the real owners of the land and people were no longer Egyptians but Macedonians and Greeks.

The Greeks had been deeply interested in Egypt ever since the time of Psametik I, first ruler of the Twenty-Sixth or Saitic Dynasty (663–525; he ruled from 663 to 609). Greek colonies were established in the Delta and flourished, in spite of Egyptian indifference or hostility.[30] During the reign of the fifth king, Ahmose II (569–525), whom the Greeks called Amasis, the Greek merchants were concentrated in a single city, Naucratis, on the Canopic mouth of the Nile in the western Delta, and that city became very prosperous. It was in all essentials a Greek city, wherein many Greek states possessed temples of their own. Amasis was kind and generous to the Greeks and popular with them; nevertheless, every privilege that the Greeks en-

---

enough to be discussed in Pauly-Wissowa, vol. 21 (1921), 732–789. Our Cleopatra was by far the most famous. She is Cleopatra VII, daughter of Ptolemaios XII Aulētēs; she was born in 69 and ended her life in 30. When one writes "Cleopatra" without qualification, she is meant. Read Plutarch's account of her in his life of Antony.

[28] Tarn and Griffith, *Hellenistic civilisation,* pp. 46, 56. Hannibal, son of Hamilcar Barca, was the greatest Carthaginian general (247–183).

[29] According to the most common traditions, Cleopatra died from the bite of an asp (Greek *aspis*) which she held to her bosom. That was a symbolic death. The royal serpent, uraeus (Greek *uraios*), combined with the sundisk, was the symbol of Rē (the sun god); it appeared also on the headdress of Egyptian kings, just over the forehead. The last native ruler of ancient Egypt was killed by the sacred uraeus.

[30] J. H. Breasted in his *History of Egypt* (New York: Scribner, 1942), p. 579, compares those Greek colonies to the European colonies in China. "If he could have had his way the Egyptian would have banished the foreigners one and all from his shores; under the circumstances, like the modern Chinese, he trafficked with them and was reconciled to their presence by the gain they brought him."

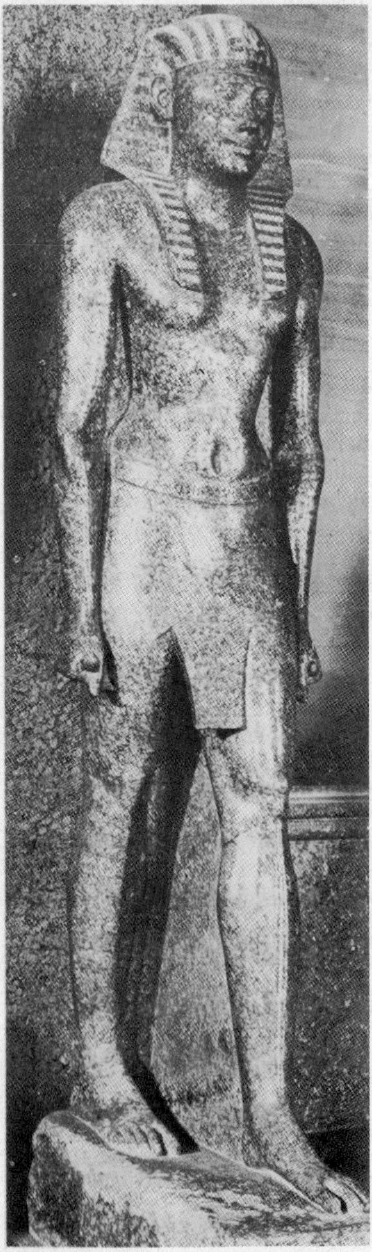

Fig. 4. Statue of Ptolemaios II Philadelphos, in the Vatican. The statue is in red granite, 2.66 m high (without the plinth, 2.40). Ptolemaios II Philadelphos (308–246), son of Ptolemaios I and Berenicē I, was the second king of the Lagid dynasty; he ruled from 285 to 246. He married Arsinoē II about 276. He is identified by two inscriptions in hieroglyphics, the shorter of which reads: "The king of Upper and Lower Egypt . . . son of Rē, Ptwlmjs. May he live forever." [Giuseppe Botti, Pietro Romanelli, *Le sculture del Museo Gregoriano Egizio* (*Monumenti vaticani di archeologia e d'arte*, vol. 9; Vatican, 1951), no. 32, pp. 24–25, pls. xxii and xxiii.]

Fig. 5. Statue of Arsinoē Philadelphos, in the Vatican. The statue in red granite is 2.70 m high (2.48 m without the plinth). Queen Arsinoē (c. 316–270), the daughter of Ptolemaios I and Berenicē I, was the sister and wife of Ptolemaios II. She is identified by two inscriptions in hieroglyphics, the shorter of which reads: "The real daughter, the real sister, the real wife, the lady of the Two Lands, 'Irsj . . . Philadelphos." [Giuseppe Botti, Pietro Romanelli, *Le sculture del Museo Gregoriano Egizio* (*Monumenti vaticani di archeologia e d'arte*, vol. 9; Vatican, 1951), no. 31, pp. 22–23, pls. xxii and xxiv.] These two statues are reproduced with the kind permission of the keepers of the Vatican Museums. It is clear that they were made at the same time as companion pieces, but the photographs made at different times and under different conditions look very different. These statues are not portraits but symbols of the Ptolemaic king and queen.

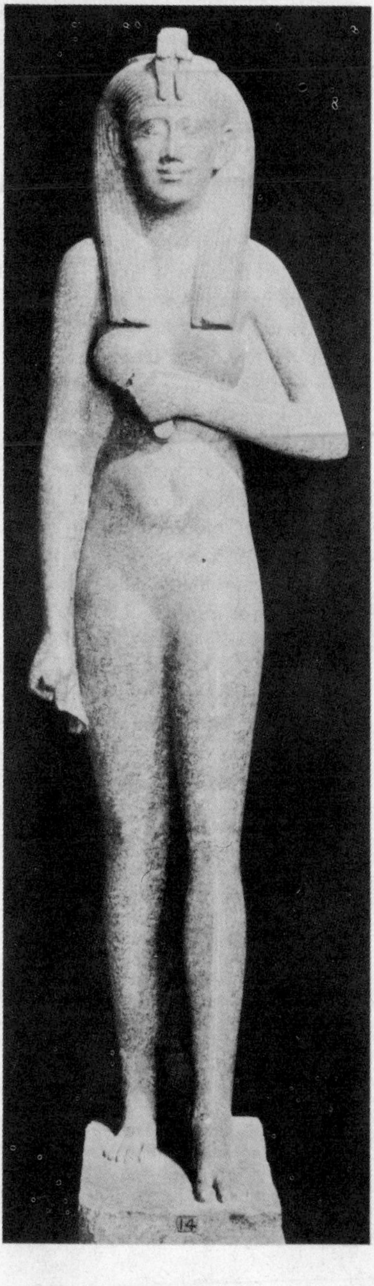

joyed was dependent upon Egyptian favor and caused considerable jealousy.

With the accession of the Ptolemaioi the situation was reversed; the Greeks were no longer guests, welcome or unwelcome, but masters. The Ptolemaioi continued the Egyptian tradition, however; they were the owners of the land and of every thing; moreover, they were sacred and divine. The king was the state. It must be added that the first Ptolemaioi, at least, were good administrators, and thanks to them Egypt was more prosperous than ever.

During the first half of the dynasty, the administration was generally efficient. Good order was kept. The annual flood of the Nile was carefully regulated, the irrigation was improved, the crops were controlled, granaries were made ready to preserve them, new animals and grains were acclimatized, the cultivated areas were increased, new crafts were introduced, coinage, trade, and banking [31] were better organized. Foreign trade was considerably extended; Egypt exported cereals, payrus, linen, glass, alabaster. One of the greatest economic innovations, probably due to Ptolemaios II Philadelphos, was the use of camels; the camels themselves may have come into Egypt before the Ptolemaioi but not long before them.[32] Ptolemaios introduced a postal service modeled on the Persian, and the camels were unsurpassable for such a purpose, being capable of great speed and endurance, or of carrying heavy loads. The one industry that the Greek rulers seem to have neglected was mining; at any rate, they did not enlarge the mineral resources and did not exploit the known ones as well as the old Pharaohs had done before them.[33] All the profits, of course, went to the king and to a small group of partners and accomplices; the peasants (then as now) received nothing but the barest needs of existence. At the beginning they did not revolt, because they were perhaps treated a little better than before and because they lacked the material and spiritual possibilities. They would have revolted only at the edge of death, and even then death was easier.[34]

[31] The inclusion of banking may astonish some readers, because they do not realize its deep antiquity. There were bankers in the Oriental empires and particularly in the Persian. Remember that Egypt had been a Persian province from 525 to 332, and the Greek conquerors were called to remove or improve Persian institutions. Hence, the Ptolemaioi inherited financial methods from both the Greek and the Persian sides. An interesting sidelight has been thrown recently on Persian banking in the Paris thesis of Guillaume Cardascia, *Les archives de Murashû. Une famille d'hommes d'affaires à l'époque Perse, 455–403* (Paris: Imprimerie nationale, 1951). The Murashū house in Nippur was one of the oldest banks in the world. A few notes on banking in Tarn and Griffith, *Hellenistic civilisation*, pp. 115–116, 250.

[32] On camels in Egypt, see Volume 1, p. 51.

[33] The study of agriculture, trade, and industries in Ptolemaic Egypt is an immense subject which has been pretty well covered by the late Mikhail Ivanovich Rostovtsev (1870–1952): *The social and economic history of the Hellenistic world* (3 vols., 1804 pp., 112 pls.; Oxford: Clarendon Press, 1941) [*Isis 34*, 173–174 (1942–43)]. Mining is dealt with in an appendix by Robert Pierpont Blake.

[34] Revolts were made very difficult and futile because of the administration, which controlled everything and was very efficient. It became less efficient, however, in the time of Ptolemaios IV Philopatōr

Inasmuch as Palestine and Egypt were united under Persian rule and continued so under the first Ptolemaioi (until 198), it is natural enough that many Jews emigrated to Egypt and the more so as that country became more prosperous and offered them greater opportunities. By the third century, however, it is probable that the majority of Egyptian Jews were already born and bred in the country; as the top management of any business was in Greek hands, the Jews were rapidly Hellenized and some of them forgot the use of Hebrew; they imitated Greek customs and adopted Greek names, preferably the names including the word *Theos* (God), such as Theodotos or Dōrothea.

The coexistence of Greek and Jewish settlements in Egypt is but one of the aspects, the main one, of a more general situation. Under Greek rule, Egypt became the most important mixing place of East and West. The Ptolemaic empire at its high tide did not include Egypt only, but also Cyrenaica, parts of Ethiopia, Arabia, Phoenicia and Coilē-Syria, Cypros, and some of the Cyclades, and it drew elements of its population from all these countries. The bulk of the people were naturally Egyptian; the top class was Macedonian and Greek; [35] there were a good many Jews but also other Orientals, Syrians, Arabs, Mesopotamians, Persians, Bactrians, Indians, and Africans like the Sudanese, Somalis, and Ethiopians. From the purely cultural point of view, the most pregnant elements in that mixture, next to the Greeks, were the Jews. We shall come back to them later apropos of the Septuagint.

The Hellenistic nations were ready to welcome foreign sages, such as the Iranian magi, the Indian gymnosophistai, and many others, because of their spiritual curiosity and even more because of a kind of religious starvation. The orientalizing Greeks were opening their hearts to the Great Mother of Phrygia, to Mithras, or to the gods of Egypt, especially Isis and Osiris. We should remember that the desire for living forms of religion had existed in Greece from early times; witness the existence and popularity of mystery cults like the Eleusinian, Orphic, and Dionysiac ones. Since the time of Aristotle and Epicuros, the old mythology had lost favor; on the other hand, the astral religion that had to some extent replaced it was too learned and too cold to satisfy the plain people. The Greeks established in Asia or Egypt were far away from their old sanctuaries, and their religious hunger caused them to be very susceptible to the Oriental mysteries. They attended or observed the festivals celebrated around them and were deeply impressed. Oriental wives were very helpful in bringing those sacred ceremonies closer

(222–05) and later. From 217 to 85 B.C. revolts increased in number, strength, and violence.

[35] The top class included a few Egyptians, chiefly high priests.

to the hearts of their Greek husbands and the number of conversions gradually increased.

The religious syncretism was especially clear and strong in Egypt. It began at the very beginning, in 331, when Alexander the Great visited the temple of Ammōn in the oasis of Sīwa [36] and was recognized by the oracle as a son of Zeus-Ammōn.[37] Egyptians had generally admitted the divine nature of their rulers and, therefore, it was natural enough for the Ptolemaic kings to assume divinity, to require and obtain divine worship for themselves. Their Greek subjects were awed by the elaborate ceremonies practiced in the Egyptian temples, and the kings were willing enough to commune with the other gods of Egypt. It was impossible for them not to share and love a religion that apotheosized them. They adopted Pharaonic manners, such as marriages between royal brothers and sisters; Ptolemaios II Philadelphos married his sister Arsinoē II. Divine kings are too exalted to marry outside of their own family.

Moreover, each Egyptian dynasty had given a new emphasis to one of the old gods or introduced a new one and in the same spirit the Ptolemaioi put in the limelight the god Sarapis. They did not actually invent him. The cult of Osiris had been gradually combined with that of the sacred bull, Apis.[38] Osiris and Apis were worshiped together in the "Sarapeion" [39] of Memphis (Ṣaqqāra).

The cult of Sarapis was typically Hellenistic, because it combined Egyptian elements with Greek ones. According to Plutarch,[40] it was formalized

[36] This is the westernmost of the Egyptian oases, some 400 miles southwest of Alexandria. It is a hard trip by motor car, and one cannot help admiring Alexander for having done it in a much harder way. The temple of Ammōn was already known to the Greeks in the seventh century. Its oracle acquired almost as much prestige and authority as those of Dōdōnē and Delphoi, and Alexander realized the political necessity of consulting. it The main book on Sīwa is by C. Dalrymple Belgrave, *Siwa, the oasis of Jupiter Ammon* (London, 1923). There is very little left of the temple. Good photographs of the ruins appear in Robin Maugham, *Journey to Siwa* (London: Chapman and Hall, 1950), pls. 13, 15, 21, 25. It is said that sal ammoniac (chloride or hydrochloride of ammonium) was first obtained by distillation of camel's dung near that temple. We are on safer ground when speaking of ammonites (fossil cephalopods); their name is certainly derived from Ammōn because of their resemblance to the ram's horn. The ram was the sacred animal of the sun-god, Amon-rē. Zeus-Ammōn was his Hellenized epiphany.

[37] Was Alexander recognized by the oracle? That is doubtful, or rather it all depends on the interpretation of the oracle's words by Alexander's staff. The oracle may have greeted Alexander with the word *Ō paidion* (O son) or *Ō pai Dios* (O son of Zeus); the two salutations might easily be confused; the second might be conventional or it might be understood literally.

[38] The dead Apis bull was identified with Osiris and was worshiped as an infernal deity. Osorapis was thus comparable to, or identical with, Haidēs or Plutōn.

[39] The name Sarapis is derived from the combination Osiris-Apis or Osorapis. Sarapis and Sarapeion (for the temple) are the Greek names; Serapis and Serapeum, the Latin transcriptions.

[40] *De Iside et Osiride*, 28. Manethōn and Timotheos were both advisers to Ptolemaios Sōtēr. Plutarch calls Timotheos *exēgētēs* (interpreter), for he was an interpreter of Eleusinian mysteries. According to old traditions, the hero, Eumol-

by a priest of Hēliopolis, Manethōn (III–1 B.C.), in collaboration with Timotheos, who was a priest of Dēmētēr, Dēmētrios of Phalēron, whom Sarapis had cured of his blindness, wrote hymns in his praise. The assimilation of Sarapis with Zeus is proved by many Greek inscriptions meaning "There is one Zeus Sarapis," somewhat like the Muslim acclamation, "There is no god but God."

The Hellenistic nature of that Egyptian cult is put beyond doubt by the fact that the liturgical language was Greek, and the art (except the hieroglyphics) was more Greek than Egyptian, or wholly Greek.

The oldest "Sarapeiôn" was the temple of Osorapis with the subterranean tombs of Apis bulls, in Ṣaqqāra. These tombs were discovered by Auguste Mariette in 1851, and the oldest of them date back to Amenhotep III (1411–1375), the Greek Memnōn. Another Sarapeion was built nearby by Nektanebis II (358–341). These two prove the antiquity and long continuity of the Osorapis cult.

Sarapeia were established during the Hellenistic period in the main Egyptian cities. The Sarapeion of Abuqīr (at the seashore, east of Alexandria) was visited by many pilgrims in search of health. The most important Sarapeion was naturally in Alexandria; it was situated on the hill upon which can still be seen "Pompey's Pillar." [41] That column may have been a part of the Sarapeion; it was preserved, or erected here, by order of Theodosios (emperor 379–395) or of the fanatical bishop of Alexandria, Theophilos,[42] in order to commemorate the destruction of the Sarapeion in A.D. 391 and the victory of Christianity.

By that time, however, the cult of Sarapis had petered out. That cult was essentially Ptolemaic, and in Roman times it had been largely replaced by the Isidic one. Theophilos' victory was less a victory over Sarapis than one over paganism in general.

"ALEXANDRIA NEAR EGYPT"

The Greek colonies where the Hellenistic culture of Egypt developed under the patronage of the Ptolemaioi constituted but a very small part of the whole country. In a sense, this was the continuation of an old usage, for under the Twenty-Sixth Dynasty the king Ahmose II (Amasis) had founded

---

pos, was the founder of those mysteries and the first priest of Dēmētēr. The priests who succeeded him were supposed to be his descendants and were called Eumolpidai. Timotheos was an Eumolpidēs. See Pauly-Wissowa, series 2, vol. 12 (1937), 1341.

[41] So called because of the medieval legend that it marked the tomb of Pompey the Great, Pompeius Magnus (106–48 B.C.). Pompey was murdered as he was landing on the Egyptian shore. The Arabs call it simply al-'Amūd (post, column).

[42] Theophilos was bishop of Alexandria from 385 to 412. It is said that he had obtained from the emperor a commission to demolish the pagan temples of Alexandria, not only the Sarapeion but also the Mithraion, and others. It is not certain, however, that the emperor gave him that power, but Theophilos was despotic and fanatical to the point of unscrupulousness.

the city of Naucratis and obliged all the Greek merchants to reside there and nowhere else. Alexander created a new city which was named after himself, Alexandria; Ptolemaios Sōtēr founded the city of Ptolemais Hermiu in Upper Egypt; and there were other Greek settlements.

While the country was dominated by the kings almost in the way a landlord dominated his own estates, the Greek colonies achieved a modicum of administrative independence according to Greek traditions.

A good many cities were said to have been founded by Alexander the Great or in his memory, and they all bore the name Alexandria. Some seventeen have been identified, practically all of them in Asia, many beyond the Tigris; two were founded on the Indus, and a third, Alexandria-Bucephala, on the Jhelum;[43] there was one beyond the Jaxartes, called Alexandria Eschatē.[44] Most of those cities have ceased to exist or have become insignificant. On the other hand, the only city that Alexander founded in Egypt, in 332, quickly assumed a very great importance under Ptolemaic patronage and has remained to this day one of the greatest cities of Western Asia and the leading harbor of the eastern Mediterranean Sea.

It is said that Alexander founded Alexandria, but that can only mean that he gave general directions for the establishment of a new city at the western end of the Nile delta. He could not do more because he left Egypt soon afterward. The real founder of the city was Ptolemaios Sōtēr. When he began his administration of Egypt, Alexandria was still too undeveloped to be used as a capital, and the government was first seated at Memphis. Sometime after Alexander's death (in Babylon, 323), Ptolemaios Sōtēr secured his body and had it brought to Memphis. As soon as Alexandria was sufficiently built up, it became the capital of the Ptolemaic kingdom, and Alexander's remains were removed to it. A temple was erected to receive them; this was called the *sēma*; it is probable that the kings of the Ptolemaic dynasty were eventually buried in the same sacred enclosure, and the *sēma* thus became a kind of national mausoleum. No trace remains of it and its location is uncertain.[45]

Strangely enough, this metropolis of Egypt was not in Egypt but outside of it. Its ancient name in Greek or Latin was *Alexandria ad Aegyptum*, which means "Alexandria near Egypt." This was geographically incorrect; Alexandria was in the northwest part of the country but not by any means at the northwest end. The temple of Ammōn which Alexander had visited was far

[43] The Jhelum river is the ancient Hydaspēs, one of the five rivers of the Panjāb. Bucephalos was Alexander's horse. See Volume 1, p. 491.

[44] Meaning the farthest one. The Jaxartēs (or Syr Daria) is the eastern one of the two rivers running into the Aral Sea, the other being the Oxos; Sogdianē is enclosed between the two rivers.

[45] *Sēma* means a sign, an omen, later a tumulus to mark a tomb by. The word "semantics," so much used today, is derived from the same root. *Sēma* was sometimes opposed to *sōma*, the material body. The Alexandrian *sēma* was probably located near the present mosque of Nabi Daniel. Excavations in that neighborhood might increase our knowledge.

to the west of it. The words "near Egypt" expressed rather a political fact. Alexandria was not an autochthonous capital but the seat of royal and colonial administration. It is as if one said "Hong Kong near China" or "Goa near India," because the first of those cities is inhabited by a vast majority of Chinese and a small minority of Englishmen; it is in China yet out of it; the second is inhabited by an overwhelming number of Indians and relatively few Portuguese; it is in India yet out of it.

Alexandria was inhabited by a small ruling class, Macedonian and Greek,[46] and by a very large number of native Egyptians. In addition, there was a considerable colony of Jews (Palestine remained a part of the Ptolemaic kingdom until c. 198) and an indeterminate number of other Orientals (Syrians, Arabs, Indians). Come to think of it, Alexandria must have been a city comparable to New York, the two ruling elements being in the first place Greek and Jewish and, in the second, British (or Irish) and Jewish. And just as New York is a dazzling symbol of the New World, so was Alexandria of the Hellenistic culture.

The comparison is valid in another respect because, if one takes into account the different speeds of sailing and the resulting shrinking of the seas, the relation of the new harbor to the harbors of ancient Greece was not very different from that of the New York harbor to the English ones. To sail from the Peiraieus (Athens' harbor) to Alexandria was almost as much of a voyage as sailing to-day from the Mersey to the Hudson. From the anthropological point of view, it is a little misleading, for Alexandria was not only Jewish but African and Asiatic to a high degree.

In that sense, Alexandria was the true daughter of a great king, for Alexander had introduced into the world a new idea of incalculable pregnancy; the Greek conception of the city-state had been replaced by that of a cosmopolitan world, whose ethical and religious heterogeneity was unified by a lay culture.

Alexandria was not only a metropolis; it was a cosmopolis, the first of its kind.[47] The Greeks were great architects, not only of temples but of whole cities; the material and spiritual principles of town planning were explained as early as the middle of the fifth century by Hippodamos of Milētos.[48] This is one of the aspects of the Greek genius. They did not allow new cities to grow at random in the way our American cities do. It is claimed that the streets of Boston were outlined by the cows going to pasture and returning to the stables. The planning of Alexandria was less casual.

[46] Plus the chief Egyptian priests who controlled the souls and collaborated with the material rulers.

[47] The word *cosmopolis* was not used by the Greeks in this sense, but Diogenēs of Sinōpē, the Cynic, was the first to use the word *cosmopolitēs*. When he was asked where he came from, Diogenēs answered, "I am a citizen of the world" (*cosmopolitēs*). This may have impressed Alexander, if he heard of it, but, even if Diogenēs originated the idea, he could not advertise it and enforce it as the emperor did. Diogenēs Laërtios, vi, 63; Volume 1, p. 489.

[48] Volume 1, pp. 295, 570.

Alexander entrusted it to Deinocratēs of Rhodos, who was the most eminent architect of his time. It was Deinocratēs who had designed the new temple of Artemis in Ephesos [49] and he had conceived the idea of cutting one of the peaks of Mount Athos into the shape of a gigantic statue of Alexander.[50] He was still living under Ptolemaios II, and it is told of him that he had planned in memory of Arsinoē (the king's wife) a temple the roof of which was armed with loadstones so that the queen's statue appeared to be suspended in the air.[51]

The city was built on a narrow piece of land limited on the north side by the Mediterranean Sea and on the south by Lake Mareōtis. It was set out around two very large avenues, a very long one (the Canopic road) extending from east to west, and a shorter one perpendicular to it. The center of the city was at or near the crossing of those main avenues. Other streets parallel to them formed a chessboard pattern. The city was divided into five sections designated by the first five letters of the Greek alphabet, which are also the first five numerals. A very large part (say a quarter or a third of the whole) was occupied by the royal palaces, a vast collection of temples and parks. The *sēma*, the Museum, the Library, and no doubt the barracks of the royal guards were all in that royal enclosure called the Brucheion. More temples and public buildings stood on the Canopic road. On the eastern hill, now called Kum al-dīk, was a great park, the Paneion (Pan's sanctuary). On another hill, at the southwest of the ancient city, was the Sarapeion. There were stadiums and hippodromes. Two vast cemeteries extended to the east and west ends and suburbs were gradually established eastward in the plain of Hadra and on the hills of Ramleh.[52] The harbors will be described presently.

The outline and details are difficult to establish with certainty, because the Greek city is like a palimpsest that has been effaced and rewritten by the Christians, then again by the Muslims. Excavations are made impossible by the richness of the modern city and the sacredness of many Islamic edifices or enclosures.

[49] The old temple of Ephesos was built in the sixth century; it was set afire by Hērostratos of Ephesos, who wished "to immortalize himself," and succeeded in doing so. According to the legend, the fire occurred on the very night of Alexander's birth in 356.

[50] The realization of that grandiose idea was not even begun. Thanks to his conception, Deinocratēs might be called the forerunner of the Danish sculptor, Bertel Thorvaldsen (1768–1844), who designed the colossal lion of Lucerne in memory of the Swiss guards massacred in 1792, and of the American sculptor, Gutzon Borglum (1871–1941), who carved figures of American presidents on the rocks of Mount Rushmore in the Black Hills of South Dakota.

[51] Pliny, *Natural history*, xxxiv, 42 or 147. A similar story was told later about the coffin of the Muslim Prophet (ṣl'm). That is a bit of ancient magnetic folklore.

[52] For more details on the old city, see E. Breccia, *Alexandrea ad Aegyptum* (Bergamo, 1914), the excellent Baedeker (ed. in English; Leipzig, 1929), and Edward Alexander Parsons, *The Alexandrian Library* (Amsterdam: Elsevier 1952) [*Isis* 43, 286 (1952)], including many maps.

THE HARBORS OF ALEXANDRIA AND THE LIGHTHOUSE

The selection of Alexandria's site for the main city of Greek Egypt was exceedingly wise. We must assume that Alexander had been guided in this by the Greek merchants who flourished in Naucratis and had obtained a good knowledge of the different localities of the Nile delta. The location was not unknown before Alexander. The island Pharos in the harbor, to which we shall return presently, was already mentioned in the *Odyssey* (IV, 355) as being a day's sailing from Aigyptos. The poet probably meant a day's sailing from the Canopic Nile, for the island was hardly more than a mile from the shore. There was a fishing village [53] on the site, but no town. Why did Alexander select that isolated spot at the western end of the Delta? One reason may have been that the harbors east of it [54] were always in danger of being choked by river alluvia; Alexandria's indirect connection with the Nile saved her from that peril.

The new city was placed between the sea and the lake Mareōtis, which afforded communication with the Nile. There were thus two harbors, one at the sea side, north of the city, and the other at the lake side, south of it. Strabōn (I–2 B.C.) records that more business came to Alexandria from the Nile than from the sea, and that is very plausible. Paris is today one of the largest harbors (if not the very largest) of France, though it depends exclusively on river and canal traffic; remember that the Nile is one of the greatest rivers of the whole world.

The sea harbor faced the island or Pharos, the existence of which was probably a determining factor in the selection of the site. The original plan of the city implied the building of a mole seven stadia in length [55] connecting Pharos with the shore and thus creating two distinct sea harbors, the eastern or Great Harbor, protected by another mole on its eastern side, and the western harbor or Eunostos (*Eunostu limēn*), the harbor of happy return.[56]

When the Nile was high, it filled the lake instead of creating marshes as it did elsewhere. Hence, the air of the city, placed between the sea and the lake and away from marshland, was relatively pure; it was cooled by the Etēsian winds blowing from the north. Another great advantage was the

[53] Rhacotis, opposite the island of Pharos. The locality may have been selected by Cleomenēs of Naucratis, who was Alexander's intendant in Egypt; Cleomenēs was a clever financier, but his extortions were carried so far that he was put to death by order of Ptolemaios Sōtēr.

[54] The Canopic branch reached the Mediterranean at Abuqīr, east of Alexandria; other branches at Rashīd (Rosetta) and farther east. Naucratis was on the Canopic arm but at some distance from the sea.

[55] The mole was 600 cubits long, 20 wide, 3 above sea level. At very high tides the sea might cover it a little, reaching the ankles of pedestrians. As the island was higher than the shore, the mole was attached to it by means of a sloping bridge made of sixteen arches of decreasing height.

[56] A long description of the harbors was given by Strabōn in his *Geography*, XVII, 1, 6–8. He observed the salubrity of Alexandria.

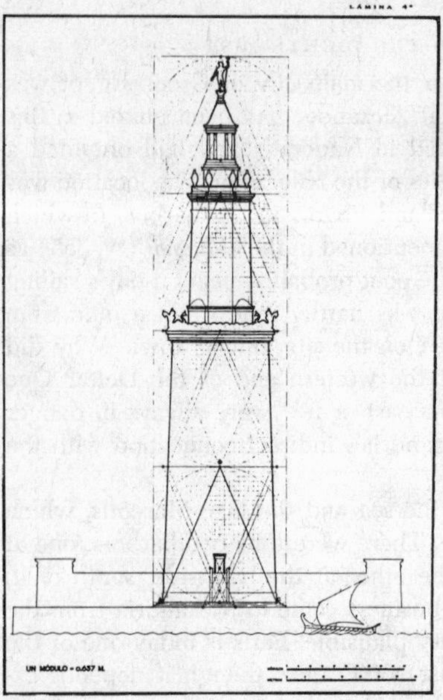

Fig. 6. Reconstruction of the Alexandria lighthouse (Pharos), by M. L. Otero (*Andalus 1*, plate 4a [1934]).

absence of malaria. It has been argued that the decadence of Greece was partly caused by the increasing occurrence of malarial fevers; the Delta, or at any rate the western part of it, was happily free from that insidious plague.[57]

The island Pharos provided a northern screen to both harbors. A great lighthouse was built on it, which every homebound sailor could see from far off. He saw not the island but the lighthouse and called it "Pharos." [58] We shall do the same from now on.

The Pharos was built upon the easternmost end of this island during the rule of Ptolemaios II Philadelphos, *c.* 270, by the architect Sōstratos of Cnidos. It excited the admiration of every traveler, not only in antiquity but also in medieval times, for it continued to exist until the fourteenth century. There are a number of references to it in medieval literature (chiefly Arabic), but the only detailed description we owe to a Hispano-Muslim scholar, Yūsuf Ibn al-Shaikh of Malaga (1132–1207), who lived in Alexandria in 1165. That description was included in his Kitāb alif-bā (meaning abecedary); it is a compendium arranged in alphabetic order for the education of his son 'Abd al-Raḥīm.[59] The Pharos had suffered much in the course of time and when Ibn al-Shaikh visited it in 1165 it was

[57] On malaria in Greece, see Volume 1, pp. 341, 357.

[58] The Greek word *pharos* acquired the meaning "lighthouse" and served to designate any one of these. The word has been transmitted to many Romance languages, *phare* in French, *faro* in Italian and Spanish, and so on. The word "pharos" is also used in English to designate a light likened to a lighthouse, as a ship's lantern. Every time we use one of these words we pay tribute to Alexandria.

[59] Printed in Cairo, 1870. The description occurs in vol. 2, pp. 537–538; its importance was discovered by Miguel Asin y Palacios, who translated and discussed it in *Andalus 1*, 241–300 (1930); Asin's explanation was completed on the technical side by the architect, Modesto Lopez Otero. See also *Andalus 3*, 185–193 (1935). The most learned study is the one of Hermann Thiersch (1874–1939), *Pharos* (266 pp., 10 pls., 455 ills.; Leipzig, 1909). Thiersch's work is still very valuable, but his conclusions must be modified because of Asin's discovery. That discovery was explained in English by the late Duke of Alba and Berwick, *Proceedings of the British Academy* (*London*) *19*, 3–18 (1933) and again in the *Illustrated London News*, 27 January 1934.

no longer used as a lighthouse; it must have been in tolerably good shape, however, for not only was he able to go to the top of it and take many measurements, but he observed in the center of the highest platform a small mosque, with four doors, surmounted by a dome. He also observed a Greek inscription (on the outside wall just below the first platform) and described its general appearance but could not decipher it. It is typical of Muslim thought in the twelfth century that the Pharos was still good enough to permit the erection of a mosque at its very top but was no longer used for its original and excellent purpose.

From the Arabic description we gather that the lighthouse was erected upon a heavy stone platform established 12 cubits above sea level; it was built in three sections – lowest, middle, uppermost – which were of decreasing area and respectively square, octagonal, and cylindrical in shape. The perimeters of the three bases measured 45 x 4 = 180 paces, 10 x 8 = 80 paces, and 40 paces.[60] The lower part was 71 meters high; there were fifty chambers or recesses in its walls. The first platform was reached by an inside circular ramp,[61] wide enough for two horsemen to meet on it without trouble. To reach the second and third platforms one used stone staircases of 32 and 18 steps. The source of light was probably a fire that was kept burning at night on the top platform. The total height was at least 120 meters and probably 140.3 meters. It was a very high tower indeed, and must have been visible on land or sea from a great distance. The Greeks and Barbarians sailing to the metropolis were awed by that wonderful sight. It was considered one of the seven wonders of the world (see below). It was destroyed by an earthquake in the thirteenth century.

The Pharos was the best advertisement of Alexandrian business and the best symbol of its prosperity. That material prosperity was in great contrast with the extreme indigence of the fallāḥīn (which has continued to this day) and also with the commercial decadence of Greece and the poverty obtaining in the greatest part of it. Athens had been reduced to the status of an impoverished provinçial town, yet its spiritual prestige was as great as ever; its schools were still the leading schools of the ancient world and it was still the main center of pilgrimage for every lover of wisdom. Alexandria was very rich, or, let us say, its kings and leading merchants and financiers controlled the business of the world; the Greek spoliation of Asia and Egypt had released the enormous wealth accumulated by Oriental kings and the circulation of gold and silver had increased considerably.

In the markets of Alexandria came together the many products of Egypt (cereals, papyrus, glassware, woven and embroidered fabrics of many kinds, carpets, jewels), those of Arabia (perfumes, incense),[62] and those of the

[60] We may assume that a cubit equals about 60 cm or 23½ in., and a pace about 70 cm or 27½ in.

[61] As in the Giralda of Seville and, the Round Tower of Copenhagen.

[62] Frankincense was required in large

Mediterranean world. Archaeological discoveries have brought to light Alexandrian objects in Hungary and Russia, not to mention nearer countries, and, in Alexandria, ceramics coming from Rhodos, Thasos, Cnidos, Crete, and elsewhere. It is noteworthy that those of Rhodos are the vast majority, because Rhodos was itself one of the greatest business centers of the eastern Mediterranean. The "central bank" of Egypt was located in Alexandria. Every industry and trade was taxed, many were monopolized [63] and farmed out to royal entrepreneurs.

The Pharos was not like the proud belfries of medieval cities, a symbol of democracy; it was rather the gigantic advertisement of the most prosperous kings of the Hellenistic age.

THE SEVEN WONDERS OF THE WORLD

We may pause a moment to wonder at that expression, traces of which exist in every Western literature: "The seven wonders of the world." It is probably an old conceit,[64] yet it occurred for the first time relatively late. The first literary account is a Greek tract entitled *Peri tōn hepta theamatōn* (*De septem orbis spectaculis*) ascribed to Philōn of Byzantion. As long as the author was identified with Philōn the mechanician, who flourished in the third or second century B.C., it might be considered early, but it is certain that this Philōn, the Philōn of "the seven spectacles," flourished not before the fourth century of our era, perhaps in the fifth.[65]

It is a short and poor treatise, which contains but little information because it is rhetorical, not descriptive, and which has reached us incompletely (the end is lacking).[66] It praises the seven wonders in the following order: (1) the "hanging gardens" of Babylon, (2) the Pyramids, (3) the statue of Zeus by Pheidias, (4) the Colossos of Rhodos, (5) the walls of Babylon, (6) the Temple of Ephesos, and (7) the Mausōleion of Halicarnassos (the end of six and the whole of seven are lost). This order is stupid; the Great Pyramid was built by Cheops (Khufu, 29th century B.C.); Nos. 1 and 5, the

quantities in the temples of many gods; Tarn, *Hellenistic civilisation,* p. 260.

[63] Much detail concerning this may be found in Bernard Pyne Grenfell, *Revenue laws of Ptolemy Philadelphus* (388 pp., 13 pls.; Oxford, 1896). Extract from Grenfell's book concerning the oil monopoly in G. W. Botsford and E. G. Sihler, *Hellenic civilization* (New York, 1915), pp. 607–609. Oil was the greatest royal monopoly and the best organized, but there were many other monopolies, such as textiles and papyrus.

[64] Strabōn (I–2 B.C.) remarked in his *Geography,* XVII, 1, 33, that the Pyramids were numbered among the seven wonders (*en tois hepta theamasi*). Hence, those wonders had already been listed before his time.

[65] In my *Introduction,* I placed the mechanician tentatively in (II–2 B.C.). In the article by W. Kroll, Pauly-Wissowa, vol. 39 (1941), 53–55, he is placed at the end of the third century B.C., and Philōn of the seven spectacles in the fourth or fifth century after Christ.

[66] First edition by Leo Allatius (Rome, 1640); second by Io. C. Orelli (Leipzig, 1816). The best is the one by Rudolf Hercher at the end of his edition of Ailianos (III–1) (Paris, 1858), vol. 2, pp. 101–105. The three editions are Greco-Latin.

hanging gardens and walls of Babylon, were constructed by Nebuchadrezzar (605–561); No. 3, the Zeus, was created by Pheidias (490–432) about the middle of the fifth century; Nos. 6 and 7 date possibly from the middle of the fourth century. I say "possibly" because, in his meaningless description of the Artemision, there is nothing to indicate whether Philōn refers to the old one built in the period 575–425 and burnt by Ērostratos in 356 or to the new one begun about 350 and burnt by the Goths in A.D. 262. King Mausōlos died in 353, and his monumental tomb was erected soon afterward by Artemisia, who was his sister, wife, and successor. The latest wonder dealt with is the gigantic statue of Hēlios, 70 cubits high, which was the creation of Charēs of Lindos [67] (fl. 290 B.C.), the favorite disciple of Lysippos. It took twelve years to build it at a cost of 300 talents. The Colossos, as it was called, stood at the entrance of the harbor of Rhodos, but the tradition that its legs bestrode the mouth of the harbor is legendary. It was destroyed by an earthquake, c. 224 B.C. The fragments remained on the ground for almost nine centuries, until they were sold by a general of Mu'āwiya (caliph 661–680) to a Jew of Emesa, who carried them away on 980 camels in 672 (there are variants to this story; the number of camels in particular varies from 900 to 30,000!).[68]

To return to the seven wonders, this conceit, favored by the sacredness of the number seven, has reached us across the ages and will never die. There have always been, and there will always be, seven wonders, but the list varies from time to time.

It is very strange that Philōn omitted the Pharos, and in that he was certainly wrong, for it was the most astounding monument of its kind until modern times and its construction involved the solution of many difficult problems.[69]

The list that appears most often is the same as Philōn's except that the gardens and walls of Babylon count as one item, and the Pharos is added.[70]

[67] Lindos was one of the three ancient cities of Rhodos; the city of Rhodos founded in 408 B.C. was relatively modern. The sun god, Hēlios, was the patron of the island. Charēs was not the only Rhodian artist; Rhodos was famous as an artistic as well as a commercial center from prehistoric times. Rhodian masterpieces of the Hellenistic age may be seen in many places. For example, the "Laocoōn" and the "Biga" (chariot drawn by two spirited horses) in the Vatican, the "Quadriga of Hēlios" in San Marco, Venice, the "Farnese Bull" in the Naples Museum, and so forth. Skevos Zervos, *Rhodes, capitale du Dodécanèse* (folio, 378 pp., 687 ills.; Paris, 1920), admirably illustrated.

[68] The best source is the *Chronographia* of Theophanēs Homologētēs (IX-1), Carolus de Boor's edition (Leipzig, 1883), vol. 1, p. 345. According to Theophanēs, the fragments were of bronze, but it is hard to believe that such a large mass of bronze would have been overlooked for nine centuries.

[69] It was the first high tower in a modern sense as opposed to a pyramid or a ziggurat.

[70] I do not know who was first to include the Pharos. It is possible that the list which includes it is more ancient than Philōn's. The vitality of the Pharos list is proved by the fact that Victor Hugo reproduced it in his *Légende des siècles* (1877–1883).

Other ancient lists included the Athēnē of Pheidias, the Asclēpieion of Epidauros, the Temple of Jupiter or Capitol in Rome, the Temple of Hadrian (117–138) in Cyzicos, and even the Temple in Jerusalem.

Fate has dealt in her own capricious way with all those wonders; the only one that still exists today is the most ancient, the Great Pyramid, two millennia older than the next in age, while the youngest of them, the Colossos of Rhodos, lasted hardly more than sixty years.

The two outstanding institutions of the Alexandrian Renaissance were the Museum and the Library. Whether these were two separate institutions or only one is a moot question. Both were royal creations; both were established in the royal part of the city and were entirely dependent upon royal pleasure. Their mutual independence or interdependence is an administrative matter that does not concern us.

The remainder of Part One is mainly devoted to the Museum and the scientific activities that originated in it or received some help or inspiration from it and to the Library and the Alexandrian humanities, most of which were centered upon the Library or inspired by it.

II

THE MUSEUM

The Ptolemies were typically Greek, in that they encouraged trades and industries and loved the fruits thereof, money, but were not satisfied to accumulate it. They were ready enough to keep all the burdens of Egypt upon the shoulders of the miserable fallāḥīn, yet, at the same time, they wanted to be known as benefactors (*evergetai*). They were anxious to raise the spiritual prestige of their kingdom and to emulate in artistic splendor not only the other Hellenistic cities but Athens itself. Therefore, it was not enough for them to bring over from Macedonia and Greece merchants and administrators; they called in also philosophers, mathematicians, physicians, artists, poets. They were Greek enough to realize that prosperity without art and science is worthless and contemptible.

FOUNDATION OF THE MUSEUM. PTOLEMAIOS I SŌTĒR AND PTOLEMAIOS II PHILADELPHOS

As soon as Ptolemaios, son of Lagos had regulated the government of Egypt and completed the foundation of Alexandria, he showed deep concern not only for the material development of the city but also for its spiritual welfare. Philanthropy, as we understand it, was as remote from his mind as could be, but he was conscious of the supreme value of Hellenic culture and wanted to establish it in Egypt. His main creation to that effect was the Museum.

A Museum is a temple of the Muses (*Musai*), who were the daughters of Zeus and Mnēmosynē (memory!), and the patron goddesses of the humanities. There were nine of them, to wit: Cleiō, muse of history; Euterpē, muse of lyric poetry; Thaleia, muse of comedy and joyful poetry; Melpomenē, muse of tragedy; Terpsichorē, muse of the dance and music; Eratō, muse of erotic poetry; Polymnia, muse of hymns; Urania, muse of astronomy; Calliopē, muse of epic poetry. Apollōn, god of the lyra, was called their leader (*Musagetēs*). A good deal of mythology is stupid and dull, but these gracious inventions are extremely pleasing and help us to understand and to love the Greek genius. Note that seven of the muses were patrons of literature, chiefly poetry in all its forms, one of history, and one other (this is very curious) of astronomy. Thus, this earliest council of the humanities made room for at least one branch of science; Urania, it is true, represented not the astronomers but the glory of heaven. Cleiō and Urania, taken together, were the earliest patrons of the history of science.

A charming use of the word "museum" was made by Euripidēs, when he referred to the *museia* of birds, where they gather to sing! Temples to all the Muses or any of them existed in many parts of Greece; there was a Museum in Plato's Academy, and the same name was given to a school of arts and letters founded in Athens by Theophrastos in Aristotle's memory, but all those institutions were dwarfed by the Ptolemaic foundation, and, when we speak of antiquity, the name Museum evokes that one and no other. The Museum of Alexandria was so famous that its name became a common name in every Western language,[1] and yet we know very little of its organization.

This is what Strabōn wrote about it:

> The Museum is also a part of the royal palaces; it has a public walk, an Exedra[2] with seats, and a large house, in which is the common mess-hall of the men of learning who share the Museum. This group of men not only hold property in common, but also have a priest in charge of the Museum, who formerly was appointed by the kings.[3]

Such a description is meager enough, yet it gives some information. In the first place, the Museum was not only a royal institution but was "a part of the royal palaces." Nothing could exist in Egypt without the king's pleasure, and everything good was to the king's credit (whatever evil there was was generally credited to the people). The Museum occupied some of the buildings in the royal city close to the great harbor.[4] There was a priest who discharged the religious duties (like the president of one of our colleges conducting services in the chapel). The members of the Museum held property in common; that is possible and plausible. In short, the Museum was a group of buildings equipped for various scientific purposes; the members lived together like the fellows or tutors in a medieval college.

[1] Compare other common names: academy (Plato), lyceum (Aristotle). Each language is an archaeological collection. The name Museum lost its original meaning, however, and today it is used chiefly for buildings containing archaeological or artistic collections. In 1794, the Jardin des Plantes of Paris was renamed the Muséum d'histoire naturelle. The Paris Museum is perhaps the closest analogy to the Alexandria Museum. The largest modern museums are staffed by learned men, who give lectures and carry out various forms of research and teaching.

[2] An *exedra* is a portico or covered colonnade, usually of a curved shape and often provided with seats, for conversation in the open air and the shade. The Greeks called it also *leschē*, as at Delphoi, for example (Volume 1, p. 229).

[3] Strabōn (I–2 B.C.), *Geography* (XVII, 1, 8). Quoted from the Loeb edition and translation in 8 vols. by Horace Leonard Jones (Cambridge, 1932), vol 8, p. 35.

[4] For comparison, think of the grand seraglio in Istanbul or of the imperial city in Peking, or just imagine all of the government and public buildings of a modern capital put together in an enclosed park.

Though we know so little concerning the organization of the Museum, we can deduce considerably more from the various activities which it encouraged. It was probably more like a scientific research institute than a college; there is no evidence that it was used for teaching purposes, or, to put it otherwise, the teaching was restricted to the very best kind, the kind that is given informally by a teacher to his apprentices and assistants. We may assume that the administration was minimum and casual. There were no examinations, no degrees, no credits. The main reward was the consciousness of good work well done, and the main punishment (save expulsion from that paradise) was the consciousness of bad work badly done.

The Museum must have included astronomical instruments, and the room or building housing them might then be called an observatory; it included also a room for anatomical dissection, for physiological experiments, and around it were botanical and zoölogical gardens. We shall speak of the library (which is an essential part of every scientific institute) in Chapter X.

The Museum was founded by the first king, but it was developed mainly by his son and successor, Ptolemaios II Philadelphos. It is impossible to determine more exactly the share of each in the great undertaking, but it is certain that an immense amount of work was accomplished in the first half of the third century and this would have been impossible if the second Ptolemaios had been obliged to begin his own efforts from scratch in 285.

Such a foundation would have been impossible without the example and stimulus of Greek genius. The founders were not only the first two kings but at least two other men, without whom they themselves would have been helpless. These men were, in order of seniority, Dēmētrios of Phalēron and Stratōn of Lampsacos.

Dēmētrios of Phalēron. Both Dēmētrios and Stratōn were successors of Aristotle and more directly of Theophrastos. This reminds us of one important explanation of the Hellenistic renaissance. The Alexandrian empire was a material thing which ceased to exist when it was broken to pieces after Alexander's death; on the other hand, the Aristotelian synthesis was a spiritual reality, which was often corrected and modified in the course of time, yet is incapable of destruction. The Museum of Alexandria was a distant continuation and amplification of the Lyceum of Athens.

Dēmētrios, born in Phalēron (the oldest harbor of Athens) c. 345, was a writer and statesman who was at one time very popular and at another very unpopular in his native city; he was absolute governor of Athens, and his fastuous habits and profligacy must have turned many people against him. When the Macedonian king Dēmētrios Poliorcētēs (the besieger) "liberated" Athens in 307, the other Dēmētrios was obliged to move out. He took refuge in Alexandria, where Ptolemaios Sōtēr welcomed him. It was not the first, nor the last, time that political exiles created or improved new opportunities. Ptolemaios needed just such a man as Dēmētrios; they were likely to stimulate one another; we cannot even be sure whether the first idea of founding the Museum and Library was the king's own or that of his protégé. It does not really matter. While in Athens, Dēmētrios had been too busy with various offices and political oratory to do much literary work. It is believed that the majority of his abundant writings (all lost) were composed in Egypt. He was perhaps the first director or the founder of the Library; at any rate his own collection of books was the nucleus of it. When Ptolemaios II succeeded his father, in 285, Dēmētrios was disgraced and exiled to Upper Egypt. According to Diogenēs Laërtios (III-1), he died of an asp bite and was buried in the district of Busiris near Diospolis; [5] that would be sometime after 283.

Stratōn of Lampsacos. The other man, Stratōn, son of Arcesilaos, was born in Lampsacos (on the Asiatic side of the Hellēspontos, Dardanelles) in the last quarter of the fourth century; he thus belonged to the generation following that of Dēmētrios; he was not only, like the latter, a pupil of Theophrastos [6] but finally suceeded him. The first Ptolemaios called him to Egypt, c. 300, to be the tutor of his son, the future Ptolemaios II, and Stratōn continued to serve in that way until 294, when he was replaced by Philētas of Cōs.[7] He may have remained in Alexandria a few more years until Theophrastos' death in 288, when he was called back to Athens to direct the Lyceum. He became the head (*scholarchēs*) of it in the 123rd Olympiad (288–284), continued to preside over it for 18 years, and appointed Lycōn of Trōas as his successor. He died c. 270–268. Said Diogenēs Laërtios, "He was generally known as 'the physicist' because more than any one else he devoted himself to the most careful study of nature." [8]

Diogenēs' biographies are generally meager from the scientific point of view, yet this brief statement gives us the main clue with regard to Stratōn's personality. We must stop to consider it, for Stratōn was not simply very important in himself (as far as can be judged indirectly, for his writings are lost), but it was he who gave to the Museum its scientific tone. The orator

[5] Diogenēs Laërtios (III-1), *Lives of eminent philosophers,* v, 75–83; Loeb edition and translation by R. D. Hicks (Cambridge, 1938), vol. 1, pp. 527–537. Presumably this was Diospolis parva, near Luxor.

[6] Theophrastos was head of the Lyceum for 35 years (323–288); Dēmētrios was his pupil at the beginning of his headship, Stratōn some 20 years later.

[7] Philētas of Cōs, poet and grammarian (d. c. 280). This is another Greek flourishing in young Alexandria and taking part in the incubation of Hellenistic culture. There must have been many others, partly because the intrigues and miseries of their mother countries drove them out and partly because Alexandria was needing them and beckoning to them.

[8] Diogenēs Laërtios, v, 58–64; Loeb edition, vol. 1, pp. 508–519. Diogenēs quotes Stratōn's will *in extenso,* from the collection of such documents made by Aristōn of Ceōs, who succeeded Lycōn. Lycōn was head of the Lyceum for 44 years, from c. 268 to c. 224. Hence, Aristōn became head c. 224.

Dēmētrios and the poet Philētas would have been incapable of doing that, for they had no knowledge of science nor any interest in it, and, but for Stratōn, the Museum might have remained a school of oratory and belles lettres.

The presence of Stratōn in Alexandria between the years 300 and 294 (or 288) is thus an event of great pregnancy. We can imagine the talks between "the physicist," his patron Ptolemaios I, and his student, the future Ptolemaios II. These three men were the main founders of the Museum.

The philosophical and physical views of Stratōn are but indirectly and imperfectly known, and whatever we know relates to his teaching in Athens, after his return from Egypt. We may assume, however, that the general direction of his mind was already established while he was in Alexandria helping to shape the scientific policies of the Museum. Diogenēs ended his biography of him with the words: "Stratōn excelled in every branch of learning, and most of all in that which is styled 'physics,' a branch of philosophy more ancient and important than the others."

In other words, the scientific tendencies of the Lyceum, which Theophrastos had stressed, were stressed even more by Stratōn. The latter must have realized that, however noble our metaphysical cogitations might be, they could lead to no safe harbor; the only road to intellectual progress was scientific research. It was his curious fate to experience the transition from the Lyceum to the Museum and later from the Museum to the Lyceum. We shall see that the Museum encouraged men of science but hardly any philosopher; it was definitely (thanks to him) a school of science, not a literary or philosophical academy.

Stratōn's physics was but the continuation of the more scientific part of Aristotle's. His bent was pantheistic and materialistic, yet he seems to have objected to atomic conceptions. I imagine that many of his contemporaries were antiatomic because they were anti-Epicurean. Moreover, whatever might be the ultimate fate of atomism (twenty-two centuries later), Epicurean atomism was not a sound approach, Platonism much less so. Stratōn tried to establish physics on a positive basis and to rid it of a vain search for final causes; as far as can be judged from fragmentary data, he tried to combine idealism with empiricism in the best Aristotelian vein, to encourage induction from experience, not deduction from metaphysical postulates. Stratonian physics was an adaptation of Aristotelian physics to more detailed knowledge and practical needs; it could not be fertile because the experimental basis was still utterly insufficient. If it was he, as I believe, who advised the Museum to fight shy of philosophy, this was partly due to the endless disputes between the Academy, the Lyceum, the Garden, and the Porch, which created confusion, more heat than light.

Yet it is not true to say as Cicero did that Stratōn neglected the most important part of philosophy, ethics. At any rate, Cicero's reproach is not justified by the list of Stratōn's writings as given by Diogenēs Laërtios (v, 59–60). As head of the Lyceum, he was naturally obliged to deal with ethical and even with metaphysical problems; yet he was primarily a physicist (*physicos*) and his main creation was the Museum. That is quite enough for his immortality.

LATER HISTORY OF THE MUSEUM

The Museum continued to exist throughout Hellenistic times. The scholars and men of science attached to it received salaries from the kings and later from the Roman governors, who appointed a director (*epistatēs*) or priest (*hiereus*) as head of it. After the middle of the second century, the Museum lost much of its importance because of political vicissitudes and of the rivalries of other institutions located in Athens, Rhodos, Antioch, even Rome and Constantinople. The early emperors, chiefly Hadrian (ruled 117–138), endeavored to restore its glory but with little success. It was all but destroyed in A.D. 270, but revived. The last scientists to glorify it were Theōn (IV–2) and his daughter, Hypatia (V–1). The murder of the latter in 415 by a Christian mob was the end of a great institution which had lasted seven centuries.

To return to the early days of the Museum or to the first century of its existence, its influence upon the progress of science was considerable. It was because of its creation and because of the enlightened patronage which enabled it to function without hindrance that the third century witnessed such an astounding renaissance. The fellows of the Museum were permitted to undertake and to continue their investigations in complete freedom. As far as can be known, collective research was now organized for the first time, and it was organized without political or religious directives, without purpose other than the search for the truth.

Great scientists and scholars were free to conduct their inquiries as they thought best and the cosmopolitan atmosphere of Alexandria enabled them to take advantage of all the work done before them, not only by Greeks but also by Egyptians and Babylonians. This will be fully illustrated in the following chapters.

III

EUCLID OF ALEXANDRIA

EUCLID'S LIFE AND WORK

One of the earliest, as well as one of the greatest, men of science connected with the new metropolis, Alexandria, was Euclid (III-1 B.C.). We all know his name and his main work, the *Elements of geometry*, but we have no certain knowledge about himself. The little that we know – and it is very little – is inferential and of late publication. This kind of ignorance, however, is not exceptional but frequent. Mankind remembers the despots and the tyrants, the successful politicians, the men of wealth (some of them at least), but it forgets its greatest benefactors. How much do we know about Homer, Thalēs, Pythagoras, Dēmocritos. . . ? Nay, how much do we know about the architects of the medieval cathedrals or about Shakespeare? The greatest men of the past are unknown, even when we have received their works and enjoy their abundant blessings.

The very places and dates of Euclid's birth and death are unknown. We call him Euclid [1] of Alexandria, because that city is the only one with which he can be almost certainly connected. Let us put together all the information that has filtered down to us. He was probably educated in Athens and, if so, he received his mathematical training at the Academy, which was the outstanding mathematical school of the fourth century and the only one where he could have gathered easily all the knowledge that he possessed. When the vicissitudes of war and political chaos made it increasingly difficult to work in Athens, he moved to Alexandria. He flourished there under the first Ptolemaios and possibly under the second. Two anecdotes help to reveal his personality. It is said that the king (Ptolemaios Sōtēr) asked him if there was in geometry any shorter way than that of the *Elements*, and he answered that there was no royal road to geometry – an excellent story, which may not be true as far as Euclid is concerned but

[1] His Greek name was Eucleidēs, but it would be pedantic to use it, because the name Euclid belongs to our language; other forms of it (like Euclide) have been acclimatized to other languages.

has an eternal validity. Mathematics is "no respecter of persons." The other anecdote is equally good. Someone who had begun to study geometry with Euclid, when he had learned the first theorem, asked him, "But what shall I get by learning those things?" Euclid called his slave and said, "Give him an obol, since he must gain from what he learns." There are still many idiots today who would judge education as Euclid's student did; they want to make it immediately profitable, and if they are given their way, education vanishes altogether.

Both anecdotes are recorded relatively late, the first by Proclos, the second by Stobaios, both of whom flourished in the second half of the fifth century; they are plausible enough; they might be literally true, or, if not, they are traditional images of the man as his contemporaries had seen him or imagined him to be. The great majority of historical anecdotes are of that kind; they are as faithful as popular imagery can be.

Was Euclid connected with the Museum? Not officially. Otherwise, the fact would have been recorded; but if he flourished in Alexandria, he was necessarily acquainted with the Museum and its Library, which were the very heart of intellectual life in all its forms. As a pure mathematician, however, he did not need any laboratory [2] and he might easily have brought from Greece all the mathematical rolls that he needed; we may assume that good students would themselves copy the texts that they were required to know or were anxious to keep. A mathematician does not need close collaborators; like the poet, he does his best work alone, very quietly. On the other hand, Euclid may have been teaching a few disciples, either in the Museum or in his own home; this would have been natural and is confirmed by Pappos' remark that Apollōnios of Perga (III–2 B.C.) was trained in Alexandria by Euclid's pupils. This helps to confirm Euclid's date, for Apollōnios lived from c. 262 to 190; this would place the teacher of his teachers in the first half of the third century.

Euclid himself was so little known that he was confused for a very long time with two other men, one much older than himself, the other considerably younger. Medieval scholars insisted on calling him Euclid of Megara because they mistook him for the philosopher Eucleidēs, who had been one of Socrates' disciples (one of the faithful who attended the master's deatl in prison), a friend of Plato's and the founder of the school of Megara. This confusion was confirmed by the early printers until late in the sixteenth century. The first to correct the error in a Euclidean edition was Federico Commandino in his Latin translation (Pesaro, 1572). The other confusion was caused by the fact that Theōn of Alexandria (IV–2), who edited the *Elements*, was believed to have added the demonstrations! If

[2] If the optical, astronomical, and musical works ascribed to him are genuine, he may have needed technical assistance and instruments, and the Museum would have been the only place where such were available. There is no reference to the Museum, however, in those works.

such had been the case, he would have been the real Euclid; the error is as deep as if one claimed that Homer had conceived the *Iliad* but that Zēnodotos of Ephesos was the real composer of it.

The Elements. My comparison with Homer is valid in another way. As every body knows the *Iliad* and the *Odyssey*, so does every body know the *Elements*. Who is Homer? the author of the *Iliad*. Who is Euclid? the author of the *Elements*.

We cannot know these great men as men, but we are privileged to study and use their works – the best of themselves – as much as we deserve to. Let us thus consider the *Elements*, the earliest elaborate textbook on geometry that has come down to us. Its importance was soon realized and, therefore, the text has been transmitted to us in its integrity. It is divided into thirteen books, the contents of which may be described briefly as follows:

Books I–VI: Plane geometry. Book I is, of course, fundamental; it includes the definitions and postulates and deals with triangles, parallels, parallelograms, etc. The contents of Book II might be called "geometric algebra." Book III is on the geometry of the circle. Book IV treats regular polygons. Book V gives a new theory of proportion applied to incommensurable as well as commensurable quantities. Book VI is on applications of the theory to plane geometry.

Books VII–X: Arithmetic, theory of numbers. These books discuss numbers of many kinds, primes or prime to one another, least common multiples, numbers in geometric progression, and so on. Book X, which is Euclid's masterpiece, is devoted to irrational lines, all the lines that can be represented by an expression, such as

$$\sqrt{(\sqrt{a}+\sqrt{b})}$$

wherein a and b are commensurable lines, but $\sqrt{a}$ and $\sqrt{b}$ are surds and incommensurable with one another.

Books XI–XIII: Solid geometry. Book XI is very much like Books I and VI extended to a third dimension. Book XII applies the method of exhaustion to the measurement of circles, spheres, pyramids, and so on. Book XIII deals with regular solids.

Plato's fantastic speculations had raised the theory of regular polyhedra to a high level of significance. Hence, a good knowledge of the "Platonic bodies"[3] was considered by many good people as the crown of geometry. Proclos (V–2) suggested that Euclid was a Platonist and that he had built his geometric monument for the purpose of explaining the Platonic figures. That is obviously wrong. Euclid may have been a Platonist, of course, but he may have preferred another philosophy or he may have carefully avoided

[3] For a discussion of the regular polyhedra and of the Platonic aberrations relative to them, see Volume 1, pp. 438–439. Briefly, Plato was so impressed by the fact that there could be only five regular polyhedra that he attached a cosmologic meaning to each of them and, furthermore, established a connection between the five solids and the five elements. The Platonic theory of five solids was fantastic, the theory of five elements equally so, and the combination of both was a compound fantasy. Yet, so great was Plato's prestige that that compound fantasy was accepted as the climax of science as well as a metaphysical triumph.

philosophic implications. The theory of regular polyhedra is the natural culmination of solid geometry and hence the *Elements* could not but end with it.

It is not surprising, however, that the early geometers who tried to continue the Euclidean efforts devoted special attention to the regular solids. Whatever Euclid may have thought of these solids beyond mathematics, they were, especially for the Neoplatonists, the most fascinating items in geometry. Thanks to them, geometry obtained a cosmical meaning and a theological value.

Two more books dealing with the regular solids were added to the *Elements*, called Books XIV and XV and included in many editions and translations, manuscript or printed. The so-called Book XIV was composed by Hypsiclēs of Alexandria at the beginning of the second century B.C. and is a work of outstanding merit; the other treatise, "Book XV," of a much later time and inferior in quality, was written by a pupil of Isidōros of Milētos (the architect of Hagia Sophia, *c.* 532).

To return to Euclid, and especially to his main work, *The thirteen books of the Elements* when judging him, we should avoid two opposite mistakes, which have been made repeatedly. The first is to speak of him as if he were the originator, the father, of geometry. As I have already explained apropos of Hippocrates, the so-called "father of medicine," there are no unbegotten fathers except Our Father in Heaven. If we take Egyptian and Babylonian efforts into account, as we should, Euclid's *Elements* is the climax of more than a thousand years of cogitations. One might object that Euclid deserves to be called the father of geometry for another reason. Granted that many discoveries were made before him, was he not the first to build a synthesis of all the knowledge obtained by others and himself and to put all the known propositions in a strong logical order? That statement is not absolutely true. Propositions had been proved before Euclid and chains of propositions established; moreover, "Elements" had been composed before him by Hippocratēs of Chios (V B.C.), by Leōn (IV-1 B.C.), and finally by Theudios of Magnēsia (IV-2 B.C.). Theudios' treatise, with which Euclid was certainly familiar, had been prepared for the Academy, and it is probable that a similar one was in use in the Lyceum. At any rate, Aristotle knew Eudoxos' theory of proportion and the method of exhaustion, which Euclid expanded in Books V, VI, and XII of the *Elements*. In short, whether you consider particular theorems or methods or the arrangement of the *Elements*, Euclid was seldom a complete innovator; he did much better and on a larger scale what other geometers had done before him.

The opposite mistake is to consider Euclid as a "textbook maker" who invented nothing and simply put together in better order the discoveries of other people. It is clear that a schoolmaster preparing today an elementary book of geometry can hardly be considered a creative mathematician; he is a textbook maker (not a dishonorable calling, even if the purpose is more often than not purely meretricious), but Euclid was not.

Many propositions in the *Elements* can be ascribed to earlier geometers; we may assume that those which cannot be ascribed to others were discovered by Euclid himself, and their number is considerable. As to the arrangement, it is safe to assume that it is to a large extent Euclid's own. He created a monument that is as marvelous in its symmetry, inner beauty, and clearness as the Parthenōn, but incomparably more complex and more durable.

A full proof of this bold statement cannot be given in a few paragraphs or in a few pages. To appreciate the richness and greatness of the *Elements* one must study them in a well-annotated translation like Heath's. It is not possible to do more, here and now, than emphasize a few points. Consider Book I, which explains first principles, definitions, postulates, axioms, theorems, and problems. It is possible to do better at present, but it is almost unbelievable that anybody could have done it as well twenty-two centuries ago.

Postulates. The most amazing part of that is Euclid's choice of postulates. Aristotle was, of course, Euclid's teacher in such matters; he had devoted much attention to mathematical principles, had shown the unavoidability of postulates and the need of reducing them to a minimum; [4] yet the choice of postulates was Euclid's.

In particular, the choice of postulate 5 is, perhaps, his greatest achievement, the one that has done more than any other to immortalize the word "Euclidean." Let us quote it verbatim:

> If a straight line falling on two straight lines make the interior angles on the same side less than two right angles, the two straight lines, if produced indefinitely, meet on that side on which are the angles less than the two right angles.[5]

A person of average intelligence would say that the proposition is evident and needs no proof; a better mathematician would realize the need of a proof and attempt to give it; it required extraordinary genius to realize that a proof was needed yet impossible. There was no way out, then, from Euclid's point of view, but to accept it as a postulate and go ahead.

[4] Aristotle's views can be read in Sir Thomas L. Heath, *Euclid's Elements in English* (Cambridge, 1926), vol. 1, pp. 117 ff., or in his posthumous book, *Mathematics in Aristotle* (305 pp.; Oxford: Clarendon Press, 1949) [*Isis 41*, 329 (1950)]. A postulate is a proposition that cannot be proved or disproved, yet that one is obliged to assert or deny in order to go forward.

[5] For the Greek text and a much fuller discussion of it than can be given here, see Heath, *Euclid*, vol. 1, pp. 202–20.

The best way to measure Euclid's genius as evidenced by this momentous decision is to examine the consequences of it. The first consequence, as far as Euclid was immediately concerned, was the admirable concatenation of his *Elements*. The second was the endless attempts that mathematicians made to correct him; the first to make them were Greeks, like Ptolemy (II–1) and Proclos (V–2), the Jew, Levi ben Gerson (XIV–1), and, finally, "modern" mathematicians, like John Wallis (1616–1703), the Jesuit father, Girolamo Saccheri (1667–1733) of San Remo in his *Euclides ab omni naevo vindicatus* (1733), the Swiss,[6] Johann Heinrich Lambert (1728–77), and the Frenchman, Adrien Marie Legendre (1752–1833). The list could be lengthened considerably, but these names suffice, because they are the names of illustrious mathematicians, representing many countries and many ages, down to the middle of the last century. The third consequence is illustrated by the list of alternatives to the fifth postulate. Some bright men thought that they could rid themselves of the postulate and succeeded in doing so, but at the cost of introducing another one (explicit or implicit) equivalent to it. For example,

If a straight line intersects one of two parallels, it will intersect the other also. (Proclos)

Given any figure, there exists a figure similar to it of any size (John Wallis)

Through a given point only one parallel can be drawn to a given straight line. (John Playfair)

There exists a triangle in which the sum of the three angles is equal to two right angles. (Legendre)

Given any three points not in a straight line, there exists a circle passing through them. (Legendre)

If I could prove that a rectilinear triangle is possible the content of which is greater than any given area, I would be in a position to prove perfectly rigorously the whole of geometry. (Gauss, 1799)

All these men proved that the fifth postulate is not necessary if one accepts another postulate rendering the same service. The acceptance of any of those alternatives (those quoted above and many others) would increase the difficulty of geometric teaching, however; the use of some of them would seem very artificial and would discourage young students. It is clear that a simple exposition is preferable to one which is more difficult; the setting up of avoidable hurdles would prove the teacher's cleverness but also his lack of common sense. Thanks to his genius, Euclid saw the necessity of this postulate and selected intuitively the simplest form of it.

There were also many mathematicians who were so blind that they rejected the fifth postulate without realizing that another was taking its place. They kicked one postulate out of the door and another came in through the window without their being aware of it!

[6] He must be called Swiss because he was born in Mulhouse, Upper Alsace, and that city was a part of the Swiss confederation from 1526 to 1798; Lambert lived from 1728 to 1777 [*Isis 40*, 139 (1949)].

NON-EUCLIDEAN GEOMETRIES

The fourth consequence, and the most remarkable, was the creation of non-Euclidean geometries. The initiators have already been named: Saccheri, Lambert, Gauss. Inasmuch as the fifth postulate cannot be proved, we are not obliged to accept it; hence, let us deliberately reject it. The first to build a new geometry on a contrary postulate was the Russian, Nikolai Ivanovich Lobachevski (1793–1856), who assumed that through a given point more than one parellel can be drawn to a given straight line or that the sum of the angles of a triangle is less than two right angles. The discovery of a non-Euclidean geometry was made at about the same time by the Transylvanian, Janos Bolyai (1802–1860). Some time later, another geometry was outlined by the German, Bernhard Riemann (1826–1866), who was not acquainted with the writings of Lobachevski and Bolyai and made radically new assumptions. In Riemann's geometry, there are no parallel lines and the sum of the angles of a triangle is greater than two right angles. The great mathematical teacher, Felix Klein (1849–1925), showed the relation of all these geometries. Euclid's geometry refers to a surface of zero curvature, in between Lobachevski's geometry on a surface of positive curvature (like the sphere) and Riemann's, applying to a surface of negative curvature. To put it more briefly, Klein called the Euclidean geometry parabolic, because it is the limit of elliptic (Riemann's) geometry on one side and of the hyperbolic (Lobachevski's) geometry on the other.

It would be foolish to give credit to Euclid for pangeometric conceptions; the idea of a geometry different from the common-sense one never occurred to his mind. Yet, when he stated the fifth postulate, he stood at the parting of the ways. His subconscious prescience is astounding. There is nothing comparable to it in the whole history of science.

It would be unwise to claim too much for Euclid. The fact that he put at the beginning of the *Elements* a relatively small number of postulates is very remarkable, especially when one considers the early date, say 300 B.C., but he could not and did not fathom the depths of postulational thinking any more than he could fathom those of non-Euclidean geometry. Yet he was the distant forerunner of David Hilbert (1862–1943), even as he was Lobachevski's spiritual ancestor.[7]

Algebra. I have said so much of Euclid the geometer that I have but little space left to show other aspects of his genius as mathematician and physicist. To begin with, the *Elements* does not deal simply with geometry but also with algebra and the theory of numbers.

[7] For details, see Florian Cajori, *History of mathematics* (ed. 2; New York, 1919), pp. 326–328; Cassius Jackson Keyser, *The rational and the superrational* (New York: Scripta Mathematica, 1952), pp. 136–144 [*Isis 44*, 171 (1953)].

Book II might be called a treatise on geometric algebra. Algebraic problems are stated in geometric terms and solved by geometric methods. For example, the product of two numbers *a*, *b* is represented by the rectangle whose sides have the lengths *a* and *b*; the extraction of a square is reduced to the finding of a square equal to a given rectangle; and so on. The distributive and commutative laws of algebra are proved geometrically. Various identities, even complicated ones, are presented by Euclid in a purely geometric form, for example,

$$2(a^2+b^2)=(a+b)^2+(a-b)^2.$$

This might seem to be a step backward as compared with the methods of Babylonian algebra, and one wonders how that could happen. It is highly probable that the clumsy symbolism of Greek numeration was the fundamental cause of that regression; it was easier to handle lines than Greek numbers! [8]

Irrational quantities. At any rate, Babylonian algebraists were not acquainted with irrational quantities, while Book X of the *Elements* (the largest of the 13 books, even larger than Book I) is devoted exclusively to them. Here again, Euclid was building on older foundations, but this time the foundations were purely Greek. We may believe the story ascribing recognition of irrational quantities to early Pythagoreans, and Plato's friend, Theaitētos (IV-1 B.C.), gave a comprehensive theory of them as well as of the five regular solids. There is no better illustration of the Greek mathematical genius (as opposed to the Babylonian one) than the theory of irrationals as explained by Hippasos of Metapontion, Theodōros of Cyrēnē, Theaitētos of Athens, and finally by Euclid.[9] It is impossible to say just how much of Book X was created by Theaitētos and how much by Euclid himself. We have no choice but to consider that book as an essential part of the *Elements*, irrespective of its origin. It is divided into three parts each of which is preceded by a group of definitions. A number of propositions deal with surds in general, but the bulk of the book investigates the complex irrationals which we would represent by the symbols

$$\sqrt{(\sqrt{a} \pm \sqrt{b})},$$

wherein a and b are commensurable quantities, but $\sqrt{a}$ and $\sqrt{b}$ are incommensurable. These irrationals are correctly divided into 25 species, each of which is discussed separately. As Euclid did not use algebraic symbols, he adopted geometric representations for these quantities and his discussion of them was geometric. Book X was much admired, especially by Arabic mathematicians; it remains a great achievement but is practically obsolete,

[8] It is highly improbable that Euclid was acquainted with Babylonian mathematics. He followed his geometric genius, even as they followed their algebraic one.

[9] For the contributions of Hippasos, Theodōros and Theaitētos, see Volume 1, pp. 282–285, 437.

for such discussion and classification are futile from the point of view of modern algebra.

Theory of numbers. Books VII to IX of the *Elements* might be called the first treatise on the theory of numbers, one of the most abstruse branches of the

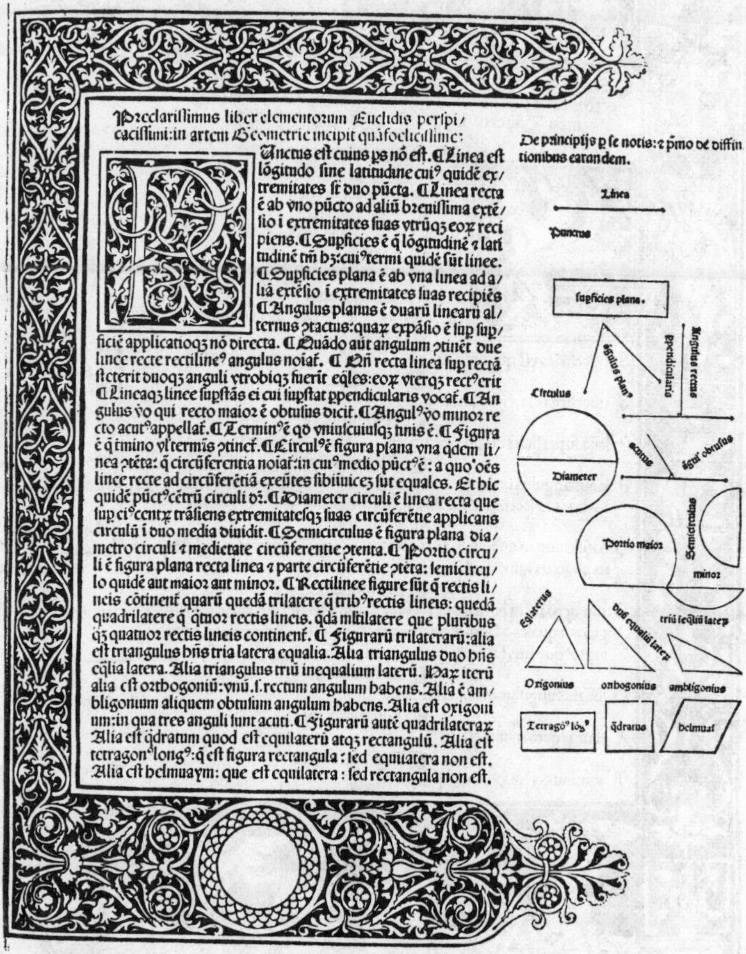
Preclarissimus liber elementorum Euclidis perspi/cacissimi: in artem Geometrie incipit quā foelicissime:

Punctus est cuius ps nō est. Linea est lōgitudo sine latitudine cui⁹ quidē ex/tremitates sīt duo pūcta. Linea recta ē ab vno pūcto ad aliū breuissima extē/sio i extremitates suas vtrūq3 eo𝔷 reci piens. Supficies ē q̄ lōgitudinē ⁊ lati tudinē tm̄ h3: cui⁹ termi quidē sūt linee. Supficies plana ē ab vna linea ad a/liā extēsio i extremitates suas recipiēs. Angulus planus ē duarū linearū al/ternus ꝯtactus: qua𝔷 expāsio ē sup sup/ficiē applicatioq3 nō directa. Quādo aūt angulum ꝯtinēt due linee recte rectiline⁹ angulus noiat. Qñ recta linea sup rectā steterit duoq3 anguli vtrobiq3 fuerit eq̄les: eo𝔷 vterq3 rect⁹ erit. Lineaq3 linee supstās ei cui supstat ppendicularis vocat. Angulus vo qui recto maior ē obtusus dicit. Angul⁹ vo minor re cto acut⁹ appellat. Termin⁹ ē qd vniuscuiusq3 finis ē. Figura ē q̄ tmino vl termīs ꝯtinet. Circul⁹ ē figura plana vna q̄dem li/nea ꝯtēta: q̄ circūferentia noiat: in cu⁹ medio pūct⁹ ē: a quo oēs linee recte ad circūferētiā exeūtes sibiiuicē sūt equales. Et hic quidē pūct⁹ cētrū circuli dr. Diameter circuli ē linea recta que sup ei⁹ cent𝔷 trāsiens extremitatesq3 suas circūferētie applicans circulū i duo media diuidit. Semicirculus ē figura plana dia/metro circuli ⁊ medietate circūferentie ꝯtenta. Portio circu/li ē figura plana recta linea ⁊ parte circūferētie ꝯtēta: semicircu/lo quidē aut maior aut minor. Rectilinee figure sūt q̄ rectis li/neis cōtinent quarū quedā trilatere q̄ trib⁹ rectis lineis: quedā quadrilatere q̄ q̄tuor rectis lineis. q̄dā mltilatere que pluribus q3 quatuor rectis lineis continent. Figurarū trilaterarū: alia est triangulus hñs tria latera equalia. Alia triangulus duo hñs eq̄lia latera. Alia triangulus triū inequalium laterū. Ha𝔷 iterū alia est orthogoniū: vnū. s. rectum angulum habens. Alia ē am/bligonium aliquem obtusum angulum habens. Alia est oxigoni um: in qua tres anguli sunt acuti. Figurarū autē quadrilatera𝔷 Alia est q̄dratum quod est equilaterū atq3 rectangulū. Alia est tetragon⁹ long⁹: q̄ est figura rectangula: sed equilatera non est. Alia est helmuaym: que est equilatera: sed rectangula non est.

Fig. 7. First edition of Euclid in any language. Translation from Arabic into Latin, revised by Giovanni Campano (Venice: Ratdolt, 1482). First page of the text proper in Harvard copy. Sarton, *Osiris* 5, 102, 130–131 (1938), including facsimiles of the same page of the *Elements* (Book III, props. 10–12) in the two incunabula editions, 1482 and 1491 (Klebs, 383).

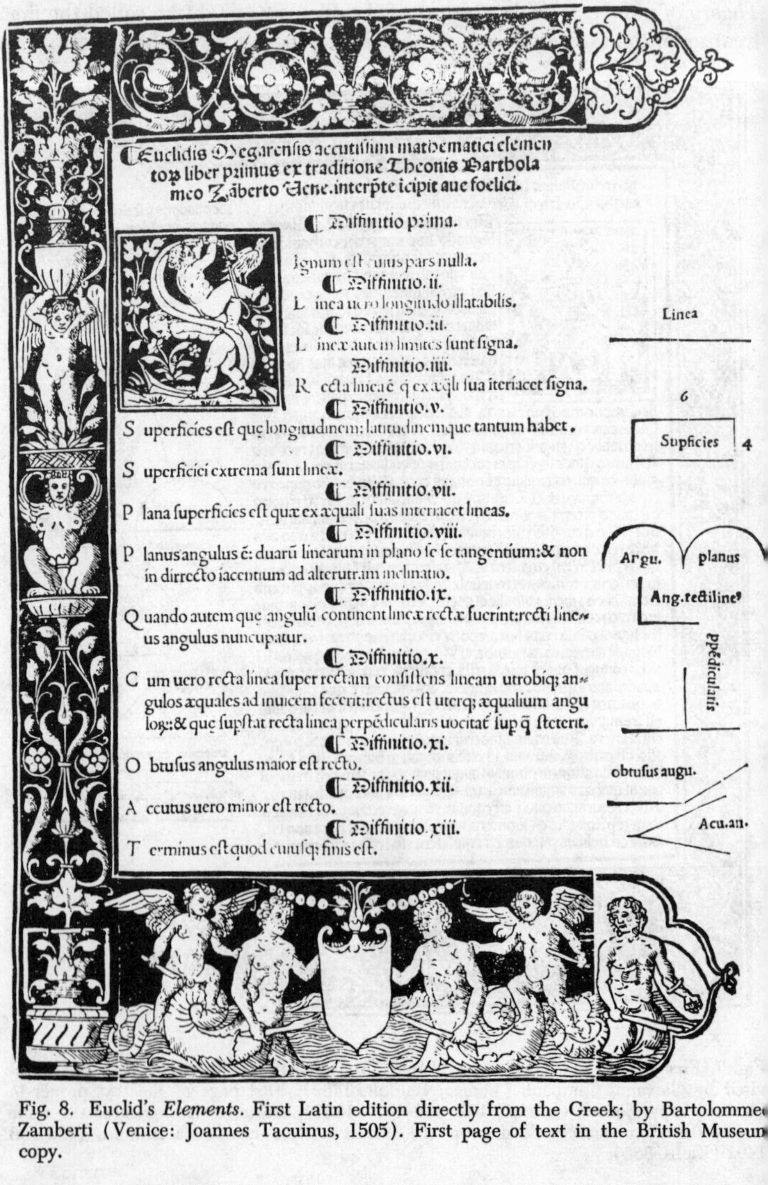

Euclidis Megarensis acutissimi mathematici elementoꝝ liber primus ex traditione Theonis Bartbolameo Zaberto Vene. interpte icipit aue foelici.

Diffinitio prima.

Signum est cuius pars nulla.

Diffinitio. ii.

Linea uero longitudo illatabilis.

Diffinitio. iii.

Lineæ autem limites sunt signa.

Diffinitio. iiii.

Recta linea ē q ex æqli sua iteriacet signa.

Diffinitio. v.

Superficies est quę longitudinem: latitudinemque tantum habet.

Diffinitio. vi.

Superficiei extrema sunt lineæ.

Diffinitio. vii.

Plana superficies est quæ ex æquali suas interiacet lineas.

Diffinitio. viii.

Planus angulus ē: duarũ linearum in plano se se tangentium: & non in directo iacentium ad alterutram inclinatio.

Diffinitio. ix.

Quando autem quę angulũ continent lineæ rectæ fuerint: rectilineus angulus nuncupatur.

Diffinitio. x.

Cum uero recta linea super rectam consistens lineam utrobiq; angulos æquales ad inuicem fecerit: rectus est uterq; æqualium angulorũ: & quę supstat recta linea perpẽdicularis uocitat̃ sup q̃ steterit.

Diffinitio. xi.

Obtusus angulus maior est recto.

Diffinitio. xii.

Accutus uero minor est recto.

Diffinitio. xiii.

Terminus est quod cuiusq; finis est.

Fig. 8. Euclid's *Elements*. First Latin edition directly from the Greek; by Bartolomme Zamberti (Venice: Joannes Tacuinus, 1505). First page of text in the British Museum copy.

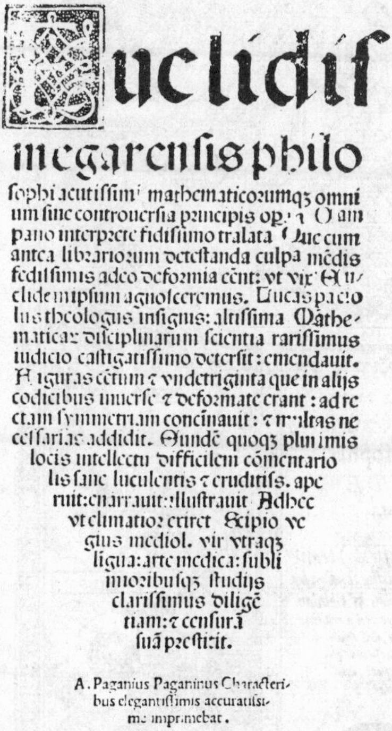

Euclidis
megarensis philo
sophi acutissimi mathematicorumque omni
um sine controversia principis op. a Cam
pano interprete fidissimo tralata Que cum
antea librariorum detestanda culpa mendis
fedissimis adeo deformia erant: ut vix Eu
clidem ipsum agnosceremus. Lucas pacio
lus theologus insignis: altissima Mathe
maticarum disciplinarum scientia rarissimus
iudicio castigatissimo detersit: emendavit.
Figuras centum et undetriginta que in aliis
codicibus inverse et deformate erant: ad re
ctam symmetriam concinnavit: et multas ne
cessarias addidit. Eundem quoque plurimis
locis intellectu difficilem commentario
lis sane luculentis et eruditiss. ape
ruit: enarravit: illustravit. Adhec
ut elimatior exiret Scipio ve
gius mediol. vir utraque
ligua: arte medica: subli
mioribusque studiis
clarissimus dilige
tiam: et censura
sua prestitit.

A. Paganius Paganinus Characteri
bus elegantissimis accuratissi
me imprimebat.

Fig. 9. Latin Euclid printed by Paganinus de Paganinis (Venice, 1509). This is a revision of Campano's text by Fra Luca Pacioli da Borgo San Sepolero. [Courtesy of Harvard College Library.] Pacioli is best known for his *Summa de arithmetica, geometria proportioni et proportionalita* (Venice: Paganinus, 1494). See *Osiris* 5, 114, 161 (1938).

ΕΥΚΛΕΙΔΟΥ
ΣΤΟΙΧΕΙΩΝ ΒΙΒΛ. ΙΕ.
ΕΚ ΤΩΝ ΘΕΩΝΟΣ ΣΥΝ
ΟΥΣΙΩΝ.
Εἰς τοῦ αὐτοῦ τὸ πρῶτον, ἐξηγημάτων Πρόκλου βιβλ. δ΄.
Adiecta præfatiuncula in qua de disciplinis
Mathematicis nonnihil.

BASILEAE APVD IOAN. HERVAGIVM ANNO
M. D. XXXIII. MENSE SEPTEMBRI.

Fig. 10. Princeps of Euclid's *Elements*. Edited by Simon Grynaeus, dedicated to Cuthbert Tunstall, printed by Joannes Hervagius (Johann Herwagen) (Basel, 1533). Title page of the copy in Harvard College Library.

mathematical tree. It would be impossible to summarize their contents, for the summary would be almost meaningless unless it covered a good many pages.[10] Let me just say that Book VII begins with a list of 22 definitions, which are comparable to the geometric definitions placed at the beginning of Book I. Then follows a series of propositions concerning the divisibility of numbers, even and odd numbers, squares and cubes, prime and perfect numbers, and so on.

Let us give a few examples. In IX, 36 Euclid proves that if $p = 1 + 2 + 2^2 + \cdots + 2^n$ is prime, $2^n p$ is perfect (that is, is equal to the sum of its di-

[10] The Greek text of Books VII to IX covers 116 pages in Heiberg's edition (Leipzig, 1884), vol. 2, and the English translation with notes 150 pages in Heath, vol. 2.

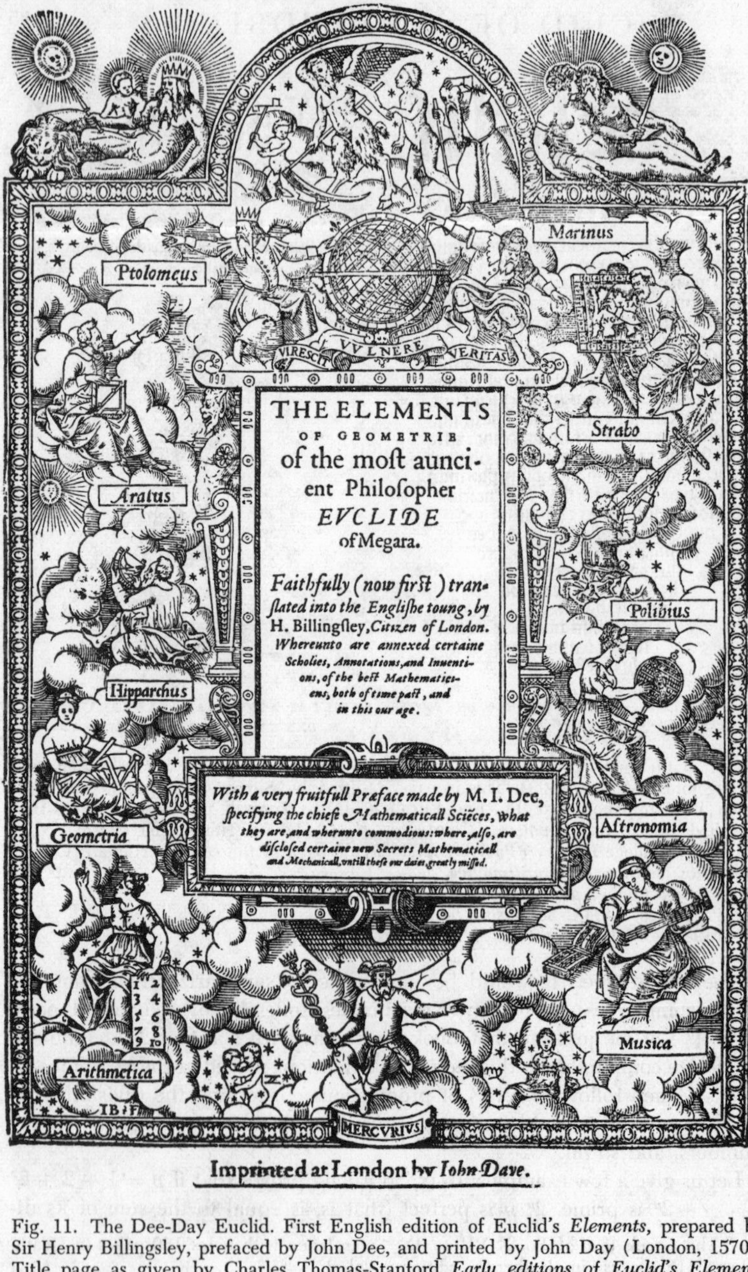

Fig. 11. The Dee-Day Euclid. First English edition of Euclid's *Elements*, prepared by Sir Henry Billingsley, prefaced by John Dee, and printed by John Day (London, 1570). Title page as given by Charles Thomas-Stanford *Early editions of Euclid's Elements* (London, 1926), pl. x.

visors). In IX, 20 we are given a very elegant demonstration that the number of prime numbers is infinite.

No matter how many primes we already know, it will always be possible to find a larger one. Consider the series of primes: $a, b, c, \ldots, l$. Their product plus 1, that is $(abc\ldots l) + 1 = P$, is either prime or not. If it is, we have found a larger prime than l; if it is not, then P must be divisible by a prime p. Now p cannot be identical with $a, b, \ldots,$ or l, because if it were it would divide their product and also unity, which is impossible.

The demonstration is so simple and our intuitive feeling *ad hoc* is so strong that one would readily accept other propositions of the same kind. For example, there are many prime pairs, that is, prime numbers packed as closely as possible, having the form $2n + 1$, $2n + 3$; examples are 11, 13; 17, 19; 41, 43. As one proceeds in the series of integers, prime pairs become rarer and rarer, yet one can hardly escape the feeling that the number of prime pairs is infinite. The proof of that is so difficult, however, that it has not yet been completed.[11]

In this field, again, Euclid was an outstanding innovator, and the few mathematicians of our own days who are trying to cultivate it recognize him as their master.

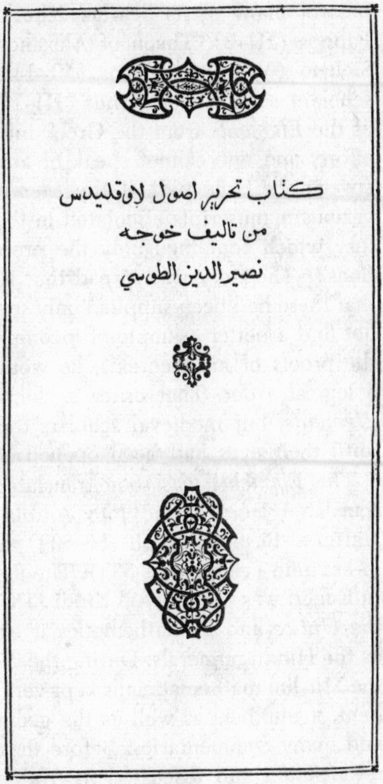
كتاب تحرير اصول لاوقليدس
من تاليف خوجه
نصير الدين الطوسي

Fig. 12. Title page of the first Arabic edition of Euclid's *Elements*, in the redaction of Nasīr al-dīn al-Ṭūsī (XIII-2); one of the first books printed in Arabic. It is a folio volume issued by the Tipographia Medicea (Rome, 1594). Upon the last page (p. 454) is a firman granted by Murād III, Ottoman sultan from 1564 to 1595. [Courtesy of History of Science Department, Harvard University.]

THE EUCLIDEAN TRADITION

The tradition concerned with the fifth postulate has already been referred to; it can be traced from the time of the *Elements* until our own. That is only a small part of the tradition, however; the Euclidean tradition, even if restricted to mathematics, is remarkable for its continuity and the great-

[11] Charles Napoleon Moore of Cincinnati offered a proof in 1944, but that proof was shown to be insufficient (*Horus: A guide to the history of science* [Wal-

ness of many of its bearers. The ancient tradition includes such men as Pappos (III–2), Theōn of Alexandria (IV–2), Proclos (V–2), Marinos of Sichem (V–2), Simplicios (VI–1). It was wholly Greek. Some Western scholars, such as Censorinus (III–1) and Boethius (VI–1), translated parts of the *Elements* from the Greek into Latin, but very little remains of their efforts and one cannot speak of any complete translation, nor of any one covering a large part of the *Elements.* There is much worse to be said; various manuscripts circulated in the West until as late as the twelfth century which contained only the propositions of Euclid without demonstrations.[12] The story was spread that Euclid himself had given no proofs and that these had been supplied only seven centuries later by Theōn. One could not find a better example of incomprehension, for if Euclid had not known the proofs of his theorems, he would not have been able to put them in a logical order. That order is the very essence and the greatness of the *Elements,* but medieval scholars did not see it, or at least did not see it until their eyes had been opened by Muslim commentators.

The *Elements* was soon translated from Greek into Syriac; it was first translated from Syriac into Arabic by al-Ḥajjāj ibn Yūsuf (IX–1) for Hārūn al-Rashīd (caliph 786–809) and al Ḥajjāj revised his translation for al-Ma'mūn (caliph 813–833). The first Muslim philosopher to be interested in Euclid was probably al-Kindī (IX–1), but his interest was centered upon the *Optics* and in mathematics it extended to non-Euclidean topics, such as the Hindu numerals. During the 250 years that followed (cent. IX to XI), the Muslim mathematicians kept very close to Euclid, the algebraist and student of numbers as well as the geometer, and published other translations and many commentaries. Before the end of the ninth century, Euclid was retranslated and discussed in Arabic by Muḥammad ibn Mūsā,[13] al-Māhānī, al-Nairīzi, Thābit ibn Qurra, Isḥāq ibn Ḥunain, Qusṭā ibn Lūqā. A great step forward was made in the first quarter of the tenth century by Abū 'Uthmān Sa'īd ibn Yaq'ūb al-Dimishqī, who translated Book x with Pappos's commentary (the Greek of which is lost).[14] This increased Arabic

---

tham, Mass.: Chronica Botanica, 1952], p. 62). The incredible complexity of the theory of numbers can be appreciated by looking at its history written by Leonard Eugene Dickson (3 vols.; Washington: Carnegie Institution, 1919–1923) [*Isis 3,* 446–448 (1920–21); *4,* 107–108 (1921–22); *6,* 96–98 (1924)]. For prime pairs, see Dickson, vol. 1, pp. 353, 425, 438.

[12] Greek and Latin editions of the propositions only, without proofs, were printed from 1547 to as late as 1587.

[13] This is Abū Ja'far (d. 872), one of the three brothers, Banū Mūsā, not Abū 'Abdallāh Muḥammad ibn Mūsā al-Khwārizmī (d. c. 850). We must assume that the latter was also a student of Euclid. See *Introduction,* vol. 1, pp. 561, 563.

[14] Pappos's authorship of the commentary is now generally accepted in spite of early doubts. The Arabic version of it was translated into German by Heinrich Suter (Erlangen, 1922) [*Isis 5,* 492 (1923)]; it was edited and Englished by William Thomson (Cambridge, 1930) [*Isis 16,* 132–136 (1931)].

ΕΥΚΛΕΙΔΟΥ

ΤΑ ΣΩΖΟΜΕΝΑ.

EUCLIDIS

QUÆ SUPERSUNT

OMNIA.

Ex Recenſione DAVIDIS GREGORII M. D.
Aſtronomiæ Profeſſoris Saviliani, & R. S. S.

OXONIÆ,
E THEATRO SHELDONIANO, An. Dom. MDCCIII.

Fig. 13. First edition of Euclid's *Opera* in Greek and Latin on parallel columns, by David Gregory (large folio; Oxford: Sheldonian Theatre, 1703.) David Gregory (1661–1708) was Savilian Professor of Astronomy in Oxford, 1691. His *Astronomiae physicae et geometricae elementa* (Oxford: Sheldonian Theatre, 1702) was the first Newtonian textbook. [Courtesy of Harvard College Library.]

Fig. 14. Frontispiece of Euclid's *Opera* edited by David Gregory (Oxford, 1703). It illustrates an anecdote told by Vitruvius (*De architectura*, first sentence of Book VI). Aristippos of Cyrēnē, one of Socrates' disciples, having been shipwrecked on the coast of Rhodos, noticed geometrical figures drawn upon the sand and exclaimed, "We may be hopeful for these are human vestiges." Many illustrations of Euclid have been given to underline his very great importance. [Courtesy of Harvard College Library.]

interest in the contents of Book X (classification of incommensurable lines) as witnessed by the new translation of Naẓīf ibn Yumn (X–2), a Christian priest, and by the commentaries of Abū Ja'far al-Khāzin (X–2) and Muḥammad ibn 'Abd al-Bāqī al-Baghdādī (XI–2). My Arabic list is long yet very incomplete, because we must assume that every Arabic mathematician of this age was acquainted with the *Elements* and discussed Euclid. For example, Abū-l-Wafā' (X-2) is said to have written a commentary which is lost.

We may now interrupt the Arabic story and return to the West. The efforts made by Western scholars to translate the *Elements* directly from Greek into Latin had been ineffective; it is probable that their knowledge of

Fig. 15. First edition of Girolamo Saccheri's famous book (Milan, 1733), containing "the vindication of Euclid and the adumbration of non-Euclidean geometry." It is very rare, but the Latin text was reprinted, Englished, and annotated by George Bruce Halsted (1853–1922) (Chicago, 1920). Saccheri may be called the forerunner of Nikolai Ivanovich Lobachevski (1793–1856).

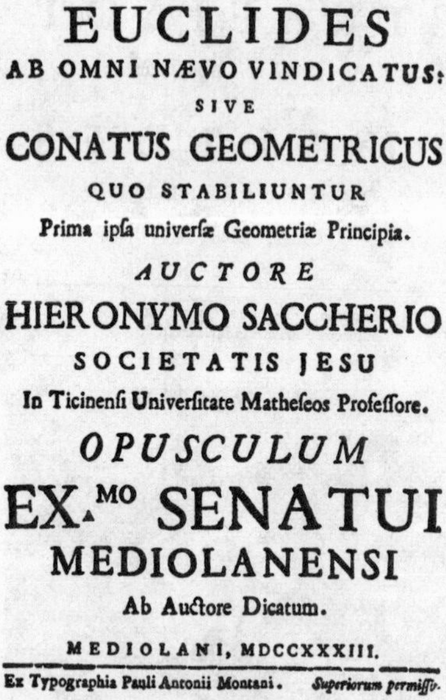

EUCLIDES
AB OMNI NÆVO VINDICATUS:
SIVE
CONATUS GEOMETRICUS
QUO STABILIUNTUR
Prima ipsa universæ Geometriæ Principia.
AUCTORE
HIERONYMO SACCHERIO
SOCIETATIS JESU
In Ticinensi Universitate Matheseos Professore.
OPUSCULUM
EX.MO SENATUI
MEDIOLANENSI
Ab Auctore Dicatum.
MEDIOLANI, MDCCXXXIII.
Ex Typographia Pauli Antonii Montani. Superiorum permissu.

Greek diminished and dwindled almost to nothing at the very time when their interest in Euclid was increasing. Translators from the Arabic were beginning to appear and these were bound to come across Euclidean manuscripts. Efforts to Latinize these were made by Hermann the Dalmatian (XII–1), John O'Creat (XII–1), and Gerard of Cremona (XII–2), but there is no reason to believe that the translation was completed, except by Adelard of Bath (XII–1).[15] However, the Latin climate was not as favorable to geometric research in the twelfth century as the Arabic climate had proved to be from the ninth century on. Indeed, we have to wait until the beginning of the thirteenth century to witness a Latin revival of the Euclidean genius, and we owe that revival to Leonardo of Pisa (XIII–1), better known under the name of Fibonacci. In his *Practica geometriae*, written in 1220, Fibonacci did not continue the *Elements*, however, but another Euclidean work on the *Divisions of figures*, which is lost.[16]

In the meanwhile, the Hebrew tradition was begun by Judah ben Solomon ha-Kohen (XIII–1) and continued by Moses ibn Tibbon (XIII–2), Jacob ben Maḥir ibn Tibbon (XIII–2), and Levi ben Gerson (XIV–1).

[15] The story is simplified for the sake of briefness; for details, see Marshall Clagett, "The medieval Latin translations from the Arabic of the *Elements* with special emphasis on the versions of Adelard of Bath," *Isis 44*, 16–42 (1953); "King Alfred and the *Elements*," *45*, 269–277 (1954).

[16] The text of that little treatise *peri diaireseōn* was restored as far as possible by Raymond Clare Archibald (1875–1955) on the basis of Leonardo's *Practica* and of an Arabic translation (*Introduction*, vol. 1, pp. 154, 155).

The Syriac tradition was revived by Abū-l-Faraj; called Barhebraeus (XIII-2), who lectured on Euclid at the observatory of Marāgha in 1268; unfortunately, this revival of the Syriac tradition was also the end of it, because Abū-l-Faraj was the last Syriac writer of importance; after his death, Syriac was gradually replaced by Arabic.

The golden age of Arabic science was also on the wane, though there remained a few illustrious Euclideans in the thirteenth century, like Qaiṣar ibn abī-l-Qāsim (XIII-1), Ibn al-Lubūdī (XIII-1), Naṣīr al-dīn al-Ṭūsī (XIII-2), Muḥyī al-dīn al-Maghribī (XIII-2), Quṭb al-dīn al-Shīrāzī (XIII-2), and even in the fourteenth century. We may overlook the late Muslim and Jewish mathematicians, for the main river was now flowing in the West.

Adelard's Latin text was revised by Giovanni Campano (XIII-2), and Campano's revision was immortalized in the earliest printed edition of the *Elements* (Venice: Ratdolt, 1482) (Fig. 7). It was reprinted by Leonardus de Basilea and Gulielmus de Papia (Vicenza, 1491). There are only these two incunabula (Klebs, 383),[17] both Latin from the Arabic. The first Latin translation from the Greek was made by the Venetian, Bartolommeo Zamberti, in 1493, and printed by Joannes Tacuinus (Venice, 1505) (Fig. 8). The next edition, also Latin, was printed by Paganinus (Venice, 1509) (Fig. 9). The Greek princeps was prepared by Simon Grynaeus, dedicated to the English mathematician and theologian, Cuthbert Tunstall, and printed by Johann Herwagen (Basel, 1533) (Fig. 10). The first English translation was made by Sir Henry Billingsley, of St. John's College, Cambridge, sometime lord mayor of London, and published with a preface by John Dee (London: John Day, 1570)[18] (Fig. 11). The princeps of the Arabic text as revised by Naṣīr al-dīn al-Ṭūsī was published by the Typographia Medicea (Rome, 1594) (Fig. 12).

The rest of the story need not be told here. The list of Euclidean editions, which began in 1482 and is not ended yet, is immense, and the history of the Euclidean tradition is an essential part of the history of geometry.

As far as elementary geometry is concerned, the *Elements* of Euclid is the only example of a textbook that has remained serviceable until our own day. Think of that! Twenty-two centuries of changes, wars, revolutions, catastrophies of every kind, yet it still is profitable to study geometry in Euclid.[19]

[17] This refers to A. C. Klebs, "Incunabula scientifica et medica," *Osiris 4*, 1–359 (1938); see Volume 1, p. 352, n.15.

[18] R. C. Archibald, "The first translation of Euclid's *Elements* into English and its sources," *American Mathematical Monthly* 57, 443–452 (1950).

[19] It is well to insist upon that, because it is decidedly not profitable to study most of the scientific classics. For example, it would be very foolish to study mathematical astronomy in Ptolemy or celestial mechanics in Newton. That would require a considerable effort and would lead to a very imperfect knowledge. It would be much easier to study modern mathematics and then modern treatises on astronomy and celestial mechanics; one's knowledge would be up to date and could be used for further progress.

BIBLIOGRAPHY

The standard edition of the Greek text of all the works, with Latin versions, is J. L. Heiberg and H. Menge, eds., *Euclidis opera omnia* (8 vols., Leipzig, 1883–1916; supplement, 1899). Volumes 1 to 4 (1883–1886) contain the thirteen books of the *Elements*; vol. 5 (1888), the so-called Book XIV by Hypsiclēs (II–1 B.C.) and Book XV by a pupil of Isidōros of Milētos in the sixth century, also abundant scholia to the *Elements*; vol. 6 (1896), Euclid's *Data* (*dedomena*) with commentary by Marinos of Sichem (V–2) and scholia; vol. 7 (1895), the *Optics* and *Catoptrics* with the commentary of Theōn of Alexandria; vol. 8 (1916), the *Phaenomena*, a treatise on spherical astronomy based upon Autolycos (IV–2 B.C.), *Scripta musica*, etc. The supplement (1899) contains the commentary of al-Nairīzī (Anaritius) to Books I to X in the Latin version by Gerard of Cremona (XII–2). This list has been given in full to illustrate that Euclid was the author not only of the *Elements* but of many other works which we had no space to discuss. More of them are mentioned in my *Introduction*, vol. 1, pp. 154–156 (1927).

Sir Thomas L. Heath, *Euclid's Elements in English* (3 vols., Cambridge, 1908; rev. ed., 3 vols., 1926) [*Isis 10*, 60–62 (1928)].

Charles Thomas-Stanford, *Early editions of Euclid's Elements* (64 p., 13 pl.; London, 1926) [*Isis 10*, 59–60 (1928)].

IV

ASTRONOMY. ARISTARCHOS AND ARATOS

ARISTYLLOS AND TIMOCHARIS

Two Greek astronomers, Aristyllos and Timocharis (III-1 B.C.), are said by Ptolemy [1] to have made astronomic observations before Hipparchos (II-2). They worked at the beginning of the third century (c. 295–283) in Alexandria; their observatory (if that is not too pretentious a term) was probably a part of the Museum. Their equipment was very simple; they probably used gnomons, sundials of some kind, and an armillary sphere, that is, a skeleton sphere made of great circles adjusted around the same center and graduated in degrees [2] (and fractions of degrees); one of the circles might be in the plane of the equator, and the other, perpendicular to it, would turn around the axis of the world; a rule or alidade would be attached to the same center in order to determine the direction of a star; with such a combination of armillaries one could measure its declination and right ascension. Timocharis' measurements were useful to Hipparchos for the determination of the precession of the equinoxes; indeed, the differences between the longitudes observed by Timocharis and by Hipparchos amounted to as much as 2°. As the period of time extending between their observations amounted to 154 or 166 years, this gave for the precession a value of 43.4″ or 46.8″ a year, a better approximation than Ptolemy's 36″ (the modern value is 50.3757″).

ARISTARCHOS OF SAMOS [3]

A far greater man than Aristyllos and Timocharis was their contemporary, Aristarchos (III-1 B.C.), whose connection either with them or with

[1] Aristyllos and Timocharis are frequently mentioned in Ptolemy's *Syntaxis*.

[2] "Graduated in degrees" is possible but uncertain. The first Greek who divided the circles of his instruments into 360 degrees is said to have been Hipparchos (II-2 B.C.), and yet he (or Ptolemy) refers to measurements in degrees by Timocharis. It may be that Timocharis' armillaries were graduated otherwise and that Hipparchos translated his measurements into degrees. The armillaries were certainly graduated in one way or another because otherwise they would have been worthless.

[3] Samos is one of the main Ionian islands, not very far northwest of Milētos.

Alexandria cannot be established. If you look at a map, you will see that it was easy enough to sail from Samos to Athens, but that sailing to Alexandria was a much longer voyage. We know that Aristarchos was a pupil of Stratōn of Lampsacos, who had been the tutor and adviser of Ptolemaios II Philadelphos and had helped the latter to create the Museum. Upon the death of Theophrastos, Stratōn succeeded him and was the head of the Lyceum for eighteen years (c. 286–268). Aristarchos could have been Stratōn's pupil either in Alexandria (before 286) or later in Athens. The second alternative seems more probable to me, and is confirmed by the failure of the astronomer Ptolemy (II–1) to mention him. The only fixed date in Aristarchos' life is 281–80, when he observed the summer solstice. If he had done that in Alexandria, Ptolemy would have mentioned him together with Aristyllos and Timocharis. As a matter of fact, we can hardly identify an astronomical school in Hellenistic times, because the observations were made not in a single place but in many – Alexandria, Athens, Sicily, Seleuceia (on the Tigris), Rhodos.

If the place of Aristarchos' activities remains undetermined, their date is sufficiently established. At the time of his observation of the summer solstice, in 281, he must have been at least twenty, hence he was born about 300 or before; on the other hand, Archimēdēs quotes him in his *Sand-reckoner*, written before 216. We may safely place him in the period III–1 B.C.

Aristarchos wrote a treatise *On the sizes and distances of the Sun and Moon* which has come down to us in its integrity. It is written in the Euclidean style and with Euclidean rigor, but is unfortunately based upon wrong data. The treatise begins with six "hypotheses":

1. The Moon receives its light from the Sun;
2. The Earth is in the relation of a central point to the sphere in which the Moon moves [this simplified view obviated the complication of a parallax];
3. At the time of Half Moon, the great circle that divides the dark portion from the bright one is in the direction of our eye [see Fig. 16].
4. At the time of Half Moon, the Moon's distance from the Sun is less than a quadrant by one-thirtieth of a quadrant [that is, equal to 87°];
5. The breadth of the Earth's shadow (at the distance where the Moon passes through it at the time of an eclipse) is that of two Moons; [4]
6. The Moon subtends one-fifteenth of a sign of the zodiac (2°).

Assumptions 4 and 6 are very inaccurate. The angle at the Earth is 89° 50′, rather than 87° – a small difference that makes a large difference in the result. The angle 89° 50′ is so close to 90° that on any drawing, however

It was already a great cultural center in the sixth century B.C., and Hērodotos considered it one of the most civilized places in the world; it gave birth or hospitality to many artists, poets, and philosophers, and to two outstanding astronomers, Aristarchos and Conōn (III–2 B.C.).

[4] As all the orbits were thought to be circular, the distances of Sun and Moon from the Earth were constant.

large, the two sides *e* and *m* would be indistinguishable from parallels and the triangle *EMS* would vanish. The second error is difficult to understand, because the approximate measurement of the apparent or angular diameter of the Moon (about 30′) was easy enough and even with poor instruments could not be so wide of the mark.

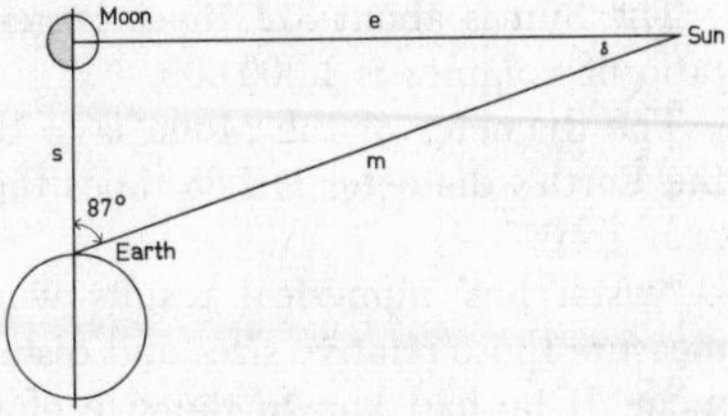

Fig. 16. Diagram to illustrate Aristarchos' hypotheses on the Moon.

Aristarchos' method was excellent, but because of the crudity of his observations the results were grossly erroneous.

The measurements that he tried to ascertain were in the nature of ratios which could be easily determined by our trigonometric methods, but trigonometry did not yet exist, and he was obliged to find those ratios by means of ingenious geometric arguments. The ratios that he found were not determined except grossly, the result x being stated in the following form:

$$a/b < x < c/d,$$

the two ratios a/b and c/d being sometimes very complex and their difference large.

For example, the final proposition (No. 18) of his treatise states that "The Earth is to the Moon in a ratio greater than that of 1,259,712 to 79,507 and smaller than that of 216,000 to 6,859." This means roughly that the ratio (of volumes) is included between 17 and 31; the true value is 49.

If the angle E equals 87°, as Aristarchos thought it did (instead of 89° 50′), then the distance of the Sun from the Earth is about 19 times the distance of the Moon (Prop. 7). The true value is 400.

Inasmuch as the apparent sizes of the Moon and the Sun are about the same,[5] he concluded (Prop. 9) that the diameter of the Sun is about 19 times that of the Moon. The true value is about 400.

The ratio of the Sun's volume to that of the Moon exceeds 5832 and is less than 8000 (Prop. 10). The true value is 106,600,000.

The radius of the Moon's orbit is 26¼ times the Moon's diameter (Prop. 11). Actually, the mean distance of the Moon is 110.5 times its diameter.

The diameter of the Sun is 6.75 times that of the Earth (Prop. 15). Actually, the ratio is 109.

[5] They are about the same, but the apparent diameter of the Moon varies from 29′26″ to 33′34″, while the mean apparent diameter of the Sun is 31′59″; the apparent diameter of the Moon may thus be smaller or greater than that of the Sun or equal to it. To put it otherwise, the apparent diameter of the Moon varies by 13.5 percent, that of the Sun by only 3.5 percent. Sōsigenēs (fl. 46 B.C.) proved the inequality of the apparent diameters by the occurrence of annular eclipses of the Sun.

The Sun is about 311 times larger than the Earth (Prop. 16). The true ratio of volumes is 1,300,000.

The diameter of the Moon is to that of the Earth as 9 is to 25, or, say, the Earth's diameter is 2.85 times the Moon's (Prop. 17). The real value is near to 3.7.

Aristarchos' numerical results were very poor, but he was the first to measure those relative sizes and distances and that was an immense achievement. If he had known the size of the Earth he could have deduced the absolute sizes of the Moon and the Sun. His results would have been grossly erroneous, but the very fact of "measuring" those celestial bodies was in his age astounding. It is possible, by the way, that he knew the size of the Earth, that is, that he knew the approximation attained by Aristotle or Dichaiarchos of Messina (IV–2 B.C.), according to whom the circumference of the Earth measured 300,000 stadia,[6] but even if he possessed that knowledge, he made no reference to it, and no use of it, in his treatise. The fact remains that thanks to Aristotle, Dicaiarchos, and Aristarchos it was made possible to measure the sizes and distances of the Sun and Moon; the actual numbers are less important than that possibility. It is as if puny man had reached those two luminaries of the day and the night.

According to Archimēdēs, it would seem that one of Aristarchos' wrong assumptions, the most shocking, was corrected by himself later in life. Instead of assuming that the apparent diameter of the Sun and Moon is about 2°, he declared it to be about 30′, which comes much closer to the truth. If so – and there is no reason for disbelieving Archimēdēs' statement – we may conclude that Aristarchos composed the treatise that has come down to us when he was relatively young.

Let me repeat that that treatise is an important landmark in the history of science, not only because it explained how to measure the distances and sizes of celestial bodies, but also because it was an anticipation of trigonometry.

And yet, however important we consider that treatise to be, it is less important than another work by Aristarchos which remained unwritten or, if it was written, was soon lost. We know it only through his younger contemporary Archimēdēs.[7] It is best to quote Archimēdēs' own words in his *Sand reckoner*; a sensitive person cannot read them without emotion when he remembers that they were written before 216 B.C.:

> You[8] are aware that universe (*cosmos*) is the name given by most astronomers to the sphere whose center is the center of the Earth, and whose radius is equal to the distance between the center of the Sun and the center of the

[6] Eratosthenēs gave a better approximation, 252,000 stadia, but that was later. Eratosthenēs was born at the time of Aristarchos' maturity.

[7] The ideas that we are going to explain now are not referred to in Aristarchos' extant treatise, and this confirms our belief that he wrote it early in life.

[8] "You" is Gelōn II, king of Syracuse, who died before 216. Archimēdēs died in

Earth. This is the common account as you have heard from astronomers. But Aristarchos of Samos brought out a book consisting of some hypotheses, wherein it appears, as a consequence of assumptions made, that the [real] universe is many times greater than the one just mentioned. His hypotheses are that the fixed stars and the Sun remain unmoved, that the Earth revolves about the Sun in the circumference of a circle, the Sun lying in the middle of the orbit, and that the sphere of the fixed stars, situated about the same center as the Sun, is so great that the circle in which he supposes the Earth to revolve bears such a proportion to the distance of the fixed stars as the center of the sphere bears to its surface.

This is stupendous, and would be incredible if we had it from another source, but we have no reason to doubt Archimedes, who was born within Aristarchos' lifetime and might have known him personally. Moreover, why would he invent such a statement? Or, if he had invented it, it would be just as stupendous.

To put it in plainer words, Aristarchos of Samos had put the center of the universe in the Sun (instead of the Earth) and assumed the daily rotation of the Earth around its own axis and the yearly rotation of the Earth around the Sun. All the planets circle around the Sun except the Moon, which alone circles around the Earth. The stars are fixed and their daily rotation is an illusion caused by the daily rotation of the Earth around its own axis in an opposite direction. The sphere of the fixed stars is so immense that the Earth's whole orbit around the Sun is like a point in comparison. The last hypothesis is the most astonishing of all, for it implied an almost inconceivable expansion of the universe. It illustrated Aristarchos' boldness. Having placed the Sun in the center of the universe, it was necessary to expand the latter immeasurably in order to account for the absence of parallactic displacements of the stars in spite of the immense size of the Earth's orbit. Aristarchos did not hesitate to accept this almost absurd consequence of the heliocentric hypothesis. It takes some effort of imagination to realize his boldness, however, because his universe was reduced to nothing by Herschel and to infinitesimal smallness by the stellar astronomy of today.

Aristarchos had conceived what we call the Copernican universe, eighteen centuries before Copernicus. The name that has been given to him in modern times, "The Copernicus of Antiquity," is fully deserved, because Aristarchos' other treatise (the one that was described above) proves that he was a conscientious astronomer. His astronomical hypothesis was not a wild one; it was justified by his experience. For example, having realized that the Sun was enormously larger than the Earth, he found it difficult to believe that the larger body was dominated by the smaller one. As to

212. The quotation is taken from the *Sand reckoner*, Heiberg's Greek-Latin edition, vol. 2 (1913), pp. 216–219; Heath's translation in his *Works of Archimedes* (Cambridge, 1897), p. 221.

the thousands of stars, why should they all turn around the Earth at an immense distance and yet with the greatest regularity? Was it not simpler to think that the Earth itself was rotating around its axis?

His hypothesis was extremely bold but not irresponsible. Moreover, it was not entirely new. An older contemporary of his, Hērakleidēs of Pontos (IV–2 B.C.), who lived just before him in Athens and whose memory must have been green in the Academy, had invented a similar hypothesis though a less complete one. Hērakleidēs had postulated the daily rotation of the Earth, and claimed that the inferior planets, Venus and Mercury, rotate around the Sun and that the Sun, Moon, and other planets rotate around the Earth. It was a kind of compromise between the geocentric and heliocentric systems, a kind of anticipation of Tycho Brahe; yet it would be less proper to call Hērakleidēs the Greek Tycho than Aristarchos the Greek Copernicus.[9]

To complete our account of Aristarchos, he was also interested in physical questions, as was natural enough for a disciple of Stratōn, and wrote a (lost) treatise on light, vision, and color. He devised a sun dial, called *scaphē* (meaning a hollow vessel, a bowl) because it was not plane, as the common dials were, but hemispherical, with a pointer where the radius would be; the direction and height of the Sun could be read off by observing the pointer's shadow with reference to lines drawn on the surface of the cavity. These are very minor achievements, however, as compared with those already described.

The Aristarchian tradition. This tradition is exceptionally interesting; one has to consider two separate traditions, the one concerning his extant treatise, the other the heliocentric hypothesis.

Let us begin with the second. In spite of the fact that the ideas of Aristarchos were almost certainly derived from those of Hēracleidēs, and were superior to them, the Heraclidean tradition was more popular and continuous. It was revived by Theōn of Smyrna (II–1), but this ends the Greek or scientific tradition. On the other hand, the Heraclidean views were mentioned by Cicero (I–1 B.C.) and Vitruvius (I–2 B.C.), and this started a Latin tradition exemplified by a remarkable group of writers, Chalcidius (IV–1), Macrobius (V–1), and Martianus Capella (V–2). Geoheliocentric views may be traced in the Hebrew writings of Abraham ben Ezra (XII–1) and Moses of Leon (XIII–2) or whoever composed the *Zohar*, and in the Latin ones of William of Conches (XII–1), Bartholomew the Englishman (XIII–1), the astrologer of Baldwin II of Courtenay (XIII–2), and Pietro d'Abano (XIV–1). The diffusion of those ideas was guaranteed by the early printed editions of Bartholomew and Pietro. The popularity of the geoheliocentric system was perhaps caused in part by the singular trajectories of the inferior planets. The views of William of Conches were typical; he did not

[9] For explanation, see Volume 1, pp. 506–508.

follow Hēracleidēs faithfully but assumed that the three orbits of the Sun, Venus, and Mercury had nearly the same radius, their centers being at short distances from each other and in line with the Earth.

The pure Aristarchian tradition was very different from the Heraclidean one. It began ominously with the charge of impiety brought against Aristarchos by his contemporary, Cleanthēs of Assos (III–1 B.C.),[10] "for moving the hearth of the universe and trying to save the phenomena by the assumption that the heaven is at rest, but that the Earth revolves in an oblique orbit, while rotating around its own axis."[11] Vitruvius thought highly of him, however, as one of the men who had an equally deep knowledge of many branches of science. "Men of this type are rare," wrote Vitruvius, "men such as were in times past Aristarchos of Samos, Philolaos and Archytas of Tarentum, Apollōnios of Perga, Eratosthenēs of Cyrēnē, Archimēdēs and Scopinas of Syracuse."[12] To return to men of science, the heliocentric views were supported by Seleucos the Babylonian (II–1 B.C.) but soon after were rejected by Hipparchos (II–2 B.C.). Hipparchos' rejection was conclusive, because he was accepted as the greatest astronomer of antiquity; it was confirmed and perpetuated by Ptolemy (II–1). It does not seem that either Hipparchos or Ptolemy paid any attention to Hēracleidēs, but they stopped the development of any system except the geocentric one. After an interval of eighteen centuries, the heliocentric views were reaffirmed by Copernicus (1543), who was well aware of the efforts made by Philolaos (V B.C.), Hicetas (V B.C.), Ecphantos (IV–1 B.C.),[13] Hēracleidēs, and Aristarchos; he was their conscious and careful revivalist.

The Heraclidean tradition was of a more literary or philosophical kind; it was almost exclusively Western, Latin and Hebrew. The Aristarchian one, on the contrary, was more scientific and Oriental, Greco-Arabic; it was defeated on purely technical grounds and was revived by Copernicus in one of the greatest scientific books of the Renaissance (1543). Then it was rejected a second time, for the best of technical reasons, by Tycho Brahe

[10] To be correct, Cleanthēs did not actually make that charge but said that it ought to be made. Cleanthēs was one of the leading Stoics, head of the Stoa from 264 to 232, the year of his death. He actually wrote a tract against Aristarchos. In their ethical zeal, the Stoics had revived some of the antiscientific prejudices of Sōcratēs. Cleanthēs' hostility to Aristarchos is revealed by Plutarch in *De facie in orbe lunae* (The face in the moon), chap. 6.

[11] According to Theōn of Smyrna (II–1), a similar charge was made implicitly by one Dercyllidas. See Eduard Hiller's edition, *Theonis Smyrnaei Expositio rerum mathematicarum ad legendum Platonem utilium* (Leipzig, 1878), p. 200.

[12] *De architectura*, I, 1; see also IX, 8. Vitruvius' selection is curious. All the men mentioned are familiar to our readers, except the last one, Scopinas of Syracuse, who is otherwise unknown.

[13] I have not spoken of these three in order not to increase unnecessarily the complexity of my story. Philolaos hailed from South Italy and the two others from Syracuse, hence they form an Italian or Western group; all were Pythagoreans. For more information about them, see my *Introduction*, vol. 1, pp. 93, 94, 118, or vol. 1, pp. 288, 290.

(1585); and it was finally established forever by Kepler (1609). The triumph of the heliocentric theory was due to something that none of the ancients, not even Apollōnios, had ever thought of, or that they would have excluded a priori – the replacement of circular orbits by elliptic ones.

The distance in time between Hēracleidēs and Aristarchos, on the one hand, and between Copernicus and Brahe, on the other, is about the same (in both groups, the younger was born at about the time of the older one's death), but their sequence was reversed, the ancient Brahe preceding the ancient Copernicus. This is easy to explain; the passage from Hēracleidēs to Aristarchos was a progress in abstraction, that from Copernicus to Brahe one in precision.

The Aristarchian tradition concerning his extant treatise is a much simpler one, being the tradition of a definite text. It was commented upon by Pappos (III–2), who insured its preservation by including it in "The little astronomy." This was a group of astronomical writings by Autolycos, Aristarchos, Euclid, Apollōnios, Archimēdēs, Hypsiclēs, Menelaos, and Ptolemy, which were transmitted together, being copied in the same rolls, and were eventually translated together by Qusṭā ibn Lūqā (IX–2) of Ba'albek. Qusṭā thus helped to create the Arabic equivalent of the "little astronomy," called *Kitāb al-mutawassiṭāt bain al-handasa wal-hai'a,* the middle books between geometry and astronomy (in the course of time a number of purely Arabic treatises were added to those which had been translated from the Greek). The main student of the *Mutawassiṭāt* was eventually the Persian Naṣīr al-dīn al-Ṭūsī (XIII–2), who paid special attention to the "Treatise on the sizes and distances of Sun and Moon"; I gather that he prepared a new edition of it, probably with commentary.

Aristarchos' treatise was included in a collection of many others, all in Latin translation, edited by Giorgio Valla (d. 1499), printed in Venice in 1488 by Ant. de Strata, then again in the same city in 1498 by Bevilaqua (Fig. 17).[14] Another and more elaborate Latin edition of Aristarchos with Pappos' commentary was issued by Federico Commandino (Pesaro, 1572) (Fig. 18). The Greek princeps (Fig. 19) was published only a century later by John Wallis (Oxford: Sheldonian Theatre, 1688). Fortia d'Urban published a Greek-Latin edition in Paris in 1810, and a French translation in Paris in 1823. A German translation was given by A. Nokk (Freiburg i.B., 1854) and a Greek-English edition by Sir Thomas Heath (Oxford: Clarendon Press, 1913).

ARATOS OF SOLOI

In order to complete our account of astronomy in the early Hellenistic days (III–1 B.C.), we must still speak of Aratos of Soloi (III–1 B.C.), a

[14] Klebs mentions only the second edition (No. 1012.1), but I have made sure

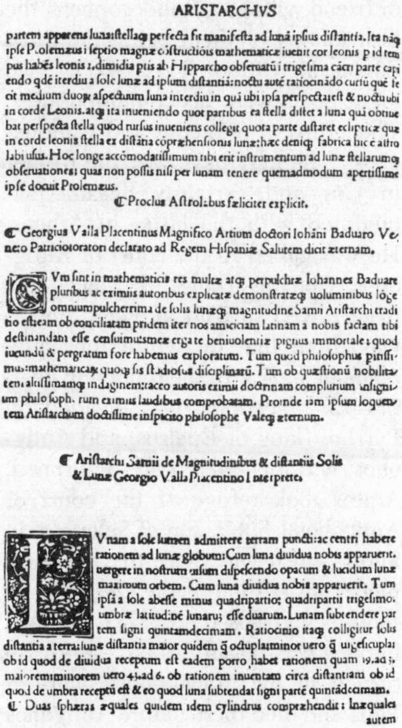
ARISTARCHVS

partem apperens lunaąstellaq̧ perfecta fit manifesta ad lunā ipsius distantia. Ita nāq̧
ipse Ptolemæus i septio magnæ cōstructiōis mathematicæ iuenit cor leonis p id tem
pus habēs leonis 2. dimidia ptis ab Hipparcho obseruatū i trigesima cācri parte capi
endo qdē iterdiu a sole lunæ ad ipsum distantiā: noctu autē ratiocinādo cursū quē fe
cit medium duoq̧ aspectuum luna interdiu in quā ubi ipsa perspectaest & noctu ubi
in corde Leonis. atq̧ ita inueniendo quot partibus ea stella distet a luna qui obtiue
bat perspecta stella quod rursus inueniens collegit quota parte distaret eclipticæ quæ
in corde leonis stella ex distātia cōpræhensionis lunæ: hæc deniq̧ fabrica hic ē astro
labi usus. Hoc longe accōmodatissimum tibi erit instrumentum ad lunæ stellarumq̧
obseruationes: quas non possis nisi per lunam tenere quemadmodum apertissime
ipse docuit Ptolemæus.

¶ Proclus Astrolabus fæliciter explicit.

¶ Georgius Valla Placentinus Magnifico Artium doctori Iohāni Baduaro Ve
neto Patricio oratori declarato ad Regem Hispaniæ Salutem dicit æternam.

CVm sint in mathematicis res multæ atq̧ perpulchræ Iohannes Baduare
pluribus ac eximiis autoribus explicatæ demonstratæq̧ uoluminibus lōge
omnium pulcherrima de solis lunæq̧ magnitudine Samii Aristarchi tradi
tio est: eam ob conciliatam pridem iter nos amiciciam latinam a nobis factam tibi
destinandam esse censuimus: meæ erga te beniuolentiæ pignus immortale: quod
iucundū & pergratum fore habemus exploratum. Tum quod philosophus pitissi
mus: mathematicaq̧ quoq̧ sis studiosus disciplinarū. Tum ob quæstionū nobilita
tem altissimamq̧ indaginem: taceo autoris eximii doctrinam complurium insigni
um philosophorum eximiis laudibus comprobatam. Proinde iam ipsum loquen
tem Aristarchum doctissime inspicito philosophe Valeq̧ æternum.

¶ Aristarchi Samii de Magnitudinibus & distantiis Solis
& Lunæ Georgio Valla Placentino Interprete.

LVnam a sole lumen admittere terram puncti: ac centri habere
rationem ad lunæ globum: Cum luna diuidua nobis apparuerit.
uergere in nostrum uisum dispescendo opacum & lucidum lunæ
maximum orbem. Cum luna diuidua nobis apparuerit. Tum
ipsā a sole abesse minus quadripartio: quadripartii trigesimo.
umbræ latitudinē lunarū esse duarum. Lunam subtendere par
tem signi quintamdecimam. Ratiocinio itaq̧ colligitur solis
distantia a terra: lunæ distantia maior quidem q̄ octupla: minor uero q̄ uigesicupla:
ob id quod de diuidua receptum est eadem porro habet rationem quam 19. ad 3.
maiorem minorem uero 43. ad 6. ob rationem inuentam circa distantiam ob id
quod de umbra receptū est & eo quod luna subtendat signi partē quintādecimam.
¶ Duas sphæras æquales quidem idem cylindrus compræhendit: Inæquales
autem

Fig. 17. First edition of the Latin translation of Aristarchos' treatise on the sizes and distances of the Sun and Moon as it occurs in the *Collectio* of Giorgio Valla of Piacenza, who was its translator (Venice: Bevilaqua, 1498). [Courtesy of the Armed Forces Medical Library, Cleveland, Ohio.]

ARISTARCHI
DE MAGNITVDINIBVS,
ET DISTANTIIS SOLIS,
ET LVNAE, LIBER
CVM PAPPI ALEXANDRINI
explicationibus quibusdam.
A FEDERICO COMMANDINO
Vrbinate in latinum conuersus, ac
commentarijs illustratus.
Cum Priuilegio Pont. Max. In annos X.

PISAVRI, Apud Camillum Franciſchinum.
M D L X X I I.

Fig. 18. First independent Latin edition of the same Aristarchian treatise. It was prepared by Federico Commandino (4 pp. plus 38 leaves; Pesaro: Camillus Franciscinus, 1572). [Courtesy of Harvard College Library.]

didactic poet, in spite of the fact that he flourished not in Alexandria, but in Cilicia and Macedonia, and was not an astronomer in the same sense as Aristarchos; his kind of knowledge was much closer to folklore and therefore very popular.

But first let us get acquainted with him. He was born in Soloi,[15] toward the end of the fourth century, perhaps as early as 315, and studied in Ephesos [16] and Athens. He was associated as disciple, auditor (*acustēs*),

that the former is not a ghost. This was the first time I discovered an omission in Klebs's excellent list.

[15] Soloi in Cilicia on the south coast of Anatolia, just north of Cypros. Soloi was the birthplace also of Chrysippos (III-2 B.C.), head of the Stoa, 233–208; "Without Chrysippos no Stoa." Soloi was rebuilt by Pompey the Great, c. 67 B.C., and then called Pompeiopolis.

[16] What kind of education would he receive in Ephesos? We may assume that philosophers and scholars had been attracted to the temple of Artemis and would

ΑΡΙΣΤΑΡΧΟΥ ΣΑΜΙΟΥ
Περὶ μεγεθῶν καὶ ἀποστημάτων Ἡλίου καὶ Σελήνης,
ΒΙΒΛΙΟΝ.
ΠΑΠΠΟΥ ΑΛΕΞΑΝΔΡΕΩΣ
Τοῦ τῆς Συναγωγῆς ΒΙΒΛΙΟΥ Β′
Απόσπασμα.
ARISTARCHI SAMII
De Magnitudinibus & Distantiis Solis & Lunæ,
LIBER.
Nunc primum Græce editus cum Federici Commandini versione Latina, notisq; illius & Editoris.
PAPPI ALEXANDRINI
SECUNDI LIBRI
MATHEMATICÆ COLLECTIONIS,
Fragmentum,
Hactenus Desideratum.
E Codice MS. edidit, Latinum fecit, Notisque illustravit
JOHANNES WALLIS, S.T.D. Geometriæ Professor Savilianus; & Regalis Societatis Londini, Sodalis.
OXONIÆ,
E THEATRO SHELDONIANO.
1688.

Fig. 19. Princeps of Aristarchos with Commandino's Latin translation and Pappos' commentary, by John Wallis (1616–1703). (Oxford: Sheldonian Theatre, 1688.) [Courtesy of Harvard College Library.]

or friend with many philosophers, the best known of them being the Stoic Zēnōn of Cition (IV–2 B.C.). Two great poets were his contemporaries, Theocritos of Syracuse and Callimachos;[17] he may have met the former in Cōs and certainly became acquainted with the latter in Athens. He was called to the court of Antigonos Gonatas (king of Macedonia from c. 283 to 239) at Pella, and it was there that he wrote the *Phainomena*, c. 275. In the following year, 274–73, Macedonia was invaded by Pyrrhos (king of Ēpeiros) and Antigonos was defeated and dethroned. Aratos took refuge at the court of Antiochos I Sōtēr, son of Seleucos, in Syria, and it was there that he completed his edition of the *Odyssey*. As soon as Pyrrhos died (272) and Antigonos was reëstablished, Aratos returned to the Macedonian court at Pella and died there before Antigonos (the latter died in 239). Aratos was a very learned man and wrote many books but his astronomical poems are the only ones which are extant.

There are two such poems, the *Phainomena* and the *Diosēmeia* (weather forecasts), the first derived from Eudoxos of Cnidos (IV–1 B.C.), the second, largely derived from Theophrastos of Eresos (IV–2 B.C.). The *Phainomena* describes the northern constellations and the Zodiac; he begins with the North Pole and the Bears and then proceeds southward, then returns to the Bears and works down again to the Zodiac. He deals altogether with thirty northern constellations and with

---

enjoy teaching young men. There might also be some kind of public education. We do not know much about that in Ephesos, but we have an extraordinary document concerning public education at the same time (III–1 B.C.) in Teōs, a city not very far from Ephesos to the northwest of it, at the seashore. You can read an English version of it in G. W. Botsford and E. G. Sihler, *Hellenic civilisation* (New York, 1915), pp. 599–601. Teōs was the birthplace of Anacreōn, a famous lyric poet of the VI/V cent.

[17] Theocritos of Syracuse, the founder of idyllic poetry, visited Alexandria c. 285. Callimachos of Cyrēnē was head of the Alexandrian library from c. 260 to 240. We shall come back to both further on.

fifteen more south of the Ecliptic; these descriptions are cómbined with mythological references. After a brief allusion to the five planets, which he does not name, he discusses five circles of the celestial sphere, to wit, the Galaxy, the Tropic of Cancer, the Tropic of Capricorn, the Equator, and the Zodiac. The end of the work (ll. 559–732) is devoted to the risings and settings of stars (*synanatolai, anticatadyseis*), that is, what stars rise with a given zodiacal sign, or set when the zodiacal sign is rising.[18]

The description of the constellations was the kind of astronomy that concerned everybody. It is still popular today and many simple people know of no other; they believe that the ability to recognize the constellations and to give them their proper names is the alpha and omega of astronomy. Yet we would not call them astronomers, any more than we would dignify with the title of botanist the people who can name every plant but have no knowledge of plant life. If the description of constellations is still popular today, we can hardly imagine what it was in ancient times. In the first place, the majority of the people saw the stars every night, while the circumstances of city life hardly give us the opportunity of being familiar with them. Moreover, the astral religion, which every one accepted, at least to some extent, gave to the constellations an awful meaning. Each was godlike. The study of heaven's luminaries was not simply astronomy, it was a survey of mythology, it was theology and religion. What a beautiful illusion that was! Think of it, the book of religion, the eternal Bible, was opened every night in the sky to everybody who cared to read it.

This state of mind justifies the sacred exordium to Aratos' poem "Ec Dios archōmestha":

> From Zeus let us begin; him do we mortals never leave unnamed; full of Zeus are all the streets and all the market-places of men; full is the sea and the havens thereof; always we all have need of Zeus. For we are also his offspring.

This is Mair's literal and accurate translation of lines 1–5, the Greek of which can be read in our facsimile of the princeps, or consider Sir D'Arcy Thompson's paraphrase:

> Let us call upon God in the beginning; let us praise His name for ever and ever. All the streets of the cities are filled with the presence of the Lord and all the marts of men; the sea also and its havens are full of His glory. Every man hath need of Him in all things, *for we are also His offspring.*

The last words are taken from the Acts of the Apostles (17:28), and that is not as arbitrary as it might seem, but on the contrary was a definite allusion of St. Paul's to Aratos. "For in Him we live, and move, and have our being, as certain also of your own poets have said, *For we are also his offspring.*" The two poets whom St. Paul had in mind were the Stoic Cleanthēs of Assos (III–1 B.C.) and Aratos.[19]

[18] This is well explained by G. R. Mair in the Loeb edition, *Callimachus, Lycophron, Aratus* (Cambridge, 1921), p. 377.

[19] This *rapprochement* was brought to

This Semitic beginning of a Greek poem is not unnatural. Aratos was born and educated in Western Asia; he had obtained some of his astronomical knowledge, whether directly or indirectly, from Babylonian sources; he must have come across many Orientals; I would not go so far, however, as to suggest that he had heard of the Psalms. That was not necessary; the author of the Psalms, Cleanthēs in his hymn to Zeus, and Aratos in his description of the glories of heaven were using similar sources, the main one of which was the religious emotion raised by the contemplation of the starry heavens.[20]

The *Phainomena* covers 730 lines, the *Weather forecasts*, 422. It is hardly necessary to point out the interest of the latter poem to every man, and especially to every farmer. The metrical form crystallized every bit of weather folklore and facilitated its remembrance. In the same way, the relative positions of the constellations were engraved in the minds of men by haunting verses.

It is hardly possible to exaggerate the importance of didactic poetry for popular education before the age of printing. Such poetry had existed in ancient Greece long before the time of Aratos – think of Hēsiodos (VIII B.C.) – but Aratos revived it and his poems were the most popular of their kind in Roman days. We shall come back to that presently; let me first remark that didactic poems continued to be written during the Renaissance and later until our own times, but they became less and less necessary and more and more artificial. The history of modern Latin literature records many of them, such as *Syphilis* by Girolamo Fracastoro (Verona, 1530), the *Anti-Lucretius* of Melchior Cardinal de Polignac (Paris, 1747). Some such poems were also published in vernaculars, for example, the *Saisons* of Jean François de Saint Lambert (Paris, 1769), and, to mention one of the most recent, *The Torch-Bearers* of Alfred Noyes (Edinburgh, 1922); only one volume of Noyes's poem has appeared, dealing with the history of astronomy from Copernicus to Sir John Herschel. This is history, history of science, and therefore more humane than astronomy itself, yet I do not see the advantage of forcing that account into a metrical yoke. Such restriction is illogical, irrelevant, and retrogressive.

The composition of such poems was a necessity in ancient times; it is today a preposterous waste of intellectual effort. A scientific poem is generally poor science and poor poetry.

The Aratian tradition. Aratos' poem seems to have pleased learned people, mathematicians and astronomers, as much as it pleased the literary ones.

---

my attention by my old friend, D'Arcy W. Thompson, in his delightful address to the Classical Association of Scotland at St. Andrews, 1935, "Astronomy in the classics," reprinted in *Science and the classics* (London: Oxford University Press, 1940) [*Isis* 33, 269 (1941–42)], pp. 79–113.

[20] Compare with Kant's famous statement in his *Kritik der praktischen Vernunft* (Riga, 1788): "Zwei Dinge erfüllen

Various commentaries were soon devoted to it, the most authoritative being that of Hipparchos (II–2 B.C.). Hipparchos' concern was the greatest homage that could be paid to Aratos. By a strange fluke, his commentary is the only work of his that has come down to us. Would that we had his own astronomical treatise instead of it! As the *Phainomena* of Eudoxos of Cnidos was available to him, he was aware that Aratos had simply versified Eudoxos' prose and he compared the two texts. Aratos' poem reproduced some of Eudoxos' mistakes and added new ones; its popularity dangerously increased the currency of those errors. This awakened the solicitude of the great astronomer. Let us quote his own words:

> Several other writers have compiled commentaries upon the *Phainomena* of Aratos, but the most careful exposition of them is that of Attalos, the mathematician of our own time.[21] Now the explanation of the meaning of the poem I do not regard as requiring great range; for the poet is simple and concise, and also easily to be understood even by readers who are only moderately informed. But to be able to distinguish in reading what he says about the heavenly bodies, which of his statements are consistent with the observed phenomena and which are erroneous, may well be considered to be a most useful accomplishment and one especially appropriate to a trained mathematician.
>
> Observing, then, that in very many of the most useful details Aratos is not in agreement with the phenomena which really occur, but that in almost all of those points not only the other commentators but Attalos also agrees with him, I have determined in view of your[22] enthusiasm for learning and looking to the benefit of all, to set out the details which seem to me to have been incorrectly given. I proposed to myself this task, not because I was desirous of getting credit for myself out of criticism of others (this would indeed be a vain and ungenerous motive; I hold on the contrary that we should be grateful to all and sundry who undertake laborious personal work for the common benefit of all), but in order that neither you nor any other enthusiasts for learning should fail to get the true view of the phenomena occurring in the universe, as is naturally the case with many persons nowadays; for the charm of poetry invests its content with a certain plausibility, and almost all who expound this particular poet associate themselves with his statements.[23]

This longish quotation has been given because it shows that Hipparchos was not a vainglorious scholar, but a truth-loving one, a good as well as a great man.

After Hipparchos, the Greek tradition petered out. There was a com-

---

das Gemüth mit immer neuer und zunehmender Bewunderung und Ehrfurcht, je öfter und anhaltender sich das Nachdenken damit beschäftigt: der bestirnte Himmel über mir und das moralische Gesetz in mir."

[21] This Attalos is otherwise unknown which is strange for one called by Hipparchos "the mathematician of our time" (*Attalos ho cath' ēmas mathēmaticos*).

[22] "You" is Hipparchos' friend, Aischriōn, to whom his book was addressed.

[23] *Hipparchi in Arati et Eudoxi Phaenomena libri tres*, I, I, 3–8, pp. 4–7 in Karl Manitius' Greek-German edition (Leipzig, 1894). English version by T. L. Heath, *Greek astronomy* (London, 1932), p. 116 [*Isis* 22, 585 (1934–35)].

ΑΡΑΤΟΥ ΣΟΛΕΩΣ ΦΑΙΝΟΜΕΝΑ.

Fig. 20. Aratos. Princeps of the *Phainomena* in the *Scriptores astronomici veteres* (Venice: Manutius, 1499) (Klebs, 405.1). The first page of the Greek text on folio 313r; lines 1–9, in the upper left-hand corner, "From Zeus let us begin . . ." A large blank is left to permit the insertion of an ornamental epsilon by the limner. The same volume contains three Latin versions of the *Phainomena* and the commentary ascribed to Theōn of Alexandria. [Courtesy of Harvard College Library.]

mentary by Achilleus Tatios (III–1), and some scholia are ascribed to Theōn of Alexandria (IV–2).

The lasting tradition was not Greek but Latin, this being due mainly to Cicero (I–1 B.C.), who translated the *Phainomena*; much of that translation (475 verses) is extant. Virgil (I–2 B.C.) was influenced by Aratos in the writing of his *Georgica*. Ovid (43 B.C.–A.D. 17) wrote of him an extravagant eulogy, "Aratos will live as long as the Sun and the Moon" (*cum sole et luna semper Aratus erit*). New translations were prepared by the Roman general Germanicus Caesar (15 B.C.–A.D. 19) and by Avienus (IV–2). Thus, the Latin Middle Ages were well acquainted with him.

Aratos' popularity is proved by the incunabula: three in Latin, one in Greek! The first two date from 1474, the one by an anonymous printer in Brescia, the other, together with the second edition of the *Astronomicon* of Manilius (I–1), by Rugierus and Bertochus in Bologna. The third was Avienus' translation printed by Strata (Venice, 1488). The fourth was in the collection *Scriptores astronomici veteres*, by Manutius (Venice, 1499) (Fig. 20). This fourth edition includes three different Latin translations plus the Greek text, and also Theōn's scholia.

Klebs, Nos. 77.1, 661.2, 137.1, 405.1.

Immanuel Bekker, *Aratus cum scholiis* (Berlin, 1828); Ernst Maass,

Arati Phaenomena (Berlin, 1893); *Commentariorum in Aratum reliquiae* (822 pp.; Berlin, 1898).

Karl Manitius, *Hipparchi in Arati et Eudoxi commentaria* (410 pp.; Leipzig, 1894), Greek and German.

Of the English translations it will suffice to mention that of the *Phaenomena* by G. R. Mair included in the edition of Callimachos and Lycophron by A. W. Mair (Loeb Classical Library, London, 1921), with Greek text on opposite pages. For the *Weather forecasts* see the verse translation by Edward Poste (London, 1880) or the literal one by C. Leeson Prince (Lewes, 1895).

Aratos' poem might be called a treatise of astronomical mythology, but it would be confusing to call it astrological. There was already much astrology at the beginning of the Hellenistic age, but that concerned religion rather than science and will be discussed briefly in Chapters XI and XIX. Aratos' purpose was descriptive and didactic; he did not try to explain any form of divination, except the short-range weather signs that the farmers of every land cannot help interpreting.

The astronomical investigations of mathematicians such as Archimēdēs, Conōn, Apollōnios, Eratosthenēs will be described in Chapters V and VI, and Chaldean and Egyptian astronomy in Chapter XIX.

V

ARCHIMĒDĒS AND APOLLŌNIOS

Ptolemaic Egypt was the main center of Greek science, but by no means the only one; wherever Greek colonies were established in Asia, in the islands, or in Magna Graecia,[1] there were definite possibilities of scientific advance. We shall come across many examples of this, the outstanding one in the third century being that of Archimēdēs of Syracuse. It is out of the question in this book to describe, even briefly, the vicissitudes of politics and war, but the historian of science must explain how it happened that great men of science did their work in one place rather than another, and why science grew up in this or that environment. It never grew up, we should remember, in a vacuum.

Now, in order to account for the occurrence of Archimēdēs in Sicily, we must give a summary of past events. It has been set forth in Volume 1 [2] that the main tensions in the Mediterranean Sea from the twelfth century on were caused by the incessant conflicts between the Greek colonies on one hand and the Phoenician on the other. From the sixth century on, the tensions became more complex in the Western Mediterranean Sea, because of Etruscan jealousies and interferences. The two leading cities in that area were Carthage in the Semitic empire and Syracuse in the Greek one. Let us focus our attention for a moment upon both of them.

Carthage was the older settlement. It had been founded by a Tyrian swarm as early as 814, and we all know the first Queen, Dido, immortalized in the *Aeneid*. It soon became the main colony of its kind, to such an extent that one ceased to speak of Phoenicians and spoke instead of Carthaginians. They established new colonies of their own in Africa, Sicily, and Sardinia. During three centuries, Greeks fought the Carthaginians for the possession of Sicily, then the conflict was taken over by the Romans. At the end of the First Punic War (264–241), the Carthaginians had conquered Spain but lost Sicily to the Romans.[3] During the Second Punic War (218–201),

[1] For a definition of Magna Graecia, see Volume 1, p. 199.

[2] Pages 108–109, 222. For a map of the Phoenician settlements about the Mediterranean Sea, see p. 102.

[3] More exactly, western Sicily, which was constituted as the first province of Rome in 227. Eastern Sicily remained under the control of Hierōn of Syracuse, who was a friend and ally of the Romans. The

the battles took place in Spain, Italy and Sicily. One of the events was the reduction of Syracuse by the Romans in 212.[4]

Syracuse was founded eighty years later than Carthage, in 734, on the southeast coast of Sicily. Thanks to its magnificent location and to the genius of its Corinthian founders, it soon became the most important city not only of Sicily but of the whole of Magna Graecia. It was bound to antagonize Carthage and the perils of war caused the establishment of a dictatorship from 485 on. In 480 (the year of Salamis!) the tyrant Gelōn defeated at Himera the Carthaginians who had invaded Sicily. His brother and successor Hierōn increased the Syracusan empire and made of this capital one of the leading centers of Hellenism. He was a friend of letters and patronized Pindaros and Aischylos. This golden age ended with his death in 467, yet one of the most glorious events of the city was the utter defeat of the Athenian expedition in 413 (this was described by Thucydidēs in a masterly fashion). The struggle between Syracuse and Carthage continued until the Romans, taking advantage of a pro-Roman party, besieged the city and took it in 212.[5]

The two preceding paragraphs end with the year 212, which is the node as far as our own narrative is concerned.

As to their spiritual glories, Carthage was the starting point at the beginning of the fifth century of the bold navigations of Hannōn and Himilcōn, and Hērillos of Carthage, disciple of Zēnōn of Cition (IV-2 B.C.), was the founder of a Stoic sect. Syracuse was the home of two famous astronomers, Hicetas (V B.C.) and Ecphantos (IV-1 B.C.), of the great poet Theocritos (c. 310–250) and of his younger contemporary, Archimēdēs (III-2 B.C.).

ARCHIMĒDĒS OF SYRACUSE

When the Roman general Marcellus besieged Syracuse, the difficulty of his task was greatly increased by the resourcefulness of an engineer named Archimēdēs, who was killed during the sack of the city in 212. According to the legend, Archimēdēs had invented various machines for defensive

---

whole of the Hispanic peninsula, except the northern part (above lat. 41° and 42°), was an intrinsic part of the Carthaginian empire from 450 to 201 B.C.

[4] For readers who wish to know the sequel of Carthage's history, the few following data are given. The third and last Punic War (149–146) ended with the utter destruction of Carthage by Scipio Aemilianus. The site was too good to be abandoned, however; it was colonized by Caesar and by Augustus and soon gave room to one of the main cities of the Roman empire. In 439, Carthage was captured by the Vandals and was their capital until 533, when Belisarios won it back for the Byzantine empire; in 698, it was taken by the Arabs. St. Louis died there in 1270 during the eighth and last Crusade, the one undertaken by himself.

[5] Main facts in the later history of Syracuse: After 212, the whole of Sicily became a Roman province and Syracuse was the capital of the eastern half. Settlers were sent to Syracuse by Augustus in 21 B.C. In A.D. 280, Syracuse was plundered by the Franks. It was conquered by Belisarios in 535, by the Arabs in 878, and by the Normans in 1085.

purposes, catapults, ingenious hooks, and also concave mirrors by means of which he deflected the sun rays and set the Roman ships on fire. The story is told that a Roman soldier came upon him while he was absorbed in the contemplation of geometric figures drawn up on the earth. Archimēdēs shouted, "Keep off," and the soldier killed him. The account of the inventions by which he tried to save his native city fired the imagination of people not only during ancient and medieval times but even as late as the eighteenth century, and he was generally thought of as a mechanical wizard. For example, Gianello della Torre, clockmaker to Charles Quint, was called "the second Archimēdēs" and as late as the eighteenth century, the inventor, Christopher Polhem, was called "the Swedish Archimēdēs."[6] That is as silly as if we were to call Edison "the American Archimēdēs." The absurdity of such nicknames is obvious as soon as one realizes that Archimēdēs, though he may have invented various machines and gadgets was primarily a mathematician, the greatest of antiquity and one of the very greatest of all times.

Plutarch had already remarked that Archimēdēs himself did not think much of his practical inventions.

> "although they had obtained for him the reputation of more than human sagacity, he did not deign to leave behind him any written work on such subject, but, regarding as ignoble and sordid the business of mechanics and every sort of art which is directed to use and profit, he placed his whole ambition in those speculations the beauty and subtlety of which are untainted by any admixture of the common needs of life.[7]

Plutarch's suggestion is plausible, and it is typically Greek. Yet, it is certain that Archimēdēs' fame was based for many centuries not upon the immortal achievements explained in his own works but upon the legends that clustered around his name. These legends had a core of truth; he did invent machines, such as compound pulleys, an endless screw, a hydraulic screw, an orrery, burning mirrors, but these activities were secondary and marginal. The orrery was actually seen by Cicero, according to whom it represented the motions of Moon and Sun so well that eclipses could be demonstrated.

The only fact of his life that can be dated with certainty is his death during the sack of Syracuse in 212. As he was said to be 75 years old in that year, that places his birth c. 287. He was the son of the astronomer Pheidias, his early interest in astronomy and mathematics was thus natural enough. He was a kinsman and friend of Hierōn II, king of Syracuse, and of

[6] Christopher Polhem (1661–1751), for whom see *Isis 43*, 65 (1952).

[7] Extracted from the life of Marcellus by Plutarch, who describes vividly the part that Archimēdēs took in the defense of Syracuse (*Plutarch's Lives*, Loeb Classical Library, vol. 5, pp. 469–479). Archimēdēs had prepared for King Hierōn offensive and defensive engines of many kinds. For the stories concerning his death, see p. 487. Marcus Claudius Marcellus (the first of that name) was the Roman general who besieged and took Syracuse; he died in 208.

the latter's son and successor, Gelōn II.[8] According to Diodōros of Sicily (I–2 B.C.), he spent some time in Egypt and that is very plausible. Alexandria was then the center of the scientific world; Archimēdēs had no equal in Syracuse, and he would naturally wish to visit the Museum and exchange views with the great mathematicians who were flourishing in its neighborhood. It was very probably in Alexandria that he became acquainted with Conōn of Samos (III–2 B.C.), the latter's pupil, Dōsitheos of Pēlusion, and Eratosthenēs.[9] It was during his stay in Alexaxndria that he invented his hydraulic screw ("Archimēdēs' screw").[10] Though we may assume that he lived mostly in Syracuse, he helps to illustrate the prestige of the Museum.

One more story. Archimēdēs requested his friends to engrave a mathematical diagram on his tombstone. The diagram (or was it a tridimensional model?) represented a cylinder circumscribing a sphere.[11] We know this through Cicero, who, when he was quaestor of Sicily in 75 B.C., discovered Archimēdēs' tomb in a ruined state, restored it, and described it.[12] The tomb has disappeared and its exact location is unknown.

Now that we know the man Archimēdēs as much as is possible, let us consider the extant works that have immortalized him.

Archimēdēs lacked the encyclopedic tendencies of Euclid, who tried to cover the whole field of geometry; he was, on the contrary, a writer of monographs of limited scope, but his treatment of any subject was masterly in its order and clearness. As Plutarch remarked in his life of Marcellus, "It is not possible to find in geometry more difficult and troublesome questions or proofs set out in simpler and clearer propositions." That is well put. Until 1907, one might have added that Archimēdēs did not indicate how he made

[8] This is chronologically possible because Hierōn II died in 216 at the age of 92; Gelōn II, appointed king by his father, died before him. It is nevertheless difficult to understand his friendship because Hierōn was the ally of the Romans in the Second Punic War and remained loyal to them. According to one Moschiōn, Archimēdēs built a ship for Hierōn; Moschiōn's elaborate description of it was preserved by Athēnaios of Naucratis, V, 40–44. The text is a very interesting document for the history of Hellenistic technology (see Chapter VII).

[9] He dedicated one of his books to King Gelōn, two to Eratosthenēs, and no less than four to Dōsitheos. These four treatises constitute more than 70 percent of the total of his extant writings. We may thus say that Dōsitheos of Pēlusion was his best friend. Pēlusion is near the seashore, east of the Suez Canal; it was the eastern key to Egypt. It is probably identical with Sin (Ezekiel 30:15, 16).

[10] The so-called Archimedean screw is a strip bent spirally around an inclined axis and encased in a hollow open cylinder. The lower end of the cylinder dipped in water and when it was rotated raised the water up to a higher level. The device is not described in the Archimedean writings which have come down to us, but that does not prove that he did not invent it. Such inventions were often realized without being explained in a literary way.

[11] Archimēdēs had established the ratio of their volumes and surfaces (3:2). The proof is given in his treatise *On the sphere and cylinder* and also in his *Method*.

[12] Cicero, *Tusculanarum disputationum*, v, 23; English translation of the relevant text in my *Appreciation of ancient and medieval science during the Renaissance* (1450–1600) (Philadelphia: University of Pennsylvania, 1955), p. 214.

his discoveries but explained them in the most dogmatic manner, caring only for order, rigor, and simplicity. We could not say that any more because in that year Heiberg published the lost *Method*, wherein Archimēdēs told us some of his secrets. We shall come back to that presently.

A dozen works have come down to us, which we shall examine briefly, adding for each of them a few remarks that will interest every educated reader, but of necessity leaving out technical details that the nonmathematician could not appreciate even at the cost of tedious explanations. As Archimēdēs was primarily a geometer, we shall examine first his geometric works, then the others dealing with arithmetic, mechanics, astronomy, and optics.

Geometry. The longest of all Archimēdēs' writings is a treatise *On the sphere and cylinder* in two books, the Greek text of which covers (in Heiberg's edition) not more than 114 pages. In that treatise he proves a number of propositions, such as the one to which he himself attached so much value that he ordered the diagram relative to it to be engraved on his tombstone, and also the one which every schoolboy knows, that the area of the surface of a sphere is four times that of one of its great circles ($4\pi r^2$). We gather from his *Method* that he had calculated the volume of the sphere ($\frac{4}{3}\pi r^3$) before its surface, and deduced the latter from the former, but in his exposition the order was reversed. The treatise begins in Euclidean fashion with definitions and assumptions. For the determination of surfaces and volumes, he uses the method of exhaustion, very skillfully and rigorously. He solved [13] the problem, "To divide a sphere by a plane into segments the volumes of which are in a given ratio," and similar ones.

His second treatise in order of length (100 pages in Greek) is the one on *Conoids and spheroids*, dealing with paraboloids and hyperboloids of revolution and the solids formed by the revolution of ellipses about their major or minor axis. The third (60 pages) is devoted to *Spirals*. This third treatise summarizes the main results of the two preceding ones and hence is also the third in chronologic order. The spiral he dealt with is the one that is called to this day the "Archimedean spiral" and which he defined as follows: "If a straight line of which one extremity remains fixed be made to revolve at a uniform rate in a plane until it returns to the position from which it started, and if, at the same time as the straight line revolves, a point moves at a uniform rate along the straight line, starting from the fixed extremity, the point will describe a spiral in the plane." [14] That clear definition would still be used today and would lead to the equation $r = a\theta$, wherein a is a constant. (There are, of course, no equations in Archimēdēs, nor in any other ancient text; our equations hardly date back to the second

[13] More exactly, he reduced the problem to a cubic equation, which he did not solve in that treatise. In a fragment known to his commentator, Eutocios (VI-1), he solved the equation by means of the intersections of a parabola and a rectangular hyperbola.

[14] The definition occurs at the beginning

half of the sixteenth century.) He finds various areas bounded by it, and what we would call the constancy of its subnormal ($=a$). His ability to obtain those results without our analytical facilities is almost uncanny.

His fourth treatise, on the *Quadrature of the parabola*, is much shorter (27 pages), but deals with a single problem.

These four geometric treatises were all dedicated to his friend, Dōsitheos of Pēlusion, who is immortalized by them; they constitute the main bulk of Archimēdēs' available works. His other geometric treatises are much shorter and less important. There is first the *Liber assumptorum* (Book of lemmas), lost in Greek but known in a Latin translation from the Arabic, concerned with special diagrams, such as the *arbēlos* (or shoemaker's knife). The *arbēlos* is bounded by three half circles whose diameters *AC*, *AB*, *BC* are collinear and coterminous (Fig. 22). The area of the circle whose diameter *BD* is perpendicular to these is equal to the area included between the three semicircles.

The *Measurement of the circle* (perhaps a fragment of a larger treatise) leads to a good approximation of π, namely, $3\frac{1}{7} > \pi > 3\frac{10}{71}$ ($3.142 > \pi > 3.141$). Archimēdēs had obtained that result by comparing the areas of two regular polygons of 96 sides, inscribed in and circumscribed about the same circle. It is difficult to know how he arrived at his approximations, for example,

$$\frac{1351}{780} > \sqrt{3} > \frac{265}{153}.$$

It may have been derived from the so-called Heronian formula,

$$a \pm \frac{b}{2a \pm 1} < \sqrt{(a^2 \pm b)} < a \pm \frac{b}{2a},$$

where a^2 is the nearest square number to the number whose root is desired. In this case, $\sqrt{3} = \sqrt{(4-1)}$, that is, $a = 2$, $b = 1$.

The *Stomachion* (or *loculus Archimedius*), another fragment, is a kind of geometric puzzle, somewhat like the Chinese tangram but more complex. The problem that it treats is to divide a parallelogram into 14 parts submitted to various relations.

According to Pappos,[15] Archimēdēs had described 13 semiregular poly-

of the treatise *De lineis spiralibus*. See Fig. 21. The spiral is generated by the point *A*, if the distance *OA* ($= r$) and the angle θ increase at a uniform rate. The Archimedian spiral is the simplest of a family of plane curves: $r^m = a^m\,\theta$.

[15] Pappos, *Synagōgē*, v, prop. 19; Greek edition by Friedrich Hultsch (Berlin, 1876), (vol. 1, pp. 351–361); French translation by Paul Ver Eecke (Bruges, 1933), pp. 272–277.

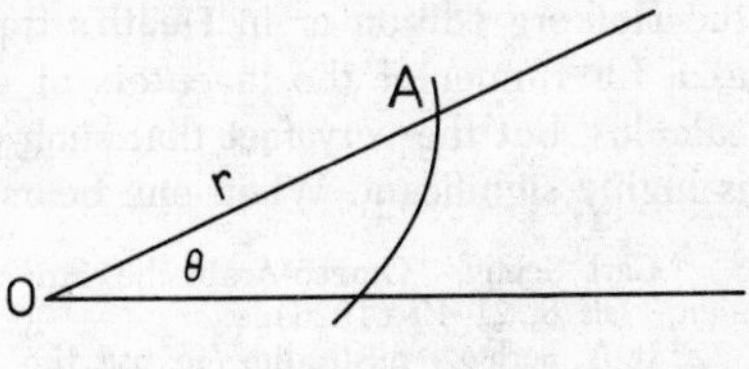

Fig. 21. The spiral of Archimedes.

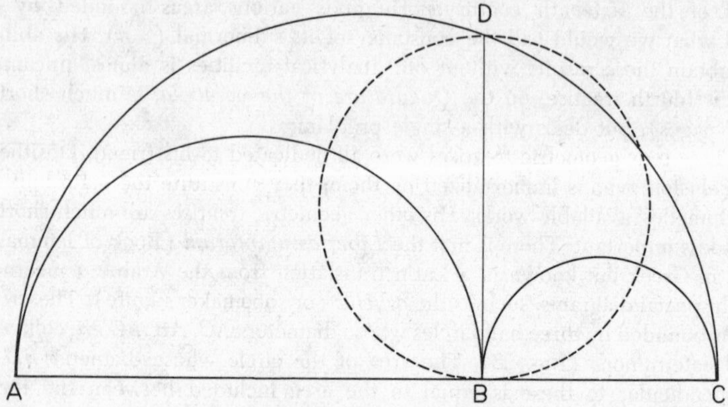

Fig. 22. Diagram of the *arbēlos*.

hedra, that is, polyhedra the faces of which are equilateral and equiangular but not similar. For example, one of them is an octahedron made of 4 triangles and 4 hexagons. The thirteenth and most complicated of these Archimedean polyhedra is one made of 92 faces, of which 80 are triangles and 12 are pentagons; it is a "snub dodecahedron," each solid angle of which is formed by 4 triangles surrounding one pentagon.

A study of his on the *Regular heptagon*, lost in Greek, was translated into Arabic by Thābit ibn Qurra (IX–2). Carl Schoy found an Arabic manuscript of it in Cairo and revealed it to the Western public in a German translation in 1926.[16]

This enumeration is more than sufficient to reveal the incredible depth and ingenuity of Archimēdēs' geometric thought. Not only did he ask questions that were original and obtain results that were almost unthinkable in his time, but he used methods that were rigorous and unique. For example, he accomplished quadratures of curvilinear plane figures and the quadrature and cubature of curved surfaces. By means of a method[17] equivalent to integration, he measured the areas of parabolic segments and of spirals, the volumes of spheres, segments of spheres, and of segments of other solids of the second degree. This cannot be explained here and now; the best way to appreciate these methods is to study his own work in the Heiberg edition or in Heath's translation. It is foolish to speak of him as a forerunner of the inventors of analytic geometry and of the integral calculus, but the very fact that such claims could have been made for him is highly significant. When one bears in mind that he had formulated and

[16] Carl Schoy, "Graeco-Arabische Studien," *Isis 8*, 21–40 (1926).

[17] It is perhaps misleading to use the word "method." He had no general method of integration but invented with great ingenuity a special way of solving each problem. Each solution was rigorous but inapplicable to other problems.

solved a good many abstruse problems without having any of the analytic instruments that we have, his genius fills us with awe.

Arithmetic. Archimēdēs' work in arithmetic and algebra was less bulky and less original. Was he at all acquainted with Babylonian methods, I wonder? [18] He might have heard of them during his stay in Alexandria; it would not have been necessary for him to hear much, the most meager suggestion would suffice to excite his mind. At any rate, it is not possible to recognize definitely Babylonian elements in his works.

Archimēdēs had been impressed by the inherent weakness of the Greek numerical system, whether it be expressed with words or with symbols. That weakness is one of the paradoxes of Greek culture; the leading mathematicians of antiquity had to be satisfied with the worst numerical system, the very basis of which was hidden by inadequate symbols.[19] His own genius was wanting in this case, for instead of inventing a better system (that was the true solution), he tried to justify Greek numerals by showing that they were sufficient to designate the very largest numbers.[20] Of course, every number system, however poor, could be justified in the same manner. He explained his views ad hoc in a treatise entitled *Archai (Principles)* or *Catonomaxis tōn arithmōn (Naming of numbers)*, which he dedicated to one Zeuxippos. That treatise is lost, but another has come down to us, the *Psammitēs* [21] (*Arenarius, Sand reckoner*), dedicated to King Gelōn, wherein an extremely large number is introduced in a very original way. "How many grains of sand could the whole universe hold?" It is clear that that question is double, for one must first determine the size of the universe; this being done, if one knows how many grains are contained in a unit of space it is easy to calculate how many the whole universe will be able to hold. It is easy provided we have the necessary number words. In the decimal system, the question would not arise, because if one understands the meaning of 10^0, 10^1, 10^2, there is no difficulty in understanding 10^n, irrespective of the size of n. Archimēdēs' solution was more complicated. The numbers from 1 to 100 million (10^8) formed his first order, those from 10^8 to 10^{16} the second order, and so on; those of the 100-millionth order ending with the number $10^{8 \cdot 10^8}$. All of these numbers form the first period; a second period may be defined in the same way, also a third, and so on up to the 10^8th period, ending with the number $(10^{8 \cdot 10^8})^{10^8}$. The decimal expression of the last number of the 10^8th period would be 1 followed by 80,000 million million zeros. The number of grains of sand in the universe is relatively small, less than 10^{63}.

[18] Volume 1, pp. 74, 118.

[19] Volume 1, pp. 206–209. The Greek system of numbers was as bad as the Semitic systems (Hebrew or Arabic).

[20] He reminds us of English mathematicians trying to justify the absurdities of English metrology.

[21] We have already spoken of the *Psammitēs* or *Sand reckoner,* which is of very great importance because it is to it alone that we owe our knowledge of the heliocentric theory of Aristarchos of Samos.

This aspect of Archimēdēs' genius is curious; instead of thinking out a numerical system that would be of use in practical life, he indulged in the conception of immense numbers – a conception that is philosophical rather than purely mathematical. It reminds us of the fancies of Buddhist cosmologists who were tormented by visions of infinity; they defined numbers (not as large as the Archimedean) and named units of increasing decimal order, up to 10^{51}, and they invented an immense period of time, *mahākalpa*, long enough for the whole drama of creation and destruction. Each *mahākalpa* follows another one. If one is able to conceive an infinity, one may conceive an infinity of infinities, and so on. At this stage of thought, this is metaphysics, not mathematics.[22]

Another treatise, called the *Cattle problem (Problema bovinum, Problēma)* and dedicated to Eratosthenēs, was devoted to a problem of indeterminate analysis. It is a problem of great complexity. One is required to find the number of bulls and cows of each of four colors; the eight unknown quantities are connected by seven single equations plus two conditions.[23]

The solution of the seven equations leads to eight numbers of 7 or 8 digits each multiplied by the same coefficient. The conditions increase that coefficient prodigiously; one of the eight unknown quantities would have more than 206,500 digits. It is strange again that Archimēdēs' interest in indeterminate analysis was combined with an Indian interest in immense numbers.

Mechanics. We now come to something that is perhaps even more remarkable than Archimēdēs' geometric investigations, namely, his creation of two branches of theoretical mechanics, statics and hydrostatics. Two of his mechanical treatises have come down to us, the *De planorum aequilibriis* and the *De corporibus fluitantibus*, both composed in Euclidean style, divided into two books, and of about equal length (50 pp. and 48 pp.). They both begin with definitions or postulates, on the basis of which a number of propositions are geometrically proved.

The first, on the *Equilibrium of planes*, begins thus:

I postulate the following:

1. Equal weights at equal distances are in equilibrium, and equal weights at unequal distances are not in equilibrium but incline towards the weight which is at the greater distance.

2. If, when weights at certain distances are in equilibrium, something be added to one of the weights, they are not in equilibrium but incline towards that weight to which the addition was made.

---

[22] For the Buddhist ideas, see William Montgomery McGovern, *Manual of Buddhist philosophy*, vol. 1, *Cosmology* (London, 1923), pp. 39 f. The modern theory of aggregates has lifted such questions from the level of futility and metaphysical verbiage to the level of science.

[23] The unknowns are W, w; X, x; Y, y; Z, z, where the capitals represent bulls, the small letters cows. Each of the four groups stands for a different color. The two conditions are that $W + Z =$ a square number and that $Y + Z =$ a triangular number.

After a few steps, he is then able to prove that "two magnitudes, whether commensurable or not, balance at distances reciprocally proportional to them." The distances to be considered are the respective distances of their centers of gravity from the fulcrum. Therefore, the end of Book I (Props. 9–15) explains how to find the centers of gravity of various figures — parallelogram, triangle, and parallel trapezium — and the whole of Book II is devoted to finding the centers of gravity of parabolic segments. The final proposition (Book II, 10) determines the center of gravity of the portion of a parabola between two parallel chords. All these propositions are geometric ones applied to statical purposes.

The treatise on *Floating bodies* is based on two postulates, the first of which is given at the beginning of Book I and the second after Prop. 7 (out of 9). They read:

Postulate 1.

"Let it be supposed that a fluid is of such a character that, its parts lying evenly and being continuous, that part which is thrust the less is driven along by that which is thrust the more; and that each of its parts is thrust by the fluid which is above it in a perpendicular direction if the fluid be sunk in anything and compressed by anything else."

Postulate 2.

"Let it be granted that bodies which are forced upwards in a fluid are forced upwards along the perpendicular [to the surface] which passes through their centre of gravity."

On the basis of Postulate 1, he proves (Prop. 2) that "the surface of any fluid at rest is a sphere the center of which is the same as that of the Earth." The main propositions of Book I, Props. 5–7, are equivalent to the famous Archimedean principle according to which a body wholly or partly immersed in a fluid loses an amount of weight equal to that of the fluid displaced. It has often been told that he discovered it when he was aware of the lightness of his own body in the water, and that he ran out of the bath shouting with joy "Heurēca, heurēca" (I have found it). This enabled him to determine the specific gravity of bodies and to solve the "problem of the crown." A golden crown made for King Hierōn was believed to contain silver as well as gold. How great was the adulteration? The problem was solved by weighing in water the crown itself, as well as equal weights of gold and silver. Book II investigates the condition of stable equilibrium of a right segment of a paraboloid of revolution floating in a fluid; there again the geometer was triumphing over the mechanician.

It would seem that Archimēdēs wrote at least one other mechanical treatise,[24] wherein he solved the problem "how to move a given weight by a given force," and proved that "greater circles overcome lesser ones when they revolve about the same center." This recalls his legendary boast to King Hierōn: "Give me a point of support [a fulcrum] and I shall move

[24] *On levers, peri zygōn; on centers of gravity, centrobarica; on equilibriums, peri isorropiōn.* These titles may refer to one treatise or to many.

the world." In order to convince the king, he managed to move a fully laden ship with no great effort by the use of a compound pulley *(polyspaston)*.

This brings us back to the mechanical inventions for war and peace which impressed posterity so profoundly that Archimēdēs' theoretical achievements were overlooked. The magnitude of his work in pure statics and hydrostatics can be appreciated in another way. Remember that Aristotelian and Stratonian physics was absolutely different from physics as we understand it today. The first physical sciences to be investigated on a mathematical basis were the rudiments of geometric optics (by Euclid and others) and, more deeply, two branches of mechanics, statics and hydrostatics. This was done by Archimēdēs, who must be called the first rational mechanician. Nor was there any other at all comparable to him until the time of Simon Stevin (1548–1620) and Galileo (1564–1642), who were born more than eighteen centuries later!

We have seen that Archimēdēs' mechanical treatises might be called geometric, and that is true of every treatise of theoretical mechanics, for mechanics is the mathematical development of certain mechanical postulates. (In the same spirit, geometry is the mathematical development of certain postulates concerning space.) It is clear that there was little difference in Archimēdēs' mind between the two fields. This impression is fortified by the study of a treatise of his which was almost completely unknown until 1906, when the illustrious Danish scholar, Heiberg, discovered it in a Constantinople palimpsest.[25] This is the *Method (ephodos) treating of mechanical problems, dedicated to Eratosthenēs*.

Few mathematicians explain their method of discovery and, therefore, their accounts are often tantalizing, for one cannot help wondering, "How did he think of that?" Their reticence may be due to a kind of coquetry but, in most cases, it is simply the fruit of necessity. The first intuition may be vague and difficult to express in scientific terms. If the mathematician follows it up, he may be able to find a scientific theory, but his way to it is tortuous and long. To describe the discovery in historical order would be equally long and tedious. It is simpler to explain it logically, dogmatically, after having thrown out everything irrelevant. The new theory then looks like a new building, after the scaffoldings and all the auxiliary

[25] Johan Ludvig Heiberg (1854–1928); portrait and biography by Hans Raeder, *Isis 11*, 367–374 (1928). A palimpsest (*palimpsēstos*, scratched or scraped again) is the name given to a manuscript written (generally on parchment) in the place of an earlier text which has been erased. The practice was caused by the high cost of parchment; monks would be tempted to erase a mathematical text which made no sense to them, and replace it with another text of greater interest. It is often possible to cause the reappearance of the erased text by means of chemicals and suitable light rays. The Archimedean text discovered by Heiberg had been erased to provide space for a euchologion (a ritual of the Orthodox Church).

constructions have been taken away, without which the building could not have been erected.

It is clear that the Euclidean mode of exposition which Archimēdēs used is as dogmatic or didactic as can be, and the order is certainly very different from the order of discovery. After having discussed the matter, probably with his friend Eratosthenēs, he wrote the *Ephodos*, for the discovery of which we must be very thankful to Heiberg, because it is one of the most revealing documents in the history, not only of ancient science, but of science in general at all times. To illustrate my bold statement, I would like to compare the *Ephodos* with a document that concerns the history of modern physiology, to wit, the *Introduction à l'étude de la médecine expérimentale* of Claude Bernard (Paris, 1865). It may seem paradoxical to compare a mathematical book written in Syracuse before 212 B.C. in the Greek language with a physiological one written more than two millennia later in French! In both cases, however, a great master tries to explain to us not his discoveries but his method of making them. Such books are uncommon in the history of science, and they are exceedingly precious.

One cannot read Archimēdēs' complicated accounts of his quadratures and cubatures without saying to oneself, "How on earth did he imagine those expedients [26] and reach those conclusions?" Eratosthenēs must have asked the same question, not only of himself but of Archimēdēs. The point is that the conclusions were reached intuitively and roughly before their validity was proved, or before it was possible to begin such a demonstration.

> It is of course easier, when we have previously acquired, by the method, some knowledge of the questions, to supply the proof than it is to find it without any previous knowledge. This is a reason why, in the case of the theorems the proof of which Eudoxos was the first to discover, namely that the cone is a third part of the cylinder, and the pyramid of the prism, having the same base and equal height, we should give no small share of the credit to Dēmocritos, who was the first to make the assertion with regard to the said figure, though he did not prove it.[27]

This statement is extremely interesting, not only in itself, but also because of the references to Dēmocritos and Eudoxos. Dēmocritos (V B.C.) discovered the volumes of cylinder, prism, and pyramid, but Eudoxos (IV-1 B.C.) was first to prove those theorems.[28] Archimēdēs remarked that Eudoxos' proof was facilitated by Dēmocritos' intuition and that some credit should be given to the latter. Now Archimēdēs was led by similar intuitions, his own. A kind of mechanical intuition, which he describes (and which makes us think of Cavalieri's),[29] enabled him to conceive the

[26] The word "expedient" is used because there was no general method, each particular problem being solved in its own way.

[27] T. L. Heath, *The Method of Archimedes* (Cambridge, 1912), p. 13. The statement occurs near the beginning of Archimēdēs' *Ephodos*.

[28] Volume 1, pp. 277, 444.

[29] Bonaventura Cavalieri (1598–1647), a disciple of Galileo, published the *Geo-*

method to be followed in a definite quadrature. He had a vision of the result before being able to prove it, indeed before attempting to do so. For further details, read the *Ephodos*, and this can be done not only in Greek or Latin but in English.

We must still say a few words about Archimēdēs' work in the fields of astronomy and optics. He wrote a (lost) book on *Sphere-making* describing the construction of an orrery to show the movement of Sun, Moon, and planets; the orrery was precise enough to foretell coming eclipses of the Sun and Moon. In the *Sand reckoner*, he described the simple apparatus (a diopter) that he used to measure the apparent diameter d of the Sun; he found that $27' < d < 32'56''$. Hipparchos referred to him and remarked that they had both made the same error in solstitial observations.[30] According to Macrobius (V–1), Archimēdēs determined the distances of the planets.

His interest in optics is proved by another lost book, *Catoptrica*, out of which Theōn of Alexandria (IV–2) quoted a single proposition: objects thrown into water look larger and larger as they sink deeper and deeper.

Considering the history of Greek astronomy and optics, it is not surprising that Archimēdēs paid some attention to those subjects. During his stay in Alexandria, he had discussed them with the disciples of Euclid and Aristarchos. Nevertheless, his own main interest was mathematical and it is admirably illustrated in the books that have come down to us.

THE ARCHIMEDEAN TRADITION

This question arises: How did Archimēdēs' works come down to us? The tradition of ancient science is almost as important as its invention, for without it the invention would have been useless.

The whole story is too complicated to be told here, because one would have to explain the tradition of the dozen items that have reached us in various ways. In order to shorten my outline, it will be convenient to number the Archimedean treatises; I follow Heiberg's order in the second Greek edition, volume 1 of which, containing the first three items, appeared in 1910, and volume 2, containing the remaining nine, in 1913:

1. *Sphere and cylinder,*
2. *Measurement of the circle,*
3. *Conoids and spheroids,*

---

metria indivisibilibus continuorum nova quadam ratione promota (Bologna, 1635), explaining the "method of indivisibles" which preceded and helped to prepare the discoveries of Newton and Leibniz. The method of exhaustion used by Eudoxos and Archimēdēs was more rigorous than Cavalieri's. Archimēdēs made his discoveries "à la Cavalieri" but was not satisfied with them until he had proved them by means of the method of exhaustion. Archimēdēs was a deeper mathematician than the Italian who followed him eighteen and a half centuries later.

[30] Ptolemy, Almagest, III, 1. *Claudii Ptolemaei opera quae exstant omnia,* vol.

4. *Spirals,*
5. *Equilibrium of planes,*
6. *Sand reckoner (Psammitēs, Arenarius),*
7. *Quadrature of the parabola,*
8. *Floating bodies,*
9. *Stomachion* (geometric puzzle),
10. *Method (Ephodos),*
11. *Book of lemmas (Liber assumptorum),*
12. *Cattle problem (Problema bovinum).*

The ancient tradition of Archimēdēs is much poorer than the Euclidean. Strangely enough, the only light in the early darkness was given by Cicero (I–1 B.C.). We know that Ptolemy (II–1) and Theōn of Alexandria (IV–2) read him, but they tell us very little. A collection of administrative documents made about the middle of the fifth century for Roman officials is preserved in the *Codex Arcerianus*, written probably in the sixth century (not later than the seventh); the scientific level of it is pretty low, yet it includes the Archimedean theorem giving the sum of the first *n* square numbers.[31]

The outstanding monument of the Greek tradition is the elaborate commentary written by Eutocios (VI–1) of Ascalon (on the Palestinian coast). It is an elaborate commentary covering items 1, 2, and 5. It fills the third volume (1915) of Heiberg's Greek edition. After that, there are no further traces of interest except that Archimedean manuscripts were copied during the Byzantine renaissance of the ninth and tenth centuries initiated by Leōn of Thessalonicē (IX–1). The prototype of the earliest manuscripts that have come to us was very probably a Byzantine one of the early ninth century. Those earliest manuscripts date back to the end of the fifteenth century and the beginning of the sixteenth; they included the old items 1, 2, 5 plus 4, 6, 7.

The prototype cannot have been later than (IX–1), for a copy of it made its way to the Dār al-islām, and was soon translated into Arabic by Qusṭā ibn Lūqā or members of his school and commented upon by Arabic mathematicians such as al-Māhānī, Thābit ibn Qurra, Yūsuf al-Khūrī, Isḥāq ibn Ḥunain, all of whom flourished in (IX–2). Some of the Arabic versions were translated into Latin. For example, item 2 *(Measurement of the circle)* was translated twice from Arabic into Latin in the twelfth century, the first time by Plato of Tivoli (XII–1) or somebody else, the second

1, *Syntaxis mathematica*, ed. J. L. Heiberg (Leipzig: Teubner, 1898–1903), pp. 194, 23: *Composition mathématique de Claude Ptolémée*, trans. N. B. Halma (Paris: Grand, 1813; facsimilé ed. Paris: Hermann, 1927), p. 153.

[31] The *Codex Arcerianus* is preserved in the Wolfenbüttel library, Braunschweig. See *Introduction*, vol. 1, p. 397. The sum of the square numbers was given by Archimēdēs in *Conoids and spheroids* (lemma to Prop. 2) and in *Spirals* (Prop. 10).

time by Gerard of Cremona (XII–2). This second translation established the text in the Latin world.[32]

A century later, a Flemish Dominican, Willem of Moerbeke (XIII–2), translated almost every Archimedean treatise directly from the Greek. The most important of his translations was that of item 8 *(Floating bodies)* because that item had been overlooked in the early Greek tradition. This translation was completed by Brother Willem at the papal court in Viterbo in 1269.[33] The Greek text of item 8 was lost and did not reappear until 1906, when Heiberg found it in the Constantinople[34] palimpsest which contained other Archimedean texts, the most precious being the *Method.*

At the time when Willem of Moerbeke was translating Archimēdēs directly from Greek into Latin, Maximos Planudēs (XIII–2) was perhaps using the Greek text for his own investigations, and the Persian Naṣīr al-dīn al-Ṭūsī (XIII–2) was revising the early Arabic versions. In the fourteenth century, a few mathematicians had access to Archimedean manuscripts — the Muslim 'Irāqī, Ibn al-Akfānī (XIV–1); Jews like Qalonymos ben Qalonymos (XIV–1), who translated them from Arabic into Hebrew, and perhaps Immanual Bonfils (XIV–2); Christians like Nicole Oresme (XIV–2) and Biagio Pelacani (XIV–2). In the fifteenth century, the number of Christians increased, the most important being Jacopo da Cremona and Regiomontanus. Leonardo da Vinci had some knowledge of him.

The *Method* (item 10) was unknown until 1906–07; it then reappeared in Greek and was soon translated into many languages. Still another item, not mentioned in the list above, the treatise on the *Regular heptagon,* was discovered by Carl Schoy in an Arabic manuscript and translated by him into German; it remained unknown until 1926. The chances of finding other unknown texts in Greek manuscripts is exceedingly small, but some may still be discovered in Arabic manuscripts, a great many of which are still uncatalogued.[35]

[32] Marshall Clagett, "Archimedes in the Middle Ages. The *De mensura circuli,*" *Osiris 10,* 587–618 (1952). Other studies by the same author on the medieval Latin tradition of Archimēdēs have appeared in *Isis* and *Osiris.* See his summary in *Isis 44,* 92–93 (1953), and "*De curvis superficiebus Archimenidis.* A medieval commentary of Johannes de Tinemue on the *De sphaera et cylindro,* " *Osiris 11,* 294–358 (1954). This John of Tinemue (?) flourished probably in the thirteenth century and his commentary was probably translated from the Arabic; *Isis 46,* 281 (1955). [See also Clagett's *Greek Science in Antiquity* (New York: Abelard-Schuman, 1955).]

[33] Viterbo (42 miles north-northwest of Rome) was part of the Patrimony of St. Peter because the "Great Countess" Matilda of Tuscany (d. 1115) had bequeathed it. Willem of Moerbeke was patronized by Clement IV (Guy de Foulques), the same who, in 1266, had ordered Roger Bacon (XIII–2) to send him copies of his writings. Clement IV died in Viterbo in 1268.

[34] For other details on the tradition of item 8 see Alexander Pogo's note, *Isis 22,* 325 (1934–35) and for the Archimedean tradition in general, see *Horus: A guide to the history of science* (Waltham, Mass.: Chronica Botanica, 1952), 18–22; *Isis 44,* 91–93 (1953). The palimpsest was found in 1899 by Papadopulos Kerameus in the Greek Patriarchate of Jerusalem, but Heiberg was first to realize its importance.

[35] For example, they may be found in composite manuscripts, which have been

The vicissitudes of these Archimedean texts are so many that one may wonder how it happened that most of them have actually reached us. Many Greek texts are lost or their rediscovery was due to a fluke, as in the case of the *Method*. Think of it, the *Method* was preserved because some ancient monks erased it; if they had not tried to destroy it, it would probably have been lost! Another case that crosses my mind as I am writing is the one of Alcman of Sardis, a lyric poet who lived in Sparta in the second half of the seventh century; one of his poems was discovered in 1855 in the packing of an Egyptian mummy! [36] Poetry, however, could be transmitted by oral tradition. That was impossible in the case of mathematics; the substance of the discoveries of mathematicians might be preserved by a succession of teachers, but the text of their works was neither remembered verbatim nor read aloud.

Traditions remained highly insecure until the text was printed. Whatever may have been the interest of a few medieval scholars in Archimēdēs, his works were never popular, and that is proved by the absence of incunabula. The first printed extract from Archimēdēs was in a collection called *Tetragonismus, id est circuli quadratura* (Venice, 1503), edited by Luca Gaurico (Fig. 23). The first important edition of his works appeared only forty years later, the Latin translation by Niccolò Tartaglia (Venice, 1543). This translation was restricted to items 5, 7, 2, and 8 (Book I only) and hence it was derived from a tradition unlike the Byzantine one (1, 2, 5 plus 4, 6, 7) and from the Moerbeke legacy. Tartaglia's edition was very imperfect but in the meanwhile another scholar, Venatorius, who was more of a philologist, was studying a manuscript belonging to Pope Nicholas V (1447–1455), and the translation of it made by James of Cremona and corrected by Regiomontanus. Making use of these manuscripts, Venatorius published the princeps (Basel, 1544), which included Latin translations and Eutocios' commentaries in Greek and Latin (Fig. 24). Tartaglia and, better than he, Venatorius revealed Archimedean geometry to the Renaissance mathematicians; by the end of the sixteenth century, there were enough of these not only to appreciate Archimēdēs but also to discuss his main difficulties.

The Greek text of 1544 was translated into Latin by Federico Commandino of Urbino (Venice, 1558) (Fig. 25), and the hydrostatics was translated into Latin by the same (Bologna, 1565). The two books on statics were published in Latin by Guido Ubaldo del Monte (Pesaro, 1588).

Curiously enough, the statics had been published in French (before Latin) by Pierre Forcadel of Béziers (2 vols.; Paris, 1565) (Fig. 26). These

---

imperfectly analyzed by nonmathematical Arabists.

[36] It was found near the second pyramid of Ṣaqqāra. It is a papyrus written in the first century of our era, preserved in the Louvre. The text is a fragment of an ode for parthenia (*melē*), that is, odes to be sung by maidens to the flute, with dancing. This particular ode was written by Alcman for the festival of the Dioscuroi, Castōr and Polydeucēs (Castor and Pollux).

Tetragonismus idest circuli quadratura per Cā
panū archimedē Syracusanū atq3 boetium ma
thematicae perspicacissimos adinuenta.

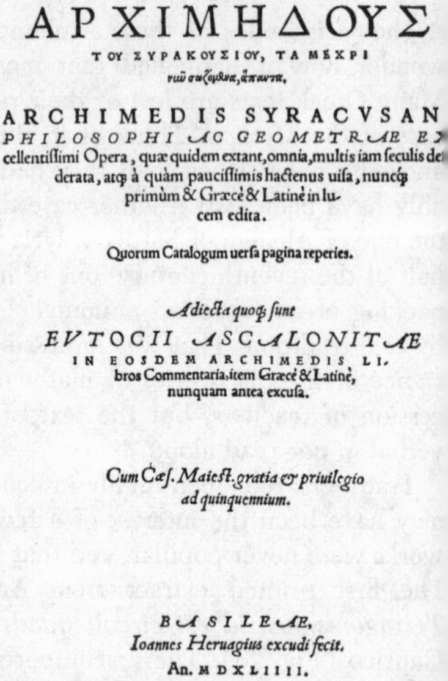
ΑΡΧΙΜΗΔΟΥΣ
ΤΟΥ ΣΥΡΑΚΟΥΣΙΟΥ, ΤΑ ΜΕΧΡΙ
νῦν σωζόμενα, ἅπαντα.

ARCHIMEDIS SYRACVSAN
PHILOSOPHI AC GEOMETRAE EX
cellentissimi Opera, quæ quidem extant, omnia, multis iam seculis de
derata, atq; à quàm paucissimis hactenus uisa, nuncq;
primùm & Græcè & Latinè in lu-
cem edita.

Quorum Catalogum uersa pagina reperies.

Adiecta quoq; sunt
EVTOCII ASCALONITAE
IN EOSDEM ARCHIMEDIS LI-
bros Commentaria, item Græcè & Latinè,
nunquam antea excusa.

Cum Cæs. Maiest. gratia & priuilegio
ad quinquennium.

BASILEAE,
Ioannes Heruagius excudi fecit.
An. M D X L I I I I.

Fig. 23. *Tetragonismus, id est circuli quadratura per Campanum Archimedem Syracusanum atque Boetium mathematicae perspicacissimos adinventa* (32 leaves, 20 cm; Venice: Sessa, 1503). This was the first Archimedean text to appear in printed form. It concerns the quadrature of the parabola and of the circle (leaves 15r–31r). Preface by Luca Gaurico (1475–1558) of Gifoni (Naples). The book includes also the "quadratures" of Euclid and Boethius (VI–1). [Courtesy of Harvard College Library.]

Fig. 24. The Archimedean princeps. First edition of the Greek text of Archimēdēs' works; it includes also a Latin translation, and the commentaries of Eutocios (VI–1) in Greek and Latin. The whole was edited by Thomas Gechauff, called Venatorius (folio, 31 cm; Basel: Joannes Hervagius [Johann Herwagen], 1544). It is divided into four parts which are generally, but not always, bound together. Parts 1 and 2 are dedicated to the Senate of Nürnberg. Part 1 (148 pp.) includes the Greek text of Archimēdēs; part 2 (169 pp.), the Latin translation; part 3 (67 pp.), Eutocios' commentaries in Greek; part 4 (70 pp.), their translation into Latin. [Courtesy of Harvard College Library.]

volumes were read by Stevin whose own investigations in statics appeared in 1586, before the Latin publication of those of Archimēdēs.

Before the end of the century, all the works of Archimēdēs were known in Europe (except the two that were discovered only in our day). They helped to create or at least to inspire the mathematical innovations of the seventeenth century.

Modern editions. J. L. Heiberg edited the Greek text in 1880–81, and revised it (3 vols.; Leipzig, 1910, 1913, 1915). Volume 3 contains Eutocios commentaries and tables. New edition (3 vols., 1930). English translation by T. L. Heath (512 pp., Cambridge, 1897), plus Supplement containing

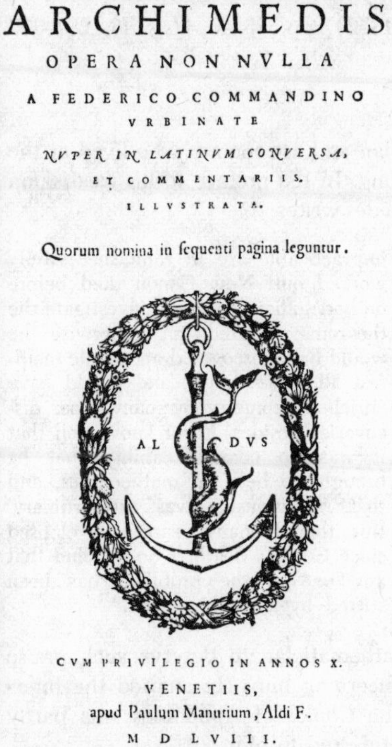

ARCHIMEDIS

OPERA NON NVLLA

A FEDERICO COMMANDINO

VRBINATE

NVPER IN LATINVM CONVERSA,

ET COMMENTARIIS

ILLVSTRATA.

Quorum nomina in ſequenti pagina leguntur.

AL DVS

CVM PRIVILEGIO IN ANNOS X.

VENETIIS,

apud Paulum Manutium, Aldi F.

M D LVIII.

Fig. 25. Latin translation of Archimēdēs (six treatises) by Federico Commandino (1509–1575) of Urbino (folio, 27.5 cm; Venice: Paulus Manutius, 1558). It is divided into two parts, the first of which contains the Archimedean text, and the second the commentaries of Eutocios and his own. The first part is dedicated to Cardinal Ranuccio Farnese, the second to another Farnese. Commandino's translation is important because of its influence in introducing the Archimedean renaissance. [Courtesy of Harvard College Library.]

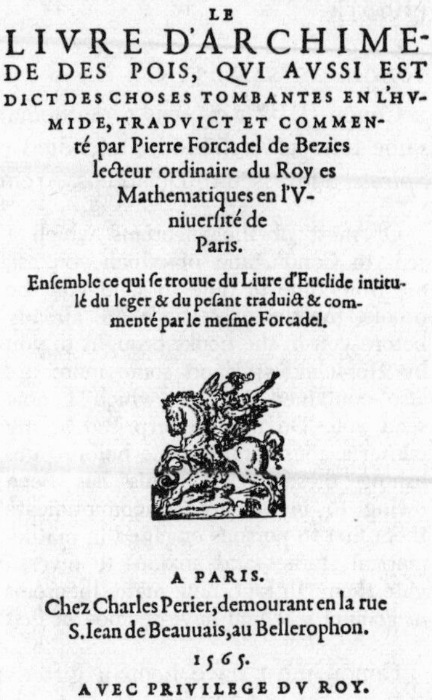

LE

LIVRE D'ARCHIME-
DE DES POIS, QVI AVSSI EST
DICT DES CHOSES TOMBANTES EN L'HV-
MIDE, TRADVICT ET COMMEN-
té par Pierre Forcadel de Bezies
lecteur ordinaire du Roy es
Mathematiques en l'V-
niuerſité de
Paris.

Enſemble ce qui ſe trouue du Liure d'Euclide intitu-
lé du leger & du peſant traduict & com-
menté par le meſme Forcadel.

A PARIS.

Chez Charles Perier, demourant en la rue
S. Iean de Beauuais, au Bellerophon.

1565.

AVEC PRIVILEGE DV ROY.

Fig. 26. French translation of Archimēdēs' hydrostatics by Pierre Forcadel (19.5 cm, 35 pp.; Paris: Charles Périer, 1565). The copy of this little book used by me belonged to Pierre Duhem. Forcadel also published a French translation of the statics (same printer, same year), which I have not seen. This was the first translation of Archimēdēs' statics in any language. The Latin translation of it by Guido Ubaldo del Monte appeared only twenty-three years later (Pesaro, 1588). [Courtesy of Harvard College Library.]

the *Method* (51 pp., 1912). French translation by Paul Ver Eecke (Brussels, 1912).

A short mechanical treatise ascribed to Archimēdēs, *Liber Archimedis de insidentibus aquae*, was edited by Maximilian Curtze, *Bibliotheca Mathematica* (1896), pp. 43–49 (*Introduction*, vol. 3, p. 735). It is of Archimedean derivation but late medieval (say XIV-1). New edition by Ernest A. Moody and Marshall Clagett, *The medieval science of weights* (Madison: University of Wisconsin Press, 1952), pp. 35–40 [*Isis* 46, 297–300 (1955)].

CONŌN OF SAMOS

Conōn (III-2 B.C.) was a mathematician and astronomer who lived at the same time as Archimēdēs and died young. In the preface to his treatise on *Spirals*, addressed to Dōsitheos, Archimēdēs writes:

Of most of the theorems which I sent to Conōn, and of which you ask me from time to time to send you the proofs, the demonstrations are already before you in the books brought to you by Hēracleidēs;[37] and some more are also contained in that which I now send you. Do not be surprised at my taking a considerable time before publishing these proofs. This has been owing to my desire to communicate them first to persons engaged in mathematical studies and anxious to investigate them. In fact, how many theorems in geometry which have seemed at first impracticable are in time successfully worked out! Now Conōn died before he had sufficient time to investigate the theorems referred to; otherwise he would have discovered and made manifest all these things, and would have enriched geometry by many other discoveries besides. For I know well that it was no common ability that he brought to bear on mathematics, and that his industry was extraordinary. But, though many years have elapsed since Conōn's death, I do not find that any one of the problems has been stirred by a single person.[38]

Conōn must have been a gifted mathematician to deserve such praise and one would like to know more concerning him. He studied the intersections of conics, and Book IV of the *Conics* of Apollōnios was partly based upon his work; Pappos (III–2) referred to him.

He wrote seven books on astronomy which were partly derived from Chaldaean (or Egyptian) observations, and he may have been the man who transmitted those observations to Hipparchos.

He compiled a new calendar or astronomical table (*parapēgma*) which gave the risings and settings of stars and weather forecasts. This table was based upon observations made in Sicily and South Italy, and this suggests that he may have associated with Archimēdēs in Syracuse as well as in Alexandria.

In any case, he must have flourished in Alexandria, for he named a constellation Comē (or Plocamos) Berenicēs in honor of Berenicē, the queen of Ptolemaios III Evergetēs.[39] It was told by the poets that she had con-

[37] This Hēracleidēs is otherwise unknown. The name was fairly common. The eponyms, the Hēracleidai were the descendants of Hēraclēs (Hercules), who united with the Dorians, had conquered the Peloponnēsos, some 80 years after the destruction of Troy.

[38] T. L. Heath (ed.), *The works of Archimedes* (Cambridge: The University Press, 1897), 151.

[39] It is the small constellation which we call Coma Berenices (Berenice's Hair), north of Virgo and between Boötes and Leo. Queen Berenicē was the daughter of Magas, king of Cyrēnē. She was put to death by her own son, Ptolemaios IV Philopatōr, in 221, soon after his accession.

secrated her hair to ensure the safe return of her husband, warring in Syria. A pretty story!

It is sufficient fame for a mathematician to have been praised in the prefaces written by Archimēdēs to his *Spirals* and by Apollōnios to Book IV of his *Conics* and to have been frequently mentioned in the *Almagest*. Yet, very few people knew that. Conōn's popular fame was based upon the verses of the Greek poet Callimachos (his contemporary) and of the Latin poet Catullus (c. 84–54).[40]

APOLLŌNIOS OF PERGĒ

There is only one other Greek geometer who can be compared with Archimēdēs and that is his younger contemporary, Apollōnios (III–2 B.C.). Some historians would say that the latter was second to Archimēdēs, but that kind of enumeration is obnoxious. They were two giants, not simply as compared with men of antiquity, but even with men of all times. To say that one was greater than the other does not make sense, if one remembers that genius cannot be measured.

Apollōnios was about twenty-five years younger than Archimēdēs, and we may assume that, without being his disciple, he was thoroughly familiar with all his works. His genius developed in another direction, however. Archimēdēs was always interested in measurements, such as quadratures, and he achieved very skillful integrations of plane or three-dimensional surfaces bounded by curved lines, and also of solids. One might call him, with proper caution, one of the ancestors of the infinitesimal calculus. On the other hand, Apollōnios' field of predilection was the theory of conic sections, which he did not measure but of which he tried to understand the forms and situations, and the various relations that might distinguish each kind of conic section or might occur when two of the same kind or of different kinds intersected. To put it tersely, one might call Archimedean geometry the geometry of measurements, and the Apollonian, the geometry of forms and situations. These two kinds of geometry, we should always remember, are not mutually exclusive, but may and do overlap. It is a difference of emphasis, Archimēdēs' on measurements and Apollōnios' on forms.

Apollōnios was born in Pergē, in Pamphylia,[41] probably c. 262. We do not know the name of his parents, but he had a son bearing his own name (Apollōnios the younger). Being very intelligent, he was sent early to

[40] We have only a fragment of Callimachos' poem, *Coma Berenices*, No. 110 in Rudolfus Pfeiffer's edition (2 vols.; Oxford: Clarendon Press, 1949), vol. 1, p. 112. This poem was imitated in Latin by Catullus (No. 66).

[41] Pamphylia is a small country along the middle of the south coast of Asia Minor; it is just west of Cypros. The story of its political vicissitudes is too complicated to be told here. In Apollōnios' time it was part of the kingdom of Pergamon, and this helps us to understand his history.

study in Alexandria and he flourished in that city under Ptolemaios III Evergetēs (247–222) and Ptolemaios IV Philopatōr (222–205). He paid a visit to Pergamon during the rule of Attalos I Sōtēr (241–197). During the rule of Ptolemaios IV, Greek power in Egypt was going down; during the rule of Attalos I, the kingdom of Pergamon was going up.[42] The date and place of Apollōnios' death are unknown and we have no idea of where or how he spent the end of his life; in that he was less fortunate than Archimēdēs, whose death in 212 was a kind of heroic climax.

Though Apollōnios composed almost as many books as Archimēdēs, he is more like Euclid in that one of his books was so much more important than the others that these could be (and are generally) overlooked. Just as Euclid is preëminently the author of the *Elements*, so Apollōnios is known as the author of the *Conica*.

The *Elements* is a textbook on plane and solid geometry; the *Conica* is also a textbook, but it deals exclusively with conic sections. Half of it is a survey and a systematic restatement of results obtained by earlier mathematicians; a large part of the work was either completely new or else consisted of known propositions explained in a new way and set in a new context which enhanced their pregnancy. Apollōnios' predecessors were many: Menaichmos (IV–2 B.C.), Aristaios (IV–2 B.C.), Euclid, and Archimēdēs.[43]

It is remarkable that, in spite of the fact that Apollōnios spent most of his life in Alexandria, his magnum opus was dedicated to Pergamenians, and this reminds us of the sad fact that his life ended in complete obscurity. Did he get into trouble with the Museum, or more probably with that debauchee and criminal, Ptolemaios IV Philopatōr? Books I to III of the *Conica* were dedicated to Eudēmos of Pergamon[44] and the remainder to Attalos I, king of Pergamon from 241 to 197. Apollōnios wrote a special preface to each of Books IV, V, VI, VII (and VIII?) and each dedication is as short as can be: "Apollōnios to Attalos, greeting." This reminds us of Archimēdēs' dedication of the *Sand reckoner* to the king of Syracuse; it is almost casual: "There are some, king Gelōn, who think that the number of the sand is infinite in multitude . . ." Now Gelōn and

[42] The prosperity of Pergamon was facilitated by Roman protection and that protection worked so well that in 133 B.C. the third Attalos bequeathed his kingdom to Rome! Greek Egypt declined during the second and first centuries but was not absorbed by Rome until 30 B.C. Ptolemaic Alexandria lasted a century longer than her rival, Attalid Pergamon.

[43] For the early history of conics, see Volume 1, pp. 503–505.

[44] This Eudēmos, a mathematician otherwise unknown, died before Apollōnios wrote the preface to Book IV of the *Conica*. He should not be confused with other Eudēmoi: Eudēmos of Cypros, disciple of Plato; the mathematician Eudēmos of Rhodos (IV–2 B.C.); Eudēmos of Alexandria (III–1 B.C.). The name Eudēmos (good people) was fairly common; 20 of them are dealt with in Pauly-Wissowa, but not this particular one (Vol. 11, pp. 894–905).

Attalos were autocrats, holding and using the power of life and death, but the intellectual freedom and essentially democratic spirit of the Greeks (even those of the Hellenistic age) were such that it seemed perfectly simple to address the king like any other man.[45] A comparison of those dedications with the extravagant and loathsome ones addressed by Renaissance scholars to petty dukes and lords is greatly to the credit of the ancients.

The *Conica* was divided into eight books of which the last is lost. Their general purpose is so well explained in the preface to his corrected version of Book I, that it is best to reproduce it, and the more so because it will give the reader an idea of Apollōnios' style, which is free from any kind of affectation and excellent.

Apollōnios to Eudēmos, greeting.

If you are in good health and things are in other respects as you wish, it is well; with me too things are moderately well. During the time I spent with you at Pergamon I observed your eagerness to become acquainted with my work in conics; I am therefore sending you the first book, which I have corrected, and I will forward the remaining books when I have finished them to my satisfaction. I dare say you have not forgotten my telling you that I undertook the investigation of this subject at the request of Naucratēs the geometer,[46] at the time when he came to Alexandria and stayed with me, and, when I had worked it out in eight books, I gave them to him at once, too hurriedly, because he was on the point of sailing; they had therefore not been thoroughly revised, indeed I had put down everything just as it occurred to me, postponing revision till the end. Accordingly I now publish, as opportunities serve from time to time, instalments of the work as they are corrected. In the meantime it has happened that some other persons also, among those whom I have met, have got the first and second books before they were corrected; do not be surprised therefore if you come across them in a different shape.

Now of the eight books the first four form an elementary introduction. The first contains the modes of producing the three sections and the opposite branches [of the hyperbola], and the fundamental properties subsisting in them, worked out more fully and generally than in the writings of others. The second book contains the properties of the diameters and the axes of the sections as well as the asymptotes, with other things generally and necessarily used for determining limits of possibility (*diorismoi*),[47] and what I mean by diameters and axes respectively you will learn from this book. The third book contains many remarkable theorems useful for the syntheses of solid loci and for *diorismoi*; the most and prettiest of these theorems are new, and it was their discovery which made me aware that Euclid did not work out the synthesis of the locus with respect to three and four lines, but only a chance portion of it, and that not successfully; for it was not possible for the said synthesis to be completed without the aid of the additional theorems discovered by me. The fourth book shows in how many ways the sections of cones

---

[45] I had been wondering whether the Attalos to whom Apollōnios dedicated the second half of his *Conica* was really the king? I think it is so because any other Attalos would have required a definition.

[46] This Naucratēs is otherwise unknown.

[47] *Diorismos* means delimitation, definition; in the plural it also means the conditions of possibility of a problem.

can meet one another and the circumference of a circle; it contains other things in addition, none of which have been discussed by earlier writers, namely the questions in how many points a section of a cone or a circumference of a circle can meet a double-branch hyperbola, or two double-branch hyperbolas can meet one another.

The rest of the books are more by way of surplusage (*periusiasticōtera*): one of them deals somewhat fully with *minima* and *maxima*, another with equal and similar sections of cones, another with theorems of the nature of determination of limits, and the last with determinate conic problems. But of course, when all of them are published, it will be open to all who read them to form their own judgement about them, according to their own individual tastes. Farewell.

Let us quote also the preface to Book IV, addressed to Attalos:

Apollōnios to Attalos, greeting.

Some time ago I expounded and sent to Eudēmos of Pergamon the first three books of my conics which I have compiled in eight books, but, as he has passed away, I have resolved to dedicate the remaining books to you because of your earnest desire to possess my works. I am sending you on this occasion the fourth book. It contains a discussion of the question, in how many points at most it is possible for sections of cones to meet one another and the circumference of a circle, on the assumption that they do not coincide throughout, and further in how many points at most a section of a cone or the circumference of a circle can meet the hyperbola with two branches [or two double-branch hyperbolas can meet one another]; and, besides these questions, the book considers a number of others of a similar kind. Now the first question Conōn expounded to Thrasydaios without, however, showing proper mastery of the proofs, and on this ground Nicotelēs of Cyrēne,[48] not without reason, fell foul of him. The second matter has merely been mentioned by Nicotelēs, in connexion with his controversy with Conōn, as one capable of demonstration; but I have not found it demonstrated either by Nicotelēs himself or by any one else. The third question and the others akin to it I have not found so much as noticed by any one. All the matters referred to, which I have not found anywhere, required for their solution many and various novel theorems, most of which I have, as a matter of fact, set out in the first three books, while the rest are contained in the present book. These theorems are of considerable use both for the syntheses of problems and for *diorismoi*. Nicotelēs indeed, on account of his controversy with Conōn, will not have it that any use can be made of the discoveries of Conōn for the purpose of *diorismoi*; he is, however, mistaken in this opinion, for, even if it is possible, without using them at all, to arrive at results in regard to limits of possibility, yet they at all events afford a readier means of observing some things, e.g. that several or so many solutions are possible, or again that no solution is possible; and such foreknowledge secures a satisfactory basis for investigations, while the theorems in question are again useful for the analyses of *diorismoi*. And, even apart from such usefulness, they will be found worthy of acceptance for the sake of the demonstrations themselves, just as we accept many other things in mathematics for this reason and for no other.[49]

[48] Nicotelēs of Cyrēnē is otherwise unknown; he is different from the Cyrenaic philosopher of the same name who flourished with his brother, Anniceris, under Ptolemaios I.

[49] These two prefaces are quoted from

There is no preface to Book III; the prefaces of Book II to Eudēmos and of Books V, VI, VII to Attalos are very short.

The contents of the *Conica* might be summed up as follows:

I. Generation of the three conic sections.
II. Asymptotes, axes, diameters.
III. Equality or proportionality of figures determined by portions of transversals, chords, asymptotes, tangents; foci of the ellipse and hyperbola.
IV. Harmonic division of straight lines. Relative positions of two conics, their intersections; they cannot cut one another in more than four points.
As Appollōnios put it in his preface to Book I, Books I to IV are an elementary introduction, while the following contain additional theorems for advanced students.
V. Maxima and minima. (This is generally considered to be his masterpiece.) How to find the shortest and longest lines to be drawn from a given point to a conic. Evolutes, centers of osculation.
VI. Similarity of conics.
VII, VIII. Conjugate diameters.

Menaichmos and Aristaios generated the conics by means of a plane cutting a right circular cone, the plane being perpendicular to the generating line of the cone. According to whether the cone's angle was acute, right, or obtuse, the section was elliptic, parabolic, or hyperbolic. Now Apollōnios showed that the three kinds of conics could be obtained as sections of the same cone, and he thus facilitated a better understanding of their unity;[50] all the conic sections belong to a single family, divided into three groups. Menaichmos' names for each group (acute-angled, right-angled, obtuse-angled) were no longer applicable to the curves generated in the new way. The names familiar to us were introduced by Apollōnios, *elleipsis*, or falling short of areas (ellipse), *parabolē*, or application of areas, and *hyperbolē*, or exceeding of areas. (If p is the parameter, $y^2 < px$, $y^2 = px$, and $y^2 > px$ in the three cases respectively.) His recognition of the two branches of the hyperbola as a single curve enabled him to show the analogies of all conic sections.

Apollōnios could construct a conic by means of tangents (III, Props. 65–67). He could also construct one defined by five points, though his construction is not stated explicitly.

A discussion of the large number of propositions of the *Conica* would be endless, but it is interesting to indicate singular omissions. Apollōnios does not speak at all of the directrix.[51] He knew the focal properties of the ellipse

---

Heath's translation in his *History of Greek mathematics* (Oxford, 1921), vol. 2, pp. 128–131.

[50] Analytic geometry expresses that unity in a simpler way. The conic sections are represented by equations of the second degree with two unknown quantities.

[51] And yet Euclid knew the relation of

and the hyperbola, but did not realize the existence of a focus in the parabola.

Such lacunae may seem almost incredible to the reader, because he has been introduced to the subject in an entirely different way. Apollōnios spoke of the foci of central conics at the end of his Book III, but our young students hear of them at the very beginning of their course. An ellipse is defined to them as the locus of a point E, the sum of whose distances a and b to two given points F_1 and F_2 is constant, $a + b = k$; the points F_1 and F_2 are the foci. The parabola is defined as the locus of a point P equidistant from a fixed point F (called the focus) and from a given straight line d (called the directrix).

As the modern student is introduced to the conics by means of analytic geometry, his approach is essentially different from that of Apollōnios, which was purely geometric; hence, his fundamental ideas are different. Yet the ancient and modern mathematicians were bound to discover finally the same results, and they did so to a large extent.

It would be foolish to study the conics at present in Apollōnios, because the modern methods (whether of analytic geometry or of projective geometry) are much simpler, easier, and deeper, but the ingenuity that enabled him to discover so much with imperfect tools is truly admirable. One can repeat apropos of him what was said above apropos of Archimēdēs; such achievements pass our imagination, they are almost weird.

Many mathematicians are mentioned in the prefaces of Archimēdēs and Apollōnios. I have already named a few of them, not that I expect the reader to remember them (I do not remember them myself), but they illustrate the relative abundance of mathematical curiosity in the third century. Besides the three kings, Hierōn II and Gelōn II of Syracuse and Attalos I of Pergamon,[52] the others are Dōsitheos, Zeuxippos, Conōn of Samos, Eudēmos of Pergamon, Naucratēs, Philōnidēs,[53] Thrasydaios, Nicotelēs of Cyrēnē. Such a list is tantalizing, because one would like to be better acquainted with them. The men to whom those two giants dedicated their books or whom they mentioned were not common men.

The other works of Apollōnios, lost in the Greek original, are known to us only through the collection of Pappos (III-2), and one of them was

focus to directrix. According to Pappos (Book VII; Hultsch, p. 678; Ver Eecke, p. 508), he showed that the locus of a point whose distance from a given point is in given ratio to its distance from a given straight line is a conic. It is an ellipse, parabola, or hyperbola according as the given ratio is $<$, $=$, or $>$ 1.

[52] Would modern kings be sufficiently interested in mathematicians to encourage the dedication of their books? It is true that Queen Victoria favored Charles Lutwidge Dodgson, but that was not because of his mathematics but because of *Alice's adventures in wonderland* (1865).

[53] Apollōnios introduced this Philōnidēs to Eudēmos at Ephesos. Like every good Greek who could afford it, they had probably made a pilgrimage to the temple of Artemis.

preserved in Arabic. This is *The cutting off of a ratio (Logu apotomē)*, eventually translated into Latin by Edmund Halley. The others are entitled *The cutting off of an area (Chōriu apotomē)*, *The determinate section* (*Diōrismenē tomē*), *Tangencies* (*Epaphai*), *Plane loci*, and *Inclinations (Neuseis)*. The contents of these six works are more or less known because of Pappos' analysis and quotations. Still other books can be ascribed to Apollōnios on weaker evidence: a comparison of the dodecahedron with the icosahedron; a study of fundamental principles; the *cochlias* (cylindrical helix), proving that it is homoeometric;[54] unordered irrationals; burning mirrors; quick delivery *(ōcutocion)*, giving a better approximation of π than Archimēdēs' but one that is less suitable for practical purposes.

It was natural enough that Apollōnios should devote a part of his attention to astronomical problems. The outstanding problem with which Greek astronomers had been struggling for two centuries was the finding of a kinematic explanation of planetary motions that would tally with the appearances and "save" them (*sōzein ta phainomena*), for example, one that would account for the apparent retrogressions of the planets. The first solution, that of the homocentric spheres, had been invented by Eudoxos of Cnidos (IV–1 B.C.) and gradually improved by Callippos of Cyzicos (IV–2 B.C.), Aristotle, and Autolycos of Pitanē (IV–2 B.C.).[55] It produced admirable results but failed to "save" all the phenomena. Something else had to be found, especially for the inferior planets. The founder of the geoheliocentric system, Hēracleidēs of Pontos (IV–2 B.C.), invented the theory of epicycles to account for the apparent motions of Mercury and Venus. In order to account for the apparent motions of the superior planets (Mars, Jupiter, and Saturn), Apollōnios generalized the use of the theory of epicycles and introduced or helped to introduce a third kind of theory, the theory of eccentrics. According to Ptolemy,[56] Apollōnios invented or perfected those two theories; Hipparchos and Ptolemy used them exclusively and rejected the theory of homocentric spheres. In later times, the latter theory was revived and the history of medieval astronomy is to some extent a protracted struggle between epicycles and homocentrics, or between Ptolemaic and Aristotelian astronomy.[57]

If we compare Aristarchos of Samos with Copernicus, then we may call Apollōnios the precursor of Tycho Brahe, though that title may be given also, with less justice, to Hēracleidēs himself.

[54] Equal in all of its parts.

[55] Autolycos showed that the theory of homocentric spheres was not compatible with the differences in the apparent sizes of Sun and Moon and with the variations in the brightness of the planets (Volume 1, p. 512).

[56] Almagest, XII, 1; *Claudii Ptolemaei opera quae exstant omnia,* vol. 2, *Opera astronomica minora,* ed. J. L. Heiberg (Leipzig: Teubner, 1907), pp. 450 f.; *Composition mathématique,* vol. 2, *Composition mathématique ou astronomie ancienne,* trans. N. B. Halma (Paris: Eberhart, 1816; facsimilé ed. Paris: Hermann, 1927), pp. 312 f. Full discussion by Otto Neugebauer, "Apollonius' planetary theory," *Communications on pure and applied mathematics* 8, 641–648 (1955).

[57] For an outline of that struggle, see

In any case, Apollōnios would deserve a very high place in the history of science even if his *Conics* had been lost. He paved the mathematical way for Hipparchos and Ptolemy and made the composition of the *Almagest* possible. It is paradoxical that his main contribution to mathematical astronomy, the theory of conics, was not exploited until more than eighteen centuries later, by Johann Kepler.

THE APOLLONIAN TRADITION

As far as the theories of epicycles and eccentrics are concerned, enough has been said already, by referring to the use that Hipparchos and Ptolemy made of them. The rest is identical with the Ptolemaic tradition itself.

We shall thus focus our attention at present upon the *Conics.* Thanks to its logical strength, clearness, and comprehensiveness, that treatise was recognized at once as the standard one in its sphere (just as Euclid's *Elements* was in another), and it was eagerly studied by the Greek epigoni. As in the case of the Archimedean tradition, we do not know what happened during the first centuries (say from the second B.C. to the third after Christ, a pretty long period). The first commentators were Pappos (III–2), thanks to whom the substance of many of Apollōnios' minor works was preserved, Theōn of Alexandria (IV–2), his famous daughter Hypatia (V–1), and finally Eutocios (VI–1).[58] After that, the story of the Archimedean tradition is repeated.

The lost prototype of the extant manuscript[59] was probably copied during the Byzantine renaissance, set under way by Leōn of Thessalonicē (IX–1), the fruits whereof appeared before the end of the ninth century, not in Byzantion, however, but in Islamic countries. Books I to IV of the *Conics* (*Kitab al-makhrūṭāt*) were translated into Arabic by Hilāl ibn al-Ḥimsī (IX–2), and Books V to VII by Thābit ibn Qurra (IX–2). It would thus seem that Book VIII was already lost; did Apollōnios complete it? In the following century Arabic mathematicians, such as Ibrāhīm ibn Sinān (X–1) and al-Kūhī (X–2), were already writing commentaries and discussing Apollonian problems, and a better translation of the *Conics*, together with a commentary on Books I–V, was prepared by Abū-l-Fatḥ Maḥmūd ibn Muḥammad (X–2) of Iṣfahān.

Many Greek books are known to us only through Arabic translations, the originals being lost, but this is the outstanding case. There is no other book of comparable importance the preservation of which we owe to the Arabic detour. Another treatise of Apollōnios (on *The cutting off of a ratio*)

---

my *Introduction,* vol. 2, pp. 16–19; vol. 3, pp. 110–137, 1105–1121.

[58] Eutocios' commentary was very elaborate. In Heiberg's Greco-Latin edition of Apollōnios, it covers 194 pp., vol. 2, pp. 168–361.

[59] The best extant manuscript of the *Conics* dates back only to the twelfth or thirteenth century, but a manuscript of Eutocios' commentary is as early as the tenth century. The Greek manuscripts of the *Conics* are restricted to Books I to IV; Books V to VII are available in Arabic manuscripts.

was, as we have already mentioned, preserved in the same way; Edmund Halley published a Latin translation of it from the Arabic in Oxford in 1706 (Fig. 27).

The Latin tradition began only in the twelfth century with a translation from the Arabic ascribed to Gerard of Cremona (XII–2), and the Hebrew tradition only in the fourteenth century with Qalonymos ben Qalonymos (XIV–1), who translated extracts from Arabic into Hebrew (this is not certain). We may overlook other details of the medieval tradition.

The weakness of that tradition is illustrated (as in the case of Archimēdēs) by the lack of incunabula. The first printed edition (Fig. 28) of the *Conics* (restricted to Books I–IV), was the Latin translation as published by Giovanni Battista Memo (Venice, 1537), but this was soon replaced by the much better one of Federico Commandino (Bologna, 1566), including the lemmas of Pappos, the commentary of Eutocios, and elucidating notes (Fig. 29).

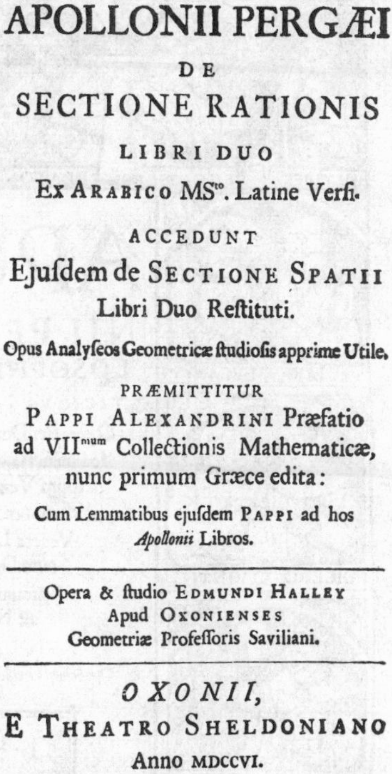
APOLLONII PERGÆI
DE
SECTIONE RATIONIS
LIBRI DUO
Ex ARABICO MS^to. Latine Verſi.
ACCEDUNT
Ejuſdem de SECTIONE SPATII
Libri Duo Reſtituti.
Opus Analyſeos Geometricæ ſtudioſis apprime Utile.
PRÆMITTITUR
PAPPI ALEXANDRINI Præfatio
ad VII^mum Collectionis Mathematicæ,
nunc primum Græce edita:
Cum Lemmatibus ejuſdem PAPPI ad hos
Apollonii Libros.

Opera & ſtudio EDMUNDI HALLEY
Apud OXONIENSES
Geometriæ Profeſſoris Saviliani.

OXONII,
E THEATRO SHELDONIANO
Anno MDCCVI.

Fig. 27. Two other Apollonian treatises first published by Edmund Halley (20 cm, 230 pp.; Oxford, 1706). Dedicated to Henry Aldrich, dean of Christ Church, Oxford. [Courtesy of Harvard College Library.]

As Books V–VII are available only in Arabic, their publication (or rather the Latin translation) did not occur until a century later. It was based upon the Arabic version as revised in 982 by Abū-l-Fatḥ al-Iṣfahānī and was prepared by the Lebanese Maronite Abraham Echellensis (= Ibrāhīm al-Ḥaqilānī) together with Giacomo Alfonso Borelli (Florence, 1661).

The Greek princeps we owe to the genius of Edmund Halley (Fig. 30), a splendid folio edition containing the Greek of Books I–IV, plus the Latin translation (revised by him from new Arabic manuscripts) of Books V–VII, a conjectural restoration of Book VIII, and the commentaries of Pappos and Eutocios (Oxford, 1710).

The mathematicians of the Renaissance could study the theory of conics in Memo's edition of 1537, or better in Commandino's of 1566. From 1566 on, they had a good knowledge of Books I–IV. In addition, they could use the restoration of Book V (maxima and minima) attempted by Francesco

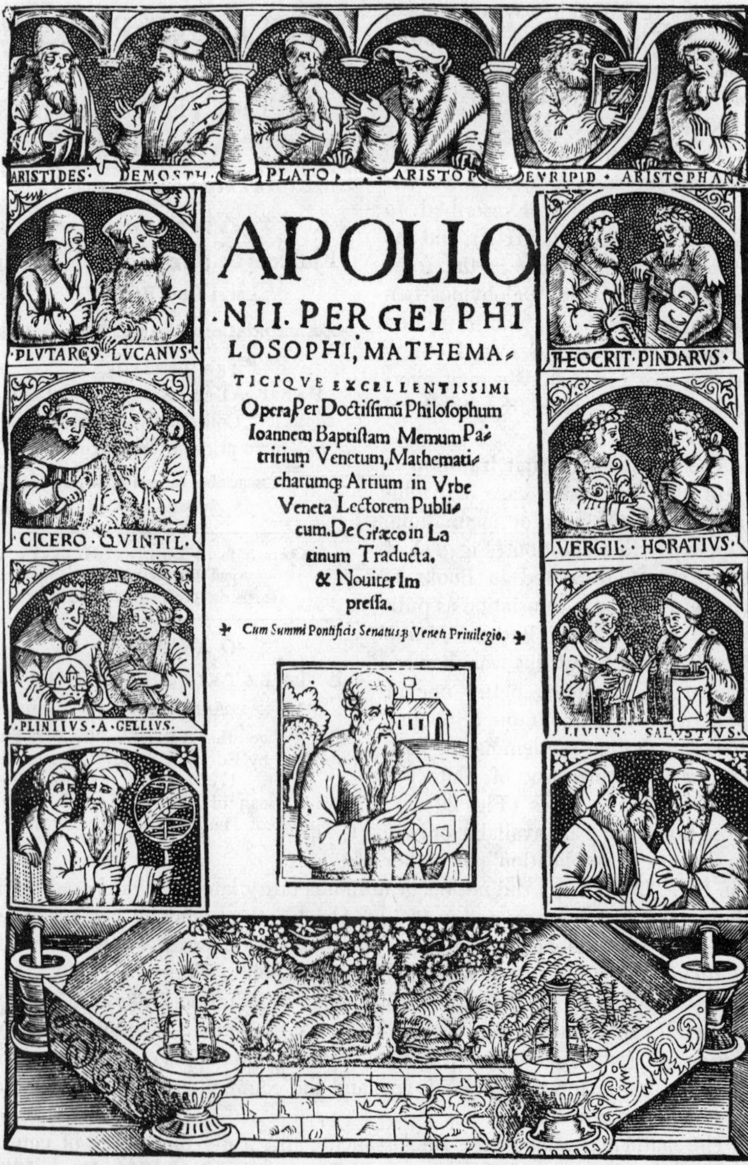

Fig. 28. First printed edition of Apollōnios, being the Latin translation of the *Conics*, Books I–IV (folio, 30 cm, 89 leaves; Venice: Bernardinus Bindonus, 1537), by Giovanni Battista Memo, patrician of Venice. This was edited after Memo's death by his son who had not enough mathematical knowledge to do it well. The book was dedicated to Cardinal Marino Grimani, patriarch of Aquileia. [Courtesy of Harvard College Library.]

APOLLONII
PERGAEI CONICORVM
LIBRI QVATTVOR.
VNA CVM PAPPI ALEXANDRINI
LEMMATIBVS, ET COMMENTARIIS
EVTOCII ASCALONITAE.
SERENI ANTINSENSIS
PHILOSOPHI LIBRI DVO
NVNC PRIMVM IN LVCEM EDITI.
QVAE OMNIA NVPER FEDERICVS
Commandinus Vrbinas mendis quamplurimis expurgata è Græco conuertit, & commentariis illustrauit.

CVM PRIVILEGIO PII IIII. PONT. MAX.
IN ANNOS X.

BONONIAE,
EX OFFICINA ALEXANDRI BENATII.
M D LXVI.

Fig. 29. Second Latin edition of Apollōnios' *Conics*, I–IV, by Federico Commandino, together with the lemmas of Pappos (III-2), the commentaries of Eutocios (VI-1), and the two books on conics by Serēnos (IV-1). (Two parts, folio, 27.5 cm, 3 + 114 leaves, 1 + 35 leaves; Bologna: Alexander Benatius, 1566.) Part 2 contains Serēnos. Each part was dedicated to a different member of the Guido Ubaldo family, dukes of Urbino. [Courtesy of Harvard College Library.]

APOLLONII PERGÆI
CONICORUM
LIBRI OCTO,
ET
SERENI ANTISSENSIS
DE SECTIONE
CYLINDRI & CONI
LIBRI DUO.

OXONIÆ,
E THEATRO SHELDONIANO, An. Dom. MDCCX.

Fig. 30. Princeps of Apollōnios, edited from the Greek manuscripts by Edmund Halley (1656–1742). Splendid folio (40 cm; Oxford, 1710) divided into three parts bound together. The first (254 pp.) contains Books I–IV in Greek and Latin, with Pappos' lemmas and Eutocios' commentaries. The second (180 pp.), Books V to VII translated from Arabic into Latin, plus restitution of Book VIII. The third (88 pp.), the treatises on sections of the cylinder and the cone by Serēnos (IV-1), in Greek and Latin. Each of the parts is dedicated to a different person. The beautiful copperplate frontispiece is the same used in the Greek-Latin Euclid (Oxford, 1703), shown in Fig. 13. [Courtesy of Harvard College Library.]

Maurolico of Messina on the basis of Pappos, and they could use also the *Libellus* of Johannes Werner (Nürnberg, 1522). This was the first book on conics to appear in Europe; note that it was printed before Apollōnios.

The knowledge of conics was applied by Johann Kepler (1609) to celestial mechanics. Just as Archimēdēs excited Descartes (1637), Apollōnios excited Girard Desargues (1636) and indirectly Pascal (1637).[60] Many other mathematicians of the seventeenth century investigated his writings: Fermat, Franz van Schooten, James Gregory, Adrianus Romanus, Princess Elizabeth (Descartes's disciple). A complete list would be very long. The works of Archimēdēs and Apollōnios acted like strong ferments from the end of the sixteenth century throughout the seventeenth. The first to put together the accumulated knowledge on conics was Philippe de La Hire, professor at the Collège de France, in three treatises (Paris, 1673, 1679, 1685).[61]

After that, the Apollonian tradition was lost in the new geometry, like a river in the ocean.

Recent editions. All the Greek texts were edited, together with the ancient commentaries, by J. L. Heiberg (2 vols.; Leipzig, 1891–1893). English translation by T. L. Heath (426 pp.; Cambridge, 1896). French translation by Paul Ver Eecke (708 pp., 419 figs.; Bruges, 1924).

For Books v–viii of the *Conics*, Halley's edition (Oxford, 1710) has not yet been superseded.

The history of mathematics is continued in Chapter XVIII.

[60] *Isis 10*, 16–20 (1928); *43*, 77–79 (1952).

[61] La Hire's treatises of 1673 and 1679 were in French; the third and most important in Latin, *Sectiones conicae in novem libros distributae* (Paris, 1685).

VI

GEOGRAPHY AND CHRONOLOGY IN THE THIRD CENTURY
ERATOSTHENĒS OF CYRĒNĒ

Though Archimēdēs and Apollōnios were interested in astronomy and physics, they were primarily mathematicians. The case of their contemporary, Eratosthenēs, is very different. His mathematical work is original, but of secondary importance in his own life; he was primarily a geodesist and a geographer, but he was also a man of letters, a philologist, an encyclopedist or polymath.

ERATOSTHENĒS OF CYRĒNĒ

Eratosthenēs son of Aglaos was born in Cyrēnē in the 126th Olympiad (276–273), c. 273; he was educated in Athens, and finally was called to Alexandria by Ptolemaios III Evergetēs (ruled 247–222), and spent in that city the rest, more than half, of his life. He died there at the age of 80, c. 192. We must try to see him in a triple background, Cyrēnē, Athens, Alexandria.

He received his first education in his native city from the grammarian Lysanias and the poet Callimachos.[1] Cyrenaica, just west of Egypt, was a cultured and ancient nation, founded by citizens of Thēra (Santorin) and Crete, c. 630; [2] the elite of its people were completely Hellenized. It was often called Pentapolis because its main cities were five in number: Cyrēnē, Apollōnia, Ptolemais, Arsinoē, and Berenicē. In particular, the capital, Cyrēnē, was one of the most cultured cities of the Hellenistic world. A num-

[1] Lysanias of Cyrēnē wrote studies on Homer and on the iambic poets. For Callimachos, see Chapter X.

[2] The founder assumed the title of king (*battos* in Libyan). The early kings were called Battos or Arcesilas. Early Cyrēnē was a center of Hellenism placed upon the north African coast between the Phoenician Tripolis (Syrtica Regio) to the west and Egypt to the east. Alexander the Great made Cyrēnē an ally and Cyrēnē remained enfeoffed to the Macedonian kings of Egypt, with periods of rebellion, until one of the last Ptolemaioi bequeathed it to Rome in 96 B.C. After 22 years of chaos, it became a Roman province, to which Crete was added in 67 B.C.

ber of distinguished people hailed from there: Aristippos, disciple of Sōcratēs and founder of the Cyrenaic school; his daughter Arētē, who succeeded him as head of the school; her son and successor, Aristippos II, nicknamed Mētrodidactos (mother taught); Anniceris, who modified the teachings of that school so much that it was called Annicerian; Callimachos and Eratosthenēs, both of whom we shall soon learn to know more intimately; furthermore, Carneadēs, second founder of the New Academy, and Apollōnios Cronos, the dialectician.[3]

During Eratosthenēs' youth, Magas, governor of Cyrenaica in the name of his uterine brother, Ptolemaios II Philadelphos, revolted against him and assumed the title of King (he died in 258). Yet Cyrēnē remained subordinated, politically and culturally, to Ptolemaic Egypt.

Athens, like Cyrēnē, was struggling to recover its political independence, but in spite of repeated failures it was still the educational and philosophical center of the Greek-speaking peoples. It was thus natural for Eratosthenēs to proceed to Athens in order to complete his education. He studied there at the feet of Arcesilaos of Pitanē (Mysia), founder of the New Academy,[4] of Aristōn of Iulis (Ceōs),[5] head of the Lyceum, and of the Cynic Biōn.[6] It should be noted that his studies were mainly philosophical, but mathematics and science never ceased to be taught in either the Academy or the Lyceum.

After the middle of the century Eratosthenēs' education was complete; a few philosophical or literary books had attracted some attention to his name and about 244 he answered a call of Ptolemaios III Evergetēs. He remained in Egypt at least fifty years, under the rule of three kings, Evergetēs, Philopatōr (whose tutor he was), and Epiphanēs (ruled from 196 to 181). We need not describe his Egyptian milieu because this has been done in previous chapters. His life was spent in active studies in three great centers of Hellenism, Cyrēnē, Athens, and Alexandria; it is as if one of our contemporaries were spending his in Oxford, Paris, and New York.

Soon after his arrival in Alexandria he began his tutorship of Philopatōr[7] and was appointed a fellow of the Museum (the tutorship of a prince and

[3] These glories of Cyrēnē are enumerated after Strabōn, *Geography*, XVII, 3, 22 (Loeb Classical Library, vol. 8, p. 205). See also Volume 1, pp. 282, 588.

[4] Also called the Second or Middle Academy. For the post-Platonic history of the Academy, see Volume 1, pp. 399–400.

[5] Not to be confused with the Stoic Aristōn of Chios, disciple of Zēnōn of Citiōn. It is remarkable that Eratosthenēs does not seem to have paid attention to the Stoa. For Aristōn of Chios, see Volume 1, p. 604; for the history of the Lyceum, p. 493. Aristōn of Chios flourished c. 260; Aristōn of Ceōs a generation later, c. 230.

[6] Could this be Biōn of Borysthenēs (Dnieper) who flourished in III-1 B.C. and was a popular philosopher or "Wanderprediger"? See von Arnim, in Pauly-Wissowa, Vol. 5 (1897), pp. 483–485.

[7] We must imagine that that tutorship was nominal; it does not seem to have improved Philopatōr, whose dissipation and crimes are as discreditable to Eratosthenēs, as those of Nero (d. A.D. 68) were to be to Seneca the Philosopher. Seneca was murdered by Nero's order in 63, but Eratosthenēs survived Philopatōr's criminal life. It must be added that Philopatōr was a patron of arts and sciences.

fellowship were in many cases correlative appointments). He was then or later a senior (or *alpha*) fellow. Upon the death of Zēnodotos (c. 234) he became chief librarian.

His education in the three cities was to a large extent philosophical and literary, yet he was a member of the Lyceum and the Museum, and was thus submitted to the influence of Aristotle, Theophrastos, and Stratōn. As a member of the Museum and Library he could not help having a share in every scientific project in addition to his own scientific investigations, to be described presently.

The earliest documents concerning him are three epigrams preserved in the *Greek Anthology*.[8] The first was composed by himself and placed by him at the end of his letter to Ptolemaios Evergetēs on the duplication of the cube; [9] the second was addressed by Archimēdēs to his friend Eratosthenēs; the third was composed by Dionysios of Cyzicos (Propontis). The first two are contemporary, the third a bit later, yet Hellenistic.[10]

Eratosthenēs received two nicknames that are significant with regard both to himself and to his time. He was called *bēta* and *pentathlos*. The first means number two or second-rate; the second name was given to athletes who had distinguished themselves in the five games,[11] and metaphorically to men who tried their hand at everything (Jacks-of-all-trades). From the social point of view these names are witnesses to the growing specialism of the Hellenistic age; not only were scientists and scholars specialized in this or that branch of knowledge but they were already beginning to despise their colleagues whose intellectual ambition was less exclusive than their own and who tried to understand as much of the world as they could. Eratosthenēs was a polymath by temperament, but also by training; his geographic investigations had been preceded by philosophical and literary studies; moreover, he was a victim of the endless opportunities that were opened to him as librarian in chief of the greatest library of antiquity.

The first nickname, *bēta*, shows that the scientists and scholars of that age were already very jealous of one another and all too ready to deflate

[8] Anthologies (*anthologia* = bouquet) of Greek poems were collected at various times from the fourth century on. The main one is the *Palatine anthology* collected by Constantinos Cephalas c. 917; it was re-edited in 1301 by Maximos Planudēs (XIII-2); see my *Introduction*, vol. 2, p. 974. Modern editions of the *Anthologia Palatina* generally contain a Planudian supplement.

[9] For the history of that problem, which Eratosthenēs himself connected with Dēlos (hence its alternative name, Delian problem), see Volume 1, pp. 278, 440, 503.

[10] Frederic Dübner, *Epigrammatum Anthologia palatina* (Greek-Latin ed., 3 vols.; Paris, 1864–1890). The three Eratosthenian epigrams are in vol. 3, I, ep. 119; VII, ep. 5, and vol. 1, VII, ep. 78. English translations by W. R. Paton in *The Greek anthology* (Loeb Classical Library, 5 vols., 1916–1918).

[11] The five games (pentathlon) were *halma* (jumping), *discos* (discus throwing), *dromos* (running), *palē* (wrestling, *lucta*), *pygmē* (boxing, *pugnus*). The last might be replaced by *acontisis* or *acōn* (javelin throwing).

those whose superiority they misunderstood and resented.[12] Now the professional mathematicians might consider him as not good enough in their field and be displeased with the abundance and variety of his nonmathematical interests. As to the men of letters and philologists, they could not appreciate his geographic purposes. Eratosthenēs might be second-rate in many endeavors, but he was absolutely first-rate in geodesy and geography; he was indeed the earliest outstanding geographer and is to this day one of the greatest geographers of all ages. This his critics could not even guess, and therefore they pooh-poohed him. There was among them a man of genius but as he was working in a new field they were too stupid to recognize him. As usual in such cases, they proved not his second-rateness but only their own.

PRE-ERATOSTHENIAN GEOGRAPHY

In order to understand Eratosthenēs' contributions, a flashback to earlier geographic endeavors is necessary. Not only had a large amount of geographic knowledge been accumulated by the middle of the third century B.C., but that knowledge was of many kinds. For example, knowledge of human geography had been gathered by historians like Hērodotos and Ctēsias in the fifth century, Ephoros in the fourth, Megasthenēs (III–1 B.C.), by travelers and explorers such as Hannōn (V B.C.), Xenophōn (IV–1 B.C.), Pytheas and Nearchos (IV–2 B.C.), Patroclēs c. 280. The last-named is not as well known as the others. He was an officer of the Seleucidae (c. 280) who explored the southern parts of the Hyrcanian (Caspian) Sea and believed that it was connected with the Arabian Sea.[13] This was a traveler's tale, perhaps of Chinese origin, but travelers' tales, however wild, might include fragments of geographic knowledge and acted as ferments.

Another kind of information was provided by the writers of land itineraries, coasting voyages (*periploi*), traveling sketches (*periēgeis*, *periodoi*), by the compilers of empirical maps, charts, or schemas (*pinaces*, *tabulae*).

Still another kind was more theoretical and more ambitious, as exemplified in the works of Anaximandros and Hecataios, both Milesians of the sixth century, or with more precision by Eudoxos of Cnidos (IV–1 B.C.), by

[12] There was perhaps an added sting in that nickname, because Eratosthenēs was an alpha fellow of the Museum. His enemies might say, "In spite of his being an alpha he is really a bēta."

[13] Aristotle and Alexander were aware of the existence of two interior seas, the Hyrcanian (our Caspian) and the Caspian (our Aral), but Alexander had wondered whether the Caspian was not connected with the Arabian Sea. That was also Patroclēs' idea. As to the Aral, it vanished from knowledge; the ancients believed that the two rivers, Jaxartēs (Sayḥūn, Sīr Daryā) and Oxos (Jayḥūn, Amū Daryā), flowed not into the Aral but into the Caspian. There may have been communications between these two lakes in very ancient days. The Araxēs of Hērodotos may have been one of those rivers, or the Volga, which actually flows into the Caspian. Such confusions were unavoidable as long as one depended not on astronomical coördinates but only on travelers' guesses. See H. F. Tozer and M. Cary, *History of ancient geography* (Cambridge, 1935), pp. 135–136, xviii.

Dicaiarchos of Messina (IV–2 B.C.) – this one is often mentioned as Eratosthenēs' forerunner – and by Timosthenēs, admiral of the fleet of Ptolemaios Philadelphos, author of a treatise on harbors and student of winds.[14]

The spherical shape of the Earth had been recognized by the early Pythagoreans and had remained a Pythagorean tenet, but it does not follow that all the geographers accepted it. For many of them, travelers and writers of itineraries, it was of no practical importance. It became crucial, however, as soon as attempts were made to develop mathematical geography and to draw a world map. One of Eratosthenēs' main achievements was precisely to establish the mathematical geography of the spherical Earth.

ERATOSTHENĒS' GEOGRAPHIC WORKS

Eratosthenēs' works were many but none has come down to us in its wholeness and most of them are known only in the form of fragments the genuineness of which cannot always be ascertained. Hence the interpretation of them is full of conjectures and the cause of endless controversies. The main user of his geographic works was Strabōn (I–2 B.C.), who criticized his facts and methods and quoted him when he had to state his disagreement, but seldom when he agreed with him. Sometimes Eratosthenēs is named (*Eratosthenus apophaseis*, *Eratosthenēs phēsi*); more often he is not.

The main works, to be discussed presently, are, in probable chronologic order, *On the measurement of the Earth* (*Anametrēsis tēs gēs*), *Geographic memoirs* (*Hypomnēmata geōgraphica*), *Hermēs*, a geographic poem.

In view of Eratosthenēs' considerable fame in antiquity, how is it that his works were allowed to disappear? They were absorbed and improved by his followers, especially Strabōn and Ptolemy. One of his early critics, Hipparchos, suffered the same fate and for the same reason. The results of ancient geography and astronomy were put together by Ptolemy and the works of Eratosthenēs and Hipparchos were replaced by Ptolemy's *Geōgraphicē hyphēgēsis* and *Almagest*.

THE MEASUREMENT OF THE EARTH

Eratosthenēs is supposed to have written a treatise on "geometry" (meaning "measurement of the Earth"), but this is not certain. The treatise is not referred to in his *Hypomnēmata* (see below); it is mentioned by Macrobius (V–1), a late witness. The subject itself is treated in the second part of the

[14] In Homer's time four winds were already recognized — Boreas, Euros, Notos and Zephyros — corresponding more or less to the four cardinal points (N, E or SE, S, W or NW). Aristotle introduced eight more (*Meteorologica* 2, 6) but his wind points were not like the vertices of a regular polygon but were arranged in groups of three to each right angle. H. F. Tozer and M. Cary, *History of ancient geography* (Cambridge, 1935), pp. 194, xxiv. The traditional division, however, was octagonal. It is represented in the Horologion of the Syrian Andronicos Cyrrhestēs, the so-called Temple of the Winds in Athens (first century B.C.).

Hypomnēmata, but that treatment might be a summary of the "geometry."

It is certain, however, that Eratosthenēs measured the Earth, and his measurement was astoundingly accurate.

His method consisted in measuring the distance between two places located on the same meridian. If the difference of latitude of these two places is known, it is easy to deduce the length of 1° or of the whole meridian. I do not say 360° because Eratosthenēs divided a great circle into 60 parts; Hipparchos was probably the first to divide it into 360°.

Eratosthenēs' was not the first estimate; according to Aristotle the circumference of the Earth amounted to 400,000 stadia; according to Archimēdēs, 300,000 stadia; according to Eratosthenēs, 252,000 stadia.[15] It is said by Cleomēdēs that his result was $50 \times 5000 = 250{,}000$ stadia, but he made various measurements and accepted 252,000 as the final result. These measurements were not accurate in the modern sense; they were approximations and the final result was probably made more acceptable for non-experimental reasons ($252 = 2^2 \times 3^2 \times 7$).

In order to determine the latitude, Eratosthenēs used a *gnōmōn* or a *sciothēron*.[16] In Syēnē,[17] at the time of the summer solstice there was no shadow at all and he concluded that that place was located on the Tropic of Cancer; Syēnē and Alexandria were, he believed, on the same meridian, their difference of latitude was 7°12′ (1/50 of a great circle) and the distance between them amounted to 5000 stadia. Thus the length of the circumference was 250,000 stadia, a result which he corrected eventually to 252,000. These assumptions were not quite correct. The differences in longitude and latitude of the two places are 3°4′ (instead of 0°) and 7°7′[18] (instead of 7°12′); the distance 5000 stadia is obviously an approximation in a round number. The distance was measured by a *bēmatistēs* (a surveyor trained to walk with equal steps and to count them). It is clear that Eratosthenēs was satisfied with approximations: the original figures, 1/50 of the circumference and 5000 stadia, are too good to be true.

It is said that he determined the position of the Tropic by means of a deep well; the sun at noon on the summer solstice would light up the well right down to the water and cast no shadow on the walls. That is not impossible, though a well would hardly be a better instrument than the *sciothēron*. The "well of Eratosthenēs" is not in Syēnē proper but in Ele-

[15] It does not follow that those estimates were in the ratios 400:300:252, for the stadia might be different in each case.

[16] A *sciothēron* is a kind of sundial. It had the shape of a bowl (*scaphē*) with a *gnōmōn* standing in the middle of it (like the radius of a hemisphere). Lines drawn inside of the bowl would enable the observer to measure the length of the *gnōmōn's* shadow immediately.

[17] Syēnē (Arabic, Aswān) in Upper Egypt on the Nile, just below the First Cataract. Its latitude is 24°5′ and the obliquity of the ecliptic was then 23°43′. It is probable that Eratosthenēs assumed the obliquity to be 24°, yet even then Syēnē was a little above the Tropic of Cancer.

[18]

Alexandria:	27°31′N,	31°12′E;
Syēnē:	30°35′N,	24°5′E;
differences:	3°4′,	7°7′.

phantine, an island in the Nile (Jazīrat Aswān), opposite Syēnē just below the First Cataract; this makes no difference.[19] The well which can be seen to this day in Elephantine is probably the nilometer (*miqyās*) described by Strabōn.

If we admit the measurement 252,000 stadia our difficulties are not yet over, for how long was a stadium? There were differences between the various stadia in different times and places, and ancient geographers were hardly conscious of them.[20] Perhaps the most acceptable solution of that insolvable puzzle is the one given by Pliny (XII, 53), according to whom one schoinos equals 40 stadia. On the other hand, according to Egyptologists the schoinos equals 12,000 cubits and the Egyptian cubit equals 0.525 meter. If so, the schoinos equals 6300 meters and Eratosthenēs' circumference equals 6300 schoinoi or 39,690 kilometers.[21] That result was almost unbelievably close to the real value (40,120 km), the error being not much above 1 percent.[22] On that basis, the Eratosthenian stadium equaled 157.5 m, shorter than the Olympic stadium (185 m) and than the Ptolemaic or Royal one (210 m).

There were about 9.45 Eratosthenian stadia to a mile; according to another interpretation, his stadium was even smaller, 10 to a mile.[23] The other stadia were larger (9, 8⅓, 8, 7½ to a mile). The smallest of these (9 to a mile) would give a circumference of 41,664 km (too large by less than 4 percent) and the others would increase the error. That does not matter very much, however. Eratosthenēs' achievement lies in his method; whichever the stadium used it would give a not implausible value of the Earth's size. That was a great mathematical achievement.

Not only was the spherical shape of the Earth vindicated, but the sphere

[19] Howard Payn, "The well of Eratosthenēs," *Observatory* 37, 287–288 (1914), with photograph of the well. Criticism by J. L. E. Dreyer, *ibidem*, 352–353). For comparison, Aydin Sayli, "The observation well," *Actes du VIIe Congrès international d'Histoire des Sciences* (Jerusalem, 1953), pp. 542–550. Elephantine (called Yebu in Egyptian, Jazīrat Aswān in Arabic) was an important military and religious center in Pharaonic times, it was also an important business center for trade with Ethiopia. What is more remarkable is that it was a Jewish center where an abundance of Aramaic papyri of the fifth century B.C. have been discovered; *Encyclopaedia Judaica*, vol. 6 (1930), pp. 446–452. Jewish colonies were established in Egypt long before Hellenistic times.

[20] There were all kinds of differences in weights and measures, calendars, chronological scales, even numerals, and the great majority of learned people were blessedly unaware of them. For a discussion of the stadia, see Aubrey Diller, "The ancient measurements of the Earth," *Isis 40*, 6–9 (1949). For numerals, see Sterling Dow, "Greek numerals," *American Journal of Archaeology 56*, 21–23 (1952).

[21] The coincidence of the two numbers 6300 is curious: 1 schoinos = 40 stadia = 12,000 Egyptian cubits = 6300 meters; and 40 stadia are included 6300 times in 252,000 stadia.

[22] 39,690 km = 24,662 mi. The corresponding diameter is 7,850 mi, only 50 mi less than the true value of the polar diameter, and 77 mi less than the equatorial diameter.

[23] On that basis (10 stadia to a mile), the circumference of the Earth would equal 37,497 km, more than 6 percent too small.

was measured. The correctness of his results was partly accidental, for it was based upon very inadequate measurements.

Eratosthenēs' main geographical work was the *Hypomnēmata geōgraphica*. As far as can be deduced from the fragments and from Suidas' description it was divided into three parts: (1) Historical introduction; (2) Mathematical geography; measurement of the Earth and of the inhabited portion thereof, *hē oicumenē* (*gē*); (3) Mapping and description (*periēgēsis*) of the countries. As the table of contents has not survived, the ascription of this or that fragment to part 2 or 3 is sometimes arbitrary, but that is of no importance.

The historical account (part 1) went back to Homer and Hesiod and explained the geographic views that preceded and gradually prepared the conception of a spherical Earth. It reviewed old ideas on the size of the Earth, the proportion of land areas to sea areas, the shape and size of the *oicumenē*, the circumambient Ocean, the Nile so vastly different from the other rivers and its mysterious inundations. Aristotle and Eratosthenēs were the first to give the true explanation of these – tropical rains of spring and early summer in the extremely distant highlands whence the Nile waters come.

Part 2 was a mathematical geography based upon the hypothesis of sphericity. It contained perhaps a summary of his older treatise on "geometry." The geographic zones [24] were established and measured. This hinged on the measurement of the obliquity of the ecliptic, for which Eratosthenēs' estimate was probably the same as Euclid's, 24°; [25] the tropical region was then 48° broad, being limited by the tropical circles of Cancer and Capricorn. The two polar circles were 24° distant from the poles, and the temperate zones occupied the spaces lying between the arctic and tropical zones. He described the main physical characteristics of each zone.

He realized that the mountains were too small, the valleys too shallow, and the catastrophies (floods, earthquakes, volcanic explosions) too weak to affect the sphericity of the Earth. According to Theōn of Smyrna (II-1), he thought that the highest mountains were only 10 stadia high (1/8000 part of the diameter), but even if he had known much higher mountains his judgment of their relative smallness would still hold.

The *oicumenē* known to Eratosthenēs extended in breadth from the latitude of Thulē, revealed to him by Pytheas, which he placed near the Arctic Circle, to the Indian Ocean and Taprobanē (Ceylon), and in length

[24] His notion of zones was thus essentially different from the earlier fifth-century one of Parmenidēs of Velia (Elea, Hyelē) and of Dēmocritos of Abdēra anterior to the discovery of the obliquity of the ecliptic (Volume 1, pp. 288, 292). The obliquity of the ecliptic, we should remember, is not constant throughout the centuries. It is now about 23°28′, whereas it was 23°43′ in Eratosthenēs' time.

[25] The result 24° was very acceptable to ancient astronomers because 24° was the angle subtended by the side of a regular polygon of 15 sides.

from the Atlantic Ocean to Central Asia and the Bay of Bengal. This gave a rectangle of about 38,000 by 78,000 stadia, that is, twice as long as it was broad; the estimate of length, however, was exaggerated by at least one-third. The coexistence of tides everywhere confirmed the hypothesis of a circumambient Ocean.

The views of Aristotle and Timosthenēs on winds have been mentioned above. It is possible that Eratosthenēs was acquainted with them and with those of Biōn the astronomer.[26] He himself wrote a treatise or chapter on winds (*peri anemōn*),[27] and established a new diagram of the winds, or wind rose. It had eight sectors: *aparctios* (N), *boreas* (NE), *euros* (E), *euronotos* (SE), *notos* (S), *lips* (SW), *zephyros* (W), *argestēs* (NW). (There are variants of those names, and the history of each of them is fairly complex.) Note that among the names given only one, *euronotos* (east-south), is constructed in the modern way. He distinguished between universal winds (*catholicoi*) and local ones (*topicoi*).

The third part of the *Hypomnēmata* dealt with mapping and descriptive geography. It may seem strange to have the treatment of maps here and not in the mathematical section, but the mathematical principles of map making were not yet understood. This weak point in Eratosthenian knowledge was sharply criticized by Hipparchos, but Hipparchos' criticisms and new theories were lost as well as those of Marinos of Tyre (II-1) and did not emerge and survive except many centuries later in Ptolemy's *Geography*. Eratosthenēs rejected the continental division (Asia, Europe, Africa) and divided the inhabited world by means of two perpendicular lines or bands, crossing each other in Rhodos (there was an old observatory on its highest mountain, Atabyrion); the horizontal one (above 35°N) passed near the Pillars of Hēraclēs (Gibraltar), followed the length of the Mediterranean Sea, and then, a little higher, of the Taurus chain; the vertical line followed roughly the Nile. That was very rough indeed and therefore it is better not to give to those perpendicular lines and to the lines parallel to them the names of latitude and longitude. These conceptions had not yet been formulated with sufficient clearness and rigor, and no wonder, for it was not yet possible to determine latitudes with great precision and longitudes with any. These two lines or bands were simply two lines of reference, permitting a rough classification of countries in four sectors. He did not try any mathematical definition of the countries, but a purely human one. Egypt is the land of the Egyptians. What is exceedingly typical of the post-Alexandrian age, he refused to speak of Greeks and Barbarians. There are among the latter some very civilized nations, such as the Indians, the

[26] Biōn ho astrologos, Strabōn I, 2, 21 (Loeb Classical Library, vol. 1, p. 106). This was perhaps Biōn of Abdēra who flourished c. 400. See Hultsch, in Pauly-Wissowa, Vol. 5 (1897), 485–487.

[27] Many texts edited by Georg Kaibel, "Antike Windrosen," *Hermes 20*, 579–624 (1885).

Romans, and the Carthaginians; on the other hand, there are some contemptible people among the Greeks.

His map was not based on an astronomical network (circles of latitude and longitude) but on a number of *sphragides* vaguely placed in each of the four mean sectors.[28] It is easy to understand Hipparchos' scorn. A *sphragis* (seal) or *plinthion* (small brick) meant to Eratosthenēs a distinctive shape, the general appearance of each country being likened to a familiar object. The idea was not new. The *actai* of Hērodotos [29] were of a similar kind. It is a popular idea rather than a scientific one. Spain was compared to an ox hide, Italy to a foot and leg, Sardinia to a human footprint, and so on. He may have been inspired by the constellations, the general shape of which was easy to observe. Note that we ourselves think of foreign countries in terms of *sphragides*; we "see" India, Indo-China, Spain or Italy. Our best references are in those terms. The most exact way of defining the position of a star is to give its coördinates, but in most cases it will be more helpful to be told that it is in this or that part of a known constellation; then we know immediately where it is. In the same way we would be very embarrassed to have to indicate the degrees of latitude and longitude that would frame Italy, but we "see" Italy, we see the boot.

I cannot help wondering, however, how such a conception could grow in ancient minds. We know the Italian boot well enough because we have seen it in atlases and maps from childhood on, but if we had no maps, what then? How could Eratosthenēs conceive the general shape of Iran? In default of astronomical coördinates all he knew to guide him were travelers' reports, the distances and relative orientation of definite places. That was not very much.

On the other hand, Eratosthenēs had accumulated much information concerning the natural products of each country and the people living in them. The bulk of that is preserved in Strabōn but we are unable to recognize the Eratosthenian elements, except in the not infrequent cases when Strabōn advertises the errors of his predecessor and criticizes him.

To sum up, Eratosthenēs had a pretty good knowledge of human geography, his knowledge of descriptive geography was empirical and poor, but he was the first to put together all the methods and facts that had accumulated down to his time. Above that, he was the first theorist and synthesist of the spherical Earth, the first mathematical geographer.

[28] According to Tozer and Cary, *History of ancient geography*, p. 181, Eratosthenēs conceived various parallels corresponding to the Cinnamon Region, Meroē, Syēnē, Alexandria, Rhodos, the Troad, Olbia near the mouth of the Borysthenēs (Dnieper), Thulē, and various meridians corresponding to the Pillars of Hēraclēs, Carthage, Alexandria, Thapsacos on the Euphrates (near the westernmost part of it), the Caspian Gates, the mouth of the Indus, the mouth of the Ganges. May be, but Eratosthenēs' knowledge of such matters was vague. He realized that some places were located at about the same latitude or longitude; but it would be incorrect to speak of definite geographic coördinates.

[29] Hērodotos, IV, 37–39.

ASTRONOMY

According to Galen,[30] an unexpected witness in astronomy, Eratosthenēs' "geometry" dealt with "the size of the equator, the distance of the tropic and polar circles, the extent of the polar zone, the size and distance of the Sun and Moon, total and partial eclipses of these heavenly bodies, changes in the length of the day according to different latitudes and seasons." This shows that Eratosthenēs had not restricted himself to *geōdaisia* (geodesy, itself a part of astronomy) but had contemplated the main astronomical problems of his time.

He estimated the distances of the Moon and Sun from the Earth to be 780,000 and 804,000,000 stadia. According to Macrobius (V–1), he said that the measure of the Sun was 27 times that of the Earth; does "measure" mean volume? Then the diameter of the Sun was thrice that of the Earth. Such measurements are mentioned for the sake of curiosity; the main addition to knowledge was the bold idea of making such measurements, and the real innovator was not Eratosthenēs but Aristarchos of Samos.

Eratosthenēs was naturally interested in the calendar. He wrote a treatise on the eight-year period, *octaetēris*, and did not consider the treatise *ad hoc* of Eudoxos of Cnidos (IV–1 B.C.) as genuine.

He was probably consulted by Ptolemaios III Evergetēs in 238 when a reform of the calendar was being deliberated by a synod of the Egyptian clergy. The reform was accepted by the synod on March 7, 238 and is generally called the decree of Canōpos.[31] It is known through various inscriptions, especially one written in three scripts (hieroglyphic, demotic, and Greek) found at Kūm al-Ḥiṣn in 1881 and preserved in the Cairo Museum.

MATHEMATICS

The most remarkable mathematical achievement ascribed to him is the invention of the famous "sieve of Eratosthenēs," [32] for the purpose of finding the prime numbers. Suppose that the integers are written out in a series; cancel the even ones, then those divisible by 3, 5, 7, 11, and so on. The remaining integers will be primes. That is simple and easy but would not lead one very far. Some of our contemporaries have been able to discover prime numbers so large that "sieving" would have to be carried out by legions of men working during an incredibly long time, even if their sieve were replaced by a machine that would cut off automatically all the multiples of the successive primes.[33] Just try to solve a relatively simple prob-

[30] Galen, *Institutio logica* (*Eisagōgē dialecticē*), ed. Carolus Kalbfleisch (98 pp.; Leipzig, 1896), chap. 12, p. 26. This text is not in Kühn's edition.

[31] Canōpos is near the mouth of the westernmost branch of the Nile, a little east of Alexandria. It was Alexandria's playground.

[32] The sieve or *coscinon* was a tool familiar to farmers and craftsmen, also to diviners. A *coscinomantis* was a man who used the sieve for the purpose of divination.

[33] The largest prime discovered to this day is $180(2^{127} - 1)^2 + 1$, *Nature 168*, 838 (November 10, 1951). Could you

lem, to sift out the prime numbers of the first million, and you will appreciate the difficulties.

He wrote a book entitled Platōnicos, which was probably a commentary on the *Timaios* or on other Platonic dialogues. It is twice mentioned by Theōn of Smyrna (II–1) in his mathematical introduction to Plato. It discussed principles of arithmetic, geometry, and music. It told the story of the Delian problem: in order to stop a plague, the priestess of Dēlos expressed Apollōn's wish that his cubic altar be doubled. It is the problem of the duplication of the cube which had exercised the minds of many mathematicians from the fifth century on.[34] Eratosthenēs proposed a new method which he described in the letter to Ptolemaios Evergetēs ending with the epigram mentioned above.[35] This was written not long before the end of Evergetēs' rule (247–222). In order to express his gratitude to the king, Eratosthenēs caused a column to be erected upon which the epigram was inscribed as well as a drawing of the contrivance (*mesolabion*) which he had devised to solve the problem.[36] Let us pause a moment to consider that. Eratosthenēs wished to thank and flatter the king, his own Evergetēs, and the best that he could think of was to dedicate to him the solution of an abstruse mathematical problem. There have been courtesans in all ages and places, but have you ever heard of another king and another courtier who behaved like that? This happened in Alexandria near Egypt shortly before 222 B.C.

PHILOLOGY

It is very curious that Eratosthenēs, who was primarily a man of science and whose fame is built on his geography, was the first man to be called a *philologos* (or perhaps *criticos*, *grammaticos*). Of course he was not by any means the first to deserve that name, but why was it given first to him, who was essentially something else? It is as if Newton were called the theologian, or Ingres, the violinist. The name would have been more fitting for other librarians, whose philological interests were supreme and exclusive.

It is probable that Eratosthenēs' appointment (c. 234) as chief librarian of the Museum was clinched because the need of a librarian familiar with mathematics and science was beginning to be felt. An *alpha* fellow of the Museum seemed to be a good choice. Yet men of science were

---

prove that that number is prime? H. S. Uhler, "Brief history of the investigations on Mersenne numbers and the latest immense primes," *Scripta mathematica 18*, 122–131 (1952). According to *Larousse Mensuel* (Paris, August 1955), p. 691, the largest prime number then known was $(2^{2281} - 1)$, obtained by electronic computer.

[34] Volume 1, pp. 278, 440, 503.

[35] See note 10.

[36] In order to solve the equation $x^3 = 2a^3$, one has to find two mean proportionals in continued proportion between a and $2a$, that is, such that $a/x = x/y = y/2a$. The *mesolabion* (or mean-finder) was a mechanical device for doing that.

still relatively rare and the great majority of learned men were philologists or men of letters and nothing else. They were unable to appreciate the new kind of learning which Eratosthenēs was representing, and therefore they did not call him *geographos* or *mathematicos* but *philologos*.

Their naming him *philologos* was not arbitrary, however, for he deserved that name from his school years in Cyrēnē and Athens when he studied belles-lettres and philosophy. Later his duties as librarian could not help but aggravate his philologic as well as his encyclopedic tendencies. Was he not in charge of all the books and of all the scholars who visited the Library? And were not the overwhelming number of books literary or philosophical and the great majority of scholars men of letters rather than of science?

His masterpiece in philology was an elaborate study of the old Attic comedy [37] (*Peri tēs archaias comōdias*) which was much used by Aristophanēs of Byzantion (II–1 B.C.) and by Didymos of Alexandria (I–2 B.C.).

It is doubtful whether Eratosthenēs prepared a corrected edition of Homer (*diorthōsis Homēru*), but he studied Homer as did every educated Greek. Homer, we should remember, was honored by them almost like a superhuman being. The *Iliad* and the *Odyssey* were read in the same spirit as the people of other nations read their sacred books. Criticism of these writings was almost as shocking to the Greeks as criticism of the Qur'ān would be to Muslims. To a man like Strabōn, Homer was the founder (*archēgetēs*) of Greek culture. It would seem that Eratosthenēs was particularly interested in Homeric geography, which was admirable in some respects (accuracy of local epithets) but less so in others. Was his criticism too sharp and indiscreet? Was his estimate of Homeric geography published in a separate treatise or only in the first part (historical) of his *Hypomnēmata*? We do not know for certain, but it seems probable that the *Hypomnēmata* contained only a summary of a more elaborate study; that summary was preserved by Strabōn.[38]

Another query rises in my mind. Was not Eratosthenēs' study of Homeric geography the seed of his geographic investigations? That is quite possible. He would not be the first man of science whose vocation was determined by romantic circumstances. A vocation is always an act of faith much anterior to the knowledge that would justify it. It is pleasant to think of Homer guiding the steps of the first mathematical geographer.

Eratosthenēs comes very close to us in another way. He was a historian, he wrote a history of philosophy, and part 1 of his *Hypomnēmata* was a history of geography.

[37] The "old comedy" was largely anterior to the fourth century. The only playwright some of whose plays have survived entire is Aristophanēs of Athens (c. 450–385), but there is an abundance of fragments from other plays.

[38] Strabōn, *Geography*, 1, 2, 3–22.

He was not the first historian of science, yet one of the very first.[39]

In the geographic field, one of his fundamental problems was the determination of localities. He could not really solve it because the latitude of a place was not easy to measure, and the longitude, exceedingly difficult. The corresponding problem in the historical field is the determination of dates in a single time sequence. Each country or city had its own means of registering deeds with reference to local standards, but it was very difficult if not impossible to harmonize different chronologies. Eratosthenēs tried to establish a scientific chronology from the war of Troy to his own day and wrote two treatises on the subject, the one entitled *Chronographiai* and the other *Olympionicai*. This second was a list of Olympic victories. Both dealt with the Olympic scale introduced by Timaios about the beginning of the third century. Timaios has established concordances between the kings and *ephoroi* of Sparta, the Athenian *archontes*, the priestesses of Argos,[40] and the Olympic victories. As the scope of those famous games was international (in the Greek world, at least), their enumeration provided an international frame of reference. Instead of saying that an event occurred in the seventh year of the rule of a local king or tyrant in Rhodos, Samos, or elsewhere, one might say that it occurred in the first, second, third, or fourth year of this or that Olympiad. The *Olympionicai* were superseded by the similar work of Apollodōros of Athens (II–2 B.C.). I do not know how much Eratosthenēs added to Timaios and how much Apollodōros to Eratosthenēs, because all those treatises are lost. The best information *ad hoc* is given by Clement of Alexandria,[41] who flourished a few centuries later.

The third century was a time when didactic poetry flourished. There were always epic and lyric poets, but the main demand of the reading people was for knowledge, easy knowledge expressed in verse. The reader has already been introduced to two didactic poets, two Greeks of Asia, Aratos of Soloi and Nicandros of Colophōn. Eratosthenēs wrote many poems, a short epic poem, *Anterinys*, describing Hesiod's death and the punishment of his murderers, an elegy, *Ērigonē*, celebrating Icaros and his daughter Ērigonē, and others, but we are more interested in two didactic poems,

[39] For earlier ones, beginning with Eudēmos of Rhodos (IV–2 B.C.), see Volume 1, p. 578.

[40] The *ephoroi* (singular, *ephoros*) or overseers of Sparta, a body of five magistrates controlling even the kings. The *archontes* were the chief magistrates of Athens, nine in number; the first, or leader, was called *the* archōn, also *archōn epōnymos*, because he gave his name to the current year. The priestesses of Argos (northeast Peloponnēsos) were in the service of Hēra, goddess of marriage and women (Juno to the Romans).

[41] Titus Flavius Clemens (c. 150–c. 214), born in Athens, converted to Christianity, head of the Catechetical School of Alexandria which provided Christian education (as against the pagan education of the Museum and Serapeum). It was a school for Christian neophytes, or catechumens (Galatians 6:6).

Hermēs and the *Catasterismoi.* Hermēs Trismegistos was of special concern to the Greco-Egyptians because he was the Greek avatar of Thoth, Egyptian god of the sciences. The poem *Hermēs* is astronomical; the extant part (35 lines) deals with zones and is the only Eratosthenian text explaining the poet's views on the subject; these have been summarized above. The *Catasterismoi* [42] describes constellations and their mythology. This was an essential part of astronomy from the Hellenistic point of view. Still another didactic poem has already been mentioned, the epigram on the duplication of the cube. According to ancient critics who knew the whole of it, *Hermēs* was his masterpiece. Such poems satisfied the scientific curiosity as well as the love of metrical words of the Ptolemaic aristocracy; they also pleased Renaissance scholars but they are not very acceptable to modern readers, be they astronomers or poets.

THE ERATOSTHENIAN TRADITION

The activities of Eratosthenēs were very complex and each of them had its own tradition. To many ancients he was primarily the critic of Homer. To others he was the founder of mathematical geography, or of descriptive geography, or even (it must be admitted, in a very imperfect way) of map making.

His mathematical and astronomical knowledge was severely criticized by Hipparchos (II–2 B.C.) but his good fame had been supported by Archimēdēs, who dedicated to him his *Cattle problem* and his greatest work, the *Method.* Surely if the greatest mathematician of antiquity chose to honor him in such a manner, there must have been qualities in him that Hipparchos failed to see.

His descriptive geography was often corrected and completely assimilated by Strabōn (I–2 B.C.). His geodesy and geographic ideas were criticized and transmitted by Polemōn ho Periēgētēs (II–1 B.C.), Poseidōnios (I–1 B.C.), Cleomēdēs (I–1 B.C.), Strabōn (I–2 B.C.), Dionysios ho Periēgētēs (I–2), Galen (II–2), Achilleus Tatios (III–1); in the Byzantine world by Marcianos of Hēracleia (V–1), Stephanos of Byzantion (VI–1), Suidas (X–2), Tzetzēs (XII–1); in the Latin world by Vitruvius (I–2 B.C.), Pliny (I–2), Macrobius (V–1), Martianus Capella (V–2) . . . Lambert of Saint Omer (XII–1); in the Arabic one, by al-Qazwīnī (XIII–2).

This list is very impressive, much more so than it deserves to be. Many names are mentioned, because if one mentions one, one cannot leave others out. In reality, the Eratosthenian works were soon reduced to fragments, and their tradition lost in the traditions of Strabōn and Ptolemy. Many Renaissance scholars were fascinated by these fragments, however, and tried to solve the many puzzles that they created. Their faith in him was astonishing. Let me give two examples of it.

[42] Or *Astrothesia,* with the same general meaning — placing of the stars. The genuineness of that work has been questioned.

When the Dutch physicist, Willebrord Snel, wished to explain his method of measurement of a part of a meridian he published it under the title *Eratosthenes batavus. De terrae ambitus vera quantitate* (Leiden, 1617). The French humanist, Claude de Saumaise, was called by his many admirers "the prince of learning" and also the Eratosthenēs of his time.[43]

The *Geography* of Strabōn was printed six times in Latin during the fifteenth century [44] and, as Eratosthenēs is quoted in them hundreds of times, men of learning using the incunabula were well acquainted with him, but there was no separate edition or important discussion of his geography until Pascal F. J. Gossellin, *Géographie des Grecs analysée ou les systèmes d'Eratosthène, de Strabon et de Ptolémée comparés entre eux* (4to, 175 pp.; Paris, 1790) (Fig. 31).

For the modern editions see my *Introduction*, vol. 1, p. 172. The poems were edited by Eduard Hiller (140 pp.; Leipzig, 1872) and the geographic fragments, by Hugo Berger (401 pp.; Leipzig, 1880).

Alessandro Olivieri, *Pseudo-Eratosthenis Catasterismi* in *Mythographi graeci* (vol. III, fasc. 1, 94 pp.; Leipzig, 1897); divided into 44 chapters, from 1. Great Bear to 44. Galaxy; index.

NOTE ON THE OLYMPIADS

The Olympic games, which took place in Olympia (Ēlis, northwest Peloponnēsos) every fourth year, were events of international importance throughout the Greek world; we might almost say throughout the *oicumenē*, because Greek influences were felt almost everywhere. The victors of these games were international heroes; oral tradition preserved the names of the Olympic victors in chronologic order and eventually lists of Olympic victories were written down. On the other hand, local events were recorded in local annals called *horographiai*,[45] and those events were dated from the beginning of each local kingship, magistracy, or priesthood. Timaios of Tauromenion (Taormina, eastern Sicilian coast) was the first to compare local chronologies, and it occurred to him that the dates of the Olympic games would afford a common standard of international validity. His work was continued and completed by Eratosthenēs; Olympic dating was used or referred to by Polybios (II-1 B.C.), Apollodōros of Athens (II-2 B.C.), Castōr of Rhodos (I-1 B.C.), Diodōros of Sicily (I-2 B.C.), Dionysios of Halicarnassos (I-2 B.C.), but it was never popular and was not used on coins or inscriptions (except a few Olympic ones).

The origin of the Olympic games is immemorial, but the first Olympiad (776–773) was reckoned from the victory of Coroibos of Ēlis in the foot

[43] Snel van Roijen in Dutch, Snellius in Latin (1591–1626). Saumaise (1588–1653) was better known under his Latin name, Claudius Salmasius. He was half a Dutchman, being professor at the University of Leiden from 1631 to 1650.

[44] Klebs, No. 935.1–6; first in Rome, 1469.

[45] *Hōra* means a limited period of time, a season, a year, an hour (*hora* in Latin). Annals were called *hōrographia* and the annalist, *hōrographos*.

race of 776. The Olympic festival occurred in the eighth month of the Eleian calendar, corresponding to the second Attic month (Megageitniōn), July–August. Thus, Ol. 1.1 covers the period from July (or August) 776 to June (or July) 775. In general, it suffices to say Ol. 1.1 = 776 B.C., but it is well to remember that the Olympic year (or the Attic year) did not begin on January 1.[46]

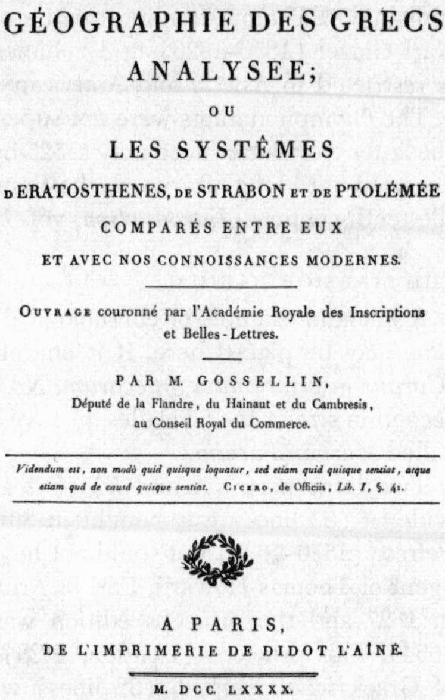

GÉOGRAPHIE DES GRECS

ANALYSEE;

OU

LES SYSTÊMES

D'ERATOSTHENES, DE STRABON ET DE PTOLÉMÉE

COMPARÉS ENTRE EUX

ET AVEC NOS CONNOISSANCES MODERNES.

OUVRAGE couronné par l'Académie Royale des Inscriptions et Belles-Lettres.

PAR M. GOSSELLIN,

Député de la Flandre, du Hainaut et du Cambresis, au Conseil Royal du Commerce.

Videndum est, non modò quid quisque loquatur, sed etiam quid quisque sentiat, atque etiam quâ de causâ quisque sentiat. CICERO, de Officiis, *Lib. I*, §. 41.

A PARIS,

DE L'IMPRIMERIE DE DIDOT L'AÎNÉ.

M. DCC. LXXXX.

Fig. 31. This book by Pascal François Joseph Gossellin of Lille (1751–1830) was the first scientific study of Eratosthenēs (29 cm, 180 pp., 8 tables, 10 maps; Paris, 1790). Subsequent investigations by Gossellin appeared under the title *Recherches sur la géographie systématique et positive des anciens* (4 vols., 29 cm, 54 maps; Paris, 1798–1813). [Courtesy of Harvard College Library.]

The dating by Olympiads was used moderately in Hellenistic times but seldom in Christian times. It was revived by Hadrian (emperor 117–138) in the year A.D. 131 (Ol. 227.3), when he dedicated the Olympieion in Athens; that year was sometimes called Ol. 1.1, which was very confusing if no explanation was added.

A list of Olympic victories compiled by the Christian chronologist Julios Africanos (III–1) was preserved by Eusebios (IV–1). It covers the period 776 B.C. to A.D. 277. The Olympic games were finally abolished in 393 by Theodosios the Great (emperor of the East 378–395).

Olympic datings were superseded by the Roman ones *ab urbe condita* and by consular dates. The foundation of Rome beginning the era was supposed to have taken place in the year corresponding to 753 B.C.[47]

Tables giving the concordance of the three chronologies (Ol., U.C., and B.C.) are available in the treatises on chronology and those on classical

[46] Christian years did not always begin on January 1. (Circumcision). The year might begin on March 1 or March 25 (Annunciation), on December 25 (Christmas), or worst of all on Easter, the date of which changes every year. The calendar style (*a navitate, ab incarnatione,* etc.) varied from time to time and from place to place; *Isis 40,* 230 (1949).

[47] There were many determinations of the Roman era, ranging from c. 870 to 729 B.C. The one generally accepted is the one suggested by Varro (I–2 B.C.), Ol. 6.3 = July 754 to July 753, and the anniversary of Rome's foundation was traditionally celebrated at the Palilia (the feast of Pales, god of shepherds), XI a. Kal. Maias = 21 April. Thus, according to tradition, Rome was born on 21 April 753. The date is as precise as it is arbitrary. F. K. Ginzel, *Handbuch der Chronologie* (Leipzig, 1911), Vol. 2, pp. 192–201.

philology.[48] The outstanding treatise on chronology is the one by Friedrich Karl Ginzel (1850–1926) in 3 volumes (Leipzig, 1906, 1911, 1914); vol. 1 is restricted to Asiatic and American chronologies.

The Olympic datings were not superseded by the Christian ones, because the latter were introduced only c. 525 by Dionysius Exiguus (VI–1) and were not used until long afterward; the Roman curia itself did not use them until the tenth century (*Introduction*, vol. 1, p. 429).

THE MARMOR PARIUM

A splendid example of chronological inscription dating from Eratosthenēs' time may be placed here. It is one of the most famous Greek inscriptions (*Corpus inscriptionum graecarum, No. 2374*). As it was found in Paros (the second in size of the Cyclades, just west of the largest, Naxos) it is generally called *Marmor Parium.*

Two large parts (*A* and *B*) of it are preserved on marble slabs 81 cm wide. *A* (92 lines) was bought in Smyrna by an agent of N. C. Fabri de Peiresc (1580–1637) but could not be delivered to him and was ceded to an agent of Thomas Howard, Earl of Arundel (1585–1646); it reached London in 1627 and the princeps edition was published by John Selden (1584–1654). This princeps (London, 1628) is itself a monument in the history of Greek scholarship. *B* (32 lines) was discovered in Paros only in 1897 and published soon afterward.

A is preserved in the Ashmolean Museum, Oxford, and *B* in the Museum of Paros.

The *Marmor Parium* is a chronology of Athens and Attica from the time of Cecrops, the legendary first king of Athens, to the archonship of Diognētos. After being translated into our era, it covers the period from 1582/1 to 264/3 B.C. It is centered upon Athenian history but registers treaties with Priēnē and Magnēsia, and so on.

The data are derived from an Atthis (or Athenian chronicle), from Ephoros of Cymē (IV–2 B.C.), from a book on inventions (*peri eurēmatōn*), and from other sources.

The text of *A* was included in the *Fragmenta historicorum graecorum*, vol. 1, pp. 533–590 (1841). The best edition of the whole is by Felix Jacoby (228 pp.; Berlin, 1904).

The history of geography is continued in Chapter XXIII.

[48] For the sake of illustration, a few events are dated below with reference to the Olympic, Roman, and Christian chronologies.

	Ol.	U.C.	B.C.
Olympic era	1.1	. . .	776
Foundation of Rome	6.4	1	753
Alexander the Great, d.	114.2	431	323
Ptolemaios II Philadelphos, d.	133.3	508	246
Archimēdēs, d.	142.1	542	212
Cato the Censor, d.	157.4	605	149
Lucretius, d.	181.2	699	55
Cicero, d.	184.2	711	43
Virgil, d.	190.2	735	19
	194.4	753	1
	195.1	754	A.D. 1

VII

PHYSICS AND TECHNOLOGY IN THE THIRD CENTURY

The history of physics is restricted to Euclid and Archimēdēs and it will not be difficult to tell it. The history of technology is more complex and also more elusive but enough will be said to give the reader some idea of the achievements and of the technical possibilities of that time. New techniques are seldom described by their inventors and frequently remain undescribed. Written descriptions and references are generally late and unconcerned with chronology. In the majority of cases, techniques can be understood and appreciated only on the basis of material objects or monuments which are seldom datable with any precision (say within a century).

As the subject cannot be covered even in brief, we must be satisfied with a few examples, and therefore it may be useful to compensate our silence with a short bibliography.

It is always useful to refer to the old books of Hugo Blümner (1844–1919), *Technologie und Terminologie der Gewerbe und Künste bei Griechen und Römern* (4 vols.; Leipzig, 1875–1887). A new, revised edition was begun but was stopped by the first World War; only vol. 1 was published, in 1912. Blümner's books deal with so large a number of topics that we cannot even enumerate them. Think of all the technical problems which men had to solve not only for industrial purposes but also for the simple needs of life.

Albert Neuburger (1867–1955), *The technical arts and sciences of the ancients* (550 pp., London, 1930), first published in German (Leipzig, 1919; again, 1921).

A number of books deal with engineering and building.

Curt Merckel, *Die Ingenieurtechnik im Alterthum* (quarto, 678 pp., 261 ill., 1 map; Berlin, 1899).

Tenney Frank (1876–1939), *Roman buildings of the Republic. An attempt to date them from their materials* (Papers and monographs from the American Academy in Rome, vol. 3, 150 pp., Rome, 1924).

Thomas Ashby (1874–1931), *The aqueducts of ancient Rome* (358 pp., 24 pl., 34 fig., 7 maps; Oxford, 1935).

Esther Boise Van Deman (1862–1937), *The building of the Roman aqueducts* (quarto, 452 pp., 60 pl., 49 fig.; Washington, 1934) [*Isis 23*, 470–471 (1935)].

Marion Elizabeth Blake, *Ancient Roman construction in Italy from the prehistoric period. A chronological study based in part upon the material accumulated by the late E. B. Van Deman* (quarto, 442 pp., 57 pl.; Washington: Carnegie Institution, 1947) [*Isis 40*, 279 (1949)].

For metallurgy, see Robert James Forbes, *Metallurgy in antiquity. A note book for archaeologists and technologists* (489 pp., 98 ill.; Leiden: Brill, 1950) [*Isis 43*, 283–285 (1952)]. Forbes's book is more concerned with high antiquity, chiefly in the Near East, and the data relative to the Hellenistic age are relatively few. See also Forbes's *Bibliographia antiqua. Philosophia naturalis* (parts I to X, Nederlandsche Instituut voor der Nabije Oosten, 1940–1950) [*Isis 36*, 208 (1946)]. In spite of its subtitle, *Philosophia naturalis*, this bibliography deals almost exclusively with technology.

Much information could be derived also from books on the history of technology in general, particularly the great *History of Technology* edited by Charles Singer and others (Oxford: Clarendon Press, 1954–). A list of these general works will be found in *Horus: A guide to the history of science* (Waltham, Mass.: Chronica Botanica, 1952), pp. 167–168.

EUCLID

Euclid is famous as a mathematician, the author of the *Elements*, but he was also a physicist, the founder of geometric optics, and treatises on music and mechanics were ascribed to him.

Of the two musical treatises one, *Harmonic introduction* (*eisagōgē harmonicē*) was more probably written by one Cleoneidēs.[1] The second, *Canonic section* (*catatomē canonos*), may be genuine.[2] Both are extant. The *Section* explains the Pythagorean theory of music. According to Proclos, Euclid had written *Elements of Music* (*hai cata musicēn stoicheiōseis*), and the *Section* is probably derived from those *Elements*.

The mechanical treatise ascribed to Euclid by the Arabs is certainly apocryphal.[3]

Euclid is said to have written two treatises on optics, the *Optica* and the *Catoptrica*. The first is genuine, the second probably apocryphal. We have the text of the *Optica*, and we have also a recension of both treatises by Theōn of Alexandria (IV–2). The *Optica* begins with definitions, or rather assumptions, derived from the Pythagorean theory that the rays of light are straight lines and proceed from the eye to the object perceived (and not in the opposite direction).[4] Euclid then explains problems of perspective. The

[1] This *Harmonic introduction* is one of the main sources for the study of the theories of Aristoxenos of Tarentum (IV–2 B.C.) in spite of its lateness. Cleoneidēs flourished at the beginning of the second century after Christ. A Latin translation, *Harmonicum introductorium*, was published by Bevilaqua (Venice, 1497) and reprinted in Giorgio Valla's *Collectio* (Venice, 1498), Klebs, Nos. 281, 1012. Greek-Latin edition by Jean Pena (Paris, 1557). French translation with commentary by Charles Emile Ruelle (Paris, 1884). Greek edition by Karl von Jan, *Musici scriptores Graeci* (Leipzig, 1895), pp. 179–207.

[2] Edited by Karl von Jan, *Musici scriptores Graeci* (1895), pp. 115–166. Both texts, the *Introduction* and the *Section*, are edited in Greek and Latin by Heinrich Menge in the *Euclidis opera omnia* (Leipzig, 1916), vol. 8, pp. 157–223.

[3] T. L. Heath, *A History of Greek mathematics* (Oxford, 1921), Vol. 1, pp. 445–446; *Introduction*, vol. 1, p. 156.

[4] This conception was bizarre, because it implied that the rays issued from the eye

Catoptrica deals with mirrors and sets forth the law of reflection. It is an elegant chapter of mathematical physics which remained almost alone of its kind for a very long period of time, but does it date from the third century B.C., or is it later, much later? The interval between Euclid and Theōn, we should remember, is very large (more than six and a half centuries).

J. L. Heiberg, *Euclidis Optica, Opticorum recensio Theonis, Catoptrica cum scholiis antiquis*, vol. 7 of *Euclidis opera omnia* (417 pp.; Leipzig, 1895). French translation of those three texts by Paul Ver Eecke, *L'optique et la catoptrique* (174 pp.; Bruges, 1938) [*Isis 30*, 520–521 (1939)]. English version of the original text of the *Optics* by Harry Edwin Burton, *Journal of the Optical Society of America* 35, 357–372 (1945).

ARCHIMĒDĒS

The mechanical treatises of Archimēdēs have been discussed in Chapter V because they illustrate his mathematical genius. He was the creator of statics and hydrostatics and we may say of mathematical physics. As we explained above, he impressed his contemporaries and the bulk of posterity not so much by mathematical creations, nor even by his physico-mathematical ones, but rather by his practical inventions. For almost two thousand years he was considered to be the archetype of the inventor and of the mechanical wizard.

ENGINEERING AND PUBLIC WORKS IN THE GREEK EAST. THE GREAT SHIPS

The outstanding building of the third century was the lighthouse built by Sōstratos of Cnidos in the harbor of Alexandria, c. 270.[5] It was built during the rule of the second king of the Lagid dynasty, Ptolemaios II Philadelphos (king 285–247). Another engineering feat illustrated his rule: the digging of a canal connecting the Mediterranean with the Red Sea. This was a very ancient undertaking, begun in the Middle Kingdom (2160–1788) and continued by Necho (king 609–593), and by Darios the Great (who was king of Persia and Egypt, 521–486).[6] It was Ptolemaios II's glory to complete the canal as much as that could be done in his time.

He also built roads, notably the one leading from Coptos (Qift) on the Nile (lat. 26° N) to Berenicē,[7] a harbor of the Red Sea. It was there that the distance between the Nile and the Red Sea, across the eastern desert, was the shortest. The road was of great importance for the trade of Egypt with Arabia and India. Berenicē was for four or five centuries the main *entrepôt* of that trade on the Western coast of the Red Sea. Its importance

---

hunted for the object and could not see it before having found it.

[5] For the description of the Pharos, see Chapter I.

[6] For the early history of the canal, see Volume 1, p. 182.

[7] Named after Berenicē, queen of Ptolemaios I Sōtēr and mother of Ptolemaios II Philadelphos.

was increased by the discovery and exploitation of gold and emerald mines in the neighborhood.

Ptolemaios II's grandson, Ptolemaios IV Philopatōr (222–205) has often been praised because of his possession of great ships, the most famous ships of antiquity. Elaborate descriptions of three of them have been preserved in the *Deipnosophistai* of Athēnaios of Naucratis (III–1).[8] Those descriptions are so interesting that it is worth while to reproduce Gulick's translation of them almost completely. The description of the first ship was borrowed by Athēnaios from a book on Alexandria written about the end of the third century B.C. by Callixeinos of Rhodos.

"Philopatōr constructed his forty-bank ship with a length of four hundred and twenty feet;[9] its beam from gangway to gangway[10] was fifty-seven feet; its height to the gunwale was seventy-two feet. From the top of the stern-post to the water-line it measured seventy-nine and a half feet. It had four steering-oars, forty-five feet long, and the oars of the topmost rowers, which are the longest, measured fifty-seven feet; these, since they carried lead on the handles and were very heavy inboard, were yet easy to handle in actual use because of their nice balance. It had a double bow and a double stern, and carried seven rams; one of these was the leader, others were of gradually diminishing size, some being mounted at the catheads.[11] It carried twelve under-girders, each of them measuring nine hundred feet.[12] It was extraordinarily well proportioned. Wonderful also was the adornment of the vessel besides; for it had figures at stern and bow not less than eighteen feet high, and every available space was elaborately covered with encaustic painting; the entire surface where the oars projected, down to the keel, had a pattern of ivy-leaves and Bacchic wands. Rich also was the equipment in armament, and it satisfied all the requirements of the various parts of the ship. On a trial voyage it took more than four thousand men to man the oars, and four hundred substitutes; to man the deck there were two thousand eight hundred and fifty marines; and besides, below decks was another complement of men and provisions in no small quantity.[13] At the beginning it was launched from a kind of cradle which, they say, was put together from the timbers of fifty five-bank ships, and it was pulled into the water by a crowd, to the accompaniment of shouts and trumpets. Later, however, a Phoenician conceived the method of launching by digging a trench under the ship near the harbour, equal in length to the ship. He constructed for this trench

[8] In Book v, 203–209; Athenaeus, *Deipnosophists*, ed. Charles Burton Gulick (Loeb Classical Library; Cambridge, 1928), Vol. 2, pp. 421–447.

[9] The Athenian trireme had a length at the water line of not over 120 feet (Gulick).

[10] There was a gangway (*parados*) running from bow to stern on each side.

[11] The rams or battering rams were beaks projecting from the prow, either above or below the water line, for cutting down an enemy's ship. The cathead is a projecting piece of timber near the bow to which the anchor is hoisted (Webster).

[12] Since the ship was 420 feet long and these cables were 900 feet, this passage would seem to prove decisively that the girders ran outside the ship from bow to stern and back (Gulick).

[13] These numbers of sailors and marines (4000, 400, 2850 and more) are incredible. There must be errors in the text, and yet the numbers are designated with the correct Greek words.

foundations of solid stone seven and a half feet in depth, and from one end of these foundations to the other he fixed in a row skids,[14] which ran transversely to the stones across the width of the trench, leaving a space below them six feet deep. And having dug a sluice from the sea, he let the sea into all the excavated space, filling it full; into this space he easily brought the vessel, with the help of unskilled men; . . . when they had barred the entrance which had been opened at the beginning, they again pumped out the seawater with engines. And when this had been done, the ship rested securely on the skids before-mentioned."

Athēnaios does not indicate his source for the second ship, but it must have been an eye-witness or a person who obtained measurements and other details from a contemporary.

"Philopatōr also constructed a river boat, the so-called 'cabin-carrier,'[15] having a length of three hundred feet, and a beam at the broadest part of forty-five feet. The height, including the pavilion when it was raised, was little short of sixty feet. Its shape was neither like that of the war galleys nor like that of the round-bottomed merchantmen, but had been altered somewhat in draught to suit its use on the river. For below the waterline it was flat and broad, but in its bulk it rose high in the air; and the top parts of its sides, especially near the bow, extended in a considerable overhang, with a backward curve very graceful in appearance. It had a double bow and a double stern which projected upward to a high point, because the waves in the river often rise very high. The hold amidships was constructed with saloons for dining-parties, with berths, and with all the other conveniences of living. Round the ship, on three sides, ran double promenades.[16] The perimeter of one of these measured not less than five furlongs. The structure of the one below decks resembled a peristyle; that of the one on the upper deck was like a concealed peristyle built up all round with walls and windows. As one first came on board at the stern, there was set a vestibule open in front, but having a row of columns on the sides; in the part which faced the bow was built a fore-gate, constructed of ivory and the most expensive wood. Entering this, one came upon a kind of proscenium, which in its construction had been roofed over. Matching the fore-gate, again, a second vestibule lay aft at the transverse side,[17] and a portal with four doors led into it. On both sides, left and right, portholes were set beneath to provide good ventilation. Connected with these entrances was the largest cabin; it had a single row of columns all round, and could hold twenty couches. The most of it was made of split cedar and Milesian cypress; the surrounding doors, numbering twenty, had panels of fragrant cedar nicely glued together, with ornamentation in ivory. The decorative studs covering their surface, and the handles as well, were made of red copper, which had been gilded in the fire. As for the columns, their shafts were of cypress-wood, while the capitals, of the Corinthian order, were entirely covered with ivory and gold. The whole entablature was in gold; over it was affixed a frieze with striking figures in ivory, more than a foot and a half tall, mediocre in workmanship, to be sure, but remarkable in their lavish display. Over the dining-saloon was a beautiful coffered ceiling of cypress wood; the ornamentations on it were sculptured, with a surface of

[14] Or "rollers" (Gulick).

[15] *Thalamēgos*. It is really the barge of state.

[16] On upper and lower deck (Gulick).

[17] That is the quarter-deck, the "side" connecting the two lateral decks (Gulick).

gilt. Next to this dining-saloon was a sleeping apartment with seven berths, adjoining which was a narrow passageway running transversely from one side of the hold to the other, and dividing off the women's quarters. In the latter was a dining-saloon, with nine couches, which was similar to the large saloon in magnificence, and a sleeping-apartment with five berths.

"Now the arrangements up to the first deck were as described. Ascending the companion-way, which adjoined the sleeping-apartment last mentioned, was another cabin large enough for five couches, having a ceiling with lozenge-shaped panels; near it was a rotunda-shaped shrine of Aphroditē, in which was a marble statue of the goddess. Opposite to this was a sumptuous dining-saloon surrounded by a row of columns, which were built of marble from India. Beside this dining-saloon were sleeping-rooms having arrangements which corresponded to those mentioned before. As one proceeded toward the bow he came upon a chamber devoted to Dionysos, large enough for thirteen couches, and surrounded by a row of columns; it had a cornice which was gilded as far as the architrave surrounding the room; the ceiling was appropriate to the spirit of the god. In this chamber, on the starboard side, a recess was built; externally, it showed a stone fabric artistically made of real jewels [18] and gold; enshrined in it were portrait-statues of the royal family in Parian marble. Very delightful, too, was another dining-saloon built on the roof of the largest cabin in the manner of an awning; this had no roof, but curtain rods shaped like bows extended over it for a certain distance, and on these, when the ship was under way, purple curtains were spread out. Next after this was an open deck [19] which occupied the space directly over the vestibule extending below it; a circular companion-way extending from this deck led to the covered promenade and the dining-saloon with nine couches. This was Egyptian in the style of its construction; for the columns built at this point bulged as they ascended, and the drums differed, one being black and another white, placed alternately. Some of their capitals are circular in shape, the entire figure described by them resembles rose-blossoms slightly opened. But around the part which is called the 'basket' [20] there are no volutes or rough leaves [21] laid on, as on Greek capitals, but calyxes of water-lilies and the fruit of freshly-budded date-palms; in some instances several other kinds of flowers are sculptured thereon. The part below the root of the capital, which, of course, rests upon the drum adjoining it, had a motif that was similar; it was composed of flowers and leaves of Egyptian beans, as it were, intertwined. This is the way in which Egyptians construct their columns; and the walls, too, they vary with alternating white and black courses of stone, but sometimes, also, they build them of the rock called alabaster. And there were many other rooms in the hollow of the ship's hold through its entire extent. Its mast had a height of one hundred and five feet, with a sail of fine linen reinforced by a purple topsail."

The third ship was not built by Ptolemaios IV but by his old contemporary, Hierōn, king of Syracuse (270–216) with the technical coöperation of no less a person than Archimēdēs (murdered in 212). Athēnaios' description of it was taken from that of Moschiōn, probably one of Hierōn's contemporaries.

[18] Or "agates"; different readings of the Greek manuscripts.

[19] A kind of "atrium" (the principal room of a Roman house, open to the sky).

[20] *Calathos,* meaning corbel, the part of a Corinthian column that spreads out between shaft and entablature.

[21] Rough leaves (*phylla trachea*), the acanthus leaves of a Corinthian capital.

"Hierōn, the king of Syracuse, he who was in all respects friendly to Rome, not only interested himself in the building of temples and gymnasia, but was also a zealous shipbuilder, constructing wheat-transports, the construction of one of which I will proceed to describe. For material he caused timber to be brought from Aetna,[22] enough in quantity for the building of sixty quadriremes. In keeping with this, he caused to be prepared dowels, belly-timbers, stanchions, and all the material for general uses, partly from Italy, partly from Sicily; for cables hemp from Iberia, hemp and pitch from the river Rhone, and all other things needful from many places. He also got together shipwrights and all other kinds of artisans, and from them all he placed in charge the Corinthian Archias as architect, urging him to attack the construction zealously; and he personally applied himself diligently to the work during the days it required. One half, then, of the entire ship he finished in six months . . . and as each part of the ship was completed it was overlaid with tiling made of lead; for there were about three hundred artisans working on the materials, not including their assistants. This part of the ship, then, was ordered to be launched in the sea, that it might receive the finishing touches there. But after considerable discussion in regard to the method of pulling it into the water, Archimēdēs the mechanician alone was able to launch it with the help of a few persons. For by the construction of a windlass he was able to launch a ship of so great proportions in the water. Archimēdēs was the first to invent the construction of the windlass. The remaining parts of the ship were completed in another period of six months; it was entirely secured with bronze rivets, most of which weighed ten pounds, while the rest were half as large again; these were fitted in place by means of augers, and held the stanchions together; fixed to the timbers was a sheath of leaden tiles, under which were strips of linen canvas covered with pitch. When, then, he had completed the outside surface, he proceeded to make complete the inner arrangements.

"Now the ship was constructed to hold twenty banks of rowers, with three gangways. The lowest gangway which it contained led to the cargo, the descent to which was afforded by companion-ways of solid construction; the second was designed for the use of those who wished to enter the cabins; after this came the third and last, which was for men posted under arms. Belonging to the middle gangway were cabins for men ranged on each side of the ship, large enough for four couches, and numbering thirty. The officers' cabin could hold fifteen couches and contained three apartments of the size of three couches; that toward the stern was the cooks' galley. All these rooms had a tessellated flooring made of a variety of stones, in the pattern of which was wonderfully wrought the entire story of the *Iliad*; also in the furniture, the ceiling, and the doors all these themes were artfully represented. On the level of the uppermost gangway there were a gymnasium and promenades built on a scale proportionate to the size of the ship; in these were garden-beds of every sort, luxuriant with plants of marvellous growth, and watered by lead tiles hidden from sight; then there were bowers of white ivy and grape-vines, the roots of which got their nourishment in casks filled with earth, and receiving the same irrigation as the garden-beds. These bowers shaded the promenades. Built next to these was a shrine of Aphroditē large enough to contain three couches, with a floor made of agate and other stones, the most beautiful kinds found in the island; it had walls and ceiling of cypress-wood, and doors of ivory and fragrant cedar; it was also most lavishly furnished with paintings and statues

[22] Aetna (*Aitnē*), the famous volcano, north of Syracuse, in the northeast of Sicily.

and drinking-vessels of every shape.

"Adjoining the Aphroditē room was a library[23] large enough for five couches,[24] the walls and doors of which were made of boxwood; it contained a collection of books, and on the ceiling was a concave dial made in imitation of the sun-dial on Achradinē.[25] There was also a bathroom, of three-couch size, with three bronze tubs and a wash-stand of variegated Tauromenian marble, having a capacity of fifty gallons. There were also several rooms built for the marines and those who manned the pumps. But beside these there were ten stalls for horses on each side of the ship; and next them was the storage-place for the horses' food, and the belongings of the riders and their slaves. There was also a water-tank at the bow, which was kept covered and had a capacity of twenty thousand gallons; it was constructed of planks, caulked with pitch and covered with tarpaulins. By its side was built a fish-tank enclosed with lead and planks; this was filled with sea-water, and many fish were kept in it. On both sides of the ship were projecting beams, at proper intervals apart; on these were constructed receptacles for wood, ovens, kitchens, handmills, and several other utensils. Outside, a row of colossi, nine feet high, ran round the ship; these supported the upper weight and the triglyph, all standing at proper intervals apart. And the whole ship was adorned with appropriate paintings. There were also eight turrets on it, of a size proportional to the weight of the ship; two at the stern, an equal number at the bow, and the rest amidships. To each of these two cranes were made fast, and over them port-holes were built, through which stones could be hurled at an enemy sailing underneath. Upon each of the turrets were mounted four sturdy men in full armour, and two archers. The whole interior of the turrets was full of stones and missiles. A wall with battlements and decks athwart the ship was built on supports; on this stood a stone-hurler, which could shoot by its own power a stone weighing one hundred and eighty pounds or a javelin eighteen feet long. This engine was constructed by Archimēdēs. Either one of these missiles could be hurled six hundred feet. After this came leather curtains joined together, suspended to thick beams by means of bronze chains. The ship carried three masts, from each of which two stone-hurling cranes were suspended; from them grappling hooks and lumps of lead could also be directed against assailants. An iron paling which encircled the ship also protected it against any who attempted to climb aboard; also grappling-cranes of iron were all about the ship, which, operated by machinery, could lay hold of the enemy's hulls and bring them alongside where they would be exposed to blows. Sixty sturdy men in full armour mounted guard on each side of the ship, and a number equal to these manned the masts and stone-hurlers. Also at the masts, on the mastheads (which were of bronze), men were posted, three on the foremast, two in the maintop and one on the mizzenmast; these were kept supplied by the slaves with stones and missiles carried aloft in wicker baskets to the crow's nests by means of pulleys.[26] There were four anchors of wood, eight of iron. The trees for the mainmast and mizzenmast

---

[23] The Greek word is *scholastērion*, a study, a place where one can study or be at leisure.

[24] A *clinē* was used as chair, couch, or bed. The word *pentaclinos* may refer to five couches or to the space needed by them; compare the Japanese use of standard straw mats, *tatami* (6 × 3 feet), to measure the area of a room. As a unit of size the mat is called *jō*; for example, a room of six *jō* is called *rokujō*, one of eight, *hachijō*.

[25] Achradinē was the "outer city" of Syracuse at the eastern end, overlooking the sea.

[26] It may surprise the reader that this

were easily found; but that for the foremast was discovered with difficulty by a swineherd in the mountains of the Bruttii;[27] it was hauled down to the coast by the engineer Phileas of Tauromenion.[28] The bilge-water, even when it became very deep, could easily be pumped out by one man with the aid of the screw, an invention of Archimēdēs. The ship was named 'Syracusia'; but when Hierōn sent her forth, he changed the name to 'Alexandris.' The boats which it had in tow were first a pinnace of three thousand talents burden; this was propelled entirely by oars. After this came fishing-boats of fifteen hundred talents burden, and several cutters besides. The numbers composing the crew were not less than . . . .[29] Next to these just mentioned there were six hundred more men at the bow ready to carry out orders. For any crimes committed on board there was a court composed of the skipper, pilot, and officer at the bow, who gave judgement in accordance with the laws of Syracuse.

"On board were loaded ninety thousand bushels of grain, ten thousand jars of Sicilian salt-fish, six hundred tons of wool, and other freight amounting to six hundred tons. Quite apart from this was the provisioning of the crew. But when Hierōn began to get reports of all the harbours, either that they could not receive his ship at all, or that great danger to the ship was involved, he determined to send it as a present to King Ptolemy of Alexandria; for there was in fact a scarcity of grain throughout Egypt. And so he did; and the ship was brought to Alexandria, where it was pulled up on shore. Hierōn also honoured Archimēlos, the poet who had written an epigram celebrating the vessel, with fifteen hundred bushels of wheat, which he shipped at his own expense to Peiraieus."

These descriptions have been reproduced verbatim in spite of the fact that much in them will seem irrelevant to the historian of technology. Such irrelevancies are typical of the age. A shipowner of Hellenistic times was not at all like a Yankee shipowner of the last century.

The references to Archimēdēs are very plausible. He was a mechanical engineer in Hierōn's service just as Leonardo da Vinci was in the service of Lodovico il Moro.

What will astonish our readers most in these descriptions is their silence concerning navigation; for example, no mention is made of the speed that those ships could attain and their dirigibility. It is probable that the three ships described by Athēnaios were more suitable for navigation on the Nile than on the Mediterranean. We know very little about the ships that trans-

---

"wheat-transport" (*ploion sitēgon*) was so heavily armed. This was a necessity because of piracy, which was from prehistoric times an endemic scourge of the Mediterranean Sea. Ships were molested not only by regular pirates but also by privateers working for one nation against others. Pompey rendered his greatest service to Rome in 67 when he defeated and destroyed the associated pirates of the eastern Mediterranean, but by and by the pirates reappeared and were checked only when Augustus established regular patrol fleets to keep them down. That Mediterranean peace was kept as long as Rome was strong enough to keep it, less than three centuries. Henry Arderne Ormerod, *Piracy in the ancient world* (286 pp.; Liverpool, 1924).

[27] Bruttii, the people of Brettia, or Bruttium, the southwest end of Italy, facing Sicily.

[28] Tauromenion, a flourishing city in east Sicily; the harbor of the Aetna region.

[29] A numeral has been lost (Gulick).

ported Egyptian grain from Alexandria to Rome, and yet they were the mainstay of Roman economic life.

The few items known to me for the study of navigation in the Mediterranean are relative to a somewhat later period but may be useful because the art of navigation was very much the same in the few centuries before Christ and after. For St. Paul's navigation see James Smith, *The voyage and shipwreck of St. Paul* (London, 1848; 3rd ed., 1866).

Lucian of Samosata (c. 120–after 180) described in his *Navigium* (*ploion*) one of the great Roman grain ships, the Isis. See Lionel Casson, "The Isis and her voyage," *Transactions and Proceedings of the American Philological Association 81*, 43–56 (1950) [*Isis 43*, 130 (1952)]; "Speed under sail of ancient ships," *ibid.* 82, 136–148 (1951). Casson concludes, "Before a favorable wind a fleet could log between two and three knots. With unfavorable winds a fleet usually could do no better than 1 to 1½ knots." [30]

Navigation in the Mediterranean Sea could be difficult, as St. Paul discovered a long time ago. As late as 1569, when shipbuilding and navigation were much improved, Venetian ships were still forbidden by law to sail back from the Near East between 15 November and 20 January.

Auguste Jal, *Archéologie navale* (Paris, 1840), vol. 2, p. 262. Lefebvre des Noëttes, *De la marine antique à la marine moderne* (Paris, 1935); *Introduction*, vol. 3, p. 157.

The Seleucid rulers of Syria emulated the Ptolemies of Egypt, and some great achievements were ascribed to them. The founder of that dynasty, Seleucos Nicatōr (312–280), was the founder also (c. 300) of the city and fortress of Seleuceia Pieria, at the seashore some four miles west of Antioch. It had been strengthened by all the technical means then available. Antiocheia (Antioch) was built by the same king and by his son Antiochos Sōtēr (280–261) and equipped with a water supply, which was gradually developed and improved during the two following centuries.

ENGINEERING AND PUBLIC WORKS IN THE ROMAN WEST

Public works of various kinds were carried out in Rome and in various provinces. For example, the earliest aqueduct, the Aqua Appia, was built in 312, and the second, the Anio vetus, in 272. The Romans were not by any means the first people to build acqueducts, but they did it very well. They applied to their construction methods invented for the building of underground sewers.

The earliest aqueduct was built by Appius Claudius, surnamed Caecus (because he became blind in his maturity); when he was appointed censor in 312 he created the aqueduct, and also the most famous of Roman roads,

[30] See also Casson's paper on "The grain trade in the Hellenistic world," *Trans. Am. Philol. Assoc.* 85, 168–187 (1954).

the Via Appia, which ran from Rome southward to Capua (and later to Brindisi).

The Aqua Appia was about 11 miles long, mostly underground, and its workmanship was crude.

Appius Claudius Caecus is the first Latin writer (in prose or verse) whose name has been transmitted to us. It is highly interesting that this first writer was also the first builder of an aqueduct and of a famous road.

During the forty years following construction of the Aqua Appia, the city of Rome increased materially in size and needed a more abundant water supply. Therefore Manius Curius Dentatus, when he became censor in 272, ordered the construction of a new and larger aqueduct, which was completed three years later. For a long time the aqueduct was called the Anio because it brought water from the upper Anio; the river Anio (Teverone) is a tributary of the Tiberis (Tevere). Later, it was necessary to call this aqueduct the *Anio vetus* to distinguish it from the *Anio novus* which was built by the emperor Claudius in A.D. 52.

The old Anio aqueduct began some 20 miles from Rome but made so many circuits that it extended to about 43 miles; much of it was underground; it was carried across some of the streams in its path on low bridges. The magnificent bridge, Ponte S. Gregorio, crossing a wide valley, the Mola di S. Gregorio, is not a part of the old aqueduct, however, but was built four centuries later under Hadrian (emperor 117–138) in order to shorten the old circuit. There are many remains of the old aqueducts in the Campagna but each of them has been restored many times and it is difficult to imagine their original state.

Such as it was, the construction of the Anio vetus was a splendid achievement, which was not repeated in the third century. Nine other aqueducts were built in the period 144 B.C. to A.D. 226; five of them will be briefly described in Chapter XX.

M. Curius Dentatus was one of the favorite heroes of the Romans; he was often praised as a symbol of ancient simplicity, frugality and disinterestedness.

The main harbor built by the Romans in the third century was probably that of Tarraco.[31] Tarraco was an old colony of Marseilles; taken by the Romans in 218 at the beginning of the second Punic War, it was used as headquarters by the two brothers Scipio, who built a fortress and a remarkable harbor. Their purpose was mainly to establish a maritime base against the Carthaginians, but the place was well chosen and Tarraco became a flourishing city. Augustus established his winter quarters there in 26 B.C. during his campaign against the Cantabri and made of it the capital of Hispania Tarraconensis.[32]

[31] Modern Tarragona, 54 miles west-southwest of Barcelona.

[32] The Hispanic peninsula was divided by Augustus into three provinces: (1)

More cities, fortresses, and harbors were built in the Mediterranean world, but there were no technological novelties. Those creations were more significant from the administrative than from the technical point of view. They illustrated the growth of Roman power and of Roman order.

The history of physics and technology is continued in Chapter XX.

Lusitania, roughly Portugal; (2) Baetica, roughly Andalusia; (3) Tarraconensis, the whole northeast, by far the largest section, twice as large as the two others put together.

VIII

ANATOMY IN THE THIRD CENTURY

The astronomical and mathematical activities have taken us sometimes away from the Museum, but anatomy brings us back to it, and all considered it is the anatomical investigations that gave to the Museum its full glory. Our knowledge of them is largely derived from Galen (II–2), who is a late witness yet had the possibility of collecting valuable evidence, not only in Alexandria, but in many other cities where anatomical traditions could still be traced back to ancient times.

The early school of Alexandria, the one that flourished under the first two Ptolemaioi (III–1 B.C.), made possible for the first time a complete survey of the structure of the human body. Anatomical investigations had been made before by Hippocratēs, his disciples, and other physicians, but never with so much continuity and so good a method. It was a period of exceptional freedom from religious prejudices, and anatomists were permitted to dissect as much as they pleased. The work done at the Museum was controlled by the kings only, and was almost unknown to the populace; hence, freedom of research was complete. Those excellent opportunities were improved by two men of genius, and the result was a golden age of anatomy. This will be better appreciated if we bear in mind that there were only two other ages comparable to it: the age of Galen (II–2), which was a rebirth, and the age of Vesalius and his successors (XVIth centry). The Alexandrian age was less a renaissance than a real beginning of systematic anatomy on a grand scale; the Vesalian renaissance was the introduction to modern anatomy.

Let us first consider the two protagonists.

HĒROPHILOS OF CHALCĒDŌN[1]

Hērophilos was born at Chalcēdōn at the end of the fourth century and was one of the men of science attracted to Alexandria by Ptolemaios Sōtēr at the beginning of the following century. He is thus one of the founders of

[1] Chalcēdōn or better, but less usual, Calchēdōn, in Bithynia, at the entrance of the Bosporos, nearly opposite to Byzantion. It was an old Greek (Megarian) colony, founded in 685 B.C. Modern Kadiköy.

the Greco-Egyptian renaissance as well as the founder of systematic anatomy. The number of his discoveries is so large and their field so comprehensive that one cannot help concluding that he undertook an elaborate survey of the whole fabric of the human body. It is clear that if a competent investigator was given a sufficient number of corpses with the freedom of dissecting them as he found necessary, he was bound to discover many things. Hērophilos and his younger assistant and successor, Erasistratos, had the same privileges as explorers who are the first to penetrate a new country.

We know little about Hērophilos' life before he answered Ptolemaios' call, except that he was a disciple of Praxagoras of Cōs, who was probably a younger contemporary of Dioclēs of Carystos (c. 340–260).[2]

According to Galen, Hērophilos was the first to undertake human dissections; we can hardly accept that statement, unless it be qualified; it may be that Galen meant public dissections (a very small public, of course) or systematic dissections carried out with assistants and students. As he was a forerunner, he had to invent the technique of dissection, and whenever a new organ was discovered, he was obliged to find a name for it. The majority of those new names have come down to us through Galen, and the Galenic writings are thus the vehicle of their first literary appearance. Hērophilōs wrote a treatise in three books on anatomy, a smaller one on eyes, and a manual for midwives (*maiōticon*).

Examples of Hērophilos' discoveries are: detailed description of the brain, distinction of cerebrum and cerebellum, meninges, calamus scriptorious (*anaglyphos calamos*), torcular (*lēnos*) Herophili; distinction between tendons and nerves (the name he gave the latter, *neura aisthētica*, implies recognition of one of their functions, sensibility; description of the

[2] That Praxagoras was Hērophilos' teacher is told by Galen; see K. G. Kühn, *Galeni opera omnia* (Leipzig, 1821–1833), Vol. 7, p. 585. The dating of Dioclēs of Carystos is the one given by Werner Jaeger (Volume 1, p. 562). Dioclēs must be placed later than I did (*Introduction*, vol. 1, p. 121) in order to make it possible for him to have been influenced by Aristotle; on the other hand, this gives very little time for three generations of students, Dioclēs, Praxagoras, Hērophilos, or four if we add Erasistratos. Our first tendency is to consider the sequence teacher-student as if it were the same as the sequence father-son, but this is not always true. Teachers are generally older than their students, but they need not be much older. In a letter which my colleague, Werner Jaeger, wrote to me (Cambridge, Mass., 4 May 1952), he states that Dioclēs' book *On diet* was written after 300. Praxagoras, Hērophilos, and Erasistratos flourished soon after, say in III–1 B.C. There is no reason to believe, says Jaeger, that Praxagoras was the disciple of Dioclēs; they were contemporaries.

To sum up, let us put it this way. The Lyceum was founded in 335. If Dioclēs flourished at the turn of the fourth century, he had plenty of time to be influenced by Aristotle. Praxagoras, Hērophilos, Erasistratos flourished in III–1 B.C.; they were contemporaries, each of them a little older than the following. Thus Praxagoras and Hērophilos were contemporaries, and so were Hērophilos and Erasistratos, but Erasistratos may have been born after the death of Praxagoras or not long before. Compare with the following situation: Aischylos, Sophoclēs, and Euripidēs were contemporaries, and so were Sophoclēs, Euripidēs, and Aristophanēs, but Aischylos and Aristophanēs were not.

optic nerves and of the eye, including the retina (his word for it was *amphiblēstroeidēs*, meaning "like a net"; the Latin-English word reproduces the same metaphor); description of the vascular system much improved; duodenum (*dōdecadactylos*, twelve fingers), a part of the small intestine close to the stomach, so called because its length is about twelve fingers' breadth; description of the liver, salivary glands, pancreas, prostate,[3] genital organs, observation of the lacteals. He made a clear distinction between arteries and veins; the arteries are six times as thick as the veins; they contain blood, not air, and are empty and flattened after death. He called the pulmonary artery the arterial vein, and the pulmonary vein the venal artery, names that remained in use until the seventeenth century.

Four forces control the organism, the nourishing, heating, perceiving, and thinking forces, seated respectively in the liver, heart, nerves, and brain. One of Aristotle's worst errors had been his localization of intelligence in the heart instead of the brain. Hērophilos rejected that error and revived the older views of Alcmaiōn (VI–B.C.), according to which the brain is the seat of intelligence.

He was eminent as a teacher as well as an investigator and founded a school which continued with decreasing vitality until the end of the Ptolemaic age.

ERASISTRATOS OF IULIS

Erasistratos was a younger contemporary of Hērophilos; it is possible that he began his career as the latter's assistant. He was born c. 304 in Iulis;[4] thus he was not an Asiatic Greek like Hērophilos but a Greek of Greece. It was therefore very natural for him to receive his education in Athens, his teachers being Mētrodōros, Aristotle's son-in-law,[5] and the Stoic Chrysippos of Soloi. He continued Hērophilos' investigations, but was more interested in physiology and in the application of physical ideas (such as atomism) to the understanding of life. He was more of a theorist than Hērophilos and had probably been more influenced by Stratōn. If we call Hērophilos the

[3] The Greek term was *adenoeideis prostatai*, meaning the glands standing in front. I do not understand the plural; there is but one prostate gland surrounding the beginning of the male urethra. Dr. Benjamin Spector, professor of anatomy in Tufts College, Boston, kindly wrote to me (23 January 1954) that the prostate may sometimes have the appearance of many glands instead of one; it is possible also that Hērophilos mistook the seminal glands for the prostate. It is interesting to note that Leonardi da Vinci neither mentioned nor figured that gland. In the *Tabulae sex* (1538) Vesalius gave no name to it, but in the *Fabrica* (1543) he called it alternately *corpus glandulosum* and *assistens glandulosis* (*assistens* is a poor translation of *prostatēs,* one who stands before and protects).

[4] Iulis was the main city of Ceōs, one of the Cyclades, close to the mainland of Attica; its modern name is Zea or Zia. Iulis was the birthplace of two great poets of the fifth century, Simōnidēs and his nephew Bacchylidēs.

[5] This Mētrodōros was a physician, disciple of Chrysippos of Cnidos (IV-1 B.C.). He was the third husband of Aristotle's daughter Pythias. Pauly-Wissowa, vol. 30 (1932), 1482, No. 26.

founder of anatomy, Erasistratos might perhaps be called the founder of physiology; he has been called also the founder of comparative and pathological anatomy (but such titles must be handled carefully).

"Comparative anatomy" was natural, because ancient physicians were obliged to dissect animals as well as men. The title of pathological anatomist was given to him because he conducted post-mortem dissections, that is, he dissected the bodies of men who had just died and whose medical history was known, and he was thus able to recognize the lesions that had caused their death.

His physiology was the first to be based upon the atomic theory, upon the theories of the Dogmatic school and the axiom of the "*horror vacui*." Many of these ideas had come to him from Praxagoras, who had been Hērophilos' teacher, but Erasistratos was more interested in them than Hērophilos had been. He tried to explain everything by natural causes and rejected any reference to occult ones.

His main anatomical discoveries concern the brain, the heart, and the nervous and vascular systems. But for his conviction that the arteries are filled with air (*pneuma zōotikon*, *spiritus vitalis*) and for his pneumatic theories in general, he might have discovered the circulation of the blood; for instance, he realized that the arteries of a living animal issue blood when incised and he suspected that the ultimate ramifications of veins and arteries were connected. He observed the chyliferous vessels in the mesentery. He realized that every organ is connected with the rest of the organism by a threefold system of vessels – artery, vein, nerve. He described correctly the function of the epiglottis (we still use the original term, *epiglōttis*) and of the auriculoventricular valves (the one on the right was called by him *triglōchin* = tricuspid). He recognized the motor and sensory nerves, distinguished more carefully between cerebrum and cerebellum, observed cerebral convolutions and noticed their greater complexity in man than in animals, traced cranial nerves in the brain itself, made experiments in vivo to determine the special functions of the meninges and different parts of the brain. He also investigated the relation of muscles to motion.

After rereading carefully this long enumeration, I must invite readers to accept its many details as carefully as I do. Our description of anatomical facts may be trusted; physiological ones require more qualifications, because one may easily misinterpret Erasistratos' ideas, which we know to be his only through Galen, and Galen's phrasing may suggest to us some ideas that did not exist in his mind, let alone the mind of Erasistratos. It is almost impossible for us to put ourselves back in their situation, and relatively easy to interpret their ideas in terms of our own knowledge.

VIVISECTION

We have said that Erasistratos made experiments in vivo to determine the functions of various parts of the brain. This implies vivisection, and it is almost certain that both he and Hērophilos carried out experiments upon the bodies of living animals. It is suspected that they did the same upon human bodies. The suspicion is based upon a text of Celsus which is so interesting that it is worth while to reproduce it verbatim.

Moreover, as pains, and also various kinds of diseases, arise in the more internal parts, they hold that no one can apply remedies for these who is ignorant about the parts themselves; hence it becomes necessary to lay open the bodies of the dead and to scrutinize their viscera and intestines. They hold that Hērophilus and Erasistratos did this in the best way by far, when they laid open men whilst alive – criminals received out of prison from the kings – and whilst these were still breathing, observed parts which beforehand nature had concealed, their position, colour, shape, size, arrangement, hardness, softness, smoothness, relation, processes and depressions of each, and whether any part is inserted into or is received into another. For when pain occurs internally, neither is it possible for one to learn what hurts the patient, unless he has acquainted himself with the position of each organ or intestine; nor can a diseased portion of the body be treated by one who does not know what that portion is. When a man's viscera are exposed in a wound, he who is ignorant of the colour of a part in health may be unable to recognize which part is intact, and which part damaged; thus he cannot even relieve the damaged part. External remedies too can be applied more aptly by one acquainted with the position, shape and size of the internal organs, and like reasonings hold good in all the instances mentioned above. Nor is it, as most people say, cruel that in the execution of criminals, and but a few of them, we should seek remedies for innocent people of all future ages.[6]

Considering the ruthlessness of that age, I am inclined to accept Celsus' statement. After all, if criminals were submitted to various tortures – and they undoubtedly were – were those early physiologists not excusable? Experiments in vivisection were less horrible than wanton torture. And yet we cannot help being horrified.[7] The Latin fathers of the church who had read Celsus, first Tertullianus of Carthage (c. 155–230) and later St. Augustine of Tagaste (V–1), did not hesitate to exploit the situation in their hatred of paganism. Pagans were so immoral that they were bad even when they tried to be good. Hērophilos' practice of embryotomy was also reproved by Tertullianus, as it would be by the Catholic doctors of today.

The main reason for disbelieving the story is that Galen does not refer to it, and yet we are indebted to him for practically all we know concerning those early anatomists. Galen's silence may be explained by his own horror. Celsus had been able to tell the story without disapproval because pagan

[6] Celsus (I–1), *De medicina. Prooemium,* quoted from the translation by W. G. Spencer (Loeb Classical Library, 1935), vol. 1, 13–15.

[7] Even as we were by the Nazi experiments on prisoners. Alexander Mitscherlich and Fred Mielke, *Doctors of infamy: the story of the Nazi medical crimes* (165 pp., 16 pls.; New York: Schuman, 1949) [*Isis 40,* 301 (1949)]. J. Schoenberg (Salonica), "Un nouveau chapitre dans l'histoire de la médecine," *Actes du VIIe Congrès d'Histoire des Sciences* (Jerusalem, 1953), p. 557-563. For vivisection in the Middle Ages, see *Introduction,* vol. 3, p. 266. It is a story told by Guibert of Nogent (XII–1).

ruthlessness had not yet been assuaged by Christian tenderness at the time he was writing; a century later, however, some progress had been made in a new direction and Galen was probably more compassionate than Celsus. In any case, the accusation of human vivisection is unproved.

EUDĒMOS OF ALEXANDRIA

The school of anatomy at Alexandria is said to have existed until the end of the Hellenistic age, but if that be true, it lost its distinction and vitality. The only anatomist who deserves to be mentioned after the two great masters is their younger contemporary Eudēmos, who flourished about the middle of the century. He made a deeper study of the nervous system, of the bones, of the pancreas,[8] of the female sexual organs, and of embryology.[9]

In short, one can trace fairly well the anatomical tradition of the first century (say 350 to 250). It is represented by the following succession: Aristotle, Dioclēs, Praxagoras, Hērophilos, Erasistratos, Eudēmos. Half of these men flourished in Alexandria and worked at the Museum.

[8] The organ was already known to Aristotle and named by him *pancreas* (*Historia animalium*, 541 b 11). We may recall for nonanatomists that the pancreas is a large gland discharging into the duodenum. The pancreas of the calf, used as food, is called a sweetbread.

[9] More details of his anatomical knowledge in Pauly-Wissowa, vol. 11 (1907), 904.

IX

MEDICINE IN THE THIRD CENTURY

Medical work has already been dealt with implicitly in the chapter on anatomy, for the anatomists were physicians, that is, they were medically trained and conscious of medical problems even if they did not practice the art. Moreover, the two traditions, anatomical and medical, are so closely interwoven, that one cannot separate them completely.

The astounding achievements of the Alexandrian anatomists described in the previous chapter were the climax of a tradition that may be summarized with the enumeration of the following names: Dioclēs of Carystos, Praxagoras of Cōs, Hērophilos of Chalcēdōn, Erasistratos of Ceōs, Eudēmos of Alexandria.

These five men followed each other closely within a century, say 340–240. As far as can be judged from a Galenic anecdote, Praxagoras was a great medical teacher. When Galen was asked to what sect he belonged, he answered, "To none," and added that he considered as slaves those who accept as final the teachings of Hippocrates, Praxagoras, or anybody else.[1] To be equated with Hippocrates by Galen was certainly a great honor.

Praxagoras, Hērophilos, and Erasistratos were primarily anatomists, but they were also physicians. Consider the pulse. In spite of the fact that Egyption physicians had taken it into account and tried to measure it,[1a] little attention was paid to it in the Hippocratic writings. As far as we know, Praxagoras was the first Greek physician to investigate the pulse and apply it to diagnosis.

Hērophilos improved that theory, using a clepsydra to measure the frequency of the pulse, from which to recognize fever. The strength of the pulse, he realized, indicated that of the heart. His pathology was empirical; he improved diagnosis and prognosis. He introduced many new drugs and recurred frequently to bloodletting. The embryo, he claimed, has only a physical, not a pneumatic, life. He invented an embryotome to cut the fetus into pieces within the womb, an instrument used by ancient obstetricians in

[1] Galen (Kühn) 19:13.

[1a] James Henry Breasted, *The Edwin Smith surgical papyrus* (Chicago, 1930), pp. 105–109 [*Isis 15*, 355–367 (1931)]. Hermann Grapow, *Grundriss der Medizin der alten Aegypter: I. Anatomie und Physiologie* (Berlin: Akademie Verlag, 1954), pp. 25, 28, 52, 69, 71.

hopeless cases. Like the older Greek physicians, he attached much importance to diet and exercise.

Erasistratos was the first physician to discard entirely the humoral theory; he was the first also to distinguish more clearly between hygiene and therapeutics and to attach greater importance to the former. Hence his insistence upon diet, proper exercise, bathing. He was opposed to violent cures, to the use of too many drugs, to excessive bloodletting (in much of this he was simply following Hippocratic conceptions). He invented the S-shaped catheter.

Our knowledge of these men is meager, yet one has the impression that their medical activities were subordinated to their scientific investigations. Excellent men of science as they were and sustained by the scientific discipline of the Museum, they must have realized that anatomical research led to tangible results while pathology and therapeutics were still unavoidably full of obscurities. They could not completely escape medical duties, and every cure was a medical experiment, but their main interest was elsewhere.

APOLLODŌROS OF ALEXANDRIA AND NICANDROS OF COLOPHŌN

Medical literature of the early Ptolemaic age was represented by the lost treatises of Apollodōros of Alexandria, which dealt the one with poisonous animals and the other (less certain) with deleterious or deadly drugs, *peri thēriōn, peri thanasimōn* (*dēlētēriōn*) *pharmacōn.* These treatises seem to have been the main source of a good many others dealing with drugs and chiefly with poisons. The ancients had a great fear of poisons to which bad luck or enmity might subject them; tyrants had special reasons to fear them and were desperately trying to find antidotes; we shall come across remarkable examples of that obsession later on.

The first to use Apollodōros' work was the poet Nicandros of Colophōn (in Ionia), who rendered the same service to agriculturists, botanists, and physicians that Aratos did to farmers and astronomers. The date of Nicandros is difficult to determine. If we place Apollodōros at the beginning of the third century, then Nicandros might be placed about the middle of the same century.[2] He would thus be a younger contemporary of Aratos and Theocritos. He was hereditary priest of Apollōn in Claros (near Colophōn). He wrote poems on many subjects, epic and erotic poems but mainly didactic ones, on husbandry (*geōrgica*) and beekeeping (*melissurgica*), prognostics (*prognōstica*, after Hippocrates), on cures (*iaseōn synagōgē*), on

[2] My dating, c. 275 B.C. (*Introduction*, vol. 1, p. 158), is perhaps too early. Nicandros has been named as one of the seven poets of the time of Ptolemaios II and is said to have flourished with Aratos at the court of Antigonos Gonatas (king of Macedonia 283–239). That is the dating which I accepted tentatively. Others would place him a century later under the last king of Pergamon, Attalos III Philomētōr (ruled 138–133). Elaborate but unconclusive discussion in Pauly-Wissowa, vol. 33 (1936), 250–265. See also *Oxford Classical Dictionary, s.v.* "Nicander."

snakes (*ophiaca*), and so on. Some of his writings may have been in prose, but everything that has come to us is in verse (hexameters). Nicandros was a typical *metaphrastēs* or interpreter, whose business it was to put available knowledge in metrical form (the task of such men was somewhat comparable to that of the popular scientific writers of today). His poems on farming and beekeeping were known to Cicero and influenced Virgil. I have not yet named his most important poems, the only two that are completely extant, on poisonous animals (*Thēriaca*, 958 verses) and on antidotes (*Alexipharmaca*, 630 verses), which were derived from Apollodōros. The *Alexipharmaca* contains (lines 74 ff.) a good clinical description of lead poisoning,[2a] with treatment. In addition to animals, 125 plants are mentioned in the two poems, and 21 poisons in the second. He was the first to refer to the therapeutic value of leeches.[3]

The popular value of such writings can hardly be overestimated, even if they were the vehicles of many errors. They carried a modicum of medical knowledge not only to physicians but also to every educated person. There were no early translations into Latin and therefore their diffusion was restricted to the Byzantine world. A Greek commentary is ascribed to Iōannēs Tzetzēs (XII–1). The Greek-sidedness of the tradition is proved by the incunabula edition. Both poems were first published in Greek (Fig. 32), together with the Greek princeps of Dioscoridēs (I–2) by Manutius (Venice, 1499; Klebs, No. 343.1).

There were quite a few later editions in Greek and Latin, but the first publication in any vernacular was the one prepared by a French physician, poet, and playwright, Jacques Grévin (born at Clermont en Beauvaisis c. 1540, died in Torino in 1570), *deux livres des venins, ausquels il est amplement discouru des bestes venimeuses, thériaques, poisons et contrepoisons . . . traduictes en vers francois* (2 parts; Anvers: Plantin, 1567–68).[4] The title was alluring to Renaissance minds.

PHILINOS OF CŌS

Philinos was a pupil of Hērophilos and hence we may assume that he flourished in the second half of the third century B.C. No writings of his have come down to us except a few fragments in Pliny and Galen.[5] He is

[2a] Plumbism or saturnism, poisoning by white lead (*psimythion*). Nicander of Colophon, *Poems and poetical fargments*, ed. and trans. A. S. F. Gow and A. F. Scholfield (Cambridge: University Press, 1953).

[3] Leeches were not used by Hippocrates. We do not know whether Nicandros succeeded in popularizing their use. They were used by Themisōn of Laodiceia (I–1 B.C.). In medieval times the word "leech" designated both the animal and the doctor. Hence, we may assume that the use of leeches had become popular; it was very popular in the nineteenth century (*Introduction*, vol. 2, p. 77).

[4] Plantin published also a Latin edition by the same Grévin (Antwerp, 1571), while Grévin's two discourses concerning the virtues of antimony were published in Paris in 1566.

[5] Given by Karl Deichgräber, *Die griechische Empirikerschule. Sammlung der Fragmente und Darstellung der Lehre* (Berlin, 1930), pp. 163–164.

ΓΕΝΟΣ ΝΙΚΑΝΔΡΟΥ.

ΝΙΚΑΝΔΡΟΥ ΘΗΡΙΑΚΑ.

Fig. 32. Princeps of Nicandros of Colophōn (III-1 B.C.). First page of the *Thēriaca.* At the top is a short biography of Nicandros. The first seven lines of the *Thēriaca* are in the middle left, a commentary being printed around them. This is part of a large folio (30.5 cm, 184 leaves) printed by Aldus Manutius (Venice, July 1499). The first and by far the largest part of this volume is the princeps of Dioscoridēs (I-2); then follow the works of Nicandros, the *Thēriaca,* the *Alexipharmaca*; finally scholia to the *Alexipharmaca* (leaves 175-184) which were perhaps printed separately; they were not included in the copy available to me. [Courtesy of Boston Medical Library.]

said to have written a criticism of the Hippocratic dictionary of Baccheios of Tanagra and notes on plants or simple drugs. He detached himself from his master, Hērophilos – for example, he rejected diagnosis based upon the pulse – and founded the so-called Empirical School of medicine, which we shall discuss in another chapter apropos of Serapiōn of Alexandria (II-1 B.C.).

ANDREAS, DISCIPLE OF HĒROPHILOS

This Andreas has sometimes been called Andreas of Carystos,[6] but this may be due to confusion with another man. We do not know for certain whence he came, but he flourished in Egypt in the second half of the third century. He was a disciple of Hērophilos and physician to Ptolemaios IV Philopatōr (ruled 222–205); he was murdered in 217 before the battle of Rhaphia[7] (where Philopatōr unexpectedly and completely defeated Antiochos the Great, king of Syria).

Many writings are ascribed to him, but none is extant. They dealt with snake bites (*peri dacetōn*), superstitions or errors (*peri tōn pseudōs pepisteumenōn*), crowns (*peri stephanōn*).[8] The most important of them seems to have been a kind of pharmacopoeia, entitled *Narthēx,* in which he described plants and roots. The title is significant; narthēx is an umbellif-

[6] Carystos in Euboia, the largest Aegean island, close to the coast of Attica. For details on this Andreas, see Deichgräber (passim) and M. Wellmann, Pauly-Wissowa, vol. 2, 2136.

[7] Rhaphia (Rafa) near the seashore on the Egyptian-Palestinian frontier, 15 miles south of Gaza, on the edge of the desert.

[8] I do not understand the real meaning of this title. *Stephanos* means that which encircles, crown, crown of victory (*palma*); *stephanē* has similar meanings, also the brim of a helmet, a helmet, woman's diadem, the brim or border of anything.

erous plant (like the carrot) which was much appreciated by the ancients because it yields a valuable drug, asafetida (antispasmodic).[9] It has a pithy stalk wherein Promētheus [10] conveyed the spark of fire from heaven to earth. Stalks of narthēx were used as rods, splints, and wands.

Our knowledge of Andreas' life and writings is derived from Serapiōn of Alexandria (II–1 B.C.), Hēracleidēs of Tarentum (I–1 B.C.), and Galen. For example, Serapiōn transmitted a *malagma* (plaster, poultice) mentioned in the *Narthēx*.

ARCHAGATHOS OF ROME

Let us now move to Rome. Its political importance was already considerable and fast growing, but as far as science and letters were concerned, Rome was still very provincial. It is not surprising that science entered the city [11] by the medical door, because sick people need doctors so badly that if they cannot find good ones they will easily become the prey of charlatans. The first Roman doctors were Greeks of both kinds, good and bad; many Greek slaves had some sort of medical knowledge and were consulted by their masters and the friends of their masters. The first Greek doctor whose name has come to us was Archagathos the Peloponnesian, who opened his *taberna* in Rome in the year 535 U.C. = 219 B.C.; [12] he was the first of a great many who flourished in the capital and in all the main cities until the end of the Roman empire. His *taberna*, that is, his office or surgery, was located near the Forum Marcelli. It is not clear whether he was used and subsidized in the Greek fashion [13] as a public physician; at any rate, he must have been moderately successful, since his name has survived. He was accepted as a Roman citizen but was accused of blasphemy and impiety because he had more confidence in therapeutics than in the protection of the household gods (*Dii penates*). That accusation has been formulated again and again, everywhere; it is clear to the superstitious mind that any medical treatment is a token of religious infidelity. The more scientific the treatment, the more sacrilegious it may seem. We have no idea how scientific Archagathos' practice was nor can we appreciate his medical knowledge, but he was a professional physician, not a magician.

The next Greek physician in Rome whose name has crossed the centuries is Asclēpiadēs the Bithynian (I–1 B.C.), but we may be sure that during the

[9] It comes from the Middle East (Afghānistān). Its Latin name is *Ferula narthex*. The *F. narthex* or *F. communis* was dealt with by Dioscoridēs (I–2) in his Book III, 91. Other *Ferulae* are described by him in the same Book III, 55, 87, 94–98.

[10] Promētheus (forethought), brother of Epimētheus (afterthought). He was credited with the discovery of many arts; he made man from clay and animated him with the artificial fire (*entechnon pyr*) stolen from Olympos.

[11] The *urbs, to asty*.

[12] That is, after the first Punic War (265–242), but before the second (218–201).

[13] An inscription in Cōs reads: "The physicians who are in the public service in the city." Wilhelm Dittenberger, *Sylloge inscriptionum graecarum* (ed. 3, Leipzig, 1920), vol. 3, p. 25, inscription 943, 1.7.

century and a half that elapsed between them many other Greeks practiced medicine in Rome. They were the only professionals, the only physicians who could bring the equivalent of a diploma or a certificate of competency obtained in the medical schools of Cōs, Athens, Alexandria, Rhodos, and other places. There were as yet no medical schools in the Latin world, though we must expect some established physicians to train their own assistants. The reaction against the Greek doctors was tremendous; it was not restricted to the ignorant but was shared by educated men who were conservative and felt it incumbent upon them to defend Roman virtues against sophisticated intruders.

We shall discuss these matters more fully in Chapter XXII apropos of Serapiōn (II-1 B.C.), Cato the Censor (II-1 B.C.), and Asclēpiadēs (I-1 B.C.). It concerns not only those men but the transmission of Greek medicine to the Romans and to ourselves.

X

THE LIBRARY

The Museum was the center of scientific research; the Library attached to it was the center of the humanities, but it was also a necessary department of the Museum itself. Therefore, it would be idle to discuss whether the Library was a part of the Museum or not. It is like the library of one of our great universities, which serves not only every department of the university but also many external needs. What is certain is that the Museum and the Library were both enclosed if not in the royal park at least in the Bruchion,[1] which was the Macedonian-Greek quarter of Alexandria, and that both were controlled by the royal will.

When the Museum was founded, it sufficed to erect a few halls and porticoes and to enroll investigators. The initial equipment was rudimentary. The growth of the Library was different. The first need was to collect manuscripts, and when these were sufficiently abundant, a building would be required to hold them and keep them in good order.

Many of the great libraries of the world grew up in the same way; that is, some of their treasures were gathered, and some of their collections well begun, before the library itself was established.

ANCIENT LIBRARIES

The Library of Alexandria was the most famous library of antiquity, but it was by no means the only one, nor the earliest. We may be certain that there were collections of papyri in Egypt and of cuneiform tablets in Mesopotamia. The most ancient libraries are lost and disintegrated (though some of their treasures may have come down to us), but archaeologists were lucky enough to discover in the ruins of Nineveh the royal library of Ashur-bani-pal (one of the last kings of Assyria, 668–626; Sardanapalos of the Greeks).[2] We may assume that libraries, private and public,[3] were not

[1] The Bruchion or Brucheion was the aristocratic quarter of the city just south of the great harbor and extending to Cape Lochias east of it. It included the royal palaces and offices, the residences of the Macedonian and Greek lords, the royal mausoleum, the Museum, and the Library.

[2] Volume 1, p. 157, with notes on Assyrian libraries of earlier times. The last king of Assyria ruled until 606 B.C.

[3] The word "public" must not be understood in the modern sense; above all, it

uncommon in the Greek-speaking world. Aristotle had a large one, and, if we may believe Strabōn, it was Aristotle himself who explained how to arrange the royal library in Egypt.[4] Other public libraries existed in Athens and later in Antiocheia, Pergamon, Rhodos, Smyrna, Cōs, and elsewhere, but the Library of Alexandria was undoubtedly the largest and eclipsed them all. In spite of the fact that it is entirely lost, we know more about it than about any other.

It was the library par excellence of classical antiquity, but strangely enough its name has not come down to us and enriched European languages as the word "Museum" has. The technical name *bibliothēcē*, which occurs in many languages, first meant a bookcase, and it also meant a collection of books in the library sense (as we would say "la bibliothèque rose"); but the use of it for a "library" came later and was not common. Polybios was first to use the word *bibliothēcē* in that sense.[5]

The characteristics of a library, we would say, are a collection of books, a building containing them, and a staff taking charge of them. That staff might be only a single person at the very beginning, but as soon as the library grew in contents and importance many employees would be needed, as well as a director or head librarian.

This introduces a very moot question: who was the first librarian?

Librarians of Alexandria. The nucleus of the Library was collected in Greece by Dēmētrios of Phalēron. He may be called the founder of the Library, though this honor might be ascribed with equal or greater justice to the first and second kings. The Library was organized by the will and at the expense of Ptolemaios Sōtēr; the organization was completed by his successor Ptolemaios Philadelphos. Hence, the fairest way of summarizing the matter would be to say that the Library was founded by Sōtēr, Philadelphos, and Dēmētrios. Was Dēmētrios the first librarian? If you please; but it would perhaps be more correct to call Zēnodotos of Ephesos the first one.[6]

should not suggest the amazing hospitality and generosity of American libraries. The terms "private" and "public" are relative. No private library was ever closed to the owner's friends; no public library was ever opened to anybody; its use might be severely restricted.

[4] Strabōn, XIII, 1, 54. That is hardly possible for Aristotle died in 322–21; yet he influenced the early librarians indirectly.

[5] Polybios (II–1 B.C.), *History*, XXVII, 4. Many Greek writers used the word *bibliothēcē* as the title of their own compilations, for example, Apollodōros of Athens (II–2 B.C.); his *Bibliothēcē* is at least a century younger, however; Diodōros of Sicily (I–2 B.C.), Phōtios (IX–2). The words *en tē basilicē bibliothēcē* (in the royal library) are used in the Septuagint (Esther 2:23).

[6] This question is connected to the one asked before. Was the Library independent of the Museum? The answer is, "If it was not independent at the very beginning, its independence grew with its own growth." A time came when the Library was a separate institution in a separate building, and then there was also a librarian or chief librarian. The same development occurs repeatedly in modern institutions (laboratories, observatories, and so on). As long as the library is small, it is taken care of by a clerk and administered by the director of the institution. When it is more

An elaborate study of the Alexandrian library was published by Edward Alexander Parson,[7] and according to him the list of librarians is:

		Tentative dates
1.	Dēmētrios of Phalēron	c.284
2.	Zēnodotos of Ephesos	284–260
3.	Callimachos of Cyrēnē	260–240
4.	Apollōnios of Rhodos	240–235
5.	Eratosthenēs of Cyrēnē	235–195
6.	Aristophanēs of Byzantion	195–180
6′.	Apollōnios Eidographos	180–160
7.	Aristarchos of Samothracē	160–145

All these men will reappear in our story later on, except Apollōnios Eidographos, a grammarian of undetermined time who occupied himself in the Library with the arrangement of Pindaros' odes.[8]

The list is uncertain in other respects. The only names about which every scholar would probably agree are Zēnodotos, Apollōnios of Rhodos, Eratosthenēs, Aristophanēs, another Apollōnios, Aristarchos. Such as it is, the list calls for two obvious remarks. First, it well illustrates Alexandrian cosmopolitanism. Second, it ends in the middle of the second century B.C. No librarian posterior to that time has ever been mentioned by anybody. We shall come back to that ominous fact presently. As far as its librarians are known (and what is a library without librarians?) the golden age of the library lasted less than one century and a half.

Growth of the Library. Thanks to the enthusiasm of its royal patrons and the ability of their first advisers, Dēmētrios and Zēnodotos, the Library grew very rapidly. The original building was already too small by the middle of the third century, and it was necessary to create a secondary library in the Sarapeion[9] (or Serapeum). Some 42,800 rolls were given or lent to the Serapeum library by the main one; this was perhaps a way of finding more room in the latter by the rejection of imperfect copies or duplicates.

The kings of Egypt were so eager to enrich their library that they employed highhanded methods for that purpose. Ptolemaios III Evergetēs (ruled 247–222) ordered that all travelers reaching Alexandria from abroad should surrender their books. If these books were not in the Library, they

---

fully grown, it requires separate housing, administration, and direction.

[7] E. A. Parson, *The Alexandrian library, glory of the Hellenic world. Its rise, antiquities and destruction* (468 pp., ill.; Amsterdam: Elsevier, 1952) [*Isis 43*, 286 (1952)]. The list of librarians is on p. 160. I copied the names but not always his dates.

[8] That is all one knows about him (Pauly-Wissowa *sub voce* Apollōnios No. 82). The name *eidographos* means a classifier of literary forms.

[9] The importance attached to that secondary library is reflected by the fact that an outstanding German journal on libraries, manuscripts, ancient literature was entitled *Serapeum* (31 vols.; Leipzig, 1840–1870). We shall use the Latin name Serapeum to designate it, as it is more familiar.

were kept, while copies on cheap papyrus were given to the owners. He asked the librarian of Athens to lend him the state copies [10] of Aischylos, Sophoclēs, and Euripidēs, in order to have transcripts made of them, paying as a guarantee of return the sum of fifteen talents; then he decided to keep them, considering that they were worth more than the money he had deposited and he returned copies instead of the originals.

The Library was the memory of the scientific departments of the Museum. The physicians needed the works of Hippocratēs and other predecessors; the astronomers needed the records of early observations and theories. One would like to know whether Babylonian and Egyptian observations were available there. How many of the earlier astronomical and astrological papyri did they have? The scientists of the Museum must know what had been done before them. It does not follow, however, that the early records were in the Library proper. The mass of those early scientific writings was not considerable and it was handier for men of science to keep them on their own bookshelves, either at home or in their laboratories. We may be sure that one of the nightmares of modern university librarians was already experienced in Alexandria, to wit, how can one reconcile the needs of the general readers with those of the special ones, and divide the books between the main library and the departmental ones?

When one passes from science to the humanities, however, the importance of the library increases immeasurably. For in the case of the humanities the library does not simply provide information, it contains the very masterpieces. The anatomist might find books in the library, but not bodies; the astronomer might find books, but not the stars, not the glory of heaven. On the other hand, if the humanist wanted to read the *Iliad* or the *Odyssey*, the songs of Anacreōn or the odes of Simōnidēs, those very treasures would be available to him in the Library and perhaps nowhere else. The Library might be called the brains or the memory of the Museum; it was the very heart of the humanities.

The Library of Alexandria was really a new start, as much as the Museum was. As much work had been done before in the field of the humanities as in the field of science, and we are fully aware, as far as the Greek world is concerned, that many books were published, sold, collected, criticized at least from the fifth century on. There had been also many libraries, large and small, private and public, but now for the first time a large number of scholars were assigned to library service.

That service was enormously more complex and difficult than that of modern librarians. To keep printed books in good order is relatively easy, for each of those books is a definite and recognizable unit. The Alexandrian librarians had to struggle with an enormous number of papyrus rolls, each

[10] I do not understand what is meant by "state copies" and who was the custodian of them. The phrase is used by H. Idris Bell, *Egypt from Alexander the Great to the Arab conquest* (176 pp.; Oxford: Clarendon Press, 1948), p. 54.

of which had to be identified first, then classified, catalogued, edited. The last word is the key to the main difficulties. The majority of the texts represented in the rolls were not standardized in any way, and their clear definition would remain almost impossible as long as they had not been thoroughly investigated, edited, and reduced to a canonic form.

To put it otherwise, the librarians of Alexandria were not simply custodians and cataloguers like those of today; they had to be, and were, full-fledged philologists. Indeed, the Alexandria Library was the nursery of philologists and humanists, even as the Museum was a nursery of anatomists and astronomers. This will be shown in some detail when we describe the activities of individual scholars.

The Library and its elaborate catalogue being lost, we have no idea of its contents, except that it was exceedingly rich and included many works that are no longer extant. The many thousands of papyri that have been discovered in Egypt and investigated in our century have revealed that the Greek population of Egypt (and the Greek-speaking Orientals) were fairly well acquainted with Greek literature. Homer was obviously the most popular author; Homeric papyri are more abundant than all the other literary papyri put together; [11] then follow, in order of decreasing frequency, Demosthenēs, Euripidēs, Menandros, Plato, Thucydidēs, Hēsiodos, Isocratēs, Aristophanēs, Xenophōn, Sophoclēs, Pindaros, Sapphō. There are very few fragments of Aristotle, but that is compensated by the discovery of a whole work of his, the *Constitution of Athens*, in a British Museum papyrus. Strangely enough, Hērodotos, who should have been of special interest to the Greeks of Egypt, was hardly represented. Not only did the papyri give us many fragments of known works but they revealed lost works like the *Athēnaiōn politeia* (just mentioned) and the medical papyrus of London, and they increased considerably our knowledge of other authors, such as Menandros, Bacchylidēs, Hypereidēs, Hērōdas, Timotheos,[12] Ephoros. "Toutes proportions gardées," the Greeks of Egypt were more literate than our American contemporaries.[13]

Rolls of papyrus. The discovery of papyrus by the Egyptians in the third millennium has been told in our first volume.[14] The essentials of papyrus

[11] Curiously enough, the *Iliad* was far more popular than the *Odyssey*. The number of *Iliad* fragments in papyri surpasses that of *Odyssey* fragments as much as Homer surpasses the other authors.

[12] Timotheos of Milētos (c. 450–360). His poem *Persai* (an account of Salamis) was discovered in the tomb of a Greek in Egypt. It is the earliest known literary papyrus; it dates from the end of the fourth century B.C., almost contemporary with the author (Berlin).

[13] Simple introductions to papyrology are available in two excellent little books: Frederick G. Kenyon (1863–1952), *Books and readers in ancient Greece and Rome* (Oxford: Clarendon Press, 1932, 1951), pp. 40–74; Bell, *Egypt from Alexander the Great to the Arab Conquest*, pp. 1–27, bibliography, pp. 152–161. For the international congresses of papyrology (1930 ff.), see *Horus: A guide to the history of science* (Waltham, Mass.: Chronica Botanica, 1952), p. 298.

[14] Volume 1, pp. 24–26. The best ancient account of papyrus is in Pliny's

manufacture remained the same in Greek and later times, but there were many obvious differences between the Egyptian and Greek papyri. The Egyptian rolls were often made of larger sheets and might be very long, sometimes exceeding a hundred feet (the record is 133 feet); the Greek rolls were smaller in size and length (say less than 50 feet) but far more numerous.

Papyrus was already an expensive material in early Egyptian times, the proof being the use of *ostraca* (potsherds such as Job used to scrape himself; Job 2:8). Nobody would have written serious matters on *ostraca* if a nice sheet of papyrus had been available to him. There is an *ostracon* in the Ashmolean Museum, Oxford, containing about nine-tenths of the *Story of Sinuhe*, one of the classics of Egyptian literature; the story was composed about the end of the twentieth century, the *ostracon* dates from the Ramesside age (roughly thirteenth to twelfth centuries). It is perhaps the largest inscribed *ostracon* in existence, but there is an abundance of shorter ones.[15]

The expensiveness of papyrus is proved also by the use of empty spaces of a roll, for example, the reverse of it, for new purposes, irrelevant to that of the original writing, and by the practice of erasing a text to make room for another (palimpsest).

We may be sure that papyrus continued to be dear in Hellenistic days, for the manufacturing of it required considerable skill and patience. The supply was a government monopoly, farmed out to individual contractors. The use of vellum began somewhat later, not before the end of the third century B.C. in Asia Minor, and as vellum was even more expensive than papyrus it did not displace it, but replaced it when papyrus was not available, which was the case in Asia, as soon as Ptolemaios Epiphanēs (ruled 205–182) had forbidden the export of it.[16]

In both cases, Egyptian and Greek, the papyrus unit was the sheet; many sheets glued together along one of their sides (generally the longer one) constituted a roll. The sheets were called *collēma*, which might be translated as something to be glued on to another of the same kind. The height of the roll was about 10 inches (it might be a little more, or much smaller) and the length seldom exceeded 35 feet. Papyrus was sold in rolls, and the

Natural history, XIII, 11–12, but it contains many errors.

[15] John W. B. Barnes, *The Ashmolean ostracon of Sinuhe* (London: Oxford University Press, 1952); *Journal of the American Oriental Society 74*, 58–62 (1954). Frans Jonkheere, "Prescriptions médicales sur ostraca hiératiques," *Chronique d'Egypte 29*, 46–61 (1954).

[16] Even in medieval times papyrus was not completely superseded by vellum. It was used for papal bulls until c. 1022; *British Museum Quarterly* 5, 27 (1931). Papyrus and vellum were finally displaced by paper under Muslim initiative; the dates of first use and of first manufacture of paper vary from country to country. That is a very complex story, for which see Thomas Francis Carter, *The invention of printing in China and its spread westwards* (New York: Columbia University Press, 1925; rev. ed. Ronald Press, 1931) [*Isis 8*, 361–373 (1926)] and my *Introduction* (by index).

writing was done on the roll (that is, the sheets were glued before the writing, not after).

Papyrus was made of strips of the pith of that plant laid together side by side, and of a second layer of strips laid at right angles to the first. Thanks to the stickiness of the pith, the two layers were stuck together by pressure. In the making of a roll all the horizontal strips or fibers were on the same side (inside or recto), all the vertical fibres outside (verso). The inside or recto was the best side for writing and in the best papyri the outside (or verso) was not used (it might be used later for the sake of economy). All the fibers were horizontal on the inside except in the last sheet, the one that remained outside when the roll was rolled up. In that sheet the arrangement of the strips was reversed and the inside ones were vertical for the sake of strength. In later times (Roman and Byzantine), that outside sheet bore various indications of administrative concern; that last sheet of the roll was the first when one unrolled it and therefore was called the first *collēma* or *prōtocollon* (hence our word protocol).

The reader may wonder how we know all that, especially if he is an old man not well acquainted with the latest discoveries. Our knowledge of (Greek) papyri is indeed fairly recent. Though some of them were discovered as early as 1778, they did not attract much attention until the end of the last century. A new scientific discipline or a new auxiliary branch of philology called "papyrology" was born in 1895–96, the very year of Röntgen's discovery. Papyrology and radiology were born in the same year! This is a very remarkable coincidence; even as the x-rays were the beginning of the new physics, papyrology was the beginning of a new history of Egypt and of classical antiquity. The papyri enabled a few scholars to see deeper into the past even as x-rays enabled others to penetrate and transcend superficial appearances.[17]

In little more than half a century of research, carried on by investigators of many countries, a large number of papyri have been found (most of them fragments, *disjecta membra*). They date from the end of the fourth century B.C. to the middle of the eighth century; most of them are in Greek, some in Latin, Coptic, or Arabic. The richest site was in Oxyrhynchos,[18] on the edge of the Libyan desert. Enough documents have come from that single locality to renew in many details our knowledge of classical antiquity and of the early Middle Ages.

How were the rolls arranged on the shelves of the Library, or what

[17] For the early history of radiology, see G. Sarton, "The discovery of x-rays with a facsimile reproduction of Röntgen's first account of them published early in 1896," *Isis 26*, 349–369 (1937).

[18] The place was named after the sacred fish of the Nile, *oxyrhynchos* ("with pointed snout"), a kind of *Mormyrus* (Arabic, *mizda*). It is in latitude 28°30′N and its present name is al-Bahnasa. Papyri could not be preserved except in dry places; hence one cannot expect to find any in the Delta.

corresponded to our shelving of books? It is impossible to say. Obviously the rolls could not be placed vertically on the shelves, as books are, but they might be placed horizontally. Even when the rolls were finally replaced by codices, it is probable that the latter were laid flat on the shelves, as is often done to this day in Oriental countries in the case of Arabic, Persian, or Chinese books.[19] Codices did not appear, however, until considerably later and were not predominant before the fifth century of our era. As Kenyon sums it up, "The thousand years of papyrus were to be succeeded by a thousand years of vellum codex, until that in turn was to give way to the paper printed book, which has so far only enjoyed half of the life of its predecessors." [20]

But we must not anticipate too much. How were the rolls arranged? As they were classified, it was necessary to group them in separate bunches. This could be done when they were placed flat on shelves, provided that they could not roll away from their companions; their rolling off could be easily avoided by adding enough vertical partitions and dividing the shelves into as many compartments or pigeonholes as might be desired.

It is probable that the more precious rolls were dealt with as the Japanese do with a *kakemono* or a *makimono*.[21] That is, the ends were reinforced, maybe, with a piece of wood which emerged on both sides and facilitated the folding in or out. A protruding label (*sillybos*) might be attached to the roller. In Roman times a number of rolls were placed in a bucket (*capsa*), which might bear a title of its own. The bucket or the pigeonhole were two equivalent solutions of the same problem and we may be sure that the one or the other was used in any large library.

We have not yet spoken of the copying itself. This was done on the ready-made roll and the superfluous part of each could be easily cut off. The scribe wrote in columns (*selis*),[22] the width of which was determined by the verses in the case of poetry; for texts in prose the columns were about 2½–3 inches wide, with intervals of ½ inch or more between them. There would be about 25 to 45 lines to a column, and 18 to 25 letters to a line; the words were not separated and there was no punctuation, except per-

[19] This was done also sometimes for Western books. We can know when it was done from the copies of old books that have their title written horizontally along the paper edges. Arabic and Chinese books often carry such titles.

[20] *Books and readers in ancient Greece and Rome*, p. 86. Papyrus rolls, VI B.C. to A.D. V; vellum codices, V–XV; printed books XV–XX. The dates concerning papyrus rolls refer to Greek papyri; Egyptian ones were far more ancient. If we took them into account, the age of papyri would extend to three millennia!

[21] Names of Japanese paintings that are rolled up. The *kakemono* is displayed by hanging lengthwise on a wall; the *makimono* is more like a papyrus roll in that it is rolled along its width; the reader unrolls it with one hand and rolls it up with the other hand.

[22] *Selis* meant originally the space between two rowing benches (*selmata*; Latin, *transtra*); later it was used to designate the space between two columns (or pages); later still, the column (or page) itself.

haps a dot or a short stroke (*paragraphos*) to indicate a pause. The end of a work would sometimes be indicated by an elaborate flourish (*corōnis*, Latin *corona*, a wreath or garland). The title, if given at all, would be put at the end of the roll, because that was the part of it which could be read as soon as one began to unroll it.

As the librarians were anxious to increase their collections, a good many rolls were copied if they could not be obtained otherwise. Certain halls of the library must have looked like a medieval *scriptorium*. It is possible that certain scribes supervised and corrected the work of the others, but it does not appear that a method or style of copying was developed, as happened later in medieval *scriptoria* such as those of Tours or Corbie, St. Albans or Bury St. Edmunds, which enables the trained paleographer not only to date a manuscript but also to say that it was written at one particular place. It is possible to distinguish Ptolemaic rolls from later ones but the specification cannot be carried further (on paleographic grounds).

The Hellenistic copyists were generally faithful, the main cause of error being the same as for modern typists, the omission of one or more lines, because the eye confuses two identical words at the beginning of two lines or the end of them (*homoioarcton*, *homoiteleuton*). Their faithfulness did not begin to compare, however, with that of the Hebrew scribes, whose duties were religious.

Size of the Library. The Library was very large, but it is impossible to know how many rolls it included. The numbers mentioned by various authors vary considerably. As the Library was steadily growing, the numbers were increasing; according to one account, there were already 200,000 rolls at the end of Sōtēr's rule; according to another, there were only 100,000 at the end of his son's rule; other accounts speak of 500,000 rolls or even 700,000 in Caesar's time. Never mind those conflicting dates. The numbers relative to definite dates may have different meanings; they may refer to works or to rolls, and there were sometimes many works to a roll or many rolls to a single work. Even today, it is difficult to answer this apparently single question, "How many books does your library contain," exactly and without ambiguities. After all, the number of books does not matter so much; the books might be very important or else trivial and worthless; they might be in perfect condition or not, there might be many imperfect or duplicate copies, or there might be few. The true richness and greatness of a library does not depend so much on the number of its books as on their quality.

It is a pity that we cannot visualize the Library. No doubt it was a fine building with elegant halls and colonnades. One would like to see the "stacks" of papyri, the desk or office where readers applied for them, the place where they were permitted to study. The halls were probably adorned with statues, bas-reliefs, or wall paintings. The most important features of a scientific institute are not the walls and fixtures, however, but the men using them; the pride of a great library is not so much its books as the distinguished scholars who are studying them, for without the latter the former are worthless.

Let us speak first of the few scholars who are named as directors of the Library or as scientific investigators charged with the organization of its contents.

ZĒNODOTOS OF EPHESOS

It would seem that some scholars combined the duties of librarians and of tutors to the royal princes. This would not be surprising, for everything in Ptolemaic Egypt gravitated around the king. The latter was not king by divine grace, he was himself divine. Stratōn was tutor to Philadelphos and when he was called to Athens to head the Lyceum, c. 288, he was replaced as tutor by the poet Philētas of Cōs. The first head of the Library,[23] Zēnodotos of Ephesos (III-1 B.C.), was a pupil of Philētas; his scholarly activities were so considerable that he probably devoted to them the whole of his time that was not eaten up by the library administration. It is highly probable, however, that the administration was still rudimentary; this was an age of administrative innocence, truly a golden age. All the chores were shared or divided amicably, without red tape, and done informally and wholeheartedly. The work was immense, for it did not suffice to put the rolls in order; each of them needed a special investigation, and not only that but the texts themselves required editing.

Zēnodotos discussed the matter with his assistants, Alexander of Pleurōn (in Aitōlia) and Lycophrōn of Chalcis (in Euboia), both Greeks of Greece proper, and they divided between them one great task, the collection and revision of the Greek poets. Zēnodotos took for himself the lion's share, Homer and other poets. He produced the first revision[24] (*diorthōsis*) of the *Iliad* and the *Odyssey*; he indicated the spuriousness of some lines but did not reject them, and introduced new readings. He compiled a Homeric glossary (*glōssai*) and a dictionary of foreign words (*lexeis ethnicai*). He was probably responsible for the division of each epic into 24 books.[25] His study of the text implied grammatical analysis and thus led to grammatical improvements. He also produced recensions of Hesiod's *Theogonia*, and corrected some poems of Pindaros and Anacrēon.

The Homeric fragments preserved in papyri reveal many variants; some

[23] The first definite head as opposed to the founder Dēmētrios. Zēnodotos lived from c. 325 to c. 234; he began his librarianship early in the rule of Ptolemaios Philadelphos (king 285–247). His edition of Homer was done before 274.

[24] I do not say the first edition. His edition was neither the first nor the last (Volume 1, p. 136).

[25] It has been suggested that the division into books was a result of the division into separate rolls, but that does not work because the average roll was large enough to accommodate two books of the *Iliad* or three books of the *Odyssey*.

rhapsodists were tempted to add lines of their own, just as a virtuoso may insert a cadenza in his rendering of a musical classic. Zēnodotos had the opportunity of comparing many Homeric rolls and his task consisted in harmonizing them.

Alexander of Pleurōn helped to classify the tragedies and satyric dramas, and was called *grammaticos* (grammarian) by Suidas (X-2). He was himself a tragic poet, included among the seven poets known as the Alexandrian Pleias.[26]

Lycophrōn of Chalcis put in order the rolls of the comic poets, and wrote an elaborate treatise on comedy (*peri cōmōdias*). We shall come back below to his work as a poet.

CALLIMACHOS OF CYRĒNĒ

Callimachos was probably born c. 310. He and Aratos were fellow students in Athens, Aratos being a few years older. Callimachos was for a time teacher of grammar at Eleusis, near Alexandria. He was introduced to the king Ptolemaios II and appointed librarian, c. 260; he held that office until his death, c. 240. By that time the Library was already so rich that it had become impossible to use it without a catalogue. Callimachos compiled one entitled *Pinaces tōn en pasē paideia dialampsantōn cai tōn ōn synegrapsan* (*Tables of the outstanding works in the whole of Greek culture and of their authors*), which was so elaborate that it filled 120 rolls. The books were divided into eight classes: (1) dramatists, (2) epic and lyric poets, (3) legislators, (4) philosophers, (5) historians, (6) orators, (7) rhetoricians, (8) miscellaneous writers. This classification is very interesting, because it reveals that the Library was essentially a literary institution. In which class were the scientific books placed? Perhaps in the fourth, or in the eighth, which was the "varia," the "glory-hole," which is necessary to complete any scheme of classification. In some of the classes the arrangement was chronological, in others by subject or alphabetical. For each book the title was given, the author's name (with discussion of authorship if necessary), the *incipit*, and the number of lines. Some of those indications were probably repeated on the label (*sillybos*) attached to each roll, for the classification of a large number of items requires some marks of identification and some labeling for each of them.

The *Pinaces* was far more than a single list, for it included historical and

[26] Pleias, plural Pleiades, the seven daughters of Atlas and the nymph Plēionē, placed among the stars. They were also called, after their father, Atlantides, and by the Romans, Vergiliae. Six Pleiades are visible to the naked eye, the seventh is not because of her shame; for she had allowed herself to be loved by a mortal. The name Pleiad was given also to the Seven Wise Men (Volume 1, pp. 167–169). The members of the Alexandrian Pleiad were, in addition to Alexander of Pleurōn, Callimachos, Apollōnios the Rhodian, Aratos, Lycophrōn, Nicandros and Theocritos (there are other selections). The name Pleïade was given to seven French poets grouped around Ronsard (1524–1585). The archaic name was typical of their archaeological tendencies.

critical remarks; thus it was a kind of *catalogue raisonné*, or it might even be called a history of Greek literature. Would that we had it, for a majority of the books that were available to Alexandrian scholars are completely lost, and many others are known only by the quotations made from them by compilers; to appreciate the value of the *Pinaces*, it will suffice to evoke the *Fihrist* of Muḥammad ibn Isḥāq ibn al-Nadīm (X–2), thanks to which we have some knowledge of a large part of the lost Arabic literature which would otherwise be as unknown to us as are a good many Greek writings.

The composition of the *Pinaces* was an immense undertaking; on the strength of it Callimachos could be called the first cataloguer (though his labor was incomparably more difficult and more original than that of modern cataloguers). It has been argued that he was not the librarian or director of the Library but its cataloguer. In the total absence of definitions of those offices, the matter cannot be profitably discussed. We must remember once for all that those early librarians were not simply librarians as with us, but men of letters, philologists, editors, lexicographers, historians, philosophers, poets. They might be each of these things, some, or all of them.

Callimachos was the teacher of the three following librarians, Apollōnios of Rhodos, Eratosthenēs of Cyrēnē (III–2 B.C.), and Aristophanēs of Byzantion (II–1 B.C.).

APOLLŌNIOS OF RHODOS

Apollōnios was a Greek Egyptian, born in Alexandria or Naucratis; he succeeded his teacher Callimachos as librarian but did not keep that office very long (say from c. 240 to 235) and went to Rhodos, where he obtained so much popularity as a teacher of rhetoric that he was naturalized, and is generally called Rhodios, the Rhodian. He finally returned to Alexandria, where he ended his days under Ptolemaios Epiphanēs (ruled 205–181). He was primarily a poet and is immortalized by his epic of the *Argonauts*. The time of his librarianship is uncertain; it may have occurred during his first stay in Alexandria (c. 240–230) or during his second stay, after Eratosthenēs' death or retirement (c. 195–192). It does not matter much, for we remember him as a poet, not as a librarian. We do not even know what he did for the Library. Was the Library already so well organized, or rather, did the kings bother so little about organization, that it was considered proper to give the directorship to a famous rhetorician and poet, a sinecure for him and an honor for the Library.[27]

ERATOSTHENĒS OF CYRĒNĒ

The first librarians, whether we count Dēmētrios with them or not, were all men of letters. Was it finally realized that the classification and investi-

[27] Librarianships were often considered as sinecures for distinguished men of letters in Europe, especially in France; for example, Leconte de Lisle (1818–94).

gation of scientific books required the care of a man of science? At any rate, the next librarian, Eratosthenēs of Cyrēnē (III-2 B.C.), was one of the greatest men of science of antiquity. He was not only a mathematician, astronomer, and geographer, but also a chronologist and even a philologist. One might even say that he was the first conscious philologist, for he was the first to assume the name *philologos*. That would be all wrong, however, for many men have deserved that name before him, and more than he, not only in Greece but also in Pharaonic Egypt, in Mesopotamia, and in India.

He completed his education in Athens but was called to Alexandria by Ptolemaios III Evergetēs (ruled 247–222) and appointed librarian c. 235; he probably remained in office until his death c. 192, at the age of 80. Two of his abundant writings were the by-products of his librarianship. One was his elaborate study of the Old Attic Comedy (*peri tēs archaias cōmōdias*) and the other his *Chronographia*, an attempt to establish the chronology of ancient Greece on a scientific basis. Callimachos and his successors were often puzzled by chronological difficulties. These difficulties were immense in antiquity because the local chronologies were independent of each other and often discordant. It was thus natural enough for a scientific librarian like Eratosthenēs to try to put some order in that chronological chaos even as he did in geodesy and the history of geography.

One might conclude that Eratosthenēs was not simply a librarian (in the sense that Apollōnios was), but that he helped to establish the chronological basis of criticism and was possibly the first classifier of scientific books.

ARISTOPHANĒS OF BYZANTION

Eratosthenēs died c. 195 and was succeeded by Aristophanēs (c. 257–c. 180), who was primarily a grammarian and lexicographer and perhaps the greatest philologist of classical antiquity. He improved the technique of textual criticism and prepared better editions of Homer, Hesiod's *Theogony*, Alcaios, Anacreōn, Pindaros, Euripidēs, Aristophanēs. He made a study of grammatical analogies or regularities, that is, he helped to organize Greek grammar and compiled a Greek dictionary (*lexeis*). Eumenēs II (197–159 B.C.) tried to steal Aristophanēs from Ptolemy Epiphanēs (205–182 B.C.) for his own library in Pergamum, whereupon Ptolemy put Aristophanēs in prison.[28]

His greatest contribution to grammar was the invention or systematization of punctuation. We are so used to reading texts which are fully punctuated that we take punctuation for granted, as we take the whole of grammar or writing itself. It is clear that punctuation is not absolutely necessary but when one has been obliged to read texts without punctuation and without

[28] F. G. Kenyon, *Books and readers in ancient Greece and Rome* (Oxford: Clarendon Press, 1931).

capitals (like Arabic) one appreciates these aids very much. It is much easier to read a text that has been carefully written, of which the words are separated, the proper words emphasized by capitals, and the sentences articulated by means of punctuation marks; it may even happen that punctuation removes ambiguities and misunderstandings. Aristophanēs was the first to understand all that clearly, but he was so much ahead of his time that those grammatical reforms were not adopted by the copyists until a long time afterward; they were still disregarded by the early printers and were not generally adopted before the middle of the sixteenth century.

The case of Aristophanēs illustrates the complexity of the services performed by the Alexandrian librarians. Librarianship in the modern sense was only a part of their job; their primary duties were philological. It was not enough to classify the books; they had to edit and rewrite them, or at least to make possible the necessary rewriting.

Aristophanēs invented not simply the ordinary marks of punctuation (similar to those that we use), but also various symbols needed in textual criticism, such as those to indicate a spurious line, missing words, metrical changes, tautology. He used such symbols in his Homeric editions. His edition of Pindaros was the first collected one, the odes being divided into sixteen books, eight on divine and eight on human themes. To all the texts edited by him he added notes (*scholia*) and sometimes introductions.[29] One of the works ascribed to him was a commentary on the *Pinaces* of Callimachos, and this confirms our belief that the *Pinaces* was far more than a catalogue and came close to being a history of Greek literature. He prepared recensions of Aischylos, Sophoclēs, Euripidēs, Aristophanēs. Finally, he compiled a dictionary or glossary (*lexeis*), a collection of grammatical analogies (or regularities) and anomalies (or irregularities), a collection of proverbs, and so on. The mass of these works is incredibly large, especially if one remembers that he was frequently a pioneer, and lacked the marvelous tools that are at the elbow of the modern philologist.

ARISTARCHOS OF SAMOTHRACĒ

The next librarian of importance, as well as the last one whose name has been transmitted to posterity, came from the little island of Samothracē, north of the Aegean Sea, close to the Thracian coast. That island was famous in antiquity because of its mysteries dedicated to the prehistoric twin gods, the Cabiri (Cabeiroi), and its name is immortalized by one of the most popular works of Hellenistic art, the Victory of Samothracē (one of the glories of the Louvre). To have given birth to one of the great philologists, Aristarchos, is another cause of pride for the little island.[30]

[29] For a fuller discussion of his philological labors, see J. E. Sandys, *History of classical scholarship* (ed. 3; Cambridge, 1921), vol. 1, pp. 126–131.

[30] It is little indeed, 68 square miles, not very much larger than Jersey in the Channel (45 square miles).

Aristarchos (II–1 B.C.) was the immediate successor or the next but one to Aristophanēs as librarian and he continued the latter's work as a literary critic and grammarian. He wrote such a large number of commentaries (*hypomnēmata*) and also critical treatises (*syngrammata*) that they filled 800 rolls (?). He was one of the first to recognize eight parts of speech – noun (and adjective), verb, participle, pronoun, article, adverb, preposition, and conjunction – and he introduced new critical symbols in his editions of the Greek poets.

From Zēnodotos to Aristarchos two parallel developments took place, textual criticism and the gradual constitution of grammar. This was not an accidental coincidence; it was impossible to discuss a text without grammatical analysis, and that analysis became sharper in proportion to the increase in subtlety of the literary criticism.

Another coincidence is even more startling, though equally natural. Anatomy and grammar, that is, the analysis of the body and that of language, developed at the same time. In both cases one must postulate a large mass of empirical knowledge, but it was in the Alexandrian age that both developments became more conscious and more systematic. It is equally difficult, or impossible, to explain the genesis of human bodies and of human language. It is marvelous that all of the intricate beauties of the Greek language, a very complex grammar as well as a rich and well-integrated vocabulary, were created to a large extent unconsciously. The main creators of Greek literature did not know grammar, but the Alexandrian philologists extracted grammar from their writings even as they extracted anatomy from their bodies. This gives the key of their achievements: neither the illustrious authors nor the grammarians created grammar, but the latter extracted it from the writings where it was implicitly contained.

Aristarchos' criticism was not only philological, it was also to some extent archaeological. He tried to find and discuss the *realia* or subject matter, the things to which the words referred.

Unfortunately, under the rules of the Ptolemaioi VI, VII, VIII conditions deteriorated in Egypt and the Library was neglected. In 145, Aristarchos was obliged to leave Alexandria and went to Cypros, where he died a few years later; it is said that he died at the age of 72 of willful starvation, because he was suffering from incurable dropsy.

The grammatical school that he had founded was continued after his death; his disciples, Apollodōros of Athens (II–2 B.C.) and Dionysios Thrax (II–2 B.C.), distinguished themselves, but the Library itself seemed to have fallen asleep. It is possible that the kings, who were facing increasing difficulties and troubles, lost their interest and reduced their support of it.

LATER HISTORY OF THE LIBRARY

Readers may be curious to know right now what happened to the Library after the middle of the second century B.C. The fact that one cannot name any librarians after Aristarchos of Samothracē is already sufficient proof of the decadence of the Library, which was but one aspect of the decadence of Hellenistic Egypt.

At the time of Caesar's siege of Alexandria, in 48 B.C., the Library was still exceedingly rich. As he could not man the Egyptian fleet riding in the harbor, which might be taken by the Egyptian commander Achillas and used against him, Caesar set fire to it. The conflagration extended to the wharves and is said to have destroyed part of the Library. This is difficult to believe, because the main Library was sufficiently distant from the harbor and docks, and the Serapeum was very far away on a hill. It is possible, however, that a quantity of books had been taken to the waterside to be shipped to Rome, and that it was those books that were destroyed.

This may explain why Marcus Antonius, the triumvir, gave to Cleopatra in 41 B.C. some 200,000 volumes taken from the Library of Pergamon. That story is far from certain, but it is plausible. If the Library had been diminished by Caesar's action, it would have been natural enough for the queen to complain and for Marcus Antonius to give her a rich compensation at the expense of his enemies.

The Library was still very important at the beginning of the Roman rule, when the Romans thought of themselves as liberators of Egypt. This is not proved, however, by the account of Josephus Flavius (I–2),[31] who does not speak of the Library as it was in his own time. During the rule of Aurelian (emperor, 270–275) the greater part of the Bruchion was destroyed. Did that involve the destruction of the main Library? At any rate, the Serapeum continued to exist.

It is possible also that books of either library or of both had been sequestered by the Roman authorities and taken to the capital. Conquerors have perpetrated such dilapidations in our own century; it was much easier to get away with them at the beginning of our era. The main enemies of the Library, however, were not the Romans but the Christians. Its decline was accelerated in proportion as Alexandria was more effectively controlled by bishops, whether Orthodox or Arian.[32] By the end of the fourth century, paganism was ebbing out of Alexandria; the Museum (if it still existed) and the Serapeum were its last refuges. The old Christians and the proselytes hated the Library, because it was in their eyes a citadel of disbelief and immorality; it was gradually undermined and brought into decay.

The Library was now concentrated in the Serapeum and the latter was finally destroyed under Theodosios the Great (emperor, 379–395), by order of Theophilos (bishop of Alexandria, 385–412), whose antipagan fanaticism was extreme. Many of the books may have been salvaged but, according to Orosius (V–1), the Library was virtually nonexistent in 416.

[31] *Antiquitates judaicae*, XII, 2. This chapter deals mainly with the Septuagint.

[32] Arianism was the imperial orthodoxy from 337 to the Council of Constantinople in 381.

The story has often been told that when the Muslims took Alexandria in 640, then again in 645 and sacked it, they destroyed the Library.[33] The *khalīfa* 'Umar is supposed to have said: "The text of those books is contained in the Qur'ān or not: If it is, we do not need them; if it is not, they are pernicious." That story is unproved. There was not much if anything left of the original library to be destroyed. The Christian fanatics had argued in the same vein as their Muslim emulators. Moreover, the pagan books were far more dangerous to the Christians, who could easily read them, than to the Muslims, who could not read them at all.

[33] For details and bibliography, see *Introduction*, vol. 1, p. 466.

XI

PHILOSOPHY AND RELIGION IN THE THIRD CENTURY

It is useful to discuss philosophy and religion in the same chapter, because they were often intertwined. The teachings of the Stoa were religious as well as philosophic, and the astral religion was derived from philosophy and science.

In spite of its political downfall and of its poverty, Athens was still the focus of philosophical teaching. Therefore, discussion of Hellenistic philosophy should begin with an account of Athenian conditions. The four main schools were the Academy, the Lyceum, the Garden, and the Porch, and to them must be added the unorganized efforts of the Cynics and the Skeptics.[1]

THE ACADEMY

After Plato's death in 347, the Academy was led by his nephew Speusippos (347–339), Xenocratēs (339–315), Polemōn (315–270), and Cratēs of Athens (270–268/4). These five men, all of them Athenians except Xenocratēs, who came from Chalcēdōn (near the entrance of the Bosphoros), were the directors of the original or Old Academy.

Upon Cratēs' death in 268/4, Arcesilaos of Pitanē (Aiolis, Mysia) was president of the school and gave such a new turn to it that it was called the New Academy. He was involved in polemics with the Stoics; in opposition to their dogmatism he revived the skeptical tendencies of Sōcratēs, Plato, and even Pyrrhōn; in opposition to their overemphasis on ethics he laid stress on clear thinking and logical skepticism. This was consistent with the scientific temper of the age. The skepticism of the New Academy was increased by Arcesilaos' successor Lacydēs of Cyrēnē (241–224/2). The first presidents were patronized by the kings of Pergamon, Arcesilaos by Eumenēs (d. 241) and Lacydēs by Attalos I Sōtēr (ruled 241–197). Attalos was a great patron of arts and letters; he gave to Lacydēs a new teaching garden (*lacydeion*) and invited him to come to Pergamon; Lacydēs declined the invitation very gracefully.

[1] These schools have been sufficiently characterized in Volume 1.

Lacydēs was succeeded by Tēleclēs (224/2–216), Evandros the Phocian (216–), and Hēgesinus of Pergamon. It is possible that the rule of the last named began only in the second century.

The names of the presidents of the Academy have been mentioned to evidence the continuity of that institution, and also its gradual deterioration. The first successors of Plato, Speusippos and Xenocratēs, were distinguished philosophers and mathematicians. Those who ruled in the third century, Polemōn, Cratēs, Arcesilaos, Lacydēs, Tēleclēs, Evandros, are almost completely forgotten; their names ring no bells in our memory.

SCHOOLS OF MEGARA AND CYRĒNĒ

Before speaking of the other Athenian schools it is well to say a few words about the provincial schools of Cyrēnē and Megara.[2] The latter had been founded by Eucleidēs of Megara (c. 450–380), who was a pupil of Sōcratēs. We know little about it; it was inspired by Parmenidēs and the Eleatics and did not last more than two generations of teachers. Eucleidēs was succeeded by Stilpōn (c. 380–300) of Megara, who seems to have been an outstanding teacher, for under his leadership the Megarian school enjoyed considerable popularity. Stilpōn was a disciple of Diogenēs the Cynic as well as of Eucleidēs, and he added Cynical tendencies to the latter's teachings; his own influence was due to his personality rather than to any doctrinal originality. One of his pupils, Menedēmos, established a new school of philosophy in his own town, Eretria (in Euboia, an island close to Attica); he was a teacher and friend of Antigonos Gonatas. The Eretrian school did not last very long; we can name only one disciple, one Ctēsibios; its teachings were criticized by the Stoic Sphairos of Borysthenēs (fl. at least to 221). It is probable that the Megarian school did not even last as long as that.

As to the Cyrenaic school, it was founded by one of Sōcratēs' immediate disciples, Aristippos of Cyrēnē, a rationalist and a hedonist. His teaching was a development of Epicureanism. It was continued by his daughter Aretē, her son Aristippos the Younger, Antipatros of Cyrēnē, Theodōros the Atheist (a strange combination of names!), Hēgēsias, and Anniceris the Younger. It was all over before the end of the third century, but the individual teachers influenced other philosophers. The views of the three last named were divergent. It is possible that the name of the school should not be used in this case except in a loose way.

These facts are not important except in so far as they illustrate the love of philosophy, which was so diffuse in the Greek population that the Athenian schools were not enough to satisfy it. Provincial schools were needed in Megara, Eretria, Cyrēnē, and probably other places. I do not know any other example of such abundance in the whole world. It was due

[2] Megara was on the isthmus separating the Gulf of Corinth from the Saronic Gulf. Upon the analogy of Central America one might call that region, placed between northern Greece and the Peloponnēsos, "Central Greece."

in part to the lack of a dominating religion and to the nonconformism, natural to the Hellenes, at once their strength and their weakness.

THE LYCEUM. THE PORCH. THE GARDEN

The Lyceum was more fortunate than the Academy in that its founder was succeeded by two men of great genius. Aristotle's leadership of it had lasted only thirteen years (335–323) but Theophrastos of Eresos ruled for 38 years (323–286) and Stratōn of Lampsacos, organizer of the Alexandrian Museum, for 19 (286–268). The next *scholarchēs*, Lycōn of Trōas, who headed the Lyceum for 44 years (268–225), was relatively unimportant.

Lycōn was succeeded by Aristōn of Iulis (Ceōs), thanks to whom Diogenēs Laërtios obtained the biographies, bibliographies, and wills of the first four headmasters. Aristōn was less a philosopher than a man of letters; he followed the path that Theophrastos had opened with his *Charactēres*, and the example given by the Cynic-Academician Biōn of Borysthenēs (c. 325–255). The golden age of the Lyceum had lasted less than seventy years (335–268).

Note that while the Old Academy was essentially an Athenian school the Old Lyceum was directed by foreigners; Aristotle was a Macedonian, Theophrastos a Lesbian, Stratōn a Mysian, and Lycōn a Troan (the last three came from the same region, northwest Anatolia). The last president, Aristōn, however, was almost like an Athenian, for his native island, Ceōs, is very close to Attica.

The most influential school of all was the Porch or Stoa. As far as ethics and politics were concerned, the importance of Stoicism can hardly be exaggerated; in a chaotic and immoral age it was the best defense of personal and civic virtues. It laid stress on conscience and duty, belief in Providence, submission [3] to one's Destiny, harmonization of one's life with the Universe (or with Nature), obedience (*eupeitheia*) to God, *ataraxia* (freedom from perturbation), *eudaimonia* (harmony of man's will with God's), *autarceia* (self-sufficiency), and also on equal participation, fellowship of men, justice, and brotherhood (*coinōnia*).[4] Stoicism was the highest ethical doctrine of the ancient world; it continued to dominate and to strengthen the best minds until the end of paganism.

[3] Submission, in Greek *eupeitheia* (obedience). There is no Greek word expressing the idea as forcibly as the Arabic word *islām*.

[4] Our knowledge of early Stoic terminology is incomplete because only fragments of Zēnōn and Cleanthēs have come down to us. Zēnōn and Cleanthēs used the words *eudaimonia* and *pronoia*; Zēnōn spoke also of *eupeitheia* (obedience), *apatheia* (insensibility to suffering), and *homonoia* (*concordia*). Marcus Aurelius used *apatheia, ataraxia* (freedom from passion), and various derivatives of *coinos* (common); he probably invented *coinonoēmosyne* (feeling of brotherhood). The word *aphilochrēmatia* (contempt of money) occurs in Plutarch. Many of the Stoic words (such as *ataraxia*) were used also by Epicureans, who shared the quietism of the Stoics.

Unfortunately, the Stoics paid little, if any, attention to science and favored divination (*manteia*) and astrology; on the ethical plane their doctrines were too abstract, cold, impersonal; this explains the ultimate victory of Christianity over Stoicism, for the Christians put a new emphasis on love, charity, and mercy.

The first teacher was Zēnōn of Cition (IV–2 B.C.), probably of Phoenician parentage, who lived until 264 and hence belongs to our century as well as to the fourth. Among his disciples were Persaios of Cition and Sphairos of Borysthenēs. His first successors as head of the Porch were Cleanthēs of Assos (III–1 B.C.) and Chrysippos of Soloi (III–2 B.C.). Cleanthēs was not only a philosopher, who helped to define the Stoic doctrine, but an inspired poet, author of the greatest religious hymn in Greek.[5] He ruled the Porch from 264 to 232, and Chrysippos from 232 to 207. As Cleanthēs was a poet, his philosophy was more sentimental than Zēnōn's; he considered the Universe a living being, God its soul, the Sun its heart. Yet he insisted that there can be no virtue without disinterestedness, and how can a man of feeling be disinterested? Stoic disinterestedness was admirable but the insensibility that was inseparable from it, much less so.[6] As to Chrysippos, his additions to Stoic philosophy were so many and so deep that it was said that "Without Chrysippos no Stoa."

Chrysippos wrote a large number of books, his successor, Zēnōn of Tarsus, very few, but by that time (the end of the third century) the popularity of Stoicism was so great that Zēnōn had many disciples. He was probably an inspired teacher, but his success was largely due to his gathering the harvest that his predecessors had sown. All these Stoic teachers were Asiatics except Sphairos, who was a Scythian!

The Garden was like the Porch in many respects; their resemblance was perhaps due to common Oriental origins and even more so to the similarity of their functions. As far as can be judged from the fragments relative to it and to its founder, the Garden was more informal and simpler than the other schools; the life was generally frugal but animated and quickened with regular feast days which brought the fellows more closely together; women were admitted to the fellowship (this we are sure of because many contemporaries were scandalized by that bold innovation and made evil innuendoes). The first teacher, Epicuros, came from Samos, the second, Hermarchos, from Mytilēnē (Lesbos). Epicuros' Athenian teaching began in 307 and he lived until 270. One can quote only two other headmasters

[5] A hymn to Zeus (*Hymnos eis Dia*), 38 lines. It is a beautiful amplification of "Thy will be done."

[6] Stoic disinterestedness was defined by the words *apatheia*, *ataraxia*, *aphilochrēmatia* (Plutarch). The conflict between insensibility and disinterestedness appears frequently and the line between them is difficult to draw. For example, saints have often been accused of being insensible or insensitive. The same reproach was rightly made to Stoics, even to the greatest of them.

in the third century, Polystratos (assisted by Hippocleidēs) and one Dionysios (fl. c. 200). Polystratos may have been a direct pupil of Epicuros; some writings of his have come down to us.[7] The other men are practically unknown.

CYNICS AND SKEPTICS

In order to complete our picture of philosophy in the third century, we must still say a few words of tendencies that were never represented by a definite school but remained unorganized and personal. Organization and institutionalization are causes of strength but also of weakness. The corporate power and glory of an institution impresses little people; it does not much influence original minds. Thus it happened that the Cynics and Skeptics had many disciples here and there in spite of the fact that one can hardly speak of a Cynical or Skeptical school. Cynicism and skepticism are states of mind, inherent in some men in all places and times. The first to express those states of mind, however, were Greeks, and they did so as early as the fourth century.

The first Cynic was Antisthenēs, one of Sōcratēs' pupils, but the most famous one was Diogenēs of Sinōpē, he who challenged Alexander the Great. Among later disciples let us recall the names [8] of Cratēs the Theban, the girl Hipparchia and her brother Mētroclēs of Marōneia (in Thrace), and Onēsicritos of Astypalaia (one of the Dōdecanēsoi). Antisthenēs was the only philosopher among them; the others were rather like friars or saints, trying to live a simple life and despising worldliness and wordiness.

The first formal Skeptic, Pyrrhōn (c. 360–270), came from Ēlis.[9] He was immortalized by his disciple, Timōn of Phlius *ho sillographos*, and has found many friends and imitators down to Montaigne and later. Every man of science is somewhat of a cynic, because he does not accept words and conventions at their face value, and of a skeptic, because he refuses to believe anything without adequate proof.

Cynicism and Skepticism favored quietist tendencies, even as Stoicism and Epicurism did. It is not strange that so many philosophers of various sects were united on this, the need of dispassionateness and disinterestedness, even of indifference, for the world around them was cruel and no peace was possible without withdrawal from the circumambient chaos. Peace was nowhere to be found save in one's own soul.

[7] Edited upon the basis of Herculaneum papyri by Carolus Wilke, *Polystrati Epicurei peri alogu cataphronēseōs libellus* (58 pp.; Leipzig, 1905), on irrational criticism of others.

[8] For details see Volume 1, 584–586. All the Greek philosophical schools are defined and described in it, because all of them were legacies of the fourth century.

[9] Ēlis in northwest Peloponnēsos; I do not know whether the city Ēlis is meant or the district. Olympia, where the Olympic games were celebrated, was in the same district, south of the city Ēlis. Phlius (district and city) is in the northeast Peloponnēsos. *Sillographos* means writer of *silloi* (a type of satiric poem).

ROYAL PATRONS

While most of the scientific work was done in Alexandria, almost every philosopher of distinction flourished outside of Egypt. The Ptolemaic kings did not favor philosophy, and I can hardly think of a philosopher patronized by one of them, except one like Eratosthenēs, who was primarily a man of science, or one like Timōn of Phlius, who shone as a man of letters.

The kings of other Hellenistic countries were more hospitable to the lovers of philosophy. Eumenēs I, king of Pergamon (263–241), favored the Academician Arcesilaos of Pitanē, and his successor, Attalos I Sōtēr (241–197), Lacydēs of Cyrēnē. Sphairos of Borysthenēs, a Stoic, was befriended by the king of Sparta, Cleomenēs III (king 235–222), and helped him in his attempt to foment a social revolution; after having failed in 222 Cleomenēs took refuge with his patron, Ptolemaios Evergetēs, but was imprisoned by Evergetēs' successor, Philopatōr, and committed suicide (220/19). Was Sphairos with him in Egypt? The main patron of philosophy, however, was Antigonos Gonatas,[10] who helped the Cynic Biōn of Borysthenēs, the Stoics Zēnōn of Cition and Persaios, and also Menedēmos of Eretria. This Antigonos was himself a philosopher and a patron of arts and letters, who tried to reëstablish Macedonian fame.

STOICISM. TYCHĒ

The most effective of those philosophies was Stoicism, under whose guidance the Greeks could become good men as well as good citizens and the city could be purified and strengthened. As one of their main tenets was to live in conformity to Nature and to Reason, one might have expected them to favor the impartial study of Nature, but they were unfortunately sidetracked. In order to obey God, let us know his will by means of divination (*manteia*); the most respectable form of divination was astrology. They favored the astral religion and the astrological superstitions derived from it.

The Stoa was encouraged in its delusion by Greek mythology (which had never been forgotten or superseded), by the Babylonian, or rather Chaldean, ideas that became a part of Greek lore under Seleucid patronage, and by similar ideas fostered in Egypt and Hellenized under the Ptolemies.

The purely Greek ingredients were the goddess Tychē (Fortuna) and the idea of *moira* or *aisa* (fate).[11] As the ideas were refined by the mythologists, there were three *Moirai*, that is, three women ordering our fate, Clōthō

[10] Antigonos II Gonatas, king of Macedonia from 283 to 239. William Woodthorpe Tarn, *Antigonos Gonatas* (513 pp., Oxford, 1913). Gonatas favored not only philosophers but also poets like Aratos of Soloi and historians like Hierōnymos of Cardia.

[11] *Moira*, from *meiromai*, to receive as one's portion; *hē eimarmenē moira*, or *hē peprōmenē moira*, that which is allotted. In both phrases the word *moira* is generally left out.

who spins the thread of life, Lachesis who distributes the lots, and Atropos, the inflexible, who cuts the thread.[12]

This is a good example of mythological elaboration of an abstract idea. The idea of *moira* was analyzed in a poetic way, each part being represented by a woman, Clōthō, Lachesis, or Atropos. This became an inexhaustible source of inspiration to poets and sculptors. Further discussion was superfluous; each artist was ready to evoke the general idea of fate or some aspect of it, the spinning out, distribution, and finally Atropos' cut, the inevitable end of every human fate, *atra mors* (black death).[13] The allegory was accepted by everybody with various degrees of literalness or symbolism. The most fascinating aspect of mythogony is its anonymous nature. Who invented the *Moirai* or the other gods and goddesses? It is impossible to know; mythology is an essential part of folklore. Who named Clōthō? Who named the common plants and animals? The gods and goddesses who symbolize many aspects of life and thought were invented anonymously and mysteriously, even as the great majority of words and the grammatical paradigms were invented.

The Greek genius was prolific in the invention of myths because it was essentially poetic. This peculiarity is better understood when one compares it with the Semitic genius. The Muslims were more fatalistic than the Greeks and they often expressed the idea of *moira* by means of equivalent words (*qisma*, *"kismet,"* or *naṣīb*) but they did not imagine women to symbolize that idea and nipped in the bud the poetic and artistic developments of it that give us so much pleasure in Greek arts and letters.

ASTROLOGY

The technical elements of astrology, the details of star worship, came from Babylonia and from Egypt. The twelve houses of the Zodiac had each properties of their own, and so did the thirty-six decans (or decades) of the Egyptian year. The main interpreters (*hermēneis*) of Fate, however, were the seven "planets" (Hēlios, Selēnē, Hermēs, Aphroditē, Arēs, Zeus, Cronos, or Sun, Moon, Mercury, Venus, Mars, Jupiter, Saturn). An elaborate correspondence was established between human events on one side and astral, planetary, events on the other, or, to put it otherwise, between the microcosmos and the macrocosm.[14] The fact that there were seven planets,

[12] According to *Plato*, the Moirai were daughters of Anankē (necessity). In Latin they were called *Parcae* and their individual names were *Nona*, *Decuma*, and *Morta*.

[13] Stephen d'Irsay, "Notes to the origin of the expression *Atra mors*," *Isis 8*, 328–332 (1926). I wonder whether the ominous meaning of *ater* was not colored by remembrance of Atropos; d'Irsay does not refer to this, however.

[14] These astronomical and astrological ideas were already old in the third century B.C. The parallelism between microcosm and macrocosm may be of Iranian or Babylonian origin; it can be traced back in Greece to Plato and Dēmocritos (Volume 1, pp. 177, 216, 421, 602).

neither more nor less, was given a mystical importance. The sacredness of the number seven is perhaps a Babylonian concept. "To the seven planets were assigned their own colors, corresponding to the seven stages of a Babylonian temple, their own minerals, plants and animals; the seven vowels of the Greek alphabet became their signs; and from them came that persistent use of 'seven' which survives in our (Hellenistic) week and appeared in the seven sleepers, the seven wonders of the world, the seven ages of man (which Shakespeare took from astrology), the seven stoles of Isis, the seven-stepped ladder of Mithras, the seven joys of the righteous in the Salathiel apocalypse,[15] the seven angels and vials of *The Revelation*, the seven gates of hell and the seventh heaven."[16] The earliest Greek document on this is the pseudo-Hippocratic treatise *De hebdomadis*, of the sixth century if not earlier. A curious relic of that superstition was given by Hegel in his *Dissertatio philosophica de orbitis planetarum* (1801), wherein he "proved" that there could not be more than seven planets![17]

How did it happen that astrology established itself so firmly in Egypt at the very time of Aristarchos and Aratos? The parallel development of scientific astronomy and astrology is due to the inheritance of a double tradition favoring astrological fancies. There was the Greek tradition which began with the *Timaios* and more emphatically with the *Epinomis*.[18] One might almost claim that Greek astrology was a fruit of Greek rationalism. At any rate, it received some kind of justification from the notion of cosmos, a cosmos which is so well arranged that no part is independent of the other parts and of the whole. Was this not proved by the tides, caused by Moon and Sun, by the menstruation of women, by the farmers' moonlore, by the general belief in lunacy.[19] The very fact that one sees the stars established the existence of some connection between them and us. The basic principle of astrology, a correspondence between stars and men, enabling the former to influence the latter, was not irrational. That principle, fortified by Greek science, came from Iran and from Persian Babylonia; the Ptolemaic astrologers received additional inspiration from their Chaldean (neo-Babylonian) contemporaries.[20] The two traditions were Greco-Baby-

[15] The Salathiel apocalypse is more commonly called "the Second Book of Ezra the Prophet," or 2 Esdras (4 Esdras in the Vulgate). It is not extant in the original Aramaic nor in Greek (except for a fragment discovered in an Oxyrhynchus papyrus), but only in old Latin and various Oriental versions. It was written in the period 66–250. It includes six visions of Salathiel (or Shealtiel) which occurred some 30 years after the destruction of Jerusalem in 586 (that is, in 556). Analysis in Robert H. Pfeiffer, *History of New Testament times* (New York: Harper, 1949) [*Isis 41*, 230 (1950), pp. 81–86].

[16] The quotation is borrowed with kind permission from W. W. Tarn, *Hellenistic civilisation* (London: Arnold, ed. 3, 1952), p. 346.

[17] For the *De hebdomadis*, see Volume 1, p. 215. For Hegel, see *Horus: A guide to the history of science* (Waltham, Mass.: Chronica Botanica, 1952), p. 37.

[18] This has been discussed at some length in Volume 1, pp. 451–454.

[19] G. Sarton, "Lunar influences on living things," *Isis 30*, 495–507 (1939).

[20] This was very natural, for the Per-

lonian and pure Babylonian; both were inculcating at one and the same time a science, astronomy, and a theology or religion, the astral religion, and the popularity of astrology among all classes was due to that very combination.

The great confusion of ideas concerning astrology to this very day is due to the fact that, whatever the purpose and aberrations of the astrologers might be, their technical basis was astronomical. If the fate of a man depended upon the positions of planets and stars at the hour of his birth (or conception), it was necessary to determine those positions as exactly as possible, and that was a purely astronomical problem. The confusion was much greater in those days because of the mixture of science with religion.

The astrologers were of two kinds, the more scientific kind who called themselves mathematicians (*mathēmaticoi*) and the more religious kind, priests or diviners, the *hōroscopoi*.[21] Those priests were either Greeks or Hellenized Egyptians, and they did not restrict themselves to astrology but practiced many other forms of divination (*manteia*, *manticē*, *technē*).

One can infer the existence of many astrological treatises written in Egypt during the third century B.C., but most of them have disappeared. The most ancient perhaps was a text ascribed to Hermēs Trismegistos[22] (thrice-greatest), a Latin translation of which was discovered by Wilhelm Gundel in a very late Latin manuscript (British Museum, Harleianus 3731, dated 1431); there is no earlier version of it except that its most important chapter was translated into French (Picard) by Arnaud de Quinquempoix (XIV–1),[23] for the Queen of France, Marie of Luxemburg.[24] The *Liber Hermetis* is obviously the relic of a Greco-Egyptian treatise; it contains Egyptian elements and expressions of Persian ancestry. It discusses the 36 decans, some 72 stars of the *sphaera graecanica*, and others of the *sphaera barbarica*.[25]

---

sians ruled in Babylonia and Egypt at about the same time, beginning in 538 or 525 and ending in both countries with Alexander's conquest in 331. After a few years of anarchy, Babylonia was ruled by the Seleucids (312–171), then by the Parthians (171 B.C.–A.D. 226), the Sassanids (226–646), and finally by Muslims. Babylonian astrology began in Persian times, the very sophisticated kind of astronomy is Seleucid. See Chapter XIX.

[21] The *hōroscopos* is the man who observes the natal hour (for it is not only the day, but also the hour that matters). The operation was called hōroscopēsis. Hence, our word horoscope, which is an operation, not a person.

[22] Hermēs, son of Zeus and Maia, was the god of occult learning. He was the equivalent of the Egyptian god Thoth, and was called Mercury by the Romans. Our word "hermetic" refers to occult learning, and also strangely enough to airtight closing! Alchemy was often called the hermetic art, and one also spoke of hermetic medicine.

[23] *Introduction*, vol. 3, p. 453. The Latin text has been edited in exemplary manner by Wilhelm Gundel in *Abhandlungen der bayerischen Akademie der Wissenschaften* (phil. hist. Abt., part 12, 386 pp., Munich, 1936). Analysis by Claire Préaux, *Chronique d'Egypte 12*, 112–115 (1937).

[24] Marie was the queen of Charles IV le Bel. She died in 1324. The French translation was thus more than a century older than the Latin Harleianus text, dated 1431.

[25] The *sphaera graecanica* included the stars known to Aratos and Hipparchos; the

The original book of Hermēs Trismegistos is undatable. We are on firmer ground with Bērōssos (III–1 B.C.), who was perhaps the main importer of Chaldean astrology from Babylonia to the West.[26] Note that his history of Babylonia was dedicated to the Seleucid Antiochos I Sōtēr (king 280–261). He is said to have established a school of astrology at Cōs. This is extremely interesting because it confirms the cultural importance of that island, strategically located at the crossing of roads connecting Greece, Egypt, Anatolia, and Syria.[27] Hippocratēs was born there and it became the site of one of the earliest medical schools; it is not astonishing to hear that it was also the cradle of the earliest school of astrology. Students could easily reach Cōs from the three continents; in that very small island medical students could go astray and sit for a change at Bērōssos' feet; this would help to explain the astrological conceits found in later medical writings, such as those of Galen (II–2).

Sudinēs (or Sudinos) of Pergamon was probably a pupil of that school, if not of Bērōssos himself. From Cōs to Pergamon was not a long voyage. Sudinēs illustrates Greco-Babylonian syncretism, for he wrote a commentary on Aratos but was famous for centuries because of his lunar tables which were of Chaldean origin. He flourished in Pergamon under Attalos I Sōtēr (king 241–197), who conquered a good part of the Seleucid territory and may have enlisted or kidnapped Chaldean astronomers.

Let me mention a few more astrologers of the third century B.C. Vitruvius refers to two other disciples of Bērōssos, Antipater and Achinapolos, but their writings are lost. They were the ones who insisted that the horoscope should be based upon the conception, not the birth, of an individual; that was a right idea, but how did they think of carrying it out?[28] There are Greek fragments of a hermetic text called *Salmeschniaka*, of Egyptian origin (c. 250?). Apollōnios of Myndos (on the Carian coast, very near Cōs) and Epigenēs of Byzantion may belong to the same age and be disciples of the Bērōssian school. Apollōnios and Epigenēs discussed the Chaldean theories

sphaera barbarica, other stars known to non-Greek astronomers. The early Egyptians divided the equatorial region into 36 decans of 10° each; the Babylonians divided the ecliptical (zodiacal) belt into 12 houses or signs of 30° each. As the equatorial and ecliptical belts overlapped, it was not difficult for star groups to pass from one system into the other. See Volume 1, pp. 27, 29, 119.

[26] Chaldean astrology had already penetrated the Greek world before Bērōssos. There were traces of it in the treatise of Theophrastos on signs (*peri sēmeiōn*). According to Proclos, "Theophrastos tells us that his Chaldean contemporaries possessed an admirable theory which predicted every event, the life and the death of every human being." It did not simply foretell general effects such as good and bad weather (*Procli in Platonis Timaeum commentaria*, ed. Ernest Diehl (Leipzig, 1906), vol. 3, p. 151.

[27] Volume 1, chap. 15, "Coan archaeology," pp. 384–391.

[28] Of course, they might subtract nine months from the date of birth, but that was very arbitrary. There is a cuneiform horoscope in the British Museum using both the actual date of birth, 15 December 258 B.C., and the date of conception derived from it, 17 March 258. Frederick H. Cramer, *Astrology in Roman law and politics* (Philadelphia: American Philosophical Society, 1954), p. 14.

of comets and disagreed about them. According to Epigenēs, the Chaldeans considered comets as fiery clusters of whirling air; according to Apollōnios, the Chaldeans considered them as planets whose orbits might be computed. The Apollonian hypothesis was approved by Seneca (I–2), whose account ended with the prophetic words: "One day a man will be born who will discover the orbits of comets and the reason why their paths are so different from those of other planets. Let us be satisfied with the discoveries already made, so that future generations may also add their mite to the truth." [29] These astounding remarks take us a little out of the Hellenistic age but not very far out; Seneca wrote them about A.D. 63.[30]

A large part of the astrological literature of the Middle Ages is derived ultimately from the Hellenistic books of Hermēs and others; this is probably true also of the Latin books translated from the Arabic.

The most remarkable feature of Ptolemaic astrology is its complete lack of interest in a man's life after death. Those texts are intensely religious yet avoid eschatological considerations. In that respect they are very different from the Indian and Christian astrological writings.[31]

The popularity of astrology in learned circles was much increased by Stoic approval. This was natural in a way because of the Stoics' conception of the universe, of its wholeness and of man's integration, harmonization, and "sympathy" with it; [32] they were ready to accept the Babylonian "correspondence" and the correlativeness of microcosm and macrocosm. Add to that their belief in divination, and astrology was justified. Their great difficulty was to reconcile Fate with Providence (*moira* with *pronoia*) and fatalism with freedom and duty. The Christian theologians were to be bothered by the same contradiction throughout the centuries.[33]

The Epicureans have often been accused, rightly, of hedonism and,

[29] Seneca, *Quaestiones naturales*, VII, 3. The whole of chapter VII is devoted to comets; it is in the middle of it (VII, 25, 4–5) that Seneca expressed his prophetic views on the progress of science. He expressed similar views in a letter to Lucilius (No. 64, quoted in *Introduction*, vol. 2, p. 484).

[30] Seneca's prophecy began to be realized by Regiomontanus, who investigated the orbit of the comet of 1472, and by Tycho Brahe, who investigated that of 1577. Cometary knowledge grew very slowly. Giovanni Alfonso Borelli suggested in 1666 that cometary orbits were parabolic and this was confirmed by Georg Samuel Dörfel in 1681, apropos of the comet of 1680. Many orbits are indeed parabolic; others are elliptic but more often than not with great eccentricity. Edmund Halley (1656–1742) cleared the matter up in his "Astronomiae cometicae synopsis," *Phil. Trans. 24*, 1882 (1705), separately published in English (Oxford, 1705), wherein he established the periodic return of the same comet, "Halley's comet," in 1531, 1607, and 1682, and predicted another return in 1758. It did return in 1759, and again in 1835 and 1910. One may say that Halley was the first to fulfill Seneca's prophecy, after a delay of 1641 years!

[31] Franz Cumont, *L'Egypte des astrologues* (254 pp., Bruxelles: Fondation égyptologique, 1937) [*Isis* 29, 511 (1938)].

[32] The Greek technical terms are *symphōnia* (Plato, Aristotle) and *sympatheia* (Aristotle, Plutarch).

[33] See the discussion of astrology in my *Introduction, passim.* The Church repudiated astrology in theory, but was repeatedly obliged to compromise with it in practice.

wrongly, of "immorality." In this respect their morality was definitely superior to the Stoic. They refused to compromise with superstition and irrationality; they rejected astrology.

ORIENTAL RELIGIONS

Astronomy was the scientific basis of astrology, while the astral religion provided its justification. That religion was acceptable to the learned people, but was utterly insufficient even for them. Their religious feelings, however, were satisfied by the mythological poetry, and their rites, liturgies, and ceremonies, by the sacred mysteries, such as the Orphica and the Dionysia. This reminds us that Dionysos [34] was one of the most popular gods of the Hellenistic world. He was orientalized under the name Sabazios, a Phrygian god who was identified with the Cyrios Sabaōth of the *Septuagint* and called the Most High God (*Theos Hypsistos*). This is but an example among many of the orientalization of religion which was growing vigorously not only in Egypt and Asia but in the Greek lands and even in the Roman territories of Western Europe. An enumeration of the foreign gods, Macedonian, Anatolian, Persian, Syrian, Mesopotamian, would be very long. In spite of the striving after one God, Hellenistic syncretism and the blind cult of Tychē (Fortuna) were destroying religion.[35]

We have already spoken of the Hellenized Egyptian gods in the first chapter, because they were the symbols and palladium of the Ptolemaic dynasty and of Ptolemaic culture. These gods did not concern Egypt alone but were carried by Greeks to Greece, even to Dēlos, and by the Romans to the western Mediterranean Sea. In the temple of Dēlos, the Egyptian triad was composed of Sarapis, Isis, and Anubis,[36] but the most popular triad or trinity was Sarapis, his wife Isis, and their son Hōros (Harpocratēs). Sarapis and Isis were saviours; greatest of all was Isis, upon whom all the religious aspirations of the Mediterranean world were gradually focused, as is shown by her innumerable titles and names. The people in distress (and who was not?) did not want only a saviour, but a celestial mother and helper (*paraclēta*). The elaborate and impressive cult of Isis paved the way for that of the Virgin Mary, Our Lady.

[34] In Latin, Dionysos was called Bacchus, and the Dionysia, Bacchanalia. The Latin name is really of Greek, Lydian, origin, Bacchos.

[35] Hellenistic syncretism was carried so far that the people worshiped not only foreign gods but various combinations of them. For example, Stratonicē, the queen of Antiochos I Sōtēr (Seleucid, 281–261), enriched the temples of Apollōn at Dēlos, of the Syrian goddess Atargatis at Hierapolis, and of the Egyptian Anubis at Smyrna. Was she considering them as different epiphanies of the one god, or was she simply playing safe?

[36] Anubis was a god of the dead, concerned with their interment and safe passage to the nether world. The Greeks sometimes identified him with Hermēs (Hermanubis). The jackal was sacred to him, the falcon to Hōros. The iconography of Isis is as complicated as her cult, which was spread everywhere and lasted until the end of the fourth century after Christ. The destruction of the Serapeum of Alexandria by Bishop Theophilos in A.D. 391 was the end of the Egyptian religion in the Christian world.

ISRAEL

There was one Oriental religion that the Greeks did not and could not absorb, the religion of Israel. This was not caused by lack of physical contact, for there were a great many Jews in the eastern Mediterranean world and the Near East. Remember that the Jews of Israel had been deported to Babylonia by Nebuchadrezzar in 597 and 586; a great many returned fifty years after, or still later, and the temple of Jerusalem was rebuilt under Persian rule (520–516). Many Jews did not return from Babylonia, however, or did not reach Jerusalem but settled in various parts of Anatolia and Syria.

As to Egypt some Jewish settlements, especially in the island of Elephantine (Aswān, Upper Nile) were very ancient, say seventh to fifth centuries. From 323 to 198 Palestine was part of the Ptolemaic kingdom, and hence Jews could pass easily from Jerusalem to Alexandria. Yet it is probable that a good part of the Jewish population of Egypt was native born.

The Jews were soon divided into two hostile groups. Those who inclined to Hellenism adopted the Greek language and Greek manners and sometimes assumed Greek names, while those who were more faithful considered the others as renegades and "collaborators" and spoke Hebrew, or rather Aramaic.[37] The Hellenizing Jews were the aristocratic party of their sect in the Seleucid and Ptolemaic kingdoms; their ideas are reflected in Ecclēsiastēs (the Preacher, Qōheleth), written between 250 and 150, and Ecclēsiasticos (The wisdom of Ben Sira), written c. 180.[38] They spoke Greek as well as Aramaic and their knowledge of Hebrew was small, in extreme cases vestigial. Their Hellenization did not generally imply the abandonment of their religion; they attended synagogues where the service had to be conducted in the Greek language. Their own Hebrew was peppered with Greek words. Such assimilation to the ruling people is to a certain extent unavoidable.

About the end of the third century, chiefly under the rule of Ptolemaios IV Philopatōr (222–205), the syncretic tendencies of the Greeks were perhaps imitated by some Hellenized Jews, both groups (Greek and Jewish) being fooled and led astray by wrong analogies. Ptolemaios IV dreamed of one God, Dionysos, who was identified with Sabazios and Sabaōth and even with Sarapis. This could not please many people, especially not the Jews, not even those who called Adonai, Theos Hypsistos.

There remained a good number of Jews, especially among the common people, whose orthodoxy was strong enough, or whose ignorance was deep enough, to protect them from Greek contamination. Their knowledge of Greek thought was small and often wrong. For example, they considered

[37] Aramaic (an old form of Syriac) was the lingua franca of the Persian empire and continued to be used extensively in the Near East by Jews and others. *Introduction,* vol. 3, p. 356.

[38] Robert H. Pfeiffer, *Introduction to the Old Testament* (New York: Harper, 1941) [*Isis 34,* 38 (1942–43)], pp. 724–731.

Epicuros as an atheist and mocker, and used the adjective Epicurean as an insult;[39] they have done so ever since. But we must not anticipate.

As the language of the orthodox Jews was Aramaic, they needed an interpretation of the Scriptures in that language. That interpretation (Aramaic, *targum*, "Chaldee paraphrase") was oral and hence is difficult to date. It was practised from the end of the sixth century (end of the Babylonian exile) to the end of the third century or later. In the meanwhile, Hebrew scribes (*ṣopherim*) were trying to establish the Hebrew text; their work was very slow and the text was not completely established until the second century of our era. The written targums (as opposed to the oral ones already mentioned) are also post-Christian (first to fourth century and later). A Hebrew Pentateuch was written in Samaritan characters for the use of the Samaritan community in the third century B.C.[40] Finally, a Greek translation of the Old Testament was begun in the same century, the so-called Septuagint; it will be discussed below in the chapter devoted to Orientalism in the Museum.

[39] Volume 1, p. 597. As often happens, the historical origin of the insult was gradually forgotten. Simeon ben Ẓemaḥ Duran (1361–1444) was the first in medieval times to rediscover that Epicuros was a Greek philosopher! (Letter from Solomon Gandz, dated Atlantic City, N. J., 16 Dec. 1952.)

[40] The Samaritan schism occurred in the century 432–332, hence their Pentateuch may have been written a little before 300. The Samaritan script is a modification of the old Phoenician alphabet discarded by the Jews for their Torah soon after 200 B.C. Pfeiffer, *Introduction to the Old Testament*, pp. 101–104 and *passim*; my *Introduction*, vol. 1, p. 15. A small remnant of the Samaritan community still exists at Nāblus or Shechem, near Mount Gerizim, their sanctuary.

XII

KNOWLEDGE OF THE PAST IN THE THIRD CENTURY

THE EARLY HISTORIANS OF ALEXANDER THE GREAT AND THE ALEXANDER ROMANCE

The greatest hero of the Hellenistic age was Alexander the Great, who died in Babylon in June 323. At the beginning of the third century there were still many people who had known him and were ready to worship him as the *"vieux grognards"* worshiped Napoleon. It had been his privilege to be educated by Aristotle and therefore he was not simply a soldier and conqueror but a humanist. When he was in Trōas, he visited the tomb of Achilleus and envied him his Homer, who was the author of his immortal fame.[1] He was determined to have sufficient witnesses of his heroic deeds in order to insure his own immortality. Not only did he have a secretary or head of the historical department, Eumenēs of Cardia, but he surrounded himself with men of letters and philosophers. His Asiatic campaign was comparable in that respect to Bonaparte's Egyptian one. The two conquerors, separated by more than twenty-one centuries, were surprisingly alike in their love of arts and letters, their dramatic sense, and the care they took to prepare their posthumous glory.

During the Asiatic campaign, Alexander had around him men like Cleitarchos of Alexandria, Ptolemaios son of Lagos, Aristobulos of Cassandreia, Callisthenēs[2] of Olynthos, Anaxarchos *ho Eudaimonicos* (the optimist) and his pupil Pyrrhōn the skeptic, and Onēsicritos of Astypalia and Nearchos the Cretan, who were respectively the pilot and admiral of his fleet. All of these

[1] Wrote Cicero, *Pro Archaia*, x. When Alexander stood near Achilleus' tomb in Sigeum, he said, "O fortunate young men who found a Homer to praise your virtue! for if the *Iliad* did not exist even the mound covering his body would have lost its name." Sigeion (Yenisheri) is the promontory near which the Greek fleet and camp were placed according to Homer.

[2] Callisthenēs was Aristotle's nephew. He described Alexander as a champion of Panhellenism and a son of Zeus. Yet he opposed Alexander's Oriental tendencies, for example, the introduction of *proscynēsis* (the prostration or obeisance required in the presence of Oriental kings). He was executed in 327 for disloyalty and this put an end to Aristotle's friendship with Alexander.

men wrote memoirs, of which only fragments have come to us but which were used in the extant histories.

The main history that has come down to us is the one by Arrian (II–1) of Nicomēdeia, who has the double distinction of having helped to immortalize Alexander and Epictētos. It was largely based upon the memoirs of Ptolemaios Sōtēr, founder of the Ptolemaic dynasty, who had been one of Alexander's friends and generals. Ptolemaios' contribution was perhaps the best part of the history of Alexander that has come down to us. It was written from the official journal of the expedition and from other official documents, and was inspired by his own experience. Ptolemaios is one of the first exemplars of the man of action writing down his own recollections, the forerunner in this of Caesar. In addition to Arrian's, three other histories are extant, the one by Diodōros of Sicily (I–2 B.C.) in book XVII of his *Bibliothēcē*, the *De rebus gestis Alexandri Magni* of Quintus Curtius (I–1) and another Latin history written by Justinus in the time of the Antonines (138–180) but copied from the earlier work of one Trogus Pompeius of the Augustan age. To put it in the simplest manner, Arrian's history was derived mainly from Ptolemaios and Aristobulos, while the three others derived ultimately from Cleitarchos.

To these four histories should be added, but kept separate from them, the life of Alexander by Plutarch (II–1). Plutarch was primarily a man of letters, a very great one, who used the worst as well as the best sources according to his poetic fancy and to his genius. One has the feeling that his portrait of Alexander is essentially true in spite of many minor inaccuracies.

The five ancient histories of Alexander that have come down to us are derived from some fifty that are lost. This is enough to show that the extraordinary *gesta* and the personality of Alexander attracted immediate attention and praise. Moreover, he had opened up a new cosmopolitan era, and his historians, drawn from many countries, continued the international tendencies that he had inspired to Ephoros of Cymē (IV–2 B.C.). His fame was so great that the Greek and Latin books of the historians did not suffice to alleviate the popular hunger and there grew up a vast cycle of legends. The "Alexander Romance" spread everywhere; more than 80 versions of it have been collected in 24 languages.[3]

Thanks to the popular legends, Alexander became one of the best known heroes of the world. Witness Chaucer's saying (*Monk's Tale*, 3821–3823):

The storie of Alisaundre is so comune
That every wight that hath discrecioun
Hath herd somewhat or al of his fortune.

[3] For the history see W. W. Tarn, *Alexander the Great* (2 vols.; Cambridge: University Press, 1948). Charles Alexander Robinson, Jr.: *The history of Alexander the Great* (296 pp.; Providence, R. I.: Brown University, 1953). For the legend, see Volume 1, p. 491. Iskandar-nāma, *Encyclopaedia of Islam*, vol. 2 (1921), p. 535. Pseudo-Callisthenēs, *The life of Alexander of Macedon*, trans. and ed. by Elizabeth Hazelton Haight (New York: Longmans, Green, 1955).

OTHER GREEK HISTORIANS

The other historians of the third century illustrate the same cosmopolitan and scientific tendencies. Let us consider a few of them out of many more. We should always remember that when we try to describe Hellenistic learning and literature we can only sample the subject, for though the number of writers was very large (over 1100 for the whole Hellenistic period) very few of their writings have been transmitted to us; our choice is thus very arbitrary, it was determined not by ourselves but by Fortune.

Crateros the Younger. When Alexander began his Asiatic expedition, he entrusted the regency of Macedonia to one of his countrymen and generals, Antipatros, and after Alexander's death the government of Macedonia and Greece was shared by this Antipatros with another Macedonian and friend of the conqueror, Crateros. Crateros married Antipatros' daughter Phila; Crateros the Younger (321–255) was born of this marriage, probably after his father's death. Phila then married Dēmētrios Poliorcētēs and had by him another son, the future Antigonos Gonatas.[4] Thus Crateros the Younger and Antigonos Gonatas were half brothers. These details have been given to help to explain Crateros' work. He published a collection of decrees of the Athenians (*Psēphismatōn synagōgē*),[5] some of which were derived from inscriptions; the majority could be obtained only from archives. It would be easier for a man in his position to do that than for the average historian. Crateros must have realized the fundamental importance of making such a collection for the writing of history. His understanding of that was similar to the understanding of astronomy and anatomy shown by some of his contemporaries. In every case, real knowledge became possible only after the relevant facts had been patiently collected and inserted in a suitable frame.

Philochoros of Athens. Under the influence of the Lyceum various men compiled collections of facts concerning Attica; these collections, called *atthides*, were arranged in chronologic order and dealt not so much with political or military history, but rather with cultural history as the *atthidographoi* understood it, that is, mythology and the origins of cults. The most famous of these works, entitled *Atthis*, was written by Philochoros, Athenian, who was an official soothsayer (*mantis cai hieroscopos*) in 306. His *Atthis* extended to 261, however, and he was executed soon after, presumably in old age, by

[4] The greatness of Antigonos II Gonatas was partly due to that of his mother Phila, one of the best Hellenistic queens, a magnanimous woman. Her two sons, Crateros and Antigonos, were loyal to each other and were praised by Plutarch in his essay on Brotherly Love (*Moralia*, 478–492). See her portrait by Grace Harriet Macurdy, *Hellenistic queens* (Baltimore, 1932), pp. 58–69. This Phila may be called Phila I to distinguish her from her daughter-in-law, Antigonos' wife, Phila II, whose wedding hymn (epithalamion) was written by Aratos of Soloi.

[5] Some 18 fragments edited by Karl Müller in *Fragmenta historicorum graecorum* (Paris, 1848), vol. 2, pp. 617–622. Plutarch made use of Crateros' collection.

Antigonos Gonatas because of alleged betrayal to Ptolemaios Philadelphos.[6] It contained much information on Athenian history, constitution, festivals, cults, epigrams, and was arranged in chronologic order by kings and magistrates (*archontes*). It is highly probable that similar annals were compiled in other Greek cities.

The mention of annals suggests the greater chronologic problem which has been discussed at the end of the chapter on Eratosthenēs. The need of a chronologic frame that would cover not single cities or nations but the whole *oicumenē*, or at least the Greek world, was first realized by Timaios of Taormina, who invented the Olympic dating; this was systematized by Eratosthenēs and practiced by a few historians. The majority of them ignored it; as far as these were concerned, it was simpler to stick to their local chronologies without attempting to correlate them with others.

Hieronymos of Cardia. The best historian of the period was probably Hieronymos of Cardia (in the Thracian Chersonēsos, Dardanelles), who was a friend of Eumenēs of Cardia (secretary to Philippos and Alexander) and after Eumenēs' death in 316 served Antigonos I the Cyclōps, Dēmētrios Poliorcētēs, and Antigonos II Gonatas. He wrote a history of Greece from the death of Alexander to that of Pyrrhos, king of Ēpeiros (that is, from 323 to 272); it covered the revolutionary period of the wars of succession and was perhaps entitled *Hai peri diadochōn historiai* (History of the successors). Hieronymos was a soldier, not a man of letters, but he could draw pictures and characters and was trustworthy. His work was much used by Diodōros, Plutarch, and Arrian.

Menippos of Gadara. A few short notes must suffice to characterize other historians, the purpose being to present them as a group, whose activities, inspired by the Lyceum and chiefly by Theophrastos, were significant. Dēmētrios of Phalēron wrote a history of his own brief rule (317–307; he died in 283); Dēmētrios of Byzantion described in great detail the Gallic invasion of Asia Minor; Pyrrhos (319–272) published his memoirs; Aratos of Sicyōn[7] (271–213) wrote *Hypomnēmatismoi* (a kind of autobiography); Duris (340–260), tyrant of Samos, wrote a Samian chronicle and histories of Macedonia and Greece (to 280) and, what was more novel, anecdotal chronicles of literature, music, and painting; Chamaileōn of Hēracleia Pontica composed a history of poetry; Phylarchos continued Duris's history to 219. A number of scholars prepared collections of biographies (*bioi*): Clearchos of Soloi,

[6] The two kings were ruling during the same period. Antigonos Gonatas ruled Macedonia from 283 to 239 and Attica for a part of that time; Ptolemaios Philadelphos ruled Egypt from 283 to 246.

[7] F. W. Walbank, *Aratos of Sicyon* (232 pp.; Cambridge, 1933); this is a study of political history. Sicyōn was the chief town of the very small district of Sicyōnia in northeast Peloponnēsos. It was considered one of the most ancient cities of Greece, pre-Homeric. It was one of the earliest schools of Greek painting and music. The great sculptor Lysippos was a native of Sicyōn.

Satyros, Antigonos of Carystos in Euboia (lives of philosophers). The Syrian (or Phoenician?) Cynic Menippos of Gadara was so well known for his satyrical pieces that Varro (I–2 B.C.) called his own satyrical letters *Saturae Menippeae*. This title had a strange fortune, because it was given to a political pamphlet, a burlesque composition by many authors, in prose and verse, French and Latin, against the "Ligue," favoring the rule of Henri IV (king of France 1589–1610); "*la Satire Ménippée*" [8] is a monument of the French language of the Renaissance.

This selection is arbitrary for many reasons, yet such as it is it is sufficient to illustrate the historical tendencies that are as characteristic of the Hellenistic Renaissance as were its scientific undertakings. There was a widespread need of factual information which was filled more or less well by scholars most of whom were not trained historians and were certainly much below the Thucydidean level, yet they prepared the way for Polybios (II–1 B.C.).

The most original historical work, however, has not yet been dealt with. It is reserved for a special chapter on "Orientalism in the third century B.C." It concerns historical investigations relative not to the Greek world proper but to India, Babylonia, Egypt, and Israel.

THE EARLIEST ROMAN HISTORIANS. Q. FABIUS PICTOR AND L. CINCIUS ALIMENTUS

Throughout the third century there were so many wars between the Succession Kingdoms of the Near East that a clear account of them would be difficult to make and a brief one impossible. The situation was often aggravated by the growth of Roman power and the increase of Roman intrigues between the rival Greek states. Each of these was ready enough to obtain Roman help against its rivals and the Romans were equally ready to exploit those desires and impulses and to play each nation against its neighbors. At the beginning of the century Roman intrigues were already flourishing in Sicily, Macedonia, and Greece. The first great conflict was the war with Pyrrhos, king of Ēpeiros (a non-Greek nation in northeast Greece), which lasted ten years (282–272); Pyrrhos was a resourceful captain and he won victories, but at the price of such heavy losses ("Pyrrhic victories") that he was finally brought to his knees; he was killed in 272 (aet. 46), leaving his country defeated, exhausted, utterly ruined. This enabled Rome to consolidate her power in Italy. She was held in check by Carthage, however, and another, more terrible, war was the only way out. The First Punic War (264–

[8] *La Satire Ménippée de la Vertu du Catholicon d'Espagne et de la tenue des Estats de Paris* . . . (Paris, 1593–1595). The *boute-en-train* of its many authors was Pierre Le Roy, canon of the Sainte Chapelle and almoner to the cardinal de Bourbon. It was translated into English as early as 1595. A great many French editions were published. The "texte primitif" was edited by Charles Read (Paris, 1878). New edition of it and of many other documents by Edouard Tricotel (2 vols.; Paris, 1877–1881). The word Menippeus was used by later writers: *Menippeus Rusticus* (London, 1698); Henry James, *Menippea* (Dresden, 1866).

241) ended with the subjugation of the greatest part of Italy; Rome took Sardinia in 238, Corsica in 227, the Eastern part of Sicily in 211. In the meanwhile, a Roman fleet was sent into the Adriatic to suppress the pirates of Queen Teuta.[9] This triumph pleased the Greeks so much that in 228 they admitted the Romans to the Isthmian Games and to the Eleusinian Mysteries. Hellas was thus opening her inner doors to a civilized friend, but within a couple of centuries, or less, Rome was to be mistress of the Hellenic house.

There was nothing in the Mediterranean world to stop Rome except the Carthaginian empire, and war between the two empires was again inevitable. The Second Punic War (218–201) was almost won by Carthage, thanks to the genius of Hannibal, one of the greatest generals of all time, yet the Romans triumphed in the end. At Zama [10] in 202 Scipio Africanus annihilated the Carthaginian army.[11] The Carthaginians were obliged to surrender Spain and all the islands, and to abandon a part of Africa to Masinissa, Rome's ally. Rome was now mistress of the western Mediterranean, and potential mistress of the whole Mediterranean world.

This very abbreviated sketch is unavoidably oversimplified. The point of it is simply to illustrate the enormous growth of Rome during the third century. One would expect Roman historians to appear, relate these political marvels, and ascribe them to Fortuna, who approved herself to be a national goddess.[12]

There were two early historians, Q. Fabius Pictor [13] (fl. 225–216) and L. Cincius Alimentus (praetor in Sicily in 209), who composed histories of Rome from the arrival of Aeneas (Aineias) to the Second Punic War, but both wrote *in Greek*. Rome was preparing herself to be mistress of the world, but her language, that is, in the last analysis, her civilization, was still immature and conscious of its inferiority.

The history of Hellenistic historiography will be continued in Chapter XXIV.

[9] She was queen of Illyria (north of Ēpeiros; along the eastern coast of the Adriatic), and the Roman war against her is called the First Illyrian War (229–228).

[10] In Numidia, just west of the Carthaginian border.

[11] Hannibal managed to escape, however. A few years later, Roman intrigues drove him out of Carthage. He took refuge at the court of Antiochos III the Great (king of Syria 223–187) and after Antiochos' defeat in 188, at that of Prusias (king of Bithynia), who betrayed him to the Romans. To avoid capture, he committed suicide in 183, aet. 64. A disciple of Alexander, Pyrrhos, and his father Hamilcar Barca, he was not only a great general but an outstanding leader of men; a great man.

[12] Fortuna loved Rome and the Romans loved her. Her cult was celebrated in the Latium, especially in Antium (on a promontory entering the Tyrrhenian Sea) and in Praeneste (near Rome, modern Palestrina). The oracles delivered in the temple of the latter city were called *Praenestinae sortes*.

[13] His grandfather, C. Fabius "Pictor," was so-called because of a painting of his in the temple of Salus publica (or Romana) in the Quirinal. This is the earliest Roman painting on record (c. 307–302). Salus was originally the Latin equivalent of Hygieia, but she had become gradually very much like Fortuna.

XIII

LANGUAGE, ARTS, AND LETTERS

THE EARLY GROWTH OF GREEK PHILOLOGY

The third century was a golden age of Greek philology, but the work done has already been explained in Chapter X devoted to the Library. We made it clear then that the "librarians" were not simply librarians in the modern sense, whose task consists in making definite books available to the public. For one thing, such "books" did not yet exist. The librarians were obliged to put in order a very large number of papyrus rolls.

As the rolls were collected rapidly by ambitious kings and piled up in large quantities, it was necessary to describe them and to divide them into groups. Each group, say poetry, was intrusted to a competent scholar. It was soon divided into subgroups – dramatists, epic poetry, lyric poetry, and so on – and gradually all the rolls relative to a single poet, say Homer, were separated from the others. This was only the beginning. It was necessary to distinguish various editions of the *Iliad*, each of which filled many rolls (these rolls were not always written by the same hand).[1] All the rolls pertaining to a single edition were finally put together. On the other hand, other texts were so short that many might be included in the same roll, and it was necessary to record the fact in the individual descriptions and eventually in the general catalogue.

The librarians of Alexandria (and of all the ancient libraries) were like the keepers of manuscript collections, or rather like the precursors of the modern keepers, those who prepared the first catalogues and were obliged not only to examine every manuscript but to read great parts of it, and to compare each manuscript with many others. These librarians were not only philologists in the fullest sense, but pioneer philologists.

[1] A roll of about 32 to 35 feet (which was the average length) would be long enough to contain one of the longer books of the New Testament (Matthew, Luke, or Acts), or a single book of Thucydides. Hence, no work of any considerable length could be contained in a single vehicle. This became possible only when the roll was replaced by the codex, and papyros by vellum. This explains why the collected works of most authors failed to reach us; a few rolls were transmitted while the others were lost. Frederic G. Kenyon, *Books and readers in ancient Greece and Rome* (Oxford: Clarendon Press, ed. 2, 1951), p. 64.

While so many good men – Zēnodotos of Ephesos, Alexander of Pleurōn, Lycophrōn of Chalcis, Callimachos of Cyrēnē, Eratosthenēs of Cyrēnē, Aristophanēs of Byzantion – were thus engaged in the study of the Greek language and in editions of the classics, others (as well as themselves) were enriching Greek literature with their own compositions. It must be admitted at once that with few exceptions their own gifts were very inferior to the ancient treasures. We have already spoken of the didactic poets, such as Aratos and Nicandros, who satisfied the needs of an age that was on the whole more scientific than poetic. It is noteworthy, however, that neither of them was Alexandrian. Aratos was a Cilician who spent half of his life in Macedonia and the other in Syria. Nicandros hailed from Ionia. Both were Asiatic Greeks.

MENANDROS OF ATHENS

The Alexandrian revolution did not put an end to the Athenian stage. New playwrights appeared who created the "New Comedy." Two of them were remarkable, Philēmōn and Menandros, and the second of these is one of the greatest in world literature.

Philēmōn of Soloi, born in 361 in Soloi (Cilicia), spent his life in Athens and Alexandria or at the Peiraieus, where he shared a villa with his mistress Glycera; he died at the Peiraieus while Athens was besieged in 262, at the age of 99. He wrote some 97 comedies, of which 54 are known to us by their titles. Our knowledge of them is otherwise restricted to fragments or to imitations by the Roman Plautus (254–184), who was close enough to him in point of time. He was clever in the invention of comic situations and obtained much success in Athens. He became a full citizen and won many competitions. His art was superficial, however, and he was not able to create characters.

His rival Menandros (342–291) was a true Athenian. He was born twenty years later than Philēmōn but lived fifty years less, hence the latter survived him by some thirty years. We must remember that when we think of them as contemporaries. Menandros was the real "star" (*astēr*) of the New Comedy, even if some of Philēmōn's "new" plays were anterior to his. He was the son of rich parents and had received a philosophical education, being influenced chiefly by Theophrastos and Epicuros. His fecundity was even greater than Philēmōn's, for in his much shorter life he composed over a hundred comedies (98 are known by title). His art was far superior to Philēmōn's, though the latter's plays were sometimes crowned in preference to his own. Not a single one has come down to us, but we have many fragments, the best-known play being the *Geōrgos* (a tiller of the earth) preserved in a papyrus.[2] Many plays have been transmitted by the Latin adaptations of Plautus and of the Carthaginian Terentius.

Menandros did not reach the level of Euripidēs, whom he greatly ad-

[2] Edited by Jules Nicole, *Le laboureur de Ménandre* (Geneva, 1898).

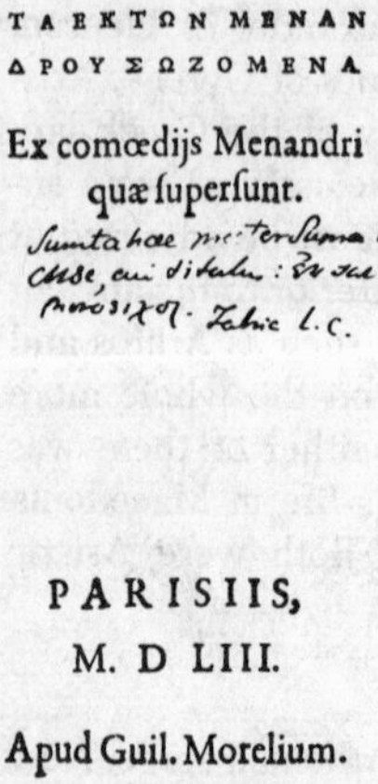
ΤΑ ΕΚ ΤΩΝ ΜΕΝΑΝ
ΔΡΟΥ ΣΩΖΟΜΕΝΑ

Ex comœdijs Menandri
quæ ſuperſunt.

PARISIIS,
M. D LIII.

Apud Guil. Morelium.

Fig. 33. Princeps of Menandros' fragments printed by Guillaume Morel (Paris, 1553) in the collection entitled *Veterum comicorum XLII quorum integra opera non extant sententiae* (small size, 15 cm, 27 leaves). [Courtesy of Harvard College Library.]

mired; yet he was a poet as well as a moralist and had a sound dramatic instinct. He did create characters, was able to diversify his language in obedience to the needs of each of them, and was sufficiently realistic. This quality was well expressed in a humorous query by Aristophanēs of Byzantion: "Which of the two, Menandros and Nature, imitates the other?" He is definitely Hellenistic, for his first play appeared on the stage in the year following Alexander's death. Many verses of his have become proverbial even in English.[3]

Menandros was invited by Ptolemaios Sōtēr to come to Alexandria, but he chose to remain in Athens. In his own time Philēmōn was sometimes preferred to him by the Athenian audiences, but his superiority was soon realized. A significant witness to that is the absence of Philēmōn papyri, while many of them contain long fragments (whole scenes) of Menandros' plays.

He was praised by Quintilian (I-2) and Plutarch (I-2) but somewhat forgotten in later times because the texts had failed to survive (except for the papyri unknown before the end of the nineteenth century). He was really one of the greatest writers of comedies, comparable to Molière.[4]

A FEW MINOR POETS

Let us speak more briefly of a few other poets. Asclēpiadēs of Samos (fl. 270) wrote love poems and epigrams. Though some epigrams (or metrical

[3] For example, "Conscience makes cowards of the bravest." The preservation of those lines was facilitated by a collection of them, put together probably in Roman times, the *Gnōmai monostichoi* (one-line sentences).

[4] The princeps edition of Menandros (1553) was included in the collection, *In veterum comicorum XLII quorum integra opera non extant sententiae* (Paris, 1553), pp. 3-56; many more editions in the sixteenth century and later. The most convenient is the Greek-English edition by Francis G. Allison, *Menander, the principal fragments* (Loeb Classical Library; Cambridge, 1929).

inscriptions) can be traced back to the seventh century, the genre obtained greater popularity (if not greater distinction) in the Hellenistic period. No Hellenistic epigrammatist attained the grace and strength of Simōnidēs (556–468) and of other poets of the fifth and fourth centuries, yet we owe to them many sensitive and curious examples of the art. Philētas of Cōs,[5] tutor to Ptolemaios Philadelphos and to Zēnodotos, was a poet as well as a grammarian and might be called the founder of the Alexandrian school of poetry. His frame was as delicate as his poetry and became legendary; it was told that he was obliged to wear leaden soles to prevent his being blown away by the wind![6]

Lycophrōn (born c. 325) of Chalcis wrote many tragedies but is chiefly remembered because of an epic poem, *Alexandra* (1474 iambics), which has the doubtful distinction of being exceedingly obscure and the more valuable one of being a witness of the impression that Roman power had made upon the Hellenistic world. The general theme of *Alexandra* is one of epic greatness: the destruction of Troy, the return of the Greeks, the struggle between Europe and Asia; above all, the Greek sufferings, felt to be a compensation for the Trojan ones (remember that Roman glory was also conceived as a vindication of Troy; Aeneas was a Trojan hero before he was a Roman one). Lycophrōn, however, was not on the level of his theme and he spoiled his poem with excessive learning and poor artistry. Its obscurity (even to its own contemporaries, not to mention ourselves) is due to bad composition, mythological confusion, and a very artificial vocabulary.[7] It is a typical example of the worst side of Hellenistic literature; it has delighted the pedants of all ages.[8]

Let us leave Lycophrōn and return to poetry. A papyrus discovered in 1890 revealed the work of the Egyptian poet Hērōdas, eight mimes describing not only lovers but pimps. Hērōdas described the bawdy life around him but he was a real artist, not a pedant.[9] He flourished in Cōs and Egypt, probably in the time of Ptolemaios Philadelphos.

In spite of his immense learning, Callimachos of Cyrēnē was also a genu-

[5] Cōs had been under Macedonian power but was "liberated" by Ptolemaios Sōtēr in 310; from that time on, it was closely connected with Alexandria. It was probably used by the Ptolemies as a summer resort. Ptolemaios Philadelphos was born there in 308. The delightful little island was illustrated in the fifth century by Hippocratēs, in the fourth century by the painter Apellēs, in the third century, by at least four poets, Philētas, Aratos, Theocritos, Hērōdas.

[6] After J. E. Sandys, *History of classical scholarship* (Cambridge, ed. 3, 1921), p. 118.

[7] Out of 3000 words, 518 are found nowhere else and 117 appear for the first time (*Oxford Classical Dictionary*). That must be a record!

[8] Convenient Greek-English edition of the *Alexandra* by A. W. Mair, *Callimachus, Lycophron, and Aratus* (Loeb Classical Library; Cambridge, 1921), pp. 477–617.

[9] First edition by Frederick George Kenyon, *Classical texts from papyri in the British Museum including the newly discovered poems of Herodas* (London, 1891). Greek-English edition with Theophrastos' *Characters* by Alfred Dillwyn Knox (Loeb Classical Library; Cambridge, 1929).

ine poet. His main work, the *Catalogue raisonné* of the Library, of which he was the director, is unfortunately lost. His other prose works are also lost but enough of his poetry has been transmitted to us to reveal his genius. We have hymns to Zeus, Apollōn, Artemis, Dēlos, Pallas, and Dēmētēr, 64 epigrams, and many fragments. His longest poetical work was an elegy, *Aitia* (*Cause*), which extended to more than 3000 lines, but very little of it is available to us. It described in the form of a dream many Greek legends and rites and was imitated in Latin by Cato the Censor (II–1 B.C.) in his *Origines* (at any rate, the title *Origines* corresponds exactly to *Aitia*). Another poem of his, *The lock of Berenice* (*Berenicēs plocamos*), had a singular fortune. It was dedicated to Berenicē, daughter of Magas, king of Cyrēnē, who married Ptolemaios III Evergetēs in 247. She hung up her hair as an ex-voto in the temple of Arsinoē Aphroditē, but it disappeared and was carried to heaven, where it became the constellation Coma Berenices (Berenice's Hair or Locks). A fine story for a poet. Only 10 lines remain of Callimachos' poem, but we have Catullus' Latin translation and it inspired Ovid. Tennyson's poem *Teresias* is derived from Callimachos' fifth hymn on the baths of Pallas. It relates the story of a young Theban, Teiresias, who happened to see Athēnē bathing; she blinded him but gave him the power of prophecy; he lived to a very old age and became one of the most famous "seers" of antiquity. Many other epigrams are delicate and sensitive, such as the one (no. VI) on the nautilus shell, dedicated to Arsinoē Aphroditē of Zephyrion.[10] Unfortunately, it helped to advertise Aristotle's mistaken account of the nautilus using its membrane as a sail and its arms as oars.[11] At his best Callimachos was very good, but he could not invite his soul often enough for he was obliged to carry an immense burden.[12]

Timōn of Phlius (northeast Peloponnēsos) was a disciple of Pyrrhōn and his prophet. A skeptic and sophist, he finally settled down in Athens where he died c. 230 aet. 90. He wrote satirical poems, or rather parodies, called *silloi*, and therefore was nicknamed *"ho sillographos."*

Euphoriōn of Chalcis studied philosophy at Athens, flourished at the court of Alexander, ruler of Euboia and Corinth, married his widow, and

[10] Arsinoē Aphroditē was the theophany of Arsinoē II (d. 270), wife of her brother Ptolemaios II Philadelphos; a temple was dedicated to her at the promontory of Zephyrion, East of Alexandria. She was a patroness of sailors (Aphroditē Euploia, Pelagia). Before her apotheosis she had proved herself to be a woman of extraordinary beauty and intelligence but as unscrupulous as the kings of her time. For more information see note 23.

[11] The Aristotelian legend refers to the "paper nautilus" (*Argonauta argo*) for which see Volume 1, p. 542. The genus *Nautilus* (*nautilos* in Greek means sailor) was so called because of that legend. The paper nautilus is not a true *Nautilus* but an *Argonauta*, a different genus of cephalopoda, related to the octopus. Would that Callimachos had known the true nautilus and its gnomonic growth, admirably explained by Sir D'Arcy W. Thompson, "La coquille du Nautile," in *Science and the classics* (London: Oxford University Press, 1940 [*Isis* 33, 269 (1941–42)], pp. 114–147.

[12] Convenient Greek-English edition of Callimachos by A. W. Mair; see note 8.

was appointed librarian at Antioch[13] by Antiochos the Great (ruler of Syria, 223–187). He probably spent the rest of his life in Antioch and was buried there (or in Apameia). Many poems are ascribed to him: epigrams, mythological pieces, *epyllia* (little epics). Very little remains, but his contemporary influence is proved by the praise and borrowings of many other poets, Greek and Latin, including Catullus and Virgil. He compiled a lexicon to Hippocratēs (lost).

The Cretan Rhianos flourished in Alexandria during the last quarter of the century. He prepared new editions of the *Iliad* and the *Odyssey*, and wrote epigrams and epics that include many geographic details; the poems are practically lost but many of those details have been preserved by Stephanos of Byzantion (VI–1) in his geographic dictionary. Rhianos' story of the Second Messenian War and of the heroism of Aristomenēs has been preserved for us by Pausanias (II–2).[14]

Cercidas of Megalopolis[15] (c. 290–220) was a Cynic, a liberal politician, and a poet. It is a great pity that his poems have vanished, because they represented a new genre, being partly devoted to the defense of the unfortunate. Cercidas was probably one of the first political poets, if not the very first.

The preceding notes, however brief, are sufficient to commemorate minor poets and to illustrate the diversity of their origins and of their gifts. We have kept for the end two longer notices on Apollōnios the Rhodian and Theocritos of Syracuse. The subject to which the former addressed himself consecrated his fame, while the latter will live forever in the hearts of men because of the genuineness of his poetry.

APOLLŌNIOS OF RHODOS

It is difficult to date Apollōnios with precision, but he was a pupil of Callimachos and that places him in the second half of the third century. He may have succeeded Callimachos as director of the Library of Alexandria (c. 240–235). The best-known event of his life was his quarrel with Callimachos, a literary battle that was gradually aggravated and poisoned by the acrimonious remarks of both rivals. Theirs was the greatest conflict

---

Very elaborate edition of Callimachos by Rudolfus Pfeiffer (Oxford: Clarendon Press, 1949, 1953).

[13] One is not astonished that there was a library in Antioch, which was a flourishing city. The Seleucid era began in 312. The founder of the Seleucid empire, Seleucos I Nicatōr (358–280) established his first capital, Seleuceia on the Tigris, in 312, and the second, Antioch on the Orontēs, c. 300. Both cities were very Greek and tried to emulate Alexandria.

[14] Messēnia is in the southwest Peloponnēsos. The second Messenian war against Sparta (685–668) was finally lost by the Messenians in spite of Aristomenēs' heroism, and Messēnia was occupied by the enemy. Aristomenēs spent the end of his life in Rhodos.

[15] Megalopolis in Arcadia, central Peloponnēsos; the Arcadians thought of themselves as the most ancient people of Greece; real Pelasgoi, they loved music and freedom. Megalopolis was a relatively new city, built upon the advice of Epameinōndas after his great victory of Leuctra (371), which put an end to Spartan hegemony.

of its kind in Hellenistic days, yet one does not know exactly what caused it. It is probable that there was no definite cause except differences of age and temperament and mutual jealousy.

Apollōnios was born in or near Alexandria, but at some time he retired to Rhodos, where he spent the end of his life. His departure may have been caused by the quarrel with Callimachos and may have abbreviated his directorship of the Library. We may assume that his main literary work was done in that island and it was there that he became famous. It is significant that he was never called Apollōnios of Alexandria but Apollōnios the Rhodian.[16]

His masterpiece was an epic poem, the *Argonautica* (Fig. 34), which has been completely preserved in spite of its relative length.[17] He was not the first to tell in verse the wonderful tale of the Argonauts, for Pindaros had done so in his fourth Pythian ode (c. 462 B.C.).

The old story may be summarized as follows. Prince Phrixos and his sister Hellē had been sacrificed to Zeus but their mother Nephelē managed to save them. In answer to her prayers a flying ram with a golden fleece took them away; Hellē fell in the sea which was named after her, Hellēspontos (Dardanelles). Phrixos reached Colchis,[18] where he was made welcome by the king, Aiētēs, who gave him his daughter Chalciopē as a wife. As to the Golden Fleece (*to chrysun cōdion*), the king caused it to be hung on an oak in a sacred wood and guarded by a sleepless dragon. Greek adventurers led by the Thessalian Iasōn undertook to capture it. Argos built for them the great ship Argō (hence her navigators were called Argonautai). Iasōn was not a common hero, for he had been nurtured by the centaur Cheirōn; he sailed accompanied by fifty adventurers as illustrious as himself (Hēraclēs, Castōr and Polydeucēs, Thēseus) and they finally reached Colchis. Thanks to the complicity of Mēdeia, another daughter of King Aiētēs, the dragon was stupefied, other obstacles were overcome, and the Golden Fleece was captured. Iasōn married Mēdeia and they returned to Greece, but they were not happy afterward.

That fantastic story may have a kernal of reality, to wit, Minoan navigations across the Black Sea. Thus, the adventures of Sindbād the Sailor in the *Thousand and one nights* were probably inspired by the navigation of Sulaimān the Merchant (IX–1) across the Indian Ocean and the China Sea.[19] The tale of the Argonautai, mixed up with an endless number of

[16] This was not exceptional in Greece or anywhere else. If one says habitually Philip the Athenian, John of Ghent, or Muḥammad al-Baghdādi, it does not follow that Philip, John, and Muḥammad were born in Athens, Ghent, or Baghdād but simply that more people associated them with those cities than with any others.

[17] It contained 5835 lines, a little less than half as many as the *Odyssey*. For the length of other epics, see Volume 1, p. 134.

[18] Colchis is a small country at the eastern end of the Black Sea; it was traversed by the river Phasis, after which the pheasant is named (*ho phasianos ornis*).

[19] *Introduction*, vol. 1, pp. 571, 636. Jean Sauvaget, *Akhbār aṣ-Ṣin wa-l-Hind* (122 pp.; Paris: Collection arabe Guillaume

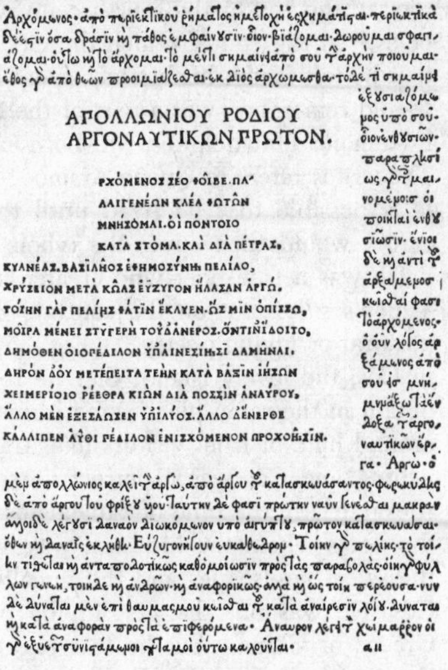
ΑΠΟΛΛΩΝΙΟΥ ΡΟΔΙΟΥ
ΑΡΓΟΝΑΥΤΙΚΩΝ ΠΡΩΤΟΝ.

ΡΧΟΜΕΝΟΣ ΣΕΟ ΦΟΙΒΕ. ΠΑ-
ΛΑΙΓΕΝΕΩΝ ΚΛΕΑ ΦΩΤΩΝ
ΜΝΗΣΟΜΑΙ. ΟΙ ΠΟΝΤΟΙΟ
ΚΑΤΑ ΣΤΟΜΑ. ΚΑΙ ΔΙΑ ΠΕΤΡΑΣ
ΚΥΑΝΕΑΣ. ΒΑΣΙΛΗΟΣ ΕΦΗΜΟΣΥΝΗ ΠΕΛΙΑΟ,
ΧΡΥΣΕΙΟΝ ΜΕΤΑ ΚΩΑΣ ΕΥΖΥΓΟΝ ΗΛΑΣΑΝ ΑΡΓΩ.
ΤΟΙΗΝ ΓΑΡ ΠΕΛΙΗΣ ΦΑΤΙΝ ΕΚΛΥΕΝ. ΩΣ ΜΙΝ ΟΠΙΣΣΩ,
ΜΟΙΡΑ ΜΕΝΕΙ ΣΤΥΓΕΡΗ ΤΟΥΔ' ΑΝΕΡΟΣ. ΟΝΤΙΝ' ΙΔΟΙΤΟ,
ΔΗΜΟΘΕΝ ΟΙΟΠΕΔΙΛΟΝ ΥΠ' ΕΝΝΕΣΙΗΣΙ ΔΑΜΗΝΑΙ.
ΔΗΡΟΝ Δ' ΟΥ ΜΕΤΕΠΕΙΤΑ ΤΕΗΝ ΚΑΤΑ ΒΑΞΙΝ ΙΗΣΩΝ
ΧΕΙΜΕΡΙΟΙΟ ΡΕΕΘΡΑ ΚΙΩΝ ΔΙΑ ΠΟΣΣΙΝ ΑΝΑΥΡΟΥ,
ΑΛΛΟ ΜΕΝ ΕΞΕΣΑΩΣΕΝ ΥΠ' ΙΛΥΟΣ. ΑΛΛΟ Δ' ΕΝΕΡΘΕ
ΚΑΛΛΙΠΕΝ ΑΥΘΙ ΠΕΔΙΛΟΝ ΕΝΙΣΧΟΜΕΝΟΝ ΠΡΟΧΟΗΣΙΝ.

Fig. 34. Princeps of the *Argonautica* by Apollōnios of Rhodos, with commentary around the text (172 unnumbered leaves; Florence: Lorenzo Francisci di Alopa, 1496); the first leaf contains a life and genealogy of the author in Greek. [From the Firmin Didot copy now in the Harvard College Library.]

other myths, was an intrinsic part of the Greek folklore and eventually of European folklore.[20]

Apollōnios' epic is divided into four books. Books I and II deal mainly with the voyage to Colchis, the main part of Book III with the love of Iasōn and Mēdeia, Book IV with the return journey. The love story is the best part of the whole work; it was the first elaborate love story of its kind and exerted a deep influence upon Roman and European letters. The abundant geographic details of Book IV are typical of a cosmopolitan age, whose geographic curiosity had been excited by Eratosthenēs.[21]

It would be tempting to write a book entitled "The Argonauts in the arts and letters," but this would take considerable pains and time, for the romantic story has inspired innumerable poets and artists.

THEOCRITOS OF SYRACUSE

We shall now end, as one should, with the best, and praise Theocritos, the greatest Greek poet of the Hellenistic age. He was born in Syracuse

Budé, 1948) [*Isis 41*, 335 (1950)]; "Les merveilles de l'Inde," *Mémorial Sauvaget* (Damas: Institut français, 1954), pp. 189–309.

[20] Its lasting popularity is proved by the creation of the order of chivalry, the Golden Fleece, in Bruges in 1429 by Philip the Good, duke of Burgundy: H. Kervyn de Lettenhove, *La Toison d'Or* (104 pp.; Brussels, 1907). The adventurers who went to California in 1848 and following years in search of gold were sometimes called argonauts. The name *argonauta* was given to the paper nautilus.

[21] The princeps of the *Argonautica*, by Lascaris (Florence, 1496), was followed by

toward the end of the fourth century, that is, during the tyranny of Agathoclēs,[22] at the end of which the city was ruined. It was not surprising then that he left Sicily; his life was spent largely in Alexandria and Cōs. Cōs, we should remember, was a part of the Ptolemaic empire, and the second king, Ptolemaios Philadelphos, was born in the island in 309. In one of his poems, Theocritos refers to Queen Arsinoē [23] as being still alive (she died in 270). It is possible that he lived until the middle of the century; his literary career would then cover the whole of the first half of the third century.

He was a true poet who created a new genre, not a secondary one like Timōn's *silloi*, but on the contrary one of the supreme kinds of poetry – pastoral or idyllic poetry [24] (Fig. 35). It was probably around Syracuse or in Cōs, the lovely island, that he received his inspiration, while he could obtain in the same place some formal training from Philētas and the poets around him, or from visitors like Aratos. The main thing, however, was his genius, and Cōs was a perfect nursery for it. He spent some time also in Alexandria in the time of Ptolemaios Philadelphos [25] and was influenced by the poets whom the Museum was nourishing, but his main teachers were the sweet landscapes and the bucolic graces, first of Syracuse and later of Cōs. He was not the first idyllic poet – there may have been others in Greece or China – but he was one of the greatest in the literature of all times and all countries. He was a poet of sunshine. Nature as reflected by his genius is not hard, as in Hesiod, nor melancholy, as in Virgil, but smiling and radiant.

---

many other editions: Venice, 1521; Paris, 1541; Geneva, 1574; Leiden, 1641; Oxford, 1777 (the two last with Latin version). Greek-English edition by R. C. Seaton (Loeb Classical Library; Cambridge, 1912).

[22] Agathoclēs, the only Hellenistic king among western Greeks, was tyrant of Syracuse from 317, proclaimed himself king of (eastern) Sicily in 304, died in 289. His rule was spoilt by incessant discord and frequent warfare. His enemies were the Carthaginians, the Greeks of western Sicily, the Romans, and also his own people, yea, his own family.

[23] Arsinoē II, daughter of Ptolemaios I and Berenicē, was perhaps the greatest of the Hellenistic queens. She married Lysimachos, one of the companions and successors of Alexander. After Lysimachos' defeat and death (281), she married her stepbrother Ptolemaios Ceraunos, and after his defeat and death (280) fled to Egypt, where she married (279) her brother Ptolemaios II Philadelphos, who was passionately devoted to her. Her power was considerable but unredeemed by any good deed. Shortly before her death in 270, she was deified under the name Philadelphos. Her influence is proved by the fact that the Faiyūm, a rich oasis in the Libyan desert, was named after her Arsinoitē nomē, and one of its cities was named Crocodeilopolis-Arsinoē. For her biography, see Auguste Bouché-Leclerq, *Histoire des Lagides* (Paris, 1903), vol. 1, pp. 164–181, and Grace Harriet Macurdy, *Hellenistic queens* (Baltimore, 1932), pp. 111–130. See also Dorothy Burr Thompson, "Portrait of Arsinoē Philadelphos," *American Journal of Archaeology* 59, 199–206, pl. 54–55 (1955), apropos of a small stone head in the Sisilianos collection in Athens, which is claimed to represent Arsinoē.

[24] The English word "idyl" is simply a transcription of the Greek *eidyllion* meaning a small *eidos*, a small form, figure, image. The verb *eidō*, to see or to know, is the same word as the Latin *video*. The word *eidyllion* does not occur in Theocritos; it was introduced later by grammarians.

[25] Praised in idyls XIV, XV, XVII. Reference to Arsinoē in XV, l. iii.

ΘΕΟΚΡΙΤΟΥ ΘΥΡΣΙΣ Η ΩΔΗ

ΕΙΔΥΛΛΙΟΝ ΠΡΩΤΟΝ.

ΘΥΡΣΙΣ Η ΩΔΗ.

Ἀδύ τι τὸ ψιθύρισμα καὶ ἁ πί-
τυς αἰπόλε τήνα,
Ἁ ποτὶ ταῖς παγαῖσι μελίσ-
δεται. ἁδὺ δὲ καὶ τὺ
Συρίσδες. μετὰ Πᾶνα τὸ δεύ-
τερον ἆθλον ἀποισῇ.
Αἴκα τῆνος ἕλῃ κεραὸν τρά-
γον, αἶγα τὺ λαψῇ.
Αἴκα δ' αἶγα λάβῃ τῆνος γέρας, ἐς τὲ καταρρεῖ
Ἁ χίμαρος, χιμάρῳ δὲ καλὸν κρῆς ἔστε κ' ἀμέλξῃς.
ΑΙ. Ἅδιον ὦ ποιμὰν τὸ τεὸν μέλος ἢ τὸ καταχὲς
Τῆν' ἀπὸ τᾶς πέτρας καταλείβεται ὑψόθεν ὕδωρ.
Αἴκα ταὶ μῶσαι τὰν οἶιδα δῶρον ἄγωνται,
Ἄρνα τὺ σακίταν λαψῇ γέρας. αἰ δέ κ' ἀρέσκῃ
Τήναις ἄρνα λαβεῖν, τὺ δὲ τὰν ὄιν ὕστερον ἀξεῖς.
Θ. Λῇς ποτὶ τᾶν νυμφᾶν λῇς αἰπόλε τᾷδε καθίξας
Ὡς τὸ κάταντες τοῦτο γεώλοφον αἵ τε μυρῖκαι,
Συρίσδεν, τὰς δ' αἶγας ἐγὼν ἐν τῷδε νομευσῶ;
ΑΙ. Οὐ θέμις ὦ ποιμὰν τὸ μεσαμβρινὸν, οὐ θέμις ἄμμιν
Συρίσδεν. τὸν Πᾶνα δεδοίκαμες. ἦ γὰρ ἀπ' ἄγρας
Τανίκα κεκμακὼς ἀμπαύεται. ἐντὶ δὲ πικρὸς
Καί οἱ ἀεὶ δριμεῖα χολὰ ποτὶ ῥινὶ κάθηται.
Ἀλλὰ τὺ γὰρ δὴ θύρσι τὰ δάφνιδος ἄλγεα ἀείδες
Καὶ τᾶς βουκολικᾶς ἐπὶ τὸ πλέον ἵκεο μοίσας.

Α·Α ii

Fig. 35. Greek edition of Theocritos together with Hēsiodos (small folio, 30 cm, 140 unnumbered leaves; Venice: Aldus Manutius, Feb. 1495). [From one of two copies in the Harvard College Library.] This is not the princeps which was published by Bonus Accursius in Milan c. 1480. The princeps also contained Hēsiodos' text; a page of it was reproduced in Volume 1 (p. 149).

According to ancient tradition, there were two other bucolic poets who followed him, Moschos of Syracuse, a grammarian, a pupil in Alexandria of Aristarchos of Samothracē (II–1 B.C.), and Biōn of Smyrna, the "neatherd," who may be placed a little later (c. 100 B.C.). Little of their work has come down to us, and that little is not pastoral. Theocritos towers above them. The simplicity, subtle beauty, and harmony of his verses cannot be described any more than one can describe music. Look yourself at the gracious images or listen to the sweet words.[26]

[26] The princeps of Theocritos was combined with that of Hesiod (Milan, 1480). A facsimile of a page appears in Volume 1, p. 149. That princeps contained 18 idyls out of 30. The Aldine edition (Venice, 1495) contained almost 29 idyls, plus fragments of Moschos and Biōn. The best edition of the bucolic poets is that by Wilamowitz-Moellendorff (Oxford, 1905). Greek-English edition of the bucolic poets by John Maxwell Edmonds (Loeb Classical Library, 1912). Arthur S. Hunt and John Johnson, *Two Theocritus papyri* (London, 1930). In the Loeb edition 380

Theocritos was a greater poet than his Hellenistic predecessors but in addition his poems have an eternal validity. Any sensitive person will understand them at once, and vibrate with him, whether he reads them in a fair translation or, preferably, in the original. By contrast, there are few people today, if any, who can read some of the ancient epigrams and poems like the *Argonautica*, not only because they are too learned but chiefly because their learning is obsolete. Until the eighteenth century and even the nineteenth, educated men were supposed to be familiar with classical mythology; that knowledge has become rare, and no one can enjoy a poem if he must consult a dictionary at almost every step in order to understand what he reads. Scholars of the Renaissance could still appreciate Apollōnios, but we can no longer. Theocritos' audience, on the contrary, is increasing, and will continue to increase. Poetry is not jeopardized by science but rather by artificial learning and pedantry.[27]

SCULPTURE

The traditions of Egyptian art were continued by the Ptolemaic kings, who favored them, but, in addition, there was some blossoming of Greek art.[28] Bryaxis, one of the sculptors employed at the Mausoleum,[29] made an "Apollōn" for the temple of Daphnē (near Antioch) and a "Serapis" for Ptolemaios Sōtēr. Greek art, however, had better chances of flourishing in the other Hellenistic kingdoms, where there was no strong competition with it as in Egypt.

Many art centers were kept alive by the rivalry obtaining between their princes. Among those centers which frequent repetitions have imprinted on my mind, let me mention Syracuse and Acragas in Sicily; Cyrēnē in Africa; Athens, Epidauros, Sicyōn, Olympia, Dēlos in Greece; Pergamon, Antioch, and Rhodos in Asia.

Lysippos of Sicyōn. Charēs of Lindos. Lysippos of Sicyōn,[30] Alexander's sculptor and the greatest one of his time, influenced the Hellenistic age in many ways. Alexander used to say that nobody should paint him but Apellēs and no one make his statue but Lysippos. The latter's activity was prodigious; some 1500 works were ascribed to him by Pliny, who no doubt

---

pages are devoted to Theocritos (30 idyls, 24 epigrams, fragments), 40 pages to Moschos, 32 pages to Biōn.

[27] The history of Hellenistic literature, Greek and Latin, will be continued in Chapter XXV.

[28] Six examples of the Egyptian art of the Ptolemaic age are given in Figs. 1–5 and 39. For other illustrations, see José Pijoán, *Summa artis* (vols. 3 and 4; Madrid, 1932); Margarete Bieber, *The sculpture of the Hellenistic age* (New York: Columbia University Press, 1955).

[29] A monument built at Halicarnassos (in Caria, southwest corner of Asia Minor) by Artemisia II to immortalize the memory of her brother and husband Mausōlos (satrap of Caria, 377–353). Many remains of it are in the British Museum.

[30] Sicyōn, in the northeast Peloponnēsos, was an artistic center from the Alexandrian age until the first century B.C.; it had a school of art and perhaps also a museum.

exaggerated; yet many of them were distributed all over Greece, teaching a new canon of the human body, slimmer than the old one, and a new technique. Lysippos published so many heads and statues of Alexander that he created the Alexandrian iconography, the Alexandrian ideal in art. His group illustrating the battle of Granicos [31] and other bas-reliefs may have been the source of inspiration for the so-called "Alexander" sarcophagus found in Sidōn (Phoenicia) and now in Istanbul. His most famous disciple was Eutychidēs of Sicyōn, immortalized by the "Tychē" group in Antioch.[32] Most of Lysippos' works were small, but one at least was gigantic, the statue of "Zeus" at Taras (Tarentum), 60 ft high. This caused another disciple of his, Charēs of Lindos, to design the "Colossos" of Rhodos (completed in 281); though the "Colossos" was destroyed by an earthquake in 225, it made such an impression on popular fancy that it was always mentioned as one of the seven marvels of the world (Fig. 36). Charēs was one of the founders of the illustrious school that flourished in Rhodos until Roman days.

A brother of Lysippos, Lysistratos of Sicyōn, was also a sculptor; he was primarily interested in making realistic portraits. He was the first to take plaster casts from the faces of his sitters. From the molds thus obtained he produced copies by pouring melted wax into them (Pliny, XXXIV, 19; XXXV, 44).

Antigonos of Carystos. Another great school was brought into being at Pergamon by Attalos I (269–197), whose victory over the Galatians (before 230) was consecrated by the title Sōtēr. He was a great patron of arts and letters and began the improvements that made of Pergamon one of the finest Hellenistic capitals. His main sculptor was Antigonos of Carystos (Euboia), whom he took away from Athens for the erection of monuments in honor of the Galatian victory. Not only did Attalos patronize the decoration of Pergamon but he ordered works of art to be made for Hellenic sanctuaries. He built a temple in Cyzicos [33] in remembrance of his wife Apollōnis, who was born in that island. She was not of royal blood but a noble woman, one of the noblest Hellenistic queens, the wife of one king of Pergamon and mother of two others. One day when she was visiting her native city with her two sons, their gentleness to her was so touching that the people of Cyzicos compared them to Bitōn and Cleobis.[34] There was

[31] Granicos, a river of Mysia, flowing into the Sea of Marmara. It was near the Granicos that Alexander defeated the last king of ancient Persia, Darios Codomannos, in 334.

[32] A gracious woman (Tychē) is seated on a hill, supported by the river Orontēs, crowned by Seleucos and Antiochos. The work is lost, but is represented by a marble copy in the Vatican. This was Antioch's Fortuna, the city's own goddess, and similar monuments were set up in other cities.

[33] Cyzicos was an island in the Sea of Marmara, not one of the Princes' Isles, and is no longer an island today. The site, now called Kapidagi, is a promontory on the south coast of the Sea of Marmara.

[34] Bitōn and Cleobis were famous, because they loved their mother Cydippē so well. She was a priestess of Hēra in

Fig. 36. Fanciful image of the "Colossos" of Rhodos, taken from the album, *Monuments de Rhodes*, by B. E. A. Rottiers (Brussels, 1828). It was a bronze statue representing the sun-god Hēlios (Sol), patron of Rhodos. It was built to immortalize the Rhodians' heroic defense of their city in 305 against Dēmētrios Poliorcētēs, was designed by Charēs of Lindos, completed in 281, and destroyed by an earthquake in 225 B.C. According to Strabōn (*Geography* XIV, 2, 5), quoting an iambic verse of an unnamed poet, the colossus was "seven time ten cubits in height." Seventy cubits equal almost thirty-one meters; a monument of such height and shape was very fragile. See also Pliny, *Natural History* XXXIV, 18.

also a school of mosaic, the leader of which was Sosos of Pergamon; he invented new types of mosaic pavements which were frequently copied in Hellenistic and Roman times.

A sculptor of Bithynia (south and southeast of the Sea of Marmara) named Doidalsēs [35] created the "Zeus Stratios" of Nicomēdeia (known only through coins) and the "Bathing (or Crouching) Aphroditē," represented by copies (Louvre).

"Victory of Samothracē." The masterpiece of the third century was the "Victory," discovered in 1863 in the Temple of the Cabeiroi in Samothracē [36] and now one of the glories of the Louvre. Scholars are not agreed about its date, but it is certainly not anterior to the third century. It may have been set up by Antigonos Gonatas to commemorate his naval victory over Ptolemaios II off Cōs c. 258, or it may commemorate a victory of the Rhodian fleet at the turn of the century.

The movement of the victorious woman is sublime in its grace and simplicity. Few Greek statues have delivered the Greek message of beauty to

---

Argos and prayed the goddess to give them the greatest gift; they both died in Hēra's temple that very night.

[35] The name Doidalsēs is not Greek but Bithynian, attested by inscriptions; Pauly-Wissowa, vol. 9 (1903), 1266.

[36] A little island in the north of the Aegean Sea not very far from the Thracian mainland. Samothracē was the central sanctuary of the Cabeiroi, non-Hellenic gods of fertility and seafaring. The mysteries of that cult had considerable prestige.

as many grateful hearts. Yet, remember, it is a legacy not of the golden age but of the Hellenistic one.

The "Lady of Elche." Another masterpiece of this age is mentioned here not only because of its grace and mysteriousness but also because it represents the western end of the Mediterranean. The "Lady of Elche" may be called Hellenistic, because it is Greek with an unmistakable difference, but our idea of Hellenistic art has generally an Oriental undertone, and the "Lady of Elche" is undoubtedly Spanish (Fig. 37). The city of Elche[37] and the region around it had remained a center of Greek culture in the Carthaginian Spain of the fourth and third centuries. There is no doubt about the "Lady's" birthplace,[38] but scholars disagree about her age. Some would make her older than she undoubtedly is and put her back to the fifth century; others would make her much younger and place her in the Roman age, as late as the second or even the first century B.C.[39] Whatever her exact age may be, she is very beautiful with an exotic (non-Greek) flavor. It is tempting and pleasing to think of her as a contemporary of the Hellenistic princesses of Egypt and Syria.

Tanagra statuettes. Statues, whether in marble or in bronze, were very expensive; popular needs were satisfied with statuettes of baked clay which were sometimes painted. The making of such statuettes had begun very early (say in the seventh and sixth centuries) and many of them were artless, that is, they did not reveal any artistic purpose, yet they might be very attractive in a homely way. That popular art reached its climax in Tanagra[40] under the influence of Praxitelēs and his school; Praxitelēs flourished from c. 370 to c. 330, hence Tanagra figurines revealing Praxi-

[37] Elche, in Latin Ilici or Illice, was on the road from Carthago Nova (Carthagena) to Valencia. It was a Greek colony, but was besieged, in 229, by the Carthaginian Hamilcar Barca, who died there; later it was a Roman colony, free of taxes or other burdens (*colonia immunis*). Hence, Iberian, Greek, Punic, and Roman influences were strangely mixed.

[38] There are obvious affinities between the Dama de Elche and the Figura femenina de "Barro Cocido procedente del Puig d'es molins, Ibiza," and the "Gran Dama Oferente del Cerro de Los Santos, Albacete," Museo Arqueológico Nacional, Madrid. Photographs of the three ladies in *Ars Hispaniae*, vol. 1 (Madrid: Editorial Plus-Ultra, 1947), fig. 138, 257–258, 299–300.

[39] The Lady of Elche was found in 1897 and taken to the Louvre. She was returned to Spain, not to Elche, however, but to the Museo del Prado in Madrid, by Vichy France in exchange for French masterpieces. Antonio Garcia y Bellido, *La Dama de Elche y el conjunto de piezas arqueologicas reingresadas en España en 1941* (Madrid: Instituto Diego Velasquez, 1943); "El arte iberico," in *Ars Hispaniae*, vol. 1 (Madrid: Editorial Plus-Ultra, 1947). Instructive review of this volume by Rhys Carpenter in *American Journal of Archaeology* 52, 474–480 (1948). Thanks to Miss Hazel Palmer of the Boston Museum of Fine Arts for bibliographic information (letter of 17 August 1954).

[40] Tanagra in eastern Boiōtia, on the railway from Athens to Thebes, 64 km from Athens and 27 km from Thebes. It is famous not only for its figurines but also as the birthplace of the poetess Corinna, who was an elder contemporary of Pindar; Pindar lived from 518 to 438.

Fig. 37. The Lady of Elche (detail). The most beautiful monument of eastern Spain and one of the most tantalizing of antiquity. [Museo del Prado, Madrid.]

telian grace and tenderness belong to the end of the fourth century and to the third. The figurines of that golden age are as delicate and lovely as they are simple and unpretentious. They were used as offerings to the dead and a great many were excavated from the Tanagra necropolis between 1870 and 1874. Others were discovered later and elsewhere or found their way from the "antika" shops of Greece and the Near East into Western museums; as Tanagra figurines commanded very high prices, many were counterfeited by rascals of our own time. Genuine statuettes of baked clay were made in other places than Tanagra, however, and even outside Greece, for example, in Alexandria,[41] and the name "Tanagra" is given to them; it is now a generic term, not necessarily an indication of birthplace.

PAINTING. APELLĒS OF COLOPHŌN

The history of painting is more difficult to describe because no monuments are left, but inasmuch as we spoke of Lysippos of Sicyōn we must evoke the memory of his contemporary, Apellēs of Colophōn (Iōnia), who was called to Pella to be the court painter to Philip and Alexander. He painted many portraits of Alexander, chiefly one for the temple of Artemis at Ephesos in which the great king wields a thunderbolt; the most famous of his paintings, however, was one of "Aphroditē Anadyomenē" (rising from the sea), which was shown in Cōs, where it attracted the devotion of pilgrims for three centuries; it was bought from the Coans by Augustus and placed by him in the temple of Julius Caesar at Rome. Apellēs

[41] Description of (local ?) Tanagra figurines in Evariste Breccia, *Alexandria ad Aegyptum* (Bergamo, 1922).

carried his technique to the highest point and was the greatest painter of the Hellenistic age. His zeal equaled his genius and a Greek maxim, corresponding to "*Nulla dies sine linea*" (no day without work) was ascribed to him.

After Alexander's departure for Asia, Apellēs flourished in Ephesos, Rhodos, Alexandria(?), and Cōs. It is said that he died in Cōs while making a copy of his "Aphroditē"; this would be at the beginning of the third century.

We know the names of other painters of his time, and the titles of some of their works, but little else. The oldest of them all was Pamphilos of Amphipolis, who was Apellēs' teacher, and was the teacher also of Pausias and Melanthios. Pamphilos flourished in Sicyōn where he led a school of painting; he insisted upon the knowledge not only of drawing but also of arithmetic and geometry.

Pausias of Sicyōn painted in encaustic.[42] He did the portrait of Glycera, a flower girl, and many other little paintings.

Melanthios was probably the head of the school of Sicyōn after Pamphilos' death; he was a master of composition and of color.

Still another painter of their group, the most illustrious next to Apellēs himself, was Prōtogenēs of Caunos,[43] who flourished in Rhodos. Up to his fiftieth year he was hardly known and had to support himself with the decoration of ships. Thanks to Apellēs' generous praise of him, he became the most famous painter of Rhodos; when Dēmētrios Poliorcētēs besieged that city in 304, he spared it to some extent in order to save Prōtogenēs' works!

Two other painters are mentioned as Apellēs' contemporaries, the Egyptian Antiphilos, who painted portraits of Philip and Alexander, and Theōn of Samos, known because of his fanciful compositions (*phantasiai*). All of which shows that painting was as popular as sculpture.

Treatises on painting were ascribed to Apellēs, to Melanthios, and to Prōtogenēs. This confirms the fact that Sicyōn was a regular school of art. A great many objects of art were public property, which suggests that Sicyōn had a museum. After the Roman conquests it was obliged to sell those treasures in order to pay its debts. Most of them were probably taken to Rome in 58 under the aedileship of M. Aemilius Scaurus junior, Sulla's stepson. This Scaurus was a master plunderer.

All the painters mentioned in this section belong to the Alexandrian age, but some may have lived until the beginning of the third century.

The paintings taken to Rome were used to adorn the temples of the gods or the palaces of rich people. Other paintings were possibly of Etruscan

[42] Encaustic is painting by means of wax with which the pigments have been combined; the wax is fused with hot irons and applied to the surface to be embellished.

[43] Caunos on the south coast of Caria was tributary to Rhodos.

provenience, and such are far better known to us than the Greek ones. That is, all the Hellenistic paintings have disappeared, while a fair number of Etruscan paintings can be admired to this very day. Our knowledge of Greek paintings is purely literary, that is, almost worthless; our knowledge of Etruscan painting (ranging from the end of the seventh century to the end of the first century B.C., more than six centuries) is based upon the monuments.[44] There is no proof that Etruscan paintings were available in the city of Rome, and the extant specimens are mainly in Tarquinii and in other Etruscan sites. Yet they were known to Roman connoisseurs and may have inspired Roman imitations.

The earliest Roman painter was C. Fabius Pictor, who decorated the temple of Salus [45] on the Quirinal hill in Rome in 302. It was for that reason that this Fabius received the surname Pictor, which was transmitted to his descendants, for example, his grandson, Q. Fabius Pictor (III–1 B.C.), the earliest historian of Rome in prose (Greek prose).

The temple of Salus had been dedicated by the censor, C. Junius Brutus Bubulcus. It is possible that C. Fabius Pictor's painting represented Bubulcus' victory over the Samnites.[46] This may have inspired the creation of other historical paintings which became fashionable in Rome in the third century and later. This was a typically Roman procedure, the use of painting for national edification. In 263, M. Valerius Messalla exhibited in the Curia Hostilia a painting representing his victory over the Carthaginians in Sicily and over their ally, Hierōn, king of Syracuse (270–216), and the practice was imitated by other victorious generals. It does not follow that the painters were Roman; it is more likely that they were Greeks. In any case, those paintings are not remembered as works of art, but rather as examples of national ostentation.

SPHRAGISTICS. PYRGOTELĒS

When we spoke of the great sculptor, Charēs of Lindos, we remarked that he was the founder of the school of mosaic that prospered in Rhodos until late Roman days. This suggests that we might deal with other arts and crafts, but that subject is endless. Let us make an exception for engraved gems, and this takes us back to the age of Alexander.

Indeed, it might take us much deeper into the past, because the art of engraving gems was developed by Babylonians and Egyptians long before the Greeks, and by Etruscans as well. The roots of that are natural enough. Engraved gems are very rare and expensive objects which help to symbolize the king's excellence and dignity. Rings and signets were needed

[44] Massimo Pallottino, *Etruscan painting* (140 pp.; Geneva: Skira, 1952), with admirable illustrations in color.

[45] Salus was the goddess of health, prosperity, and public welfare (Salus publica, or Romana). She was worshiped publicly on 30 April in conjunction with Pax, Concordia, and Janus.

[46] Samnium is a mountainous country in central Italy, which was conquered with difficulty by the Romans in 343–290.

as material proofs of a transfer of sovereignty, as when the dying Alexander gave his ring to Perdiccas; they were needed more often as signs manual to authenticate documents or were given to ambassadors and ministers of state as proofs of their authority. In addition, all kinds of magical virtues were readily ascribed to precious gems and jewels.[47] One of the earliest engravers known to us[48] was Pyrgotelēs, in the service of Alexander the Great, who placed him on the same level as his painter, Apellēs, and his sculptor, Lysippos. It was Pyrgotelēs and he only who engraved the king's rings and seals. His importance in the king's eyes is natural enough, for he was creating the symbols and talismans of royal power.

The history of Hellenistic art will be continued in Chapter XXVII.

[47] Remember the beautiful story of Polycratēs' ring which I told in Volume 1, p. 190. Polycratēs of Samos was crucified in 522. Many other stories about gems and rings in E. A. Wallis Budge, *Amulets and superstitions* (London, 1930).

[48] The very earliest was Theodōros of Samos who engraved the Polycrātes' ring mentioned in the previous footnote and flourished c. 550–530. Another was his contemporary Mnesarchos, also of Samos, Pythagoras' father. The most illustrious engraver of the fifth century was Dexamenos of Chios. As many rings, seals, and engraved gems were produced between the time of Polycratēs and that of Alexander, there must have been many goldsmiths and engravers between Theodōros and Pyrgotelēs.

XIV

ORIENTALISM IN THE THIRD CENTURY

The most astonishing part of Hellenistic learning is the study of Oriental countries and cultures. It astonishes us less, however, as soon as we realize that it was a natural consequence of the Alexandrian invasion of Asia and of the prolonged contacts between Greeks, Egyptians, Jews, and Asiatics in the Succession States. Our account will be divided into five sections, dealing respectively with India, Egypt, Babylonia, Phoenicia, and Israel.

INDIA

Nearchos and Megasthenēs. The Cretan Nearchos (IV–2 B.C.) flourished at Amphipolis, Macedonia, and at the court of Philip. The latter banished him, but as soon as Alexander was in power he recalled Nearchos and took him on his Asiatic expedition. A fleet built in 326 by Alexander's order on the Hydaspēs [1] was entrusted to Nearchos. He sailed down the river to the mouth of the Indus and was forced to seek shelter from the southwest monsoon in a natural harbor which he called Alexandri Portus (Karachi); he then continued westward along the coast of the Ichthyophagi to the Persian Gulf. At Harmozia (Hormuz) he landed and was able to visit Alexander, who was leading his army not far from the coast. He observed pearl fisheries and a shoal of whales, sailed to the head of the Persian Gulf and up the Tigris and the Pasitigris (in Susiana), where he met Alexander's army before its arrival at Susa.

Nearchos' voyage had taken five months (September 326 to February 325). He wrote an account of it which is lost, but the substance of which has been preserved by Flavius Arrianus (II–1). After Alexander's death, Nearchos received the government of Lysia and Pamphylia under the supreme leadership of Antigonos the Cyclōps (king of Asia 311–301).

[1] Or Jhelum river; the northernmost of the Five Rivers of the Panjāb, the five tributaries of the Indus.

Alexander's brutal invasion of northern India had angered the Indians, who considered him a "demon-like outer barbarian"[2] without the slightest respect for their usages and traditions. Therefore, they would not learn anything from him even in warfare. Chandragupta[3] continued on an immense scale the tradition of a fourfold army (horse, foot, chariots, elephants) and drove the Macedonian garrisons out of the Panjāb. The founder of the Seleucid dynasty in Western Asia, Seleucos Nicatōr (king of Syria 312–281), crossed the Indus and tried to reconquer the lost territories but was beaten by Chandragupta, probably in the Panjāb, and obliged to abandon all the northern lands; in exchange, Chandragupta gave him 500 elephants which Seleucos could use against his western enemies. After the peace, Seleucos sent to Chandragupta as his ambassador Megasthenēs (III-1 B.C.), who had served in Kandahār. The date of his embassy was c. 305. We do not know how long Megasthenēs remained at the Maurya court, but it must have been long enough to collect a vast amount of information on India. His book is unfortunately lost, but essential parts of it have been preserved by Diodōros (I-2 B.C.), Strabōn (I-2 B.C.) and chiefly in the *Indica* of Flavius Arrianus. Megasthenēs realized the immense extent of India, the bigness of its main rivers the Ganges and the Indus, the fertility of its cultivated parts, the multitude of its cities. He stated that there are in all 118 Indian nations or tribes. He described the Royal Road which connected the valleys of the Indus and of the Ganges. Starting from the Indus and crossing the Panjāb, it reached the Jumna, and then continued down that river to its confluence with the upper Ganges. The road itself (as distinguished from the rivers) was lined with trees, and provided with wells, hostels, and police stations at regular intervals. The importance of his account can hardly be exaggerated, because it is the main, if not the only, Greek source on ancient India; much of it has been confirmed by Indian authorities.

It should be added that India as Megasthenēs understood it was restricted to the northern part of it, north of the Deccan. He was aware of the existence of Taprobanē (Ceylon) but thought of it as being far away south of the peninsula. He described not only the geography and climate of India but also its administration, and the religion, manners, and customs of its peoples. His account, being informed with friendliness, is pleasing to read.[4]

[2] Words used by Vincent A. Smith in *The Oxford History of India* (Oxford, ed. 2, 1923), p. 139: "a demon-like outer barbarian who hanged Brahmans without scruple and won battles by impious methods in defiance of the scriptures . . ."

[3] Chandragupta was called in Greek Sandrocottos, and his capital Pāṭaliputra, on the middle Ganges, was called in Greek Patna. He was the founder in 322 of the Maurya dynasty (322–185 B.C.). With the advent of that dynasty, Indian chronology becomes clear, if not always precise.

[4] Megasthenēs in Karl Müller, *Fragmenta historicorum graecorum,* vol. 2 (Paris, 1848), pp. 397–439, with Latin translation. Christian Lassen, *Indische Alterthumskunde* (5 vols.; Bonn, 1847–1862). See also the editions of Diodōros, Strabōn, and Arrian.

Chandragupta was succeeded by his son, Bindusāra, in 298, and Megasthenēs by another Seleucid ambassador, Dēimarchos. As the latter was the ambassador of the second Seleucid king, Antiochos I Sōtēr (ruled 281–261), that cannot have been before 281. On the other hand, Ptolemaios Philadelphos (ruled 285–246) sent to Pāṭaliputra an envoy called Dionysios. This may have been during the rule of Bindusāra or that of Aśoka, who succeeded him in 273. Unfortunately, neither Dēimarchos nor Dionysios was a writer like Megasthenēs, and the Greek information stems exclusively from the latter.

Aśoka and Buddhist expansion. The Seleucid embassies of Megasthenēs and Dēimarchos to Chandragupta and Bindusāra, and the Ptolemaic embassy of Dionysios to Bindusāra or Aśoka, had familiarized the Hellenistic world with the first three emperors of the Maurya dynasty, with India, and with its religions, Hinduism, Jainism, and Buddhism.

The Maurya empire was truly immense, and it was admirably organized. At its climax c. 250 (under Aśoka) it included the whole of the Indian peninsula (except the Tamil southern tip, below 15°N), and extended northward to Balūchistān, Afghānistān below the Hindu Kush, Kashmir, Nepal (but not Assam). Of course the imperial authority did not penetrate with equal force into every part of that endless territory and many tribes managed to enjoy their freedom in the hills and forests.

The empire created by Chandragupta (ruled 322–298), founder of the dynasty, was greater than Alexander's empire and it lasted longer. Chandragupta was a great conqueror, a very intelligent administrator but utterly unscrupulous. The policies of Maurya administration were revealed with utter cynicism by Chandragupta's minister Kauṭilya (or Cāṇakya) in the elaborate treatise Arthaśāstra,[5] which must be read in conjunction with Megasthenēs. The Arthaśāstra was partly derived from Vedic sources, that is, from the fourth Vēda, the Atharva-vēda, dealing with magic and sorcery. The bulk was probably the creation of Kauṭilya himself, an Indian Machiavelli of considerable experience. Historians of science will consult the book with profit not only to understand government and administration about the beginning of the third century B.C. but also to obtain information on Indian medicine, mining, census-taking, meteorology, shipping, surveying, and so on, and above all to observe many aspects of Indian life.

Chandragupta was a Hindu but became a Jain toward the end of his life. His son Bindusāra (emperor 298–273) continued the conquest of the

[5] Bibliography *ad hoc* in my *Introduction*, vol. 1, p. 147. R. Shama Sastry, *Index verborum* (Mysore, 1924–25). Johann Jakob Meyer, *Das altindische Buch vom Welt- und Staatsleben* (quarto 1071 pp.; Leipzig, 1926), with Sanskrit glossary. Scholars are not agreed about the date of the Arthaśāstra; I have accepted the earliest one. The dating varies from 300 B.C. to A.D. 300. Franklin Edgerton, *The elephant-lore of the Hindus* (148 pp., New Haven, 1931) [*Isis 41*, 120–123 (1950)], p. 2.

Indian peninsula and was succeeded in 273 by his own son Aśoka,[6] who ruled the empire for forty years, and will always be remembered as one of the noblest emperors in the whole past.

During his father's lifetime Aśoka had served as viceroy in Taxila and later in Ujjain.[7] Though his rule began in 273, he was not crowned until 269. The empire which he inherited was so large that there was hardly any need of increasing it and he waged only one war of aggression, the conquest of Kaliṅga (in 261) on the coast of the Bay of Bengal. He had been brought up as a Hindu, most likely a worshiper of Śiva, but his conquest of Kaliṅga brought him such deep remorse that he became a fervent Buddhist. Herein lies his importance. It was thanks to him that Buddhism ceased to be a local sect and became a national religion, nay, an international one, to this day one of the leading faiths of the world. This deserves full emphasis even in a history of science, because Buddhism was the vehicle of much science in India and east of it, just as Christianity was a vehicle of science and culture in Palestine and west of it.

We may call Aśoka the Constantine of Buddhism, or even the St. Paul of Buddhism, bearing in mind, however, that his conversion to Buddhism occurred three centuries before Paul's conversion to Christianity and his proclamation of Buddhism (if we place it c. 260 B.C.), almost six centuries before the Edict of Milan (A.D. 313). These pregnant decisions of his are well known, because they are represented by a long series of inscriptions, the most elaborate series of its kind anywhere. They date from 261 to 242 and are scattered over the whole territory of the Maurya empire; some of them are inscribed on rocks or boulders, others on highly finished columns (Fig. 38). The inscriptions are written in various dialectical forms of Sanskrit according to the regions where they are placed; the script is Brāhmī (an earlier form of the Dēvanāgarī used in Sanskrit and cognate languages), except for some of them near the northwest frontier, which are written in Kharōshṭhī script (a form of Aramaic script used in that neighborhood).

The infinite miseries stemming from the Kaliṅga war (261) which his ambition had caused filled Aśoka with distress. We must assume that his conversion to Buddhism occurred about that time and made him realize the evils he had perpetrated.[8] His catechist was Upagupta of Mathurā, fourth patriarch of the Buddhist church.

[6] The treaty of peace between Chandragupta and Seleucos, c. 302, stipulated a matrimonial alliance. Does that mean that Chandragupta married a daughter of Seleucos Nicatōr? If that wife was the mother of Bindusāra, then Aśoka had a Seleucid grandmother.

[7] Taxila is in the northwest frontier of India (now Pakistān). Alexander the Great had been there in 326 B.C. Ujjain in Central India (Mālwā, Gwalior State) is one of the oldest and holiest cities of India. Taxila became a center of Buddhism; Ujjain, of Hinduism and Sanskrit learning. There was an observatory in Ujjain and one of the greatest Indian mathematicians, Brahmagupta (VII-1), was born there in 598.

[8] Doubts have been cast upon the genuineness of Aśoka's remorse and faith. His

Fig. 38. This very elegant pillar was erected by Aśoka in 243 B.C. at Lauṛiyā-Nandan-gaṛh in Nepal. The shaft is a single piece of sandstone 32 feet, 9½ inches high, and it diminishes from a base diameter of 35½ inches to one of 22¼ inches at the top. The height of the capital including the lion, which faces the rising sun, is 6 feet 10 inches. Hence the whole monument is almost 40 feet high. The inscription is a copy of the so-called Pillar Edicts I to VI (out of seven); it is practically perfect. The whole monument is also in perfect condition, except that the lion capital was slightly damaged by a cannon shot in the time of Aurangzeb (sixth Mogul emperor of India, 1658–1707). Vincent A. Smith, *Aśoka* (Oxford, ed. 3, 1920), pp. 118, 147, 198–208. The lion is probably a symbol of Buddha, as the lion of the Śākya clan. Benjamin A. Rowland, *The art and architecture of India* (London: Penguin, 1953), p. 43 and pl. 8. [Courtesy of the Fogg Museum, Cambridge, Mass., and of Professor Rowland.]

The intense remorse with which he was smitten because of his Kaliṅga crime is expressed in one of the inscriptions, the longest Rock Edict (No. XIII), which is unique in world literature – a conqueror declaring his guilt and his sorrow:

The country of Kaliṅga was conquered when King Priyadarśin,[9] Beloved of the Gods, had been anointed eight years [that is, in 261]. One hundred and fifty thousand were therefrom captured, one hundred thousand were there slain, and many times as many died. Thereafter, now, when the country of Kaliṅga has been acquired, the Beloved of the Gods has zealous compliance with Dhaṃma, love for Dhaṃma, and teaching of Dhaṃma.[10] That is the remorse of the Beloved of the Gods on having conquered Kaliṅga . . .

. . . The Beloved of the Gods desires for all beings noninjury, self-control, equable conduct and gentleness.

And this conquest is considered to be the chiefest by the Beloved of the Gods, which is conquest through Dhaṃma. And that again has been achieved by the Beloved of the Gods here and in the bordering dominions, even as far as six hundred yōjanas[11] where (dwell) the Yavana king called Aṃtiyoka, and beyond this Aṃtiyoka, the four kings called Turamāya, Aṃtekina, Maga, and Aliksuṃdara [follows a long enumeration of Eastern nations].

. . . And this edict of Dhaṃma has been engraved for this purpose, why?, in order that whosoever may be, my sons and great grandsons, may not think of a new conquest as worth achieving, that in regard to a conquest, possible only through (the use) of arrow, they may prefer forbearance and lightness of punishment, and that they may regard that to be the (real) conquest which is conquest through Dhaṃma. That is (good) for here and hereafter. May attachment to Dhaṃma develop into attachment to all kingdoms (*chakras*). That is (good) for here and hereafter.[12]

In addition to Buddhist propaganda, this inscription contains two noble ideas: first, remorse for evil done; second, the affirmation that the only valid conquest is the one accomplished

Without ambition, war, or violence
By deeds of peace, by wisdom eminent
By patience, temperance.

---

horror at the sufferings of the victims of the Kaliṅga war has been compared with the horror professed by Napoleon III at Solferino. It is probable that Aśoka and Napoleon were both sincere. Did Aśoka use Buddhism as a protective cloak for his imperialism, even as the Russians used the Orthodox Church or Communism for theirs? That is quite possible, for the motives of men are often mixed. It is futile, however, to discuss Aśoka's motives. Thanks to him Buddhism was greatly strengthened and diffused.

[9] Priyadarśin (one of gracious mien), or more fully Devānāṃ-priya Priyadarśī Rāja, is the king's name in most of the inscriptions. His personal name Aśoka appears only in one inscription (Maski, near his southern frontier). It is significant that he called himself only *rāja*, not *mahārāja* or *rājādhirāja*, that is, king, not great king or king of kings.

[10] Dhaṃma = dharma, the Law, the Buddhist faith.

[11] *Yōjana* is a measure of length difficult to define exactly; there were a long *yōjana* and a short one (about 9 miles and 4½ miles). The word was also used to denote a day's march (c. 12 miles, but variable). Lionel D. Barnett, *Antiquities of India* (London, 1913) [*Isis* 2, 408 (1914–1919)], p. 217. The Persian *parasangēs*, equaling 30 stadia, was shorter, but it was also an itinerary distance, a stage.

[12] Quoted from the translation by D.

These lines are quoted from *Paradise regained* (III, 90–92), but note that Milton wrote them in 1671, while Aśoka's inscription was engraved a short time after 261 B.C.!

The references to Yavana (Greek) kings are enormously interesting; these kings can be identified as follows: Antiochos Theos (king of Syria 261–246), Ptolemaios II Philadelphos (king of Egypt 285–247), Antigonos Gonatas (king of Macedonia 283–220), Magas (king of Cyrenaica, d. 258), Alexander II (king of Ēpeiros 272–240). At the time when that edict was published (soon after 261) the four Hellenistic kings were alive and in power; the first of them to die was Magas, in 258. What did *they* know of Aśoka?

The government of the empire had been so well organized by its first two rulers that Aśoka could continue it very much as it was, though he probably tried to mitigate its severity and cruelty, for as he put it "all men are my children." He took pains to encourage perseverance and patience and to discourage jealousy, harshness, indolence; he appointed special ministers (*mahāmātra*) responsible for the observance of the Law (*dharma mahāmātra*). These were what we would call ministers of religion, and it is significant that their duty concerned Brahmanical sects as well as Buddhism. We are afraid that these efforts were largely futile and that his admonitions remained counsels of perfection. He could not change the nature of the Indians, and how could a benevolent autocrat control his officers in distant places? The goodness of an autocrat is always and unavoidably betrayed by the greed and cruelty of his subordinates.

The paramount duty was *ahimsā*, nonviolence, noninjury to living creatures. Aśoka forbade the killing of animals in the course of hunting or otherwise, their castration and other evils.[13]

Aśoka explained many other duties: reverence for parents and teachers, obedience to them, kindness to all, charity, and toleration; he took measures to insure the comfort of travelers, the poor, the unfortunate of every kind. The best is to let him speak in his own inscriptions:

> Everywhere in the dominions of King Priyadarśin, Beloved of the Gods, as well as of those of his frontier sovereigns, such as the Chodas, Pāṇḍyas, Sātiyaputra, Keralaputra, as far as the Tāmraparṇi, the Yona [Greek] king called Amtiyaka [Antiochos] and also those who are the neighbours of Aṃtiyaka – everywhere has king Priyadarśin, Beloved of the Gods, established medical treatment of two kinds, medical treatment for men and medical treatment for animals. Wherever medicinal herbs, wholesome for men and wholesome for animals, are not found, they have everywhere been caused to be imported and planted. Roots and fruits, wherever they are not found, have been caused to be imported and planted. On the roads wells have been caused to be dug, and trees caused to be planted for the enjoyment of man and beast.[14]

R. Bhandarkar, *Aśoka* (Calcutta, ed. 2, 1932), pp. 329–334.

[13] More details about the killing of animals in the Arthaśāstra, II, 26.

[14] Bhandarkar, *Aśoka,* rock edict II, undated, complete.

The words "medical treatment" used thrice are a translation of *chikīchha*. Other scholars have translated it "remedies" (Emile Senart) and "hospital" (Johann Georg Bühler). Hence the controversy: did Aśoka found hospitals (in which case he would probably be the first founder) or not? The quarrel is somewhat futile. It is certain that he provided accommodations for sick men and even for sick animals, but were the places set apart for the sick real hospitals? When does a house or hall set apart for the sick deserve to be called a hospital? Similar difficulties exist with reference to the beginning of every institution. Are babies ever comparable to adults? [15]

King Priyadarśin, Beloved of the Gods, honours (men of) all sects, ascetics and householders and honours (them) with gift and manifold honour. But the Beloved of the Gods does not think so much of gift and honour as – what? – as that there should be a growth of the essential among (men of) all sects. The growth of the essential, however, is of various kinds. But the root of it is restraint of speech, – how? – namely, there should not be honour to one's own sect or condemnation of another's sect without any occasion; or it may be a little on this and that occasion. On the contrary, other sects should be honoured on this and that occasion. By so doing one promotes one's own sect, and benefits another's sect. By doing otherwise one injures one's own sect and also harms another sect. For one who honours one's own sect and condemns another's sect, all through attachment to one's own sect, – why? – in order that one may illuminate one's own sect, in reality by so doing injures, more assuredly, one's own sect. Concourse (*samavāya*) is therefore commendable, – why? – in order that they may hear and desire to hear (further) one another's Dhamma. For this is the desire of the Beloved of the Gods – what? – that all sects shall be well-informed and conducive of good. And those who are favourably disposed toward this or that sect should be informed: The Beloved of the Gods does not so much think of gift or honour as – what? – as that there may be a growth of the essential among all sects and also mutual appreciation. For this end are engaged the Dharma-mahāmātras (overseers of the Sacred Law), the Vrajabhūmikas and other bodies (of officials). And this is the fruit, the exaltation of one's own sect and the illumination of Dhaṃma.[16]

This edict has been quoted *in extenso*, in spite of its redundancy (typical of Buddhist literature), because it is an astounding plea for toleration of the best kind. It is not enough to tolerate other sects than one's own; one should be ready to praise them. It took more than nineteen centuries for the Christian sects to understand that and some of them have not understood it yet.

. . . Now, for a long time past previously, there were no Dharma-Mahāmātras. Dharma-Mahāmātras were created by me when I had been conse-

---

[15] See my notes on hospitals in my *Introduction*, Vol. 2, pp. 95, 245–257; Vol. 3, pp. 293–295, 1747–1749). George E. Gask and John Todd, "The Origin of hospitals," in E. A. Underwood (ed.), *Science, medicine and history; essays on the evolution of scientific thought and medical practice, written in honour of Charles Singer* (London: Oxford University Press, 1953), vol. 1, pp. 122–130.

[16] Bhandarkar, *Aśoka*, rock edict XII, undated.

crated thirteen years [that is, in 256]. They are employed among all sects; and (also) for the establishment of Dhaṃma, promotion of Dhaṃma, and for the welfare and happiness of those devoted to Dhaṃma. They are engaged among the Yavanas, Kaṃbojas and the Gandhāras, and the hereditary Rāshṭrikas and others on the Western Coast (Aparānta); among the Brāhmaṇs and Gṛihaptis who have become hirelings, and among the helpless and the aged for (their) welfare and happiness; and also for the unfettering of those devoted to Dhamma. They concern themselves with (money) grant, the unfettering or the release of (anyone) who is bound with fetters, according as he is encumbered with progeny, is subjected to oppression, or is aged . . ."[17]

. . . Thus saith Priyadarśin, Beloved of the Gods: On the roads have I planted the banyan trees. They will offer shade to man and beast. I have grown mango-orchards. I have caused wells to be dug at every eight koses[18] and I have had rest-houses. I have made many watering sheds at different places for the enjoyment of man and beast. This (provision of) enjoyment, however, is, indeed, a trifle, because mankind has been blessed with many such blessings by the previous kings as by me. But I have done this with the intent that men may practise (such) practices of Dhaṃma. . ."[19]

It is certain that Aśoka organized Buddhist missions not only to various parts of his empire but also to the Western countries and to Ceylon. The mission to Ceylon is the only one about which we have plenty of information (from Singhalese sources). Aśoka's son, Mahendra or Mahinda, was in charge of it and was sent at the request of Tissa,[20] king of Ceylon, c. 247. Mahendra settled in the island and died there in 204; he was helped by his sister, nicknamed Sanghamitrā (friend of the order), who died in the following year. This mission was very fortunate in view of what happened in later times; while Buddhism was gradually driven out of India by Hinduism, it has never ceased to flourish in Ceylon. The ruins of the monastic city of Anurādhapura, "the Buddhist Rome," constitute the most impressive monument to the memory of Aśoka's family and of the early Singhalese converts.

Aśoka was an ardent Buddhist, anxious to proselytize yet remaining tolerant; witness Rock Edict XII, quoted above. For example, he offered gifts to the Ājīvika monks, an order closely akin to the Digambara or nude Jains.

In 249, his old teacher, Upagupta, took him on a pilgrimage to the holy places; it was probably then that he visited the sacred tree at Buddh Gayā.[21]

[17] *Ibid.*, middle part of rock edict V, 256 B.C.

[18] This is probably *krōśa*, an itinerary measure: four *krōśa* equal one *yōjana* (see note 11).

[19] Bhandarkar, *Aśoka*, a middle section of pillar edict VII, which is very long and was engraved in 242 B.C.

[20] Dēvānampiya Tissa (247–207). H. W. Codrington, *Short history of Ceylon* (London: Macmillan, rev. ed., 1939), pp. 11 f.

[21] Buddh Gayā, south of Patna in the center of Bihar. It is there that the Buddha experienced the Enlightenment under the sacred bo or pipal tree (*Ficus religiosa*). A slip of that very bo tree was taken to Ceylon by Sister Sanghamitrā and planted in the Mahāmēgha garden in Anurādhapura, c. 240 B.C.; it is to this day one of the main attractions of the pilgrims to that place.

In 240, Aśoka gathered a Buddhist council in his capital, Pāṭaliputra. According to Buddhist tradition, that was the third council; the sixth council (Chattha Sangayana) met in Rangoon in 1954–56, the year 1956 being supposed by them to be the 2500th anniversary of the Buddha's death (*Mahā parinibbhana*).[22]

It is not known for certain when and where Aśoka died. It was not many years after the council, probably in or near 232. At any rate, his rule is supposed to have ended in that year. According to Tibetan tradition, he died in Taxila. He was succeeded by two grandsons, Daśaratha and Samprati, the former ruling the eastern provinces, Magadha, of which Pāṭaliputra was probably the capital, and the latter, the western province, with Ujjain as its capital. Samprati was as devoted to Jainism as his grandfather to Buddhism. The last ruler of the Maurya dynasty was murdered in 185 B.C. by his commander in chief, who founded the short-lived Sunga dynasty (185–173). Another part of the Maurya empire, the extreme southeast of it, the deltas of the rivers Gōdāvarī and Kṛishṇā, broke off soon after Aśoka's death and was ruled for about 450 years (c. 230 B.C.–c. A.D. 225), by some thirty kings of the Āndhra dynasty.

The golden age of the Maurya dynasty lasted a little less than a century (322–232). The first three emperors ruled almost exactly during the same period as the first three Ptolemaioi (323–222). They were great patrons of art; their buildings have disappeared but some splendid examples of Asokan sculpture have come down to us, such as the lion column in Lauṛiyā-Nandangaṛh in Nepal (243 B.C.) and the four-lions capital which stood in the deer park of Sārnāth, scene of the Buddha's first preaching.[23] That art is pure and beautiful and its technique astonishingly mature and precise. The monolithic columns, some of which are over 40 ft high, were admirably dealt with, and the technique of polishing hard stone carried to an inimitable perfection.

Aśoka's main achievement, however, was the diffusion of Buddhism. He is one of the three giants of Indian culture, the two others being Akbar, third emperor of the dynasty (1542–1605), and Gandhi, creator of India's independence (1869–1948).[24] These men are as different from one another as their times, and yet they have qualities in common that illustrate the spiritual unity of India.

[22] U Hla Maung, "The sixth great Buddhist Council," *Forum, Journal of the World Congress of Faiths,* No. 20 (London, 1954), pp. 6–8. According to the tradition of the Burmese Buddhists of today, the Buddha died in 545 B.C.; the date generally accepted by Western scholars is later, 483–477 (*Introduction,* Vol. 1, p. 68). Buddhist traditions are full of inconsistencies.

[23] For discussion and illustrations, see Benjamin Rowland, *The art and architecture of India: Buddhist, Hindu, Jain* (Pelican history of art; Baltimore: Penguin Books, 1953).

[24] For Gandhi, see G. Sarton, "Experiments with truth by Faraday, Darwin and Gandhi," *Osiris 11,* 87 (1954).

Short bibliography

Vincent Arthur Smith, *Asoka, the Buddhist emperor of India* (Oxford: Clarendon Press, 1901; ed. 2, 1909; ed. 3, 1920, 278 pp.).

Jean Przyluski, "La légende de l'empereur Açoka (Açoka-avadāna) dans les textes indiens et chinois," *Annales du Musée Guimet 32* (476 pp.; Paris, 1923). The Aśokāvadāna, written in the second half of the second century B.C., is preserved in two Chinese versions composed in A.D. 300 and 512.

Devadetta Ramakhrisna Bhandarkar, *Aśoka* (University of Calcutta, 1925; ed. 2, 432 pp., 1932). My extracts from the inscriptions are quoted on the basis of Bhandarkar's second edition and translation of them.

George Peiris Malalasekera, *Dictionary of Pāli proper names* (2 vols.; London: Murray, 1937–38), vol. 1, 216–219.

EGYPT

Manethōn. During the rule of Ptolemaios Sōtēr (323, 304–283), Hecataios of Teōs wrote a romantic account of Egypt which familiarized the Greeks with the idea that the Nile Valley was the cradle of civilization.[25]

His effort was renewed a little later by a man who was far more competent than himself, Manethōn. While Hecataios was a Greek interested in Egypt, Manethōn was a Hellenized Egyptian, a native of Sebennytos (now Samannūd) in the Eastern Delta, on the Damietta branch of the Nile, a priest in the temple of Sebennytos and later a high priest in old Hēliopolis (near Cairo). Not only were some of the main historical sources available to him, but he was able to read them critically and to point out the errors of Greek historians such as Hērodotos and Hecataios. It is probable that his work was done at the request of Ptolemaios Philadelphos (285–247), who was anxious to prove that Egyptian civilization was at least as old as the Mesopotamian described by Bērōssos in the service of Antiochos I (ruled 280–261).

Manethōn was younger than Hecataios, but he was already employed by the first Ptolemaios, together with a Greek, Timotheos, who was also a priest, or a royal councilor in religious matters. The two men, Manethōn and Timotheos, organized the Greco-Egyptian cultus of Sarapis. The statement that Sarapis entered Alexandria in 286 (or 278) may refer to the inauguration of Bryaxis' statue of the god or to the beginning of the cultus.

Manethōn's main work is the *Aigyptiaca*, which is lost and known only through an early Greek epitome and fragments. It is a history of Egypt from the beginning to 323, which has been of considerable help to modern Egyptologists. The familiar division of the dynasties into Old Empire (dyn. I–VI, 3200–2270), Middle Empire (dyn. XI–XIII, 2100–1700), New Empire (dyn. XVIII–XXIV, 1555–712), Late Period (dyn. XXV–XXX, 712–332)[26] was already implied in Manethōn's work. His chronology, however defective, is extremely important, because it was derived from original

[25] Teōs is in the middle third of the Ionian coast while Milētos, where the elder Hecataios lived in the sixth century, is in the lower third. Fragments of Hecataios Abdēritēs in Müller, *Fragmenta historicorum graecorum*, vol. 2, pp. 384–396.

[26] The dates added by me are modern estimates by Georg Steindorff. Dyn. VII

documents available in the temple archives, such as the royal lists of Abydos (British Museum), Karnak (Louvre), Sakhāra (Cairo Museum), the Turin Papyrus (c. 1200 B.C.), and the Palermo Stone (c. 2600 B.C.)

He wrote other books, all concerned with Egyptian history, religion, and science. Judging from the few fragments remaining from his *Epitomē tōn physicōn* (*Epitome of physical matters*), his "physics" was mythology rather than science. He was aware of Greek cosmology and, as he was writing in Greek, his purpose was to explain Egyptian "physics" to a Greek audience. It was much easier for an Egyptian to learn Greek and read Greek authors than for a Greek to understand hieroglyphics. His religious books were used by Plutarch in his treatise on Isis and Osiris.

The Greeks of the Hellenistic period were probably more curious to read Hecataios' romance than Manethōn's chronology; on the other hand, the Jews were deeply interested in the latter's chronology because of the Egyptian antiquities of their own past. Manethōn's history was exploited first by Jewish historians like Jōsēphos (I–2) and later by Christian chronologists, like Sextos Julios Africanos (III–1), Eusebios (IV–1), Geōrgios Syncellos (IX–1), for all of them, Christian as well as Jew, were trying to establish the Biblical chronology as well as possible.[27] Jōsēphos (I–2) criticized Manethōn for confusing with the Jews "a crowd of Egyptians, who for leprosy and other maladies had been condemned to banishment from Egypt." This was the earliest statement connecting leprosy with Egypt and with the Jews.[28]

Manethōn of Sebennytos has been confused with "Manethōn" of Mendēs. The latter's real name was Ptolemaios of Mendēs; he studied Egyptian matters somewhat later, probably in the time of Augustus. The confusion may have been facilitated by the fact that Mendēs was near Damietta (thus not very far from Sebennytos), and was a sacred place, yet occupied by Greek mercenaries during the XXIXth dyn. (398–379). The god Mendēs, worshiped in Mendēs, was a ram (or he-goat) which became very popular in Ptolemaic times; a famous stele found at Mendēs expresses the devotion of Ptolemaios and Arsinoē Philadelphos to the sacred ram and recalls the privileges and festivals which the temple enjoyed.

The calendar of Sais. A Greek papyrus found in 1902 at al-Ḥiba [29] is a calendar for Sais and the district (nomē) around it, together with an astronomical introduction.

---

to X (2270–2100) constitute an intermediate period; dyn. XIV to XVII (1700–1555) constitute another, the Hyksos period.

[27] *Fragmenta historicorum graecorum*, vol. 2, pp. 495–510. The most convenient edition of Manethōn's fragments in Greek and English is that by W. G. Waddell (Loeb Classical Library; Cambridge: Harvard University Press, 1940).

[28] Manethōn (Loeb edition), p. 121. Jōsēphos, *Contra Apionem*, I, 26–31. For the origins of leprosy, see *Introduction*, Vol. 3, pp. 275 ff.).

[29] Al-Ḥibah on the Nile (about 28°50′) was the site of a Ptolemaic city. Many

The whole was written at Sais, c. 300 B.C. or soon afterward, by a follower of Eudoxos (IV-1 B.C.), for his pupils' instruction. He explains the different years in use in Egypt. The year of the calendar is an ordinary Egyptian *annus vagus* of 365 days beginning with Thoth I (the account of the first three months is missing).

The following details are recorded under the various days: (1) changes of the seasons indicated by equinoxes and solstices (the author seemed to believe that the equinoxes divided the year into two almost equal parts, of 183 and 182 days); (2) passing of the Sun at its rising from one of the 12 great constellations to another; (3) risings and settings of certain stars or constellations; (4) weather forecasts; (5) stages in the rising of the Nile; (6) Greco-Egyptian festivals celebrated in Sais; (7) the lengths of day and night. The length of the longest day is given as 14 hours, which corresponds to the latitude of Sais.

The papyrus was fairly long, but we have only 16 fragments of it. It was edited and translated by Bernard P. Grenfell and Arthur S. Hunt, *The Hibeh papyri. Part I* (London: Egypt Exploration Fund, 1906), No. 27, pp. 138–157, pl. viii.

BABYLONIA. BERŌSSOS[30]

Berōssos flourished during the rule of Antiochos I Sōtēr (king of Syria 281–262). His name is the Greek transcription of a Babylonian one; hence we may assume that he was not a Greek but rather a Hellenized native. He was born not later than 340, flourished in Babylon until the beginning at least of Antiochos' rule, then moved to Cōs, where he founded a school (Cōs was then in Ptolemaic power); the date of his death is unknown.

Antiochos Sōtēr tried to do what the first two Ptolemaioi were doing in Egypt and his method was the same. Berōssos, whose service he engaged, was a priest of Marduk in Babylon, and thus had an inside knowledge of Babylonian history and religion and was competent to exploit Babylonian (or Chaldean) sources. His work, written in Greek and dedicated to Antiochos (hence posterior to 281), was entitled *Babylōnica* (better than *Chaldaica*); it was divided into three parts (curiously enough, Manethōn's work was also tripartite). It is lost, but has been more or less reconstructed from the quotations made by Jōsēphos (I-2) and Eusebios (IV-1).

The three books dealt with the following periods: (1) from the Creation to the Flood, 432,000 years; (2) from the Flood to Nabonassar, king of Babylon in 747, 34,090 + 1,701 = 35,791 years; (3) from Nabonassar to Cyros, 209 years or to Alexander 424 years – in all 468,000 or 468,215 years.

Greek papyri were found in the Ptolemaic necropolis, all but one derived from mummy cartonnage and all of the third century B.C. Sais is far away, near Ṭanṭa in the western Delta, half way between Alexandria and Cairo.

[30] The name, of Babylonian origin, is written with one or two sigmas and also with omicron instead of ōmega. The accentuation may occur on each of the three syllables. Such vacillation of the accent is typical of foreign words.

The first book and part of the second were necessarily cosmological, hence Berōssos was called the "astrologer."

His book was the main vehicle of Chaldean astrology to Egypt and to the Hellenistic world in general, and that was, for good or evil, its main function. How much of that astronomical or astrological knowledge was purely Chaldean and how much Iranian or Greek is difficult to say. Berōssos was discussing the elements (*stoicheia*), the seven planets and their virtues, and so on.

As far as antiquity goes, Berōssos had been able to dig deeper than Manethōn, and Antiochos had beaten the Ptolemaioi. This was the first battle between the "Assyriologists" and the "Egyptologists," and the former won.[31]

A curious proof of the influence of Babylonian literature upon Greek is given by the *Iamboi* of Callimachos, which include the quarrel between the Laurel and the Olive. Now this poem of about 72 lines can be compared with a Babylonian one of the very same kind, except that the rivals are not the Laurel and the Olive but the Tamarisk and the Date. The general idea is the same; to put it in Christian terms, it is the eternal conflict between Mary and Martha.[32]

PHOENICIA

One Menandros of Ephesos, who flourished in Alexandria or in Pergamon, used Phoenician records (*anagraphai*) and wrote a history of Tyros which is lost but fragments of which are quoted by Jōsēphos (I–2) in his book *Against Apiōn*. He dealt with Hiram, king of Tyre, a contemporary of Solomon ben David, king of Isarel.[33]

ISRAEL

The outstanding achievement of Hellenistic orientalism, the *Septuagint*, was accomplished in Egypt, being initiated by the Museum and the second Ptolemaios. At the end of this survey, we realize more fully than we did at the beginning that the third century was the golden age of Hellenism and that the climax occurred in Egypt about 250 B.C.

[31] Text of Berōssos in Müller, *Fragmenta historicorum graecorum*, vol. 2. Paul Schnabel, *Berosi Babyloniacorum libri tres quae supersunt* (Leipzig, 1913); *Berossos und die babylonisch-hellenistische Litteratur* (275 pp.; Leipzig, 1923).

[32] Greek-English text of Callimachos in the Loeb edition by A. W. Mair, *Callimachus, Lycophron, and Aratus* (Loeb Classical Library; Cambridge, 1921), pp. 280–88; Babylonian-German text in Erich Ebeling, "Die babylonische Fabel und ihre Bedeutung für die Literaturgeschichte," *Mitteilungen der altorientalischen Gesellschaft 2*, part 3 (Leipzig, 1927).

[33] Müller, *Fragmenta historicorum graecorum*, vol. 4 (Paris, 1851), pp. 445–448. Isaac Preston Cory, *Ancient fragments of the Phoenician, Carthaginian, Babylonian, Egyptian and other authors*, new ed. by Edward Richmond Hodges (London, 1876), pp. 27–32. Pauly-Wissowa, vol. 29 (1931), 762. For Hiram, king of Tyre, see 1 Kings 5.

Haec tibi pentadecas tetragonon respicit illud
Hospitium petri et pauli ter quinqʒ dierum.
Namqʒ instrumētum vetus hebdoas innuit: octo
Lex noua signatur. ter quinqʒ receptat vtrunqʒ.

Vetus testamentū multiplici lingua nūc
primo impressum. Et imprimis
Pentateuchus Hebraico Gre-
coatqʒ Chaldaico idioma-
te. Adiūcta vnicuiqʒ sua
latina interpreta-
tione.

Fig. 39. Title page of vol. 1 of the Complutensian Bible, the first polyglot; printed in 1514–17. It is called Complutensian because it was printed in Alcalá de Henares (Complutum in Latin). In the middle of the title page is the coat of arms (printed in red) of Cardinal Jiménez de Cisneros (1437–1517), who patronized this magnificent publication, 600 copies being printed at his expense. The work was not actually published (distributed) until 1521, four years after his death. Four pages of errata are placed before the title page. Volume 1 contains the Pentateuch or Torah. It is a heavy folio (37 cm high, 4.5 cm thick without the covers which are heavy in proportion). There were altogether six volumes; the typography of vol. 2 and following was less complicated. [Courtesy of Harvard College Library.]

Let us first consider the situation in the orthodox Jewish world. The greatest part of the Old Testament was available. During the first part of the century, various historical books were edited, to wit, the two Books of Chronicles (65 chapters), the Book of Ezra (10 chapters), the Book of Nehemiah (13 chapters).[34] The Chronicler tells the history of Judah from Adam to the end of the Babylonian captivity (538–536); Ezra and Nehemiah continue the story from 536 to 432. The Books of Ezra and Nehemiah were derived from the memoirs of two Hebrew priests, Ezra and Nehemiah, who lived in the fifth century, at a time when the Hebrew language had not yet been replaced by Aramaic. The Book of Nehemiah was the last Hebrew work written while this language was still living.[35]

At the time of the editing of those historical books, the people were speaking Aramaic and their ignorance of Hebrew was such that it was necessary to provide them with a *targum* or interpretation in Aramaic (the "Chaldee paraphrase").

Another work of greater importance was completed in the first half of the same century. This was the Book of Proverbs (in 31 chapters) or, to give it its full title, "The proverbs of Solomon the son of David, king of

[34] In the Catholic canon (Vulgate, Douay Bible), the book of Ezra (Esdras in Greek) is called "1 Esdras" and the book of Nehemiah, "2 Esdras, alias Nehemias." The First (and Second) Books of Esdras are included in the Apocrypha by Catholics and Protestants, but Catholics call them "3 (and 4) Esdras."

[35] Robert H. Pfeiffer, *Introduction to the Old Testament* (New York: Harper, 1941) [*Isis 34*, 38 (1942–48)], p. 838. Hebrew has become a living language again in our century and was made the official language of Israel in 1948.

Trāsla.Gre.lxx.cū interp.latina. Transla.B.Hiero. Ter.Heb. Ben. Ca.i. Pritiua.heb.

Transla.Chal. Interp.chal. Pritiua chal.

Fig. 40. First page of Genesis in the Complutensian Bible. The Septuagint text is in the first column, with a Latin interlinear translation; the standard Latin translation by St. Jerome (IV–2) is in the middle column; the Hebrew original text in the right; the Chaldean paraphrase with its Latin version at the bottom of the page. There are thus six concurrent texts with notes! Observe that in St. Jerome's text (the Vulgate) there are no blank spaces; each blank is replaced by a row of zeroes. The early printers did not like blanks. The Hebrew text was printed and published before the great Hebrew Jewish Bible (4 vols.; Venice, 1524–26). [Courtesy of Harvard College Library.]

Ἡ ΘΕΙΑ ΓΡΑΦΗ ΠΑΛΑΙΑ ΤΕ ΚΑΙ ΝΕΑ.

ΓΕΝΕΣΙΣ.

ΚΕΦ. α΄. 1

Ν ἀρχῇ ἐποίησεν ὁ
θεὸς τὸν οὐρανὸν καὶ
τὴν γῆν. ἡ δὲ γῆ ἦν
ἀόρατος καὶ ἀκατα-
σκεύαστος. καὶ σκό-
τος ἐπέκειτο ἐπάνω
τῆς ἀβύσσου, καὶ
πνεῦμα θεοῦ ἐπε-
φέρετο ἐπάνω τοῦ
ὕδατος. καὶ εἶπεν ὁ θεός· γενηθήτω φῶς, καὶ ἐγένετο φῶς. καὶ εἶδεν ὁ θεὸς τὸ φῶς ὅτι καλόν. καὶ διεχώρισεν ὁ θεὸς ἀνὰ μέσον τοῦ φωτὸς, καὶ ἀνὰ μέσον τοῦ σκότους. καὶ ἐκάλεσεν ὁ θεὸς τὸ φῶς, ἡμέ-
ραν. καὶ τὸ σκότος, ἐκάλεσε νύκτα. καὶ ἐγένετο ἑσπέρα
καὶ ἐγένετο πρωΐ ἡμέρα μία. Καὶ εἶπεν ὁ θεός· γενηθήτω στερέωμα ἐν μέσῳ τοῦ ὕδατος. καὶ ἔστω διαχω-
ρίζον ἀνὰ μέσον ὕδατος καὶ ὕδατος. καὶ ἐγένετο οὕτως. καὶ ἐποίησεν ὁ θεὸς τὸ στερέωμα. καὶ διεχώρισεν
ὁ θεὸς ἀνὰ μέσον τοῦ ὕδατος ὃ ἦν ὑποκάτω τοῦ στε-
ρεώματος. καὶ ἀνὰ μέσον τοῦ ὕδατος τοῦ ἐπάνω τοῦ στε-
ρεώματος. καὶ ἐκάλεσεν ὁ θεὸς τὸ στερέωμα, οὐρανόν.
καὶ εἶδεν ὁ θεὸς ὅτι καλόν. καὶ ἐγένετο ἑσπέρα καὶ
ἐγένετο πρωΐ ἡμέρα δευτέρα. Καὶ εἶπεν ὁ θεός·
συναχθήτω τὸ ὕδωρ τὸ ὑποκάτω τοῦ οὐρανοῦ, εἰς συ-
ναγωγὴν μίαν. καὶ ὀφθήτω ἡ ξηρά. καὶ ἐγένετο οὕτως, καὶ συνήχθη τὸ ὕδωρ τὸ ὑποκάτω τοῦ οὐρανοῦ
εἰς τὰς συναγωγὰς αὐτῶν. καὶ ὤφθη ἡ ξηρά. καὶ ἐκάλεσεν ὁ θεὸς τὴν ξηρὰν, γῆν. καὶ τὰ συστήματα τῶν ὑδά-
των ἐκάλεσε θαλάσσας. καὶ εἶδεν ὁ θεὸς ὅτι καλόν. καὶ εἶπεν ὁ θεός· βλαστησάτω ἡ γῆ βοτάνην χόρτου σπεῖρον σπέρμα κατὰ γένος καὶ καθ᾿ ὁμοιότητα. καὶ ξύλον κάρπιμον ποιοῦν καρπόν. οὗ τὸ σπέρμα αὐτοῦ ἐν
αὐτῷ κατὰ γένος ἐπὶ τῆς γῆς καὶ ἐγένετο οὕτως.
καὶ ἐξήνεγκεν ἡ γῆ βοτάνην χόρτου σπεῖρον σπέρμα
κατὰ γένος καὶ καθ᾿ ὁμοιότητα. καὶ ξύλον κάρπιμον ποιοῦν καρπὸν, οὗ τὸ σπέρμα αὐτοῦ ἐν αὐτῷ κατὰ
γένος ἐπὶ τῆς γῆς. καὶ εἶδεν ὁ θεὸς ὅτι καλόν. καὶ ἐ-
γένετο ἑσπέρα καὶ ἐγένετο πρωΐ ἡμέρα τρίτη.
Καὶ εἶπεν ὁ θεός· γενηθήτωσαν φωστῆρες ἐν τῷ στερεώματι τοῦ οὐρανοῦ εἰς φαῦσιν ἐπὶ τῆς γῆς, τοῦ διαχω-
ρίζειν ἀνὰ μέσον τῆς ἡμέρας καὶ ἀνὰ μέσον τῆς νυκτός.
καὶ ἔστωσαν εἰς σημεῖα καὶ εἰς καιροὺς καὶ εἰς ἡμέρας
καὶ εἰς ἐνιαυτούς. καὶ ἔστωσαν εἰς φαῦσιν ἐν τῷ στερεώματι τοῦ οὐρανοῦ ὥστε φαίνειν ἐπὶ τῆς γῆς. καὶ ἐ-
γένετο οὕτως. καὶ ἐποίησεν ὁ θεὸς τοὺς δύο φωστῆρας
τοὺς μεγάλους. τὸν φωστῆρα τὸν μέγαν, εἰς ἀρχὰς τῆς
ἡμέρας, καὶ τὸν φωστῆρα τὸν ἐλάσσω, εἰς ἀρχὰς τῆς
νυκτὸς καὶ τοὺς ἀστέρας. καὶ ἔθετο αὐτοὺς ὁ θεὸς ἐν
τῷ στερεώματι τοῦ οὐρανοῦ, ὥστε φαίνειν ἐπὶ τῆς γῆς.
καὶ ἄρχειν τῆς ἡμέρας καὶ τῆς νυκτὸς, καὶ διαχωρί-
ζειν ἀνὰ μέσον τοῦ φωτὸς καὶ ἀνὰ μέσον τοῦ σκότους.
καὶ εἶδεν ὁ θεὸς ὅτι καλόν. καὶ ἐγένετο ἑσπέρα καὶ ἐ-
γένετο πρωΐ ἡμέρα τετάρτη. Καὶ εἶπεν ὁ θεός·
ἐξαγαγέτω τὰ ὕδατα, ἑρπετὰ ψυχῶν ζωσῶν καὶ
πετεινὰ πετόμενα ἐπὶ τῆς γῆς κατὰ τὸ στερέωμα τοῦ
οὐρανοῦ. καὶ ἐγένετο οὕτως. καὶ ἐποίησεν ὁ θεὸς, τὰ
κήτη τὰ μεγάλα, καὶ πᾶσαν ψυχὴν ζώων ἑρπετῶν ἃ
ἐξήγαγε τὰ ὕδατα κατὰ γένη αὐτῶν. καὶ πᾶν πε-
τεινὸν πτερωτὸν κατὰ γένος. καὶ εἶδεν ὁ θεὸς, ὅτι καλά, καὶ εὐλόγησεν αὐτὰ ὁ θεὸς λέγων. αὐξάνεσθε καὶ
πληθύνεσθε. καὶ πληρώσατε τὰ ὕδατα ἐν ταῖς θα-
λάσσαις, καὶ τὰ πετεινὰ πληθυνέσθωσαν ἐπὶ τῆς
γῆς, καὶ ἐγένετο ἑσπέρα καὶ ἐγένετο πρωΐ ἡμέρα πέμ-
πτη. Καὶ εἶπεν ὁ θεός. ἐξαγαγέτω ἡ γῆ, ψυχὴν ζῶ-
σαν κατὰ γένος. τετράποδα καὶ ἑρπετὰ καὶ θηρία
τῆς γῆς κατὰ γένος. καὶ ἐγένετο οὕτως, καὶ ἐποίησεν ὁ θεὸς τὰ θηρία τῆς γῆς κατὰ γένος. καὶ τὰ κτή-
νη κατὰ γένος αὐτῶν. καὶ πάντα τὰ ἑρπετὰ τῆς γῆς
κατὰ γένος. καὶ εἶδεν ὁ θεὸς ὅτι καλά. καὶ εἶπεν ὁ
θεός. ποιήσωμεν ἄνθρωπον κατ᾿ εἰκόνα ἡμετέραν καὶ
καθ᾿ ὁμοίωσιν. καὶ ἀρχέτωσαν τῶν ἰχθύων τῆς θαλάσσης καὶ τῶν πετεινῶν τοῦ οὐρανοῦ. καὶ τῶν κτηνῶν,
καὶ πάσης τῆς γῆς, καὶ πάντων τῶν ἑρπετῶν τῶν
ἑρπόντων ἐπὶ τῆς γῆς. καὶ ἐποίησεν ὁ θεὸς τὸν ἄνθρωπον. κατ᾿ εἰκόνα θεοῦ ἐποίησεν αὐτόν. ἄρσεν καὶ θῆ-
λυ ἐποίησεν αὐτούς. καὶ εὐλόγησεν αὐτοὺς ὁ θεὸς λέγων. αὐξάνεσθε καὶ πληθύνεσθε καὶ πληρώσατε τὴν
γῆν, καὶ κατακυριεύσατε αὐτῆς. καὶ ἄρχετε τῶν ἰ-
χθύων τῆς θαλάσσης καὶ τῶν πετεινῶν τοῦ οὐρανοῦ,
καὶ πάντων τῶν κτηνῶν καὶ πάσης τῆς γῆς, καὶ πάντων τῶν ἑρπετῶν τῶν ἑρπόντων ἐπὶ τῆς γῆς. καὶ εἶπεν ὁ θεός. ἰδοὺ δέδωκα ὑμῖν πάντα χόρτον σπόρι-
μον σπεῖρον σπέρμα ὅ ἐστιν ἐπάνω τῆς γῆς. καὶ πᾶν
ξύλον ὃ ἔχει ἐν ἑαυτῷ καρπὸν σπέρματος σπορίμου ὑ-
μῖν ἔσται εἰς βρῶσιν, καὶ πᾶσι τοῖς θηρίοις τῆς γῆς,
καὶ πᾶσι τοῖς πετεινοῖς τοῦ οὐρανοῦ. καὶ παντὶ ἑρ-
πετῷ ἕρποντι ἐπὶ τῆς γῆς ὃ ἔχει ἐν ἑαυτῷ ψυχὴν
ζωῆς, καὶ πάντα χόρτον χλωρὸν εἰς βρῶσιν. καὶ ἐγένετο οὕτως. καὶ εἶδεν ὁ θεὸς τὰ πάντα ὅσα ἐποίησε,
καὶ ἰδοὺ καλὰ λίαν, καὶ ἐγένετο ἑσπέρα καὶ ἐγένετο πρωΐ ἡμέρα ἕκτη.

ΚΕΦ. β΄. 2

Αἱ συνετελέσθησαν ὁ οὐρανὸς καὶ ἡ γῆ καὶ
κ πᾶς ὁ κόσμος αὐτῶν. καὶ συνετέλεσεν ὁ θεὸς ἐν τῇ ἡμέρᾳ τῇ ἕκτῃ τὰ ἔργα αὐτοῦ ἃ ἐ-

Israel; to know wisdom and instruction; to perceive the words of understanding; to receive the instruction of wisdom, justice, and judgment, and equity; to give subtilty to the simple, to the young man knowledge and discretion. A wise man will hear, and will increase learning; and a man of understanding shall attain unto wise counsels; to understand a proverb, and the interpretation; the words of the wise and their dark sayings."

The short title Proverbs (*Mushli*) is conventional and misleading; the book contains the teachings of wise men, some of which might be used as proverbs, but most of them, not. It is not simply a collection of wise teachings but a bundle of many such collections gathered at different dates. Irrespective of the dates of particular verses or groups of verses, the book as a whole cannot be older than the fourth century and the final editing occurred in the second half of the third century.[36]

The Jews who had emigrated to Egypt or were born there of emigrant parents had forgotten their Hebrew and even their Aramaic and spoke a Greek dialect (Judeo-Hellenistic Greek). Of course, the well-educated Jews spoke excellent Greek, but even they had generally neglected their mother tongue, if not their religion.

According to tradition, Dēmētrios of Phalēron suggested to Ptolemaios II Philadelphos [37] the importance of translating the Old Testament or at least the Pentateuch from Hebrew into Greek; this would be valuable to Jews who could no longer read Hebrew; it would be even more valuable to Greeks who never had been able to read it. The translation of the Holy Book of the Jews would enable their Greek patrons to understand them better. The translation was at first limited to the Torah and was approved by Eleazar, high priest in Jerusalem. It is significant that the initiative of the translation came from the Greek, not from the Jewish side. The tradition as it had crystallized by the middle of the second century B.C. is well known through the Greek letter of Aristeas to Philocratēs; [38] it was current in Alexandria and was accepted by the Church Fathers, with the exception of St. Jerome (IV–2).

[36] *Ibid.*, pp. 640–659.

[37] Dēmētrios was not on good terms with Ptolemaios Philadelphos, but he may have made this suggestion before his disgrace.

[38] Paulus Wendland, *Aristeae ad Philocratem epistula cum ceteris de origine versionis LXX interpretum testimoniis* (262 pp.; Leipzig, 1900). H. St. J. Thackeray, edition of the Greek text appended to H. B. Swete and R. R. Ottley, *Introduction to the Old Testament in Greek* (640 pp.; Cambridge, 1914). Moses Hadas, ed. and trans., *Letter of Aristeas to Philocrates* (Dropsie College edition of Jewish apocryphal literature, 234 pp.; New York: Harper, 1951) [*Isis 43*, 287 (1952)]. The most probable date of that text is "about 130 B.C."

Fig. 41. The first edition of the Septuaginta to be actually published (distributed) was issued by Aldus Manutius and his father-in-law, Andrea Torresani (Venice: Aldus, February 1518). It is a magnificent folio of 452 leaves, 34 cm high. The title page is meager and would reproduce badly. We reproduce the first page of the Greek text, which contains the first chapter of Genesis. The top ornament, title, and first capital are printed in red; the title means "The Holy Scriptures, Old and New." [Courtesy of Harvard College Library.]

The gist of the story is as follows. Ptolemaios II accepted Dēmētrios' advice and sent Aristaios and Andreas to Jerusalem on an embassy to the high priest Eleazar, begging him to lend the necessary manuscripts and to send to Alexandria six representatives from each of the twelve tribes. Eleazar was willing to oblige his king. The text communicated by him was written on skins (*diphtherai*). The seventy-two scholars were established in the island of Pharos and their translation was finished in seventy-two days. It is for that reason that the Greek translation of the Old Testament was called *Septuaginta* (Septuagint in English; seventy in round numbers for 72).[39]

The legendary nature of that account is obvious. The early part of the Septuagint, the Torah or Pentateuch, is written in a very poor Judeo-Greek; according to specialists that dialect is Egyptian rather than Palestinian. I have read only Genesis, and was horrified by the language; it would not be fair to compare that language with the best Attic one, but it is very fair to compare it with that of the Gospels written almost four centuries later. The language of the Gospels is immeasurably superior to that of Genesis. How was this permitted to happen? For there were plenty of Greeks in Alexandria who knew their language perfectly and whose collaboration might have been easily drafted by the Court or the Museum.

Such as it is, the Septuagint is very precious to us, because it was made before the standard Hebrew text was established by the Sopherim. Moreover, the oldest Greek manuscripts are older than the oldest Hebrew ones (with the exception of some of the rolls discovered in 1947 in Jordanian caves, on the eastern shore of the Dead Sea).[40] The Septuagint is so important that its testimony can never be overlooked; the Old Testament scholar must know Greek as well as Hebrew.

The Septuagint became a sacred text for the Christians.[41] There are thus two Old Testament traditions, the Christian one based upon the Septuagint and the Vulgate,[42] and the Jewish one based upon the Hebrew text established by the Sopherim (completed by the end of the second century of our era) and interpreted by the Masoretes in the tenth century.[43]

In short, we owe to the Alexandrian scholars the earliest edition of the

[39] *Hē hermēneia cata tous hebdomēconta* (*interpretatio septuaginta seniorum*), abbreviated to "*hoi O*" or "the LXX." The translation was at first restricted to the Pentateuch, but by 132 B.C. almost the whole Old Testament had been translated into Greek by Alexandrian Jews; the name *Septuaginta* was extended to mean the whole of that first Greek translation of the Old Testament; the Septuagint is almost entirely pre-Christian.

[40] See Chapter XVI.

[41] All the scriptural quotations in the New Testament and the Greek Church Fathers are derived from it. Some Jews like Philōn (I-1) and Jōsēphos (I-2) always referred to Old Testament texts in their Septuagint form.

[42] When St. Jerome (IV-2) prepared the Vulgate in Bethlehem between 386 and 404, he used the Septuagint but, realizing its shortcomings, he also used Hebrew and Aramaic sources.

[43] The early Hebrew text was purely consonantal; vowel points were added only in the seventh century. A new standard text was established three centuries later, together with its *masorah* (or interpretation); this was done by the two main masoretic schools of the tenth century, in

Pentateuch in any language; we owe to them a part of our knowledge of a text equally sacred to Jews and Christians. Our indebtedness to Hellenistic Egypt is considerable, and this part of their legacy, the Septuagint, is not by any means the least.[44]

The history of Orientalism in Hellenistic times will be continued in Chapter XXVIII.

Tiberias and Babylon. The Tiberias tradition is immortalized in the fundamental printed text of the Old Testament, edited by Jacob ben Ḥayyim ibn Adonijah (4 folio vols.; Venice, 1524–26).

[44] For more information than could be given here, see Pfeiffer, *Introduction to the Old Testament*, pp. 104–108. The first *printed* text of the Septuagint was included in the great Complutensian Polyglot, published under the auspices of Cardinal Ximenes de Cisneros (Alcala, 1514–17). The publication was delayed until 1521, however, and the first *published* edition (the princeps) was the Aldine (Venice, 1518/19), though it was printed after the Complutensian. Third edition under the auspices of Sixtus V (Sixtine edition; Rome, 1587). The Cambridge University Press published a portable edition of the Greek text (4 vols., 1887–94); this was revised at least thrice. The larger Cambridge edition, in three volumes (bound in 9 parts), appeared in 1906–1940.

PART TWO

THE LAST TWO CENTURIES BEFORE CHRIST

XV

THE SOCIAL BACKGROUND

If we consider the Hellenistic age, as is generally done, a period of three centuries, these three centuries do not coincide exactly with the three centuries before Christ, for the period is assumed to begin in 323 with the death of Alexander the Great and to end in 30 B.C. with the establishment of the Roman Empire. Both dates are a little artificial, but they are the best available, provided we do not take them too pedantically. The Alexandrian empire did not disintegrate immediately after Alexander's death and Roman imperialism began before Augustus.

Part One was devoted to the first part of that age, the Alexandrian renaissance (roughly the third century); Part Two will deal with what might be called the decadence and fall of Hellenism, the last two centuries of the pre-Christian era.

During those two centuries, the known world (the *oicumenē*), that is, the world known to the learned people, continued to be Greek or Hellenistic. The learned world was remarkably international; the culture favored by the educated people was Greek, and their best language was the Greek *coinē*; [1] the world was international-minded in its highest aspects (religion, philosophy, science, and art) and it was philanthropic in the Stoic sense, except that it was deeply vitiated by slavery, the existence of which was taken for granted, as a law of nature. The best men, those who were most free of superstitions and fanaticism, continued more or less consciously the Alexandrian and Stoic traditions of *homonoia* (unity of mankind) and *coinōnia* (communion, participation).[2] Unfortunately, tumults, revolutions, and wars and all the evils stemming therefrom never stopped anywhere for any length of time, and it was increasingly difficult even for the gentlest men and the wisest to remain very long *"au dessus de la mêlée."*

THE HELLENISTIC WORLD

The leading people in the Near East were Greek, but Greek mercenaries, officers, civil servants, and their followers had been used so much in Egypt

[1] The *coinē dialectos* was the "common language" such as was used in the Septuagint and later in the New Testament. *Hē coinē ennoia* (or *epinoia*) is "common sense."

[2] For *homonoia* and *coinōnia*, see Volume 1, p. 603.

and all over Eastern Asia, and had been so widely distributed, that the Greek communities or individuals were lost in the sea of native populations. There were not enough Greeks to Hellenize the African and Asiatic nations, and an increasing number of the younger generations had native mothers. By the end of the second century, if not before, the Hellenistic world was still Greek on the surface, but outside of the Greek continent and some of the islands it was more and more impregnated with foreign elements. The old division between Greeks and barbarians was steadily losing its validity.

Without trying to be complete and without bewildering ourselves with political details which are innumerable, let us take a bird's-eye view of that world.

Continental Greece was still fairly homogeneous; there were many Macedonians and Romans around but few Orientals; the bulk was Greek. In spite of its vicissitudes, Athens was still the sacred focus of Greek culture and Greek education; Corinth flourished until 146; many other cities were able to emerge time after time from national or local disasters.

The golden age of Ptolemaic Egypt was over, but Alexandria was still the largest center of Hellenic culture as well as the richest emporium. By the year 200, it was still the greatest city of the world,[3] though Rome would surpass it before long; by the time of Augustus the population of Alexandria was perhaps as large as a million. In the second century, Greeks, Egyptians, and Jews were already very much mixed; the dominating culture was Hellenistic; the prominent Jewish and native families used the Greek language and often assumed Greek names.[4] The most illustrious members of the Ptolemaic dynasty were the first two, who flourished in the third century B.C., and the very last, Cleopatra VII (d. 30 B.C.), one of the most extraordinary women in the whole past.[5]

The three most important islands from the cultural point of view were Dēlos, Cypros, and Rhodos. Dēlos, being a sacred place, enjoyed some kind of neutrality and was therefore a nest of political intrigues. In 167, Rome declared Dēlos to be a free harbor in order to damage the Rhodian trade. It was sacked by order of Mithridatēs in 88 and again in 69. When Pompey destroyed the pirates in 67, it enjoyed a modicum of prosperity but never recovered its former splendor.

Cypros was most of the time a dependency of Ptolemaic Egypt, whose vicissitudes it was obliged to share; it became a Roman province in 58 B.C.

Rhodos was an independent maritime power and a center of trade, art, and science. We shall come back to it many times, especially when we deal

[3] I do not say "Western world" because it must be understood once and for all that my survey does not cover India or the Far East, but is largely restricted to the Western *oicumenē*.

[4] Jews favored names derived from Theos (God), such as Theodotos and Dōrothea.

[5] A brief history of the dynasty has been given in Chapter I; we need not return to it.

with Panaitios (II–2 B.C.), Hipparchos (II–2 B.C.), and Poseidōnios (I–1 B.C.). Its trade was protected by an excellent fleet which managed to repress the pirates and to create for a time a "Rhodian peace" in the eastern Mediterranean. Its maritime law was adopted by the Antonines and was probably the source of the "Rhodian navigation law" compiled under Leōn III the Isaurian, c. 740, of medieval codes, and even of the Venetian usages in later times.[6] It controlled some land on the Asiatic shore, the Peraia (*Peraea Rhodiorum*), and the Romans increased its share in 188, but deprived the city of it some twenty years later.[7] The role played by Rhodos in the Hellenistic age has been compared with that of the Republic of Venice in the sixteenth and seventeenth centuries.

Let us pass to Asia. The main kingdom was the Seleucid, which included at first Syria, Cilicia, and Mesopotamia. The outstanding king was Antiochos III the Great (ruled 223–187), who acquired Armenia but made the mistake of underestimating Roman power. He was defeated by the Romans in a naval engagement and in two land battles at the Thermopylai (196) and Magnēsia in Lydia (190) and was obliged to sign the peace of Apameia (188), which put an end to his Mediterranean influence. The Seleucid kingdom remained powerful in Asia Minor. Another great king was his son Antiochos IV Epiphanēs (ruled 175–164), who realized that his main duty was to Hellenize Syria; he made the mistake, however, of trying to seduce the Jews from their own religious duties and caused the Maccabean revolt (168); the Jews obtained religious freedom in 164 and political independence in 142 (until the beginning of Roman rule in 63 B.C.). The last Seleucid ruler, Alexander Balas (ruled 150–145), needed Roman support to preserve what little power he still possessed. After his expulsion and death (145), the kingdom disintegrated, and it finally became a Roman province in 64 B.C.

The Seleucid capital was Antioch (Antiocheia on the river Orontēs, some 14 miles from the sea). It was one of the leading cities of the Hellenistic world, vying with Alexandria, and as cosmopolitan as the latter. Its population had grown fast because of the arrival of many Greek refugees (exiled Aetolians and Euboeans) and Jews.[8] When the Seleucid kingdom was annexed by Pompey in 64, Antioch was the capital of the Roman province of Syria. Apameia, also on the Orontēs, upstream (that is, south) from Antioch, was a natural fortress which served as military headquarters;[9] it was there that the peace of 188 was signed. It was a much smaller city than Antioch

[6] *Nomos Rhodiōn nauticos; Introduction,* vol. 1, p. 517. The main medieval code was the Catalan *Llibre del consolot de mar* of Barcelona compiled about the middle of the fourteenth century; *Introduction,* vol. 3, pp. 324–325, 1140.

[7] P. M. Fraser and G. E. Bean, *The Rhodian peraea and islands* (192 pp., ill.; London: Oxford University Press, 1954).

[8] It was natural enough for Jews to move northward along the coast, being attracted by Antiocheian prosperity. It was much easier to go from Jerusalem to Antioch than to Alexandria.

[9] It was a kind of arsenal, and it was there that the Seleucid kings kept their elephants and horses and probably their stud.

but not a mean one. The fortress was seized by the Romans only in 46. Under Augustus, it still had more than a hundred thousand inhabitants.

Smyrna in Lydia (in the same latitude as Chios) was one of the richest cities of the western coast of Asia Minor, competing with Milētos and Ephesos, but more nearly permanent than they were. Its harbor was one of the best of the Near East, and its hinterland was full of resources. It was much favored by the Romans and took sides with them against the Seleucid kingdom and against Mithridatēs of Pontos.

The city of Pergamon and a large territory around it detached from the Seleucid kingdom was developed by the Attalid dynasty. Attalos I Sōtēr was its first "king" (241–197) and the first to refuse to pay tribute to his eastern neighbors, the Galatians.[10] His son and successor, Eumenēs II (king 197–159), made of Pergamon the most sophisticated city of the Near East next to Alexandria, and the most friendly to the Romans.

The Pergamene renaissance, begun by Attalos I and carried to its climax by Eumenēs II, was almost as astonishing as the Alexandrian renaissance accomplished a century before by the first and second Ptolemaioi. While Alexandria was built near the coast, almost at sea level, Pergamon was established some fifteen miles inland upon a steep hill near the confluence of three rivers. The Attalid kings built an acropolis for themselves at the top and many public buildings on the slopes; beautiful temples and theaters could be seen from a distance one above the other upon successive terraces. The great altar celebrating their victory (c. 235) over the Galatians was completed during the rule of Eumenēs II; it represented the heroic conflict between gods (the Pergamenians) and giants (the defeated Galatians) and was one of the most remarkable monuments of the ancient world.[11] Attalid patronage caused the emergence in Pergamon of a school of art and also of literature; the library of Pergamon, of which more will be said at the end of this chapter, was the greatest library of antiquity, next to that of Alexandria.

The Attalid kings were so friendly to Rome that they were considered betrayers of Hellenism. Attalos III (ruled 138–133), the last of his dynasty, trusted the Romans overmuch and himself too little; he was apparently more interested in the cultivation of herbs and the study of poisons than in politics;

[10] The Galatians or Gauls were the offspring of real Gauls or Celts who had emigrated to Bithynia at the invitation of Nicomēdēs I (278–250) and later had moved eastward and established themselves in the central part of Asia Minor (Galatia, the main city of which was Ancyra, now Ankara, Turkey's capital). Their name is familiar to us because of St. Paul's Epistle to the Galatians. It is claimed that some of them were still speaking a Celtic language when St. Jerome (IV-2) visited them. That is hard to believe. Their common language was Greek and they were often called Gallograeci.

[11] It was well known in Europe because all the sculptural parts had been brought to Germany and the altar rebuilt in the Berlin Museum. Those parts have been taken away by the Russians and their present location is unknown. G. Sarton, *Galen* (Lawrence: University of Kansas Press, 1954), p. 9. *Bonner Berichte aus Mittel- und Ostdeutschland. Die Verluste der öffentlichen Kunstsammlungen, 1943–46* (Bonn, 1954), p. 20.

his kingdom was bequeathed by him to Rome [12] and soon after his death, in 133 (aet. c. 37), it became the province of Asia.

The temple of Artemis had given to Ephesos a considerable fame and sanctity in the Greek world. "Artemis of the Ephesians" was an oriental fertility goddess whom the Greek colonists Hellenized.[13] Her famous temple was burned down in the very night of Alexander's birth (356), but was soon rebuilt. Ephesos was a part of the kingdom of Pergamon and thus became Roman in 133; it was eventually the chief city of "Asia." The cult of Artemis and the pilgrimages to Ephesos continued until the end of paganism; [14] they were not stopped by St. Paul's Epistle nor even by the destruction of the city and the temple by the Goths in A.D. 262.

The main cities of the western Mediterranean, Syracuse and Carthage, were by this time under the Roman yoke. The surrender of Syracuse in 212 is well known to historians of science because of Archimēdēs' death, which was incidental to it. As to Carthage, it was annihilated in 146; its site was too valuable to be abandoned, however, and in the following century it accommodated a Roman colony; the new Carthage was the capital of proconsular Africa. The cultural heritage of the Punic city was small, yet it included Mago's work to which we shall come back in Chapter XXI.

This sketch is sufficient to indicate the variety and richness of the Mediterranean world, though we have spoken of only a very few cities among many; others will be referred to in the text or footnotes as we proceed.

The number of cities in the eastern provinces as well as in the western ones was considerable, though we should remember that it was much smaller in pre-Christian times than later. Consider, for example, the survey made by Arnold Hugh Martin Jones in his *Cities of the Eastern Roman Provinces* (592 pp., 8 maps; Oxford: Clarendon Press, 1937), which covers the period from Alexander's conquests to Justinian (VI–1), inclusive. His pages teem with names of cities, but many of them are Roman (Augustan or later) or even Byzantine. Nevertheless, those mentioned in our pages are only a very few of those that were already flourishing before Christ.[15]

THE GROWTH OF ROME

The outstanding feature of the period was the steady, the relentless, growth of Rome. It seems as if that growth was partly unconscious or un-

[12] Attalos' testament is textually known through an inscription found in the Pergamon theater. Edited by Wilhelm Dittenberger, *Orientis Graeci inscriptiones selectae* (Leipzig, 1903), vol. 1, no. 338, pp. 533–537. The motives of that bequest are not quite clear; Attalos III was a very strange personality. Esther V. Hansen, *The Attalids of Pergamon* (Ithaca: Cornell University Press, 1947), pp. 136–142.

[13] Her avatars were the Great Mother of Anatolia, Artemis and Diana Ephesiorum.

[14] The ruins of the second temple were discovered in 1869. St. John Ervine, "John Turtle Wood, discoverer of the Artemision," *Isis* 28, 376–384 (1938).

[15] A. H. M. Jones deals with thirteen regions or provinces, each of which boasted many cities: 1, Thrace; 2, Asia; 3, Lycia; 4,

Fig. 42. The she-wolf of the Capitol suckling the twins Romulus and Remus. According to the legend Romulus and Remus, children of a Vestal and the god Mars, were abandoned to die but were saved by the wolf who mothered them. Romulus was the founder of Rome, and the wolf was soon called *mater Romanorum*. The early Sabini and Romans had probably a wolf totem, and the Lupercalia (15 February) was perhaps their most ancient festival. This bronze wolf was made as early as the fifth century B.C. in a Greek studio of central Italy (say, Cumae) or in an Etruscan one (Veii, near Rome). The nurslings were added rather late, c. 1474, and are ascribed to Antonio Pollaiuolo (1429–1498). The nursing wolf is represented on an Etruscan stela (Bologna) and on many Roman coins. The story of this monument is extremely complicated and mysterious as was shown by Jérôme Carcopino, *La louve du Capitole* (90 pp., 6 pls.; Paris, 1925). It is one of the most impressive relics of the past, for it evokes the beginnings of Rome, Greek and Etruscan influences, and finally the Italian Renaissance. [Museo dei Conservatori, Campidoglio, Rome.]

premeditated. Rome was very ancient; according to its own calendar, the city was founded in 753, but for centuries it had been but one nation among many others. The essential difference lies in this, that Rome survived all others as if she were eternal – as indeed she is. An endless series of wars did not interrupt her growth, but emphasized its main phases: the Punic

The Gauls; 5, Pamphylia, Pisidia, and Lycaonia; 6, Bithynia and Pontos; 7, Cappadocia; 8, Cilicia; 9, Mesopotamia and Armenia; 10, Syria; 11, Egypt; 12, Cyrenaica (and Crete); 13, Cypros. See *The cities of the eastern Roman provinces* (Oxford: Clarendon Press, 1937). For the Byzantine period he lists 48 cities in Asia, 34 on the Hellespont, 28 in Lydia, 35 in Caria, 40 in Lycia, and so forth. See *The Greek city from Alexander to Justinian* (Oxford: Clarendon Press, 1940).

Wars (I, 264–241; II, 218–201; III, 149–146), the Macedonian Wars (I, 215–205; II, 200–197; III, 171–168; IV, 149–148), the Syrian War (192–189), the Jugurthine war in Africa (111–105), the Mithridatic wars (I, 88–84; II, 83–81; III, 74–64), the conquest of Gaul (58–51), the invasion of Britain (54). Add to that the civil wars, the agrarian reforms of the Gracchi (133–121), the Servile Wars in Sicily (I, 135–132; II, 103–99), the Social War (91–88), the Civil War in Rome (88–82) ending with Sulla's dictatorship, the Third Servile War in Italy (73–71), the First Triumvirate (Caesar, Crassus, Pompey, 60–51), the murder of Pompey in 48, the murder of Caesar in 44, the Second Triumvirate (Antony, Lepidus, Octavian) in 43, the Battle of Action won by Octavian over Antony in 31. After that Octavian became the emperor Augustus and a new world began – the Roman empire.

While Rome was struggling in all these wars abroad and during the domestic revolutions, the kingdoms of the Near East were fighting each other and there was always one of them that wanted Roman help against its enemies. Rome was willing enough to oblige them and to take full advantage of her alliances. The kingdoms gained or lost, Rome was sometimes beaten, but her gains were always greater than her losses. They finally went under; she grew bigger and stronger. Thus was the empire built up in spite of endless calamities.

Let us look at it a little more closely without going into too many details. It was about 212 that Rome, then already half a millennium old, was first inveigled into Hellenistic affairs. When her hands were freed after the end of the Second Punic War (201), Rhodos and Attalos I of Pergamon appealed to her, and this was the first of many arbitrations that obliged her to intervene in Eastern troubles and to make the best of them. It was not always Rome's deliberate purpose to intervene but she was entangled willy-nilly and did not scruple to improve every opportunity which her own wishes or *fortuna* opened to her. In 197, Philip V of Macedonia was defeated at Cynoscephalai (Thessaly) by Titus Quinctius Flaminius and the Aitolians, and at the Isthmian games [16] of 196, Flaminius proclaimed the freedom of Greece! (Conquerors have always liked to pose as deliverers.) In spite of the Aitolian help given in 197, the Aitolian League was subjected to Rome in 189; the Achaian League was made subservient to Rome in 183; thus were the Greek cities gradually disarmed. Antiochos III the Great, defeated at Magnēsia in Lydia by Scipio Asiaticus, was obliged in 188 to sign the peace of Apameia. Twenty years later, his son, Antiochos IV Epiphanēs, could have conquered Egypt but Rome told him to keep out. In the same year (168), the last king of Macedonia, Perseus, was defeated by Aemilius Paulus at Pydna. The Romans were now getting harder and even less scrupulous than before; their imperialistic tendencies were growing fast. In 167, they divided Mace-

[16] The Isthmian games, organized in 581, were international festivals held every other year at Corinth in honor of Poseidōn. Half a century later, the Romans would suppress not only the freedom of Corinth but its very existence.

donia into four republics, each of which was obliged to pay tribute. In 164, they restored the government of Egypt to Ptolemaios VI Philomētōr and gave Cyrēnē to his brother Ptolemaios VIII Evergetēs (who bequeathed it to them!). The first Hellenistic kingdom to become a Roman province was Macedonia, in 148. The year 146 must have seemed a very auspicious one to Roman minds; it marked the end of the Third Punic War, and the destruction of Carthage by Scipio Aemilianus and of Corinth by Mummius Achaicus. Mummius dissolved the Achaian League and sent the treasures of Corinth to Rome. This was a kind of climax in Rome's ascent to power, and she began to realize the beauty of Greek culture. Cicero thought of this period as of a golden age.

By this time Rome counted eight provinces; I, Sicilia, 241 (Syracuse included in 212); II, Sardinia, 238 (Corsica added to it c. 230); III, Hispania citerior, 205 (northwest Spain, capital Tarragona); IV, Hispania ulterior, 205 (Baetica, Andalusia); V, Gallia cisalpina, c. 191 (northern Italy); VI, Illyricum, 168 (east of the Adriatic); VII, Africa, 146; VIII, Macedonia and Achaia, 146.

In 133, Pergamon was bequeathed to Rome and became the province of Asia a few years later. In 116, Phrygia, east of Asia, was added to it. Ptolemaios Apiōn,[17] king of Cyrēnē (117–96), bequeathed his kingdom to Rome in 96 (not annexed until 75).

In the meanwhile, Mithridatēs VI Eupatōr [18] of Pontos had considerably increased his own kingdom, annexed Colchis and Armenia, and defeated the Parthians, but in 92 a treaty with Rome was forced upon him. There was a growing resentment in the eastern lands (Pontos, Parthia, Armenia, Cappadocia) with western power. Mithridatēs decided to take advantage of it, attempted to liberate "Asia," and in 88 ordered a general massacre of the Romans of Asia Minor and the islands (about 100,000 Romans perished). This started rebellions elsewhere and Rome was obliged to intervene not only for her own sake but for the defense of Hellenism, that is, of Roman Hellenism. Greece and Greek Asia never recovered from the Mithridatic wars (three wars which lasted from 88 to 64). The center of the eastern trade was partly displaced from Dēlos to Puteoli (near Naples).

To return: in 83, Tigranēs the Great (king of Armenia from 96 to 56) invaded Syria and Mesopotamia and put an end to the Seleucid dynasty. In 74, Nicomēdēs IV, the last king of Bithynia (92–74), bequeathed his kingdom to Rome. At the time of Pompey's final victory over Mithridatēs the Great in 64, Pontos became a Roman province (Bithynia being included in

[17] Apiōn was an illegitimate son of Ptolemaios VIII Evergetēs; the latter had already bequeathed Cyrēnē to Rome. The circumstances and meanings of those bequests are not clear to me.

[18] The first Mithridatēs (or Mithradatēs) had founded the Pontic dynasty in 337. He was of Persian race and his name is derived from that of Mithras. That dynasty developed at the expense of its eastern and southern neighbors, the Armenians and the Parthians (Arsacids). Conflicts with Rome began only in the first century B.C. Mithridatēs Eupatōr is often called "the Great."

it), and Syria another. At that time the whole of Asia Minor, the Greek peninsula, and Cyrenaica were controlled by Rome, the various nations being either provinces or protectorates. In the second category one may count Galatia, Cappadocia, and to some extent Ptolemaic Egypt. Crete was annexed by Rome in 66; Cypros in 58. Ten years later Caesar forced the restoration of Cleopatra VII as queen of Egypt. She committed suicide in 31. Egypt became a Roman province in 30, Galatia in 25, and Cappadocia – last of all – under Vespasian (emperor A.D. 69–79).

This enumeration, long to the point of tediousness, is nevertheless very incomplete and imperfect. Every article of it would require long qualifications. Such as it is, it suggests the relentless growth of Rome and the preparation of the formal empire.

CAESAR AND AUGUSTUS

The end of that story, which may be symbolized by two gigantic names – Caesar and Augustus – is better known to educated readers and yet a summary of it may be helpful to them. Perhaps one might add to those two names a third one – that of Pompey the Great.[19] He was the conqueror of Mithridatēs, the destroyer of the pirates, and the organizer of the Roman provinces in the East. He was defeated by Caesar at Pharsalos (Thessaly) in 48 and murdered the same year when landing in Egypt. He was a military rather than a political genius, yet he was a great administrator and his activities made possible and facilitated the constitution of the empire seventeen years after his death. Cicero paid a simple and fine tribute to him when he wrote, "*hominem enim integrum, castum, et gravem cognovi.*" [20]

Julius Caesar [21] was also a military genius, but he was much else and much more. While Pompey was a general in the field before he was twenty-five, Caesar's military career began much later, at forty-three, and (as Pascal thought) that was mighty late to begin the conquest of the world. He started the Gallic wars (58–51) at an age when Alexander was already dead and Napoleon beaten. Up to that time he had been mainly a popular agitator; now he began to lead armies and govern provinces, and he did it in the best manner. His military and administrative duties never diminished his love of literature and he was himself a man of letters of the first order (we shall come back to that in chapter 25). His prestige was largely based on that fact, for the Roman elite fully realized the superiority of mind over material power, and Caesar was both intelligent and powerful. He was one of the first victorious generals to respect and pity his fallen enemies; this does not mean that he was always clement, but he did not indulge in cruelty for its own

[19] Gnaeus Pompeius (106–48), called Magnus since 81.

[20] *Ad Atticum*, XI, 6, 5. "I have known a man who was honest, pious, and grave."

[21] Pompey (106–48) and Caesar (c. 101–44) were almost exact contemporaries, and they reached the same age, 57 or 58 years.

sake. After his victory at Pharsalos over Pompey (48), his restoration of Cleopatra, his mistress, as queen, his victory over Pharnacēs at Zēla [22] (47, "*Veni, vidi, vici*"), his victory over the remnant of the Pompeian army in Thapsus [23] (46), he celebrated four triumphs, Gallic, Alexandrian, Pontic, and African. He was dictator, held many powers and honors, controlled all the main offices. This was too much for the defenders of freedom. A conspiracy led by Marcus Brutus and Gaius Cassius was formed against him and he was murdered in the Senate on the Ides of March 44; he died at the foot of Pompey's statue.

After Caesar's murder, there was a kind of political vacuum which was gradually filled by Marcus Antonius ("Mark Antony") and by the emergence of a young man, Caesar's great-nephew, then eighteen, Gaius Julius Caesar Octavianus ("Octavian"), and in 43 the Second Triumvirate was made up of those two and M. Aemilius Lepidus. The Triumvirate strengthened itself with a widespread proscription and confiscation of lands and money; its most illustrious victim was Cicero, murdered by Antony's agents on 7 December 43.[24]

In the following year, the triumvirs celebrated Caesar's apotheosis, erected a temple to his memory in the Forum, and continued the fight against their enemies. They defeated the combined forces of Cassius and Brutus at Philippoi in Macedonia in that same year (42); [25] both committed suicide. In 41, Antony met Cleopatra at Tarsus (Cilicia), returned with her to Egypt, and formally married her in 36. Antony's subjugation by Cleopatra caused leading Romans to fear that Roman interests would be sacrificed to Oriental ones. Cleopatra posed as Isis and as a Roman empress; the Romans were more frightened by her than they had ever been by a foreigner (except Hannibal), and prophecies were scattered that, after defeating Rome, she would begin a golden age during which the West and the East would be reconciled by justice and love. Caesar, had he lived, might have helped her to conquer Rome with the help of Roman power, but Antony was not capable. Octavian defeated him at the naval battle of Action [26] in 31; in 30, Antony commited suicide; Cleopatra hoped for a while to realize her ambitions through Octavian (after Caesar and Antony had failed her), but she could not swing him and destroyed herself.

Egypt was now reduced to the status of a province, and Octavian was mas-

[22] Ta Zēla, near Amaseia, in Pontos. Pharnacēs, son of Mithridatēs the Great, was king of Pontos or of Bosporos (Kerch).

[23] In Byzacium or the Byzacena regio, the eastern part of the province of Africa. Thapsus, on the eastern coast of Tunisia, north of Mahdia, was originally a Phoenician factory.

[24] G. Sarton, "Death and burial of Vesalius and, incidentally, of Cicero," *Isis 45*, 131–37 (1954).

[25] Many of our readers remember Philippoi as the first European place where the Christian gospel was preached (by St. Paul); there are four references to it in the New Testament.

[26] Action (Actium) is at the entrance of the Ambracian Gulf on the Ionian coast of Greece. It was a sacred place because of a famous temple to Apollōn.

Fig. 43. Cleopatra VII (d. 30 B.C.) the last queen of Egypt, represented as Hathor, and her son by Julius Caesar, Ptolemy XIV, known as Caesarion (Temple of Dendera). [Courtesy of the Metropolitan Museum, New York.]

ter of the world. He promised "to restore the Republic," and he did indeed restore peace. The temple of Janus [27] was closed in 29 for the first time since 235, and the altar of peace (*ara pacis*) dedicated in 9. In the meanwhile in the year 726 U.C. (27 B.C.), Octavian became absolute emperor (*autocratōr*) and was called Augustus (*sebastos*, venerable). In 13, he was *pontifex maximus* (high priest).[28] In 2 B.C., being then consul for the thirteenth time, he was called *pater patriae*, a title which gave him great satisfaction. He died in A.D. 14 at Nola in Campania (near Naples).

We may add a few reflections on the creation of the Roman empire in 31 and Caesar's share in it as well as Octavian's. The latter did not have Caesar's genius and generosity, but he was clever, cruel, and efficient; he stood on Caesar's shoulders and managed to do what he did because of Caesar's preparation. This is not an uncommon process; mediocre men often succeed where greater ones have failed. The former succeed partly because of their shortcomings and lack of scruples; the latter may fail because of circumstances but also because of their virtues. After Pharsalos (48), Caesar was master of the Roman world but the obstinate remembrance of freedom was still too strong to be overcome; after Philippoi (42), democracy and freedom in the old sense were ended; after Action (31), that is, after a civil war of twenty years, the most terrible war of its kind, the witnesses of freedom were all dead and Octavian's hypocritical "restoration of the republic" in 27 was wel-

[27] Janus meant a door (hence *janitor*, doorkeeper). The god Janus was generally represented with two heads (*Janus bifrons, sive geminus*) facing opposite directions (just as a door has two sides). The temple of Janus was open in war and closed in peace. The mythologic development is not clear to me.

[28] A title that has been inherited by the popes.

come.[29] People were so tired of war that the evils of dictatorship were forgotten. Augustus played his role very well, using old words such as *dēmocratia*, *libertas*, and *res publica* with new meanings. No monarchy was ever more absolute than his (at least in the West); every power was concentrated in his hands, and, as his empire extended to the whole world, there was no place for exile. And yet he was always camouflaging his absolutism or complaining of it; according to his own declarations, his purpose was not at all to enslave the people but rather to rejuvenate the old ideals.

His political life is limited by two documents, both of which are extant. The first is his preamble to the edict of proscription (November 43), which was preserved by Appianos of Alexandria (II–2), and the second his political will written in A.D. 13 after fifty-six years of unlimited power. According to Suetonius (II–1), he had ordered it to be engraved on bronze tablets; these tablets have disappeared but, happily, a copy of the inscription and a Greek translation of it engraved upon the walls of the Augustan temple at Ancyra are still extant.[30] The inscription (as compared with those of Oriental rulers) is relatively modest in spite of the fact that it enumerates all his achievements, not simply military ones but also constitutional changes, economic, political, and diplomatic matters, the large number of monuments built or restored by his order, and so on.

ROMAN LIBRARIES

Among many aspects of social life we shall select the one that is closest to this book's purpose and the reader's interest – the libraries. The Library of Alexandria has been described in Chapter X. It was undoubtedly the richest library of antiquity, but it was not by any means the earliest one nor the only one. We may assume that almost every great city of the Hellenistic age had a library of its own. These libraries were generally owned by the rulers and opened to their families. The learned men, poets, and artists who graced the royal court with their presence were able to use the library, but no library was "public" in the modern sense. Each library was at best semipublic, like the Pierpont Morgan Library and many other repositories of our own time, that is, one could not make use of it without some formality but bona fide students were welcomed. After all, the general problem was the same then

[29] Octavian's illiberality did not begin immediately after Action. Cicero's letters were published by his order, or with his permission, and must have caused a deep impression. Thus Cicero, murdered in 43 by Antony, was partly rehabilitated by Octavian after 31. Or did Octavian intend to show him up?

[30] It was my privilege to examine it during my visit to Ankara in August 1952 (Ancyra = Angora = Ankara). The *Monumentum ancyranum*, as the inscription is called, was illustrated by Georges Perrot, *Exploration archéologique de la Galatie et de la Bithynie* (2 vols. folio; Paris, 1862–1872). I regret to say that it has suffered much since Perrot's time. There are many editions of the Latin and Greek texts of that inscription. For example, William Fairley's in *Translations and reprints* (Philadelphia, 1898), vol. 5, no. 1, and Jean Gage, *Res gestae divi Augusti ex monumentis ancyrano et antiocheno latinis, ancyrano et apolloniensi graecis* (Paris: Belles Lettres, ed. 2, 1950).

as now. Every collector wishes the objects of his collection to be appreciated and their value remains doubtful as long as it has not been explained and exploited by competent scholars. The collector of books needs devoted readers, and this is even more true when lack of time, attention, or skill prevents him from reading them himself.

When Antiochos the Great (223–187) developed his capital, Antioch, on the Orontēs, he was naturally anxious to have it equal Alexandria and to be equipped with the institutions that were then already considered necessary for the fame of a great city: temples, theaters, circuses, collections of works of art and of manuscripts. The Library of Antioch was placed c. 221 B.C. under the care of the poet and grammarian, Euphoriōn of Chalcis. Euphoriōn's merit cannot be judged because his works are lost, but he was imitated not only by Greek poets but also by Roman ones like Cornelius Gallus (c. 69–26); Tibullus (c. 48–19), and Propertius (c. 50–10), and Virgil referred to him. Some kind of museum and library existed in Antioch at least until the end of the Seleucid dynasty.

The library of Pergamon, founded or developed by Eumenēs II (197–159), was second only to that of Alexandria. It is said to have included some 200,000 volumes at the time of Antony's alleged gift of it to Cleopatra. As Eumenēs needed a competent librarian for his endless acquisitions, he tried to debauch Aristophanēs of Byzantion, who was librarian in Alexandria from 195 to 180 under the rule of Ptolemaios V Epiphanēs. When Ptolemaios discovered this, he ordered Aristophanēs to be imprisoned and the export of papyrus to be forbidden. The story goes that the lack of papyrus obliged the Pergamenians to find some other material and to develop (not to invent) the use of skins (*diphtherai*); that new material was eventually called "parchment" in remembrance of its origin. All of this contains parcels of truth but has been much exaggerated in the telling. It is possible that the Egyptian kings put an embargo on papyrus, not only to vex the Attalid upstarts but also to protect a dwindling supply. They probably used some kind of vellum,[31] but it is certain that the majority of rolls continued to be made of papyrus. The change from papyrus to vellum and from roll to codex [32] did not become current until Christian times (second and third centuries).[33] St.

[31] The earliest documents on vellum were discovered by Franz Cumont in the Roman fortress of Dura-Europos on the upper Euphrates; they bear dates equivalent to 190–189, 196–195 B.C. This would suggest that the use of vellum was already established in the third century. On the other hand, the word *pergamēnē* (hence parchment) is not found earlier than the Edict of Diocletian (A.D. 301); it is thus probable that skins were used much less than papyrus.

[32] These two changes were roughly simultaneous but not exclusively so. Most rolls were made of papyrus and most codices of parchment, but there were also papyrus codices and parchment rolls (the latter continue to exist to this day, in the form of furāmina and diplomas).

[33] Christianity seems to have favored the use of codices. In the third century (after Christ), and to a smaller extent in the fourth, the roll was more used for pagan works and the codex for Christian ones. Frederick G. Kenyon, *Books and readers in ancient Greece and Rome* (Oxford: Clarendon Press, ed. 2, 1951), vellum and the codex, pp. 87–120. Papyrus

Jerome (IV–2) records in one of his letters (Epist. 141) that papyrus rolls in the library of Pamphilos of Caisareia (in Palestine) were gradually replaced by vellum codices.

To return to the pre-Christian libraries, it was said that parchment was used in Pergamon; no such suggestion was ever made concerning Alexandria. As to the number of parchment manuscripts in Pergamon, we can only guess. No extant manuscript has ever been identified as being from one of those two libraries. If it be true that the Pergamon library was given by Antony to Cleopatra c. 34, then the two libraries were finally united and suffered the same fate, gradual dilapidation and destruction. We know that the Greek literature which has come down to us is only a small fraction of that which was available in ancient times.[34]

Of course, new libraries were constantly gathered in the Greek and the Roman world. There were new libraries in Pergamon in Galen's time (II–2), and he saw others in other cities; there were also booksellers in every center that he visited and he bought books from them for his own use.

What about Rome? The first important library was probably that of Lucius Licinius Lucullus (c. 117–56), the bulk of which was gathered in the East. It was available to his friends, especially the Greek ones, who flocked thither as to a museum. Cicero and Caesar had rich libraries of their own, but the "public library" of which the latter contemplated the formation did not materialize because of the abrupt ending of his life. The first public library of Rome was founded in the *Atrium libertatis* in 37 by Gaius Asinius Pollio (I–2 B.C.), a man of letters, friend of Virgil and Horace, who organized public recitations. Two more libraries were established by Augustus, one on the Campus Martius and the other on the Capitoline hill. This second one was founded in 28 and its prefect was C. Julius Hyginus (I–2 B.C.). The general arrangement of both libraries was the one devised by Caesar. There was a temple for religious services and close to it an open rectangular colonnade; the Greek books were kept on one side and the Latin ones on the other. That was a very natural classification; further details of classification and administration are unknown, nor do we have any definite knowledge of the size and contents of those libraries. There were also many private collections; many of the books available in Rome had been obtained as parts of public loot or private booty; others had been bought from impoverished owners or from regular booksellers. For example, after the siege of Athens in 84, the dictator Sulla brought to Rome what remained of Aristotle's library.[35]

It is noteworthy that the libraries planned by Caesar and realized by Augustus included a temple. That was the old idea of a museum (a temple devoted to the Muses); any collection of art and letters, any institution of learning or research, was placed under their protection. There are a great

---

codices found in Egypt are Coptic, that is, Christian.

[34] For proof, see Kenyon, *Books and readers*, p. 28.

[35] For the history of Aristotle's library, see Volume 1, pp. 476–477.

many museums in the modern world, but the Muses have generally been driven out of them by heartless and hardheaded administrators.

ARCHIVES AND ACTA DIURNA

In addition to those libraries, there were collections of archives, either in the Senate, in the *Ara pacis*, or in other public buildings. Magistrates took an oath to respect the laws of the state and had to be acquainted with the *acta* (the government's decisions). Caesar's *acta* were ratified by the Senate after his murder and magistrates took an oath to observe them (in 45); in 29 and 24 they took similar oaths to observe Augustus' *acta*. This implies that there was a definite place where the *acta* were recorded and preserved and where the persons concerned could examine them. The *acta* (or *commentarii*) *Senatus* were also preserved.

Furthermore, since Caesar's first consulship (in 59), a kind of official gazette was published daily, the *Acta diurna*. It included (1) numbers of births and deaths in Rome as registered in the temples of Venus and Libitina; (2) financial news, supply of corn; (3) wills of important people, trials, new magistrates; (4) extracts from the senatorial proceedings (*acta Senatus*); (5) court circular; (6) varia, such as prodigies and miracles, new buildings, fires, funerals, games and tales.

Copies of the *Acta diurna* were available not only in Rome but also in the provinces. Leading men and great officers of the state were not satisfied with the official gazette, however, but employed private diarists and secretaries who sent them the news and gossip of the day by messengers, and they depended also, to a large extent, upon the benevolent services of their friends. Good examples of such mutual services are given in Cicero's correspondence, part of which has come down to us.

The most important news, that which every citizen was expected to know, was advertised on notice boards (*alba*) in public places. The writing was in black on a white board but the headings were red (*rubricae*). The news published "*in albo*" could be read at leisure by any passerby and copied by him, if he were interested. Hence, a sufficient publicity was easily obtained.

XVI

RELIGION IN THE LAST TWO CENTURIES[1]

To understand the religious situation in the Hellenistic world it is perhaps best, or at least simplest, to think of it as a triangular conflict. One side of the triangle represents the purely Greek religions; the second side, the Oriental religions; the third, the Jewish. There are thus six kinds of tensions.

GREEK RELIGION

Let us place ourselves in a Greek community and consider the main tensions from the Greek point of view. A fastidious reader would interrupt us here and say, "What do you call a Greek community?" The answer is not difficult, though it cannot be precise. It is not simply a matter of race or parentage, nor of language, though the linguistic factor may be the strongest. The members of a Greek community spoke Greek and cultivated as well as they could Greek ideals; they knew Homer, Plato, Aristotle as members of an English community know Chaucer, Shakespeare, and Milton. In both cases, that knowledge would often be superficial, and intuitive rather than technical, yet it would create a strong bond of union. Their religion would be the old mythology as recreated by the Mysteries and the Festivals, but with the addition of various Oriental gods, such as Isis and Osiris, Sarapis, Mithras, the Great Mother goddess of Anatolia (Cybelē), and others.[2] The saying that the only vital things in Hellenism were philosophy and the Oriental religions is an exaggeration,[3] for many Greek gods were still alive and their cults flourishing. Moreover, some of the old gods were very real

[1] For religion in the third century B.C., see Chapter XI.

[2] Considerable details are given by Franz Cumont, *Les religions orientales dans le paganisme romain* (Paris, ed. 4, 1929) [*Isis 15*, 271 (1931)]. Cumont tells us when a definite god was introduced in Rome or in another city, how his cult was established, how it developed or was interfered with. The main sanctuary of Cybelē (or Agdistis) was in Pessinus (West Galatia, now Bahihisar). Her statue was removed to Rome in 205 B.C. (Livy, XXIX, 10).

[3] It is repeated with approval by W. W. Tarn and G. T. Griffith, *Hellenistic civilisation* (London: Arnold, ed. 3, 1952), p. 336.

to the Greek people. For example, in the healing temples, the god Asclēpios [4] actually appeared to the patients, and there were other epiphanies that were as indisputable to the witnesses as those of new Madonnas are today. We know of those epiphanies because (then even as now) they might cause the foundation of a new temple or a new cult and the fact would be recorded in inscriptions, some of which have come down to us. The old oracles were predominantly Greek (except the one of Ammōn in the oasis of Siwa) and new oracles were listened to. The old Greek mysteries, such as those of Dēmētēr, Dionysios, and the Orphic and Eleusinian mysteries, were if anything more popular in Hellenistic times than before. It is true that Oriental mysteries were added or Oriental features might be grafted upon the Greek stems, but the purely Greek mysteries [5] and festivals were just as alive as modern pilgrimage resorts. The Dionysiac mysteries were regulated by Ptolemaios IV Philopatōr (ruled 221–205) and their western form, the Bacchanalia, had to be repressed by the Roman Senate as early as 186. The Eleusinian mysteries retained their prestige until the very end of paganism. A religious revival, proved by many events, began after 146, when Greece was a Roman protectorate.[6] This suggests that calamities (such as the utter destruction of Corinth in 146) would increase piety, for the Greeks had no more hope except in their own gods. When everything else failed, religion remained the only palladium of Greek culture and of the good life.

The fact that the new temple to Apollōn in Didyma [7] remained unfinished for centuries does not prove a lack of popular religion, but of money and governmental support. Temples are not built by people but by their rulers.

What was more ominous was the continuous deflection of popular prayers toward Egyptian or Oriental gods, but the people were not aware of any infidelity; they prayed for their own salvation. Their desperation drove them to various forms of gnosis, magic, and occultism. Their religion did not become less intense but more superstitious.

Though the Jews were fairly numerous,[8] not only in their homeland but also in many Hellenistic cities, and had many commercial and political connections with the Greek people, many of them (perhaps most of them) did not compromise in religious matters and did not influence the Greek and Oriental religions. They allowed Greek to replace Aramaic as their common

[4] Asclēpios was worshiped early in Rome. After a plague that decimated the city c. 300, a sanctuary was dedicated to him c. 291. This was not a clandestine but an official foundation, authorized by the government after due consultation of the Sibylline books.

[5] For a new reassessment of the mysteries, see Raffaele Pettazzoni, "Les mystères grecs et les religions à mystères de l'antiquité. Recherches récentes et problèmes nouveaux," *Cahiers d'histoire mondiale* 2, 302–312, 661–667 (1954–55).

[6] Tarn and Griffith, *Hellenistic civilisation*, p. 39.

[7] Branchidae (Bragchidai), just south of Milētos. There was a temple to the oracle of Apollōn Didymēios. See Volume 1, p. 182.

[8] Archaeological proof of their numbers and ubiquity is given by Erwin Ramsdell Goodenough, *Jewish symbols in the Greco-Roman period* (4 vols., Bollingen Series; New York: Pantheon, 1953–54).

language and their knowledge of the sacred language, Hebrew, deteriorated. As Greek citizenship implied worship of the city gods, the Jews could not become citizens without apostasy and in almost every case they remained apart (a nation, *laos*, not a *dēmos*); there never could be a real fusion between the Jewish and Hellenistic populations as there often was between the Hellenistic and other Orientals. Jewish literature was influenced, to some extent, by Greek literature, but the opposite was hardly possible in the pre-Christian age (Philōn and Jōsēphos both belong to the first century after Christ). The Septuagint obtained an immense authority over Hellenistic Jews, but there is no trace of its influence upon the contemporary non-Jews.

The Jewish literature available in the two last centuries B.C. was quite considerable; most of it was available in Hebrew, some in Aramaic, and some in Greek. Of course, the older parts of the Old Testament (the very oldest dating back to 1200 or before, and the rest composed between 1200 and 300 B.C.) were circulated among faithful Jews. Greek literature began with Homer (say in the ninth century), but by that time some Hebrew writings were at least three centuries old. We are not concerned at present with the early strata of the Old Testament, but only with the creations that are more or less certainly posterior to 200 B.C.

A brief survey of them will give the reader an idea of the intellectual ferment obtaining among the Jews.[9] One word first: the writings of the period from 200 to 1 B.C. include not only many of those in the English Old Testament but also practically all of the Old Testament Apocrypha.[10]

HEBREW WRITINGS. THE OLD TESTAMENT APOCRYPHA

To begin with poetry, few people realize that in spite of the high antiquity of some of the Psalms (24:7–10; 45), a good many are of later creation, posterior to 400 or even to 200. The latest are the Maccabean Psalms (44, 74, 79, 83, etc.) and the Hasmonean Psalms (2, 110, etc.). The final edition of the Book of Psalms or *Psaltērion* occurred only after 200; that is, of course, the Hebrew edition entitled *tehillim*, *tillim*, or *tehillot* (praises); the Greek translation, *Biblos psalmōn*, followed soon.[11]

The Proverbs suggest the same remarks as the Psalms. None is as old as the older Psalms, yet some of them may go back to the sixth century; others

[9] My main source of information about such matters, but not by any means the only one, is the book of my friend, Robert H. Pfeiffer, *Introduction to the Old Testament* (New York: Harper, 1941) [*Isis 34*, 38 (1942–43)]; for a brief stratigraphy of the Old Testament, see pp. 21–23.

[10] The reader will bear in mind that the contents of the Old Testament are not the same in Hebrew, in the Greek Septuagint, and in the Latin Vulgate, not to mention other versions; nor are they the same in the Protestant and Catholic English translations. Thus writings that are canonical in one of these Bibles may be apocryphal in the others. For the sake of simplicity, I shall call apocryphal the writings not included in the King James Version, and not bother about the other versions.

[11] *Psaltērion* meant a musical instrument (psaltery) and *psalmos* the playing thereof or a musical composition. Later *psalmos* meant a sacred song (*mizmor*, psalm), and *psaltērion*, the collection of them (psalter).

may be early Hellenistic. In spite of its title, *Mishle Shelomoh* (Proverbs of Solomon), Solomon is no more the author than David is the author of the Psalms, and the contents are not proverbs in the usual sense. The various "teachings" added at different times and edited at the beginning of the Hellenistic age do not appear in exactly the same order in Hebrew and in the Septuagint (and other Bibles); the Greek order seems to be closer to the original than the one preserved in the Hebrew canon.

A short poem may be mentioned to end with. It is the so-called Prayer of Manasseh,[12] a penitential prayer, probably Pharisaic, written in Greek between 150 and 50. It is apocryphal but sometimes inserted in Greek Bibles after II Chronicles 33:12–13; it is referred to in v. 18. Manasseh, king of Judah (692?–639?), was for a time a prisoner in Babylon and this is a prayer which he was supposed to have said in his captivity.

Three books of the Wisdom Literature are definitely Hellenistic – Ecclēsiastēs, Ecclēsiasticos, and the Wisdom of Solomon.

Ecclēsiastēs was written in Hebrew about the turn of the third century or the beginning of the second by an author who calls himself Qoheleth (the Preacher, Ecclēsiastēs); it begins, "The words of the Preacher, the son of David, king in Jerusalem." It is an extraordinary work which the rabbis admitted in the Hebrew canon by mistake, because of the saying already quoted (1:1) and another like it (1:12), "I the Preacher was king over Israel in Jerusalem." According to Gandz, Qoheleth did not mean that he was king but headmaster (*prostatēs*); he is the only head of a secular school in Israel known to us.[13] The meaning is elucidated at the end of the book (12:9–14). He was a Hellenizer and perhaps in sympathy with the Seleucid nation, while the common people of Israel favored Egypt. There are Epicurean ideas in his book (9:7–9), but such ideas were older than Epicuros. Ecclēsiastēs is a book of self-communion, and its originality is reflected in statements calling its author "the Omar Khayyam of the Bible," the "Sphinx of the Old Testament," and calling his book "The Book of the Two Voices" (orthodox and heretic). He has been compared also with Spinoza and with Pascal, but Gandz prefers to compare him with Epicuros, and therein lies the paradox, an Epicuros in the Hebrew canon! [14] There is no proof that Qoheleth read Epicuros but that was not necessary because many Epicurean ideas were in the air.

The main message, however, was not Epicurean but *sui generis*. "Vanity of vanities, saith the Preacher; all is vanity" (12:8).

[12] Manassēs in Greek; *Proseuchē Manassē*.

[13] My good friend, Solomon Gandz, wrote letters to me concerning Qoheleth in the last year of his life. My conclusions are influenced by his.

[14] Robert Gordis, *Koheleth, the man and his word* (408 pp.; New York: Jewish Theological Seminary, 1951) [*Isis* 43, 58 (1952)]. In the Hebrew language, "Epicurean" came to mean unbeliever, infidel (Volume 1, p. 597), hence the almost incredible paradox: Epicuros canonized!

Ecclēsiasticos is apocryphal. Until 1896, it was known only through Greek and Syriac versions, but a large part of the Hebrew original has been discovered since. It was written a little later than Ecclēsiastēs, say about 180; the Greek translation was made in Egypt some fifty years later, in 132. It is called the "Wisdom of Jeshua son of Sirach," or the "Wisdom of Sirach"; the Greeks called it also *Panaretos sophia* (all virtuous wisdom) and the Talmudists, "Book of Ben Sira."

We find in it allusions to the Empedoclean doctrine of opposites (coexistence in nature of two opposite forces) and to the Aristotelian fancy that the heart is the seat of intelligence. Its author fully appreciated the work of physicians, scribes, and craftsmen and said of the latter (38:34–35), "They support the fabric of the world and their prayer is in the practice of their trade." The book ends with a review of Jewish history down to the high priest Simon (d. 199). This begins with the oft-repeated words (44:1), "Let us now praise distinguished men, our forefathers before us . . ."

Ecclēsiastēs and Ecclēsiasticos are among the golden books of universal literature, together with the *Meditations* of Marcus Antonius (II–2). The Wisdom of Solomon (*Sophia Salōmōnos*) is less important; it was written in Greek more than a century later, in the period 50 B.C.–A.D. 40, for the Jews of Egypt. One can distinguish two parts in it (1:1 to 11:5; 11:6 to 19:22) which are probably the works of different authors, done at slightly different dates. The Wisdom of Solomon was well known to St. Paul and to the authors of Ephesians, Hebrews, and Peter's first epistle.

On account of its later date it is even more typical of Hellenistic Judaism than the two previous works. It contains allusions to the four elements and to the doctrine according to which the substance of the embryo is formed out of the catamenia, which cease to be discharged during pregnancy (Aristotle, *De generatione*).

The best-remembered line is, "The souls of the upright are in the hands of God" (3:1).

The main historical book of this age is the Book of Daniēl, of which about one-half (2:4 to 7:28) was written in Aramaic, the rest in Hebrew, at the end of the reign of Antiochos IV Epiphanēs (172–164), or more exactly after the desecration of the Temple and the revolt of the Maccabees in 168. The apocalyptic visions of Daniēl (chapters 7–12) fall between 168 and 165 when the Temple was restored. They refer to the fall of Babylon in 538; remember the vision of Nebuchadrezzar (he who had taken Jerusalem in 597 and 586), driven from the sons of men to eat grass as oxen do (4:33, 5:21).[15]

[15] A very interesting contribution to Daniēl exegesis was given by Solomon Gandz, "Mene Mene Tekel Upharsin, a chapter in Babylonian mathematics," *Isis 26*, 82–94 (1936–37).

Three additions to Daniēl occurred in the Septuagint, and have passed from it into the Orthodox and Catholic canons, but not the Protestant one. They are The Song of the Three Children (*Preces trium puerorum*), The Story of Susannah, and Bel and the Dragon.

The Song of the Three Children and the Prayer of Azariah associated with it are two hymns of thanksgiving which follow 3:23 in the Greek Daniēl (*Proseuchē Azariu cai hymnos tōn triōn*). They are pieces of Jewish liturgy of which the Prayer dates from c. 170 and the Song from c. 150. They were probably written in Hebrew and translated into Greek for the Septuagint.

The Story of Susannah is placed at the very beginning of the Greek Daniēl. In spite of its being apocryphal (to Protestants), it is one of the best-known dramatic stories of the world literature and it has inspired many paintings. It shows that Jews were ahead of Greeks and Romans in defending cross-examination as a means of eliciting the truth in a legal trial. Witnesses to a capital charge must be examined separately; if they contradict each other, the charge falls; if their testimony is proved to be false, they should be executed. Susannah was written by a Pharisee in the last century B.C., in Hebrew or in Greek.

The story of Bel and the Dragon (*Bēl cai Dracōn*) is placed at the end of the Greek Daniēl. It was written c. 100 B.C., probably in Greek, to discredit idolatry. The dragon was a kind of serpent and there was definitely a snake cult in Greece, for example, in the healing temples (Asclēpieia).[16]

The three additions to Daniēl included in the Septuagint were revised, or retranslated into Greek, by Theodotion, a Jew of Ephesos (or Sinōpē?) who flourished under Marcus Aurelius (II–2). The text found in Christian Bibles (or Apocrypha) is generally derived from Theodotion, not from the Septuagint.

The Apocrypha include various books of history or religious fiction:

(1) The Book of Tōbit (and his son Tōbias), written c. 200–175, probably by an Egyptian Jew, in Greek.

(2) The Book of Judith, written in Hebrew after the Maccabean war, say c. 150, when the Pharisaic movement was growing. The Greek version is a much expanded revision of the Hebrew story.

(3) The Book of Esthēr tells the story of a Jewish girl who became the queen of Xerxēs (king of Persia 485–465) and used her wifely influence to save her people. The event is celebrated every year in the feast of Purim. The story was written in Hebrew, c. 150–125. Half a century later it was translated into Greek, and the Greek text included various additions which are, to some extent, apocryphal. They are put together in the King James version (10:4 to 16:24), but are scattered in the Septuagint. Hence, the English Book of Esthēr is materially different from the Hebrew one and also from the Greek one.

[16] For snake worship, see Volume 1, pp. 332, 335, 389–390.

(4) The first book of Esdras (or 3 Esdras) is an imaginative account of the rebuilding of the Temple, probably written in Greek, in Egypt, c. 150. It follows, to some extent, the accounts found in II Chronicles (35–36), Ezra[17] (1–10), Nehemiah (8), but contains new material, the most noteworthy being the discussion of the three guardsmen of King Dareios (3–4) as to "What is strongest?" Wine? the king? woman? but "Truth is great and supremely strong" (4:41).

In the Hebrew manuscripts the Books of Ezra and Nehemiah (Neemias in Greek) were put together until as late as 1448. In the Vulgate, these two books are called 1 and 2 Esdras and the two apocryphal texts, 3 and 4 Esdras.

The persecution of the Jews by Antiochos IV Epiphanēs, the Maccabean revolt which it caused, and the liberation of the Jewish nation under the Hasmonean dynasty have created a series of writings called the Books of Maccabees. There are five such books.

(5) The first book of Maccabees is the only one that is genuinely historical; it covers the period (175–132) and is a trustworthy source; it was written c. 90–70, probably in Hebrew; the Hebrew text is lost but we have a Greek translation. The original dating in this and the second book is according to the Seleucid era, which began in 311.

(6) The second book of Maccabees was written in Greek, probably in Alexandria, c. 50 B.C., by one Jason (Iasōn) of Cyrēnē. It is a summary of the events of (175–160), derived from I Maccabees, and has no historical value of its own.

(7) 3 Maccabees is more fictional than historical and does not deal with the Maccabean revolt but with the martyrdom of Egyptian Jews under Ptolemaios IV Philopatōr (ruled 222–205). It was written in the first century B.C., or shortly afterward.

1, 2, and 3 Maccabees exist in the Septuagint; 1 and 2 Maccabees in the Vulgate.

(8) 4 Maccabees is a philosophical discourse, using historical facts as examples, to prove the supremacy of reason and piety over passions. It was written in Greek by a Stoic Jew, probably toward the end of the pre-Christian age.

5 Maccabees is added here *pro forma*. It is a later work preserved in the Syriac Peshitta[18] and is simply a translation of book VI of the *History of the Jewish War* by Jōsēphos (I–2).

[17] Ezra is really the same name as Esdras; Esdras is Greek, while Ezra is Hebrew with the initial ʿain left out. It is better to write Ezra for the Old Testament book and Esdras I and II for the apocrypha. Esdras II is a later, post-Christian series of apocalyptic visions dating from A.D. 66 to 270; it was probably written in Hebrew and developed in Greek; it is known only in the Latin Vulgate and in Syriac, Ethiopic, Arabic, and Armenian versions. Three verses of the Greek text (15:57–59) have been discovered in an Oxyrhynchos papyrus (1910).

[18] For the Mappaqtā Peshiṭtā or "plain version," see *Introduction*, vol. 1, p. 291. The Peshiṭtā was made in Edessa about A.D. 150 or before.

Other apocryphal books of the Old Testament, written between, let us say, 150 B.C. and A.D. 50, are not even included in the Apocrypha of the King James Version of 1611. They are the Book of Jubilees, the Book of Enoch, the Testaments of the Twelve Patriarchs,[19] the Assumption of Moses, and perhaps others still. Those just named are known only in translations, Greek, Ethiopic, or Latin, but they were probably composed in Hebrew or Aramaic.

Many of the books dealt with above were written originally in Greek (which had become the common language of many Eastern Jews instead of Aramaic); that is the case for the Wisdom of Solomon, Bel and the Dragon, 1 (or 3) Esdras, 2, 3, and 4 Maccabees. Irrespective of their original language, all were soon available in Greek, and to those Greek texts must be added other parts of the Septuagint (not apocryphal in our sense) which are posterior to the third century B.C.

Note that, until very recently, none of the Hebrew writings were represented by contemporary manuscripts. It is one of the most paradoxical vicissitudes of Old Testament texts that the Greek manuscripts are many centuries earlier than the earliest Hebrew ones. Hence, students of the Old Testament are obliged to refer to the *Septuaginta*, which translated Hebrew texts older than those represented in the Hebrew Bibles. Considering the great antiquity of the Old Testament, the establishment of its Hebrew text was incredibly slow. Jewish scribes of Palestine prepared the "text of the ṣopherim" in the second century after Christ, vowels and accents were added to it in the seventh century, and a new standard text, the Masoretic, was available only in the first half of the tenth century, or rather two texts became thus available, for there were two Masoretic schools, in Tiberias and in Babylon. The leader of the Babylonian (or Oriental) tradition was Ben Naphtali, and the leader of the Palestinian (or Occidental) tradition was Ben Asher. The Palestinian tradition was supreme; it was consecrated by the fundamental printed (Hebrew) Bible (4 folio vols.; Venice, 1524–25).[20] The Hebrew text had been printed before, however, in the Complutensian Bible, under the auspices of Francisco Ximenes de Cisneros, cardinal archbishop of Toledo; this was completely printed before the Cardinal's death in 1517 but not published until 1521 (in Complutum = Alcalà de Henares).

The Greek text of the Septuagint was first printed in the Complutensian Bible (1517), but first published by the Aldine Press (Venice, 1518/19).

[19] The Testaments of the Twelve Patriarchs are two apocalyptic and Messianic hymns, the most glorious achievements of the time of Joannes Hyrcanos (134–104), untitled king of Israel. They anticipated the Sermon on the Mount and influenced New Testament writings.

[20] For the Ṣopherim and Masoretic editions, see *Introduction*, vol. 1, pp. 291, 624, and for more details, Pfeiffer, *Introduction to the Old Testament*.

THE DEAD SEA SCROLLS. THE ESSENES

In the spring of 1947 one of the most startling discoveries of our time was made accidentally by a Bedouin boy; in a natural cave on a cliff on the northwestern shore of the Dead Sea, he discovered Hebrew scrolls preserved in jars. As soon as the discovery was known to Jewish and Christian archaeologists in Jerusalem, it created considerable emotion and curiosity. The scrolls found by Bedouins were offered for sale and efforts were soon made to find more of them, and chiefly, to find them in a more scientific way. A better exploration of the caves was conducted by G. Lankester Harding, of the Jordanian Department of Antiquities, and by the Dominican father Roland de Vaux of the École Biblique of Jerusalem. They explored systematically some 267 caves in the western hills of the Dead Sea and found many thousands of fragments. They also excavated the stone ruins of an Essenian monastery at Khirbet Qumrān in the same neighborhood.

The scrolls and fragments thus far deciphered cover many parts of the Bible and include new documents, such as thirty Hymns of Thanksgiving, a commentary on Habakkuk, an apocalyptic treatise, the War of the Sons of Light against the Sons of Darkness, and an Essenian Manual of Discipline, which anticipates Christian treatises, such as the Teachings of the Apostles or Didachē.

The Essenes were a Jewish sect, organized in the form of a brotherhood or monastic order, which flourished from the second century B.C. until Christian times. Their organization was communistic and very ascetic; they were strongly interested in learning and had gathered a library. The monastery of Khirbat Qumrān [21] was built c. 136–106 and occupied until A.D. 68. It is plausible that the Hebrew scrolls and fragments found in the neighboring caves are the remnants of the Essenian library, "the Dead Sea Library." The original writings which that library contained (apocalypses and manuals of discipline) may be said to represent the period elapsing between the Old Testament and the New, the transition from Judaism to Christianity.

The fragments are innumerable, and many are extremely difficult to decipher and to identify; a long time will elapse before the fragments that belong together are properly rejoined. These preliminary tasks will occupy scholars for thirty years or more and the conclusions of today are necessarily provisional, yet some cannot help being made if only for the sake of guidance.

A vast number of books and papers are devoted to the many controversies that the Dead Sea Scrolls have already awakened, and will continue to awaken. Among the many authors I shall mention only André Dupont-Sommer, *Aper-*

---

[21] This is obviously a later name. Khirba in Arabic means ruins; the name was not given to the monastery until some time (probably a long time) after its abandonment.

çus préliminaires sur les MSS de la Mer Morte (Paris: Maisonneuve, 1950), *Nouveaux aperçus* (*ibid.*, 1953); Harold Henry Rowley, *The Zadokite fragments and the Dead Sea scrolls* (Oxford: Blackwell, 1952), with elaborate bibliography; Millar Burrows, *The Dead Sea scrolls* (New York: Viking, 1955); Edmund Wilson, *The scrolls from the Dead Sea* (New York: Oxford University Press, 1955). O. P. Barthelemy and J. T. Milik, *Discoveries in the Judaean desert*; vol. 1, "Qumran cave I" (New York: Oxford University Press, 1955).

The collection of the Dead Sea scrolls belonging to the Hebrew University in Jerusalem was edited with facsimiles and notes by the late Eleazar Sukenik, his son, General Yigael Yadin, and Dr. Avigad (2 vols.; Jerusalem: Bialik Institute and Hebrew University, 1955). English translation of the texts, edited by Eleazar Lipa Sukenik, *The Dead Sea scrolls of Hebrew University* (44 pp., 116 pls.; Jerusalem: Magnes Press, 1955). This collection includes the Book of Isaiah, the Psalms of Thanksgiving and the Book of the War of the Sons of Light against the Sons of Darkness.

In my book, *Ancient science and modern civilization* (Lincoln: University of Nebraska Press, 1954), p. 18, I suggested for a date of Dead Sea scrolls one posterior to A.D. 70. On the basis of new information, especially that concerning the Essenian library of Khirbat Qumrān, I am now more inclined to conclude that most of those Hebrew texts are pre-Christian.

Jewish communities in Oriental cities, outside of Palestine itself, were numerous and sometimes ancient. Some of the Egyptian ones dated back to the seventh century; but the largest Egyptian community was of course that of Alexandria, which was young. From 301 to 200 Judea was an integral part of Ptolemaic Egypt and the number of Egyptian Jews increased enormously during that period. Jews were abundant also in Damascus, Antioch, in various Ionian cities, in Dēlos, and elsewhere. There was a growing community in Rome, and most of the Roman Jews spoke Greek, as is proved by the very large proportion of Greco-Jewish inscriptions.[22]

Some of the Jews were faithful to their religion in a fanatical way; at the other extreme many others were anxious to be assimilated and to be considered not Jews but Greeks (that is a story applying to all times and places). They read the Scriptures, if at all, in Greek; services in their synagogues were conducted in Greek; they gave Adonai the name Theos Hypsistos (which applied just as well to Zeus); they often adopted Greek names for themselves and imitated Greek manners and customs; in Asia Minor some Jews intermarried with gentiles and adopted more or less Greco-Oriental cults; in this they were simply imitating the religious syncretism that was so fashionable among the Greeks themselves.

It is clear that Jews who were as accommodating as that could not be treated except in a friendly way. There might be social discrimination against them, but no violent anti-Semitism existed before Roman and Christian times.

[22] In the Jewish catacombs of Rome, 74 percent of the inscriptions are in Greek; Harry Joshua Leon in *Transactions of the American Philological Association 58*, 210 (1927).

JEWS VS. GREEKS

Though the Jews were fairly numerous in certain places, they were always a minority, while the Egyptians in Egypt and Asiatics in Asia Minor constituted an overwhelming majority. Those Orientals were swallowing up the Greeks, blood and brain. There never was a Semitic danger outside of Israel, but the Oriental absorption was ominous.

Said Tarn, "From the second century Hellenism was between the hammer and the anvil, the sword of Rome and the spirit of Egypt and Babylon. One man saw it, Antiochos Epiphanēs, and has been called a mad man ever since. But his attempt to unify his realm on a basis of Greek religion and culture failed; and Greek religion got no second chance." [23]

The story of Antiochos IV Epiphanēs (Seleucid king 175–163) is very instructive. In his time, the priestly aristocracy in Jerusalm and the high priest himself, one Iasōn, were Greek-minded. Being deceived by his own Hellenic dreams and by the Hellenizing tendencies of the Jews, he tried to Hellenize them completely. In 167 he was impudent enough to dedicate the Temple to Zeus; he built a citadel in Jerusalem and tried to suppress Judaism. The result was a profound reaction, the creation of the Pharisee party, and the Maccabean revolt. In 164 the Temple was rededicated to Adonai, but the war continued. In 142, the Jews expelled the Seleucid garrison and obtained political freedom. Not for very long, however; eight years later Antiochos VII Sidētēs took Jerusalem and razed the walls; Antiochos' death in 129 was the final end of Seleucid power. Joannes Hyrcanos ruled the Jews and his rule (until 104) was the golden age of the Hasmonean (or Maccabean) dynasty; unfortunately, he started to conquer his neighbors, the Samaritans and the Idumaeans, whom he Judaized by force. After that there were so many disputes and rebellions that Pompey was obliged to intervene in 63. The Romans found it expedient, however, after a while to delegate their authority to one of those Idumaeans whom Hyrcanos had "converted," the so-called Herod the Great (Hērōdēs), who was an unscrupulous tyrant, king of Judea from 37 to 4 B.C. Ten years later, A.D. 6, Judea became a Roman province and Roman control continued until 395.

Did the Jews influence the gentile community around them? The Greek translation called the *Septuaginta* had been begun at the initiative of Demētrios of Phalēron and at the request of Ptolemaios II Philadelphos (see Chapter 14); hence we must assume that some of the Greeks read the books as they became available, say the Pentateuch, but there is little evidence of that.[24] It is probable that the Septuagint was used mainly, if not

[23] Tarn and Griffith: *Hellenistic civilisation*, p. 33.

[24] My main source of information is Robert H. Pfeiffer, *History of New Testament times with an introduction to the Apocrypha* (New York: Harper, 1949) [*Isis 41*, 230–231 (1950)] and his friendly letter of 21 May 1955.

exclusively, by those Jews who could not read the Hebrew original any more, or who could not read it without help. Did the gentiles read Aristeas' letter written about the middle of the second century B.C.? Again, that cannot be proved, but those who read it found in it a subtle apology for Jewish concerns. Even without any reading there must have been among the Greeks of Alexandria (and other places) who had Jewish friends some who realized that those Jews were not simply clever businessmen, perhaps over clever, but the guardians of an old religious tradition, the inheritors of a literature that was as old and venerable as the Greek literature and perhaps older.

Aristeas, allegedly a pagan, praised the Jewish Law and rituals. A contemporary Jew, Aristobulos of Alexandria, who flourished under Ptolemaios VI Philomētōr (ruled 181–145) wrote a Greek commentary on the Pentateuch which we know only through extracts of a much later date. If his date is accepted, his lost treatise would be the first bridge between Greek philosophy and Alexandrian Jewry. He made the claim that Homer, Hesiod, Pythagoras, Plato, and Aristotle had borrowed from the Jewish tradition; this was an extravagant claim because it implied not only the great antiquity of the Old Testament but the existence of a Greek translation of it, early enough to have been used as a source by the earliest Greek authors! That fantastic theory had an extraordinary fortune, as will be shown presently.

Another story is even more startling, because it illustrates Jewish impact upon the Greeks not of Alexandria but of Rome. One Alexander, born at Milētos c. 105, was brought to Rome as a prisoner of war and freed by Sulla (c. 80), when he assumed the name L. Cornelius Alexander.[25] He was the teacher of C. Julius Hyginus (librarian of the Palatine library) and wrote so many books that he was nicknamed Alexander Polyhistor. This Alexander helped to diffuse Jewish history in Rome, and he explained that Jewish culture was the oldest, and that the best which the Greeks knew they had obtained from Jewish sources. Such saying found receptive ears and may account for Semitic tendencies among pagans or Orientals, for example, those *Sabbatistae* of Cilicia who kept the Sabbath and worshiped Adonai.

The story that Hebrew was the original language of mankind originated perhaps from the same source of Semitic enthusiasm. It was as fantastic a story as the other, because it was impossible to find any resemblance between Hebrew on the one hand and Greek or Latin on the other, yet it was also very popular.[26]

This is a very curious deformation of a real truth, the Oriental origins of

[25] That is, according to Roman usage, he assumed the *praenomen* Lucius and the *nomen gentilicium* Cornelius of the man to whom he owed his emanicipation, L. Cornelius Sulla (138–78).

[26] For the study of that conceit, Hebrew the mother of languages, see my *Introduction,* vol. 3, p. 363. Holger Pedersen, *Linguistic science in the nineteenth century* (Cambridge, 1931), pp. 7–9, 240.

Greek science and Greek wisdom. The predecessors of the Greeks were not Jews, however, but Egyptians and Babylonians.

Legend that Greek wisdom was derived from Hebrew sources. This delusion, that Greek wisdom was derived from Hebrew sources, is so curious and its popularity so unexpected that the reader will forgive me for inserting here a historical outline of its wayward approval.

The early defenders of Christianity were anxious to diminish pagan glory as far as they could, and their acceptance of the Old Testament led them to praise the old Hebrew tradition. Thus, in his *Apology* (1, 59), Justin Martyr (II–2) was willing to connect Plato with Moses. This was done more thoroughly by another early Father of the Church, Clement of Alexandria (c. 150–c. 220). In the first book of his *Strōmateis* (*Miscellanies*), Clement argued that the Old Testament is much older than any part of Greek philosophy and that the Greek philosophers must have borrowed from the Jews. In book II he explains in detail the originality and superiority of the moral teachings revealed in the Old Testament as compared with Greek philosophy.

To make a long jump, we find similar ideas in the Arabic letters of the Brethren of Purity (X–2), the *Rasā'il ikhwān al-ṣafā*. In letter 21, a Greek orator boasting of Greek wisdom and science is asked, "How did you obtain such knowledge if not from the Jews in Ptolemy's time and from the wise men of Egypt? You took that knowledge home and claimed it as your own." Such views were transmitted to the Jews when Qalonymos ben Qalonymos (XIV–1) translated that letter or treatise 21 from Arabic into Hebrew in 1316.[27] Roger Bacon (XIII–2) shared the belief of many Christian doctors that the Hebrew culture was the original.

To return to the Jews, Meir ben Aldabi (XIV–2) of Toledo assumed that Greek knowledge was ultimately of Hebrew origin. Another Castilian, Meir ben Solomon Alguadez (XIV–2), who translated the *Nicomachean Ethics* from Latin into Hebrew, sets forth in his preface that Aristotle was really explaining the precepts of the Torah.

Medieval conceits, you will object. Those conceits survived the Middle Ages, the Renaissance, and even the Enlightenment. A few examples must suffice, "I am not very hostile to Greek letters," said a preacher to Henry the Eighth, "since they are derived from the Hebrew." [28] In his *Harmonie étymologique des langues* (Paris, 1606), Etienne Guichard proved that all languages, French included, were derived from Hebrew.[29] They did even better than that in England. Zachary Bogan of Corpus Christi, Oxford, pub-

[27] This tallied with other Jewish traditions according to which Aristotle himself had obtained his knowledge from Jewish sources; Aristotle was either of the Jewish race or a Jewish proselyte; *Introduction*, vol. 2, p. 962.

[28] Francis Hacket, *Henry the VIIIth* (Garden City, 1931), p. 105.

[29] Louis Petit de Julleville, *Histoire de la langue française* (Paris, 1896), vol. 1, p. III.

lished a book the title of which is revealing, *Homerus Hebraïzōn* (Oxford, 1658), while James Duport, Master of Magdalene, Cambridge, in his *Gnomologia Homerica* (Cambridge, 1660), traced the resemblances between Homer and the Old Testament. Another Greek scholar of the following generation, Joshua Barnes (1654–1712), persuaded his wife that the *Iliad* and the *Odyssey* were the work of King Solomon.[30]

In his *L'origine et le progrès des arts et des sciences* (428 pp.; Paris, 1740), Charles Noblot showed that the Jews, not the Egyptians, were the true originators.

All this was capped by the learned Salomon Spinner, *Herkunft, Entstehung und antike Umwelt des hebraïschen Volkes: ein neuer Beitrag zur Geschichte der Völker Vorderasiens* (548 pp.; Vienna, 1933) [*Isis 24*, 262 (1935)], who elaborated the singular thesis that Clement of Alexandria had defended more than seventeen centuries before him.

Much space has been devoted to the Jews and to Hebrew writings, partly because those writings, as far as they were included in the Bible and the Apocrypha, are integral parts of our own tradition, and the others concern us also indirectly, partly because they are tangible enough. The same cannot be said of the Oriental religions (Egyptian, Iranian, Anatolian, Syrian, and others), which were many, complex, and multiform and were illustrated by monuments rather than books. As to the Greeks (and later the Romans), their own mythology and their own cults had been gradually colored with Oriental infiltrations.

The Greeks, and the Romans after them, conquered Asia and Egypt but were defeated by the Oriental gods. The Romans were gradually overwhelmed by Greek culture, but that culture was deeply orientalized. The common people of the Greek and Roman lands were dominated by superstitions; the religion of the learned men was a kind of scientific pantheism, but they themselves were not free of superstitions, for they believed in astrology and in many forms of divination.[31]

Though the Greeks were often ready to welcome Oriental gods into their own pantheon, in one case at least the opposite happened and a Greek god set out to conquer the outside world. That was Dionysios,[32] whose glory was diffused by poets and artists. He was amalgamated with a Thracolydian god, Sabazios, whose cult was popular in Phrygia, Lydia, and Pergamon, identified with the Cyrios Sabaōth of the Septuagint, and called Theos Hypsistos. He was represented in Phrygian costume, with the thunderbolt

[30] Martin Lowther Clarke, *Greek studies in England 1700–1830* (Cambridge: University Press, 1945) [*Isis 37*, 232 (1947)], p. 2.

[31] Franz Cumont (1868–1947), *Astrology and religion among the Greeks and the Romans* (New York, 1912); *Les religions orientales dans le paganisme romain* (ed. 4, 350 pp., ill.; Paris, 1929); *Lux perpetua* (558 pp., ill.; Paris: Geuthner, 1949) [*Isis 41*, 371 (1950)].

[32] Called Bacchus in Latin from the Greek Bakchos (or Iakchos).

and eagle of Zeus, and sometimes with a snake. In Egypt he was identified with Sarapis. The common people, visiting temples, praying, making offerings, attending festivals, did not worry about such ambiguities, or, more exactly, were not aware of any. They simply hoped to obtain the god's favor and protection.

One good result of Hellenistic syncretism was the absence of intolerance, at least of religious intolerance. Whatever intolerance there was, was not religious but racial and political, and in many cases was restricted to snobbishness. The Greeks, except perhaps the Jewish ones, were not exclusive.

NATIONAL CULTS

There is one aspect of religion, however, which caused exclusiveness; that was the city cult, open only to citizens, and the cult of national gods, such as was developed in various countries and finally on a very large scale by the Romans. The cult of heroes was natural to the Greeks, but that cult was not exclusive. A kind of ruler cult was introduced by Alexander the Great and this was imitated in different ways by many Hellenistic sovereigns.

The Ptolemaioi who ruled Egypt were deified after their death, and later the apotheosis occurred during their lifetime; they were then *theoi epiphaneis* (gods come to light, living gods). The first Ptolemaios to call himself Epiphanēs was the fifth, who ruled from 205 to 180. Another early example of divine ruler was Eumenēs II, king of Pergamon from 197 to 160. The Seleucid king Antiochos IV (king 175–163) was also entitled Epiphanēs, while an earlier Seleucid Antiochos II (ruled 261–247) and Ptolemaios XII (ruler of Egypt from 80 to 51) were simply called Theos (god).

This dangerous innovation was passed to the Romans. Cicero has a share of guilt in this for in the *Somnium Scipionis* (c. 51), he developed the Stoic conceit that great men may become divine after death. Divine honors were given to Caesar during the last year of his life (45–44) and were probably one of the causes of his assassination. Augustus was a divine ruler from the Hellenistic point of view and in Egypt he received the apotheosis accorded to the Ptolemaioi; his Roman titles, *Divi filius* and *Augustus*, were suggestive of divinity, and he was deified after his death. The cult of Augustus was combined with that of the goddess Roma.

Such cults became national duties for citizens, and their deliberate omission, a sufficient proof of disloyalty. The difficulties of the Jews with Roman authorities stemmed chiefly from their refusal to recognize any god but God. Roman intolerance of the Jews, therefore, was not religious but political; subjects who rejected the national cult could not be tolerated.

The real intolerance of a religion is the one that forbids its exercise and prevents its sectaries and devotees from taking part in it. Such intolerance is terrible and really intolerable, for it breaks the most sacred tradition of good people, withdraws from them their fathers' blessings, and gives them the feeling that their ancestral sacredness has been taken away from them. On account of the prevailing syncretism, this kind of intolerance was practically unknown in pre-Christian times.[33] Jews were persecuted, if at all, not because of their own religion but because of their failure to accomplish the religious part of their national duties.

[33] One exceptional example of such intolerance was given by Antiochus IV Epiphanēs, as told above.

XVII

PHILOSOPHY IN THE LAST TWO CENTURIES POSEIDŌNIOS, CICERO, AND LUCRETIUS[1]

There were schools of philosophy in many cities of the Mediterranean world – in Athens, Alexandria, Pergamon, Rhodos, Rome – and the philosophers moved as frequently from one school to another as they were to do in the Middle Ages. Not only did the teachers move but so did the students in search of wisdom. The latter were comparable to ill people who go from one health resort to another hoping to be cured. If the students did not obtain wisdom in Athens, they perhaps thought that they would find it in Alexandria or Rhodos. They might.

Students who hailed from Rome itself or from one of the western provinces had another reason, a very good one, to travel eastward. They would thus obtain a better knowledge of Greek, and would be able to speak that language with fluency and to write it more correctly. Wisdom was illusive, but the Greek language and Greek culture were tangible.

The situation will be easier to understand if we think of the many Asiatic and African students who come to America. Each of them is in quest of some technique but, in addition, they hope to obtain a better knowledge of English, and that knowledge will be a certain acquisition. They may fail to master the technique, but they will acquire an instrument of universal value, the English tongue.

In order to explain the philosophic endeavors of those days, let us make two surveys, first, of the teaching of philosophy in one place, Athens, and second, of the teaching of one kind of philosophy, the Stoic, in many places. We shall then complete the picture with the portraits of three outstanding personalities, Poseidōnios, Cicero, and Lucretius.

ATHENIAN SCHOOLS

In spite of its political downfall, Athens was still a nursery of Greek genius, and the four schools of philosophy that had become traditional

[1] For philosophy in the third century B.C., see Chapter XI.

continued to flourish – the Academy, the Lyceum, the Porch, and the Garden. We know the schoolmasters who ruled during the second and first centuries, some thirty of them almost equally divided. It is interesting to review them and to consider their great diverseness in the service of definite traditions.

It is natural to begin with the Academy, which was still the leading school; it is true that it kept the leadership in part because of its catholicism; its dogmatism was very moderate and it was perhaps more ready than the other schools to accept innovations.

We know the names of at least nine masters of the Academy during this period (maybe there were no others, for nine is not too few for two centuries). Hēgēsinus of Pergamon is the first; then comes Carneadēs of Cyrēnē (c. 213–129), who was the founder of the "Third Academy" and its *prostatēs* (or director) until 137–36. He seems to have been a good critic and rhetorician, and he became quite famous (in Rome as well as in Athens) in spite of the fact that he left no writings of his own. His Roman fame was due to a strange concatenation of accidents. The city of Orōpos, on the boundary of Boiōtia and Attica, had long been a bone of contention between them; the Athenians, having destroyed it, were fined five hundred talents by their Roman masters. They decided to send a delegation to Rome to plead their cause; this was in the year 156–55. It is typical that the members of the group were philosophers and even more typical that they were philosophers of three kinds: Carneadēs, representing the Academy; the Peripatetic, Critolaos; and the Stoic, Diogenēs of Babylon.[2] The fine was reduced, but, more important, the embassy represents the introduction of Greek philosophy at Rome.

We should be very grateful to Carneadēs because of his strong denunciations of divination in general and of astrology in particular; he provided the best arguments against the astrologers and those arguments were eventually repeated and amplified by Cicero, but they were powerless to stop the rising tide of superstition when political vicissitudes jeopardized and finally canceled freedom of thought.[3]

The successors of Carneadēs of Cyrēnē were his namesake, Carneadēs the son of Polemarchos (c. 136–131), Cratēs of Tarsos (c. 131–127), Cleito-

[2] No Epicurean was chosen as a member of that embassy. However, Athenian Epicureans found their way to Rome. Two of them were there in the following century, Phaidros and Patrōn. The selection of philosophers as members of that embassy was the more remarkable considering that the Senate had passed a decree a few years before (in 161) excluding all foreign teachers of philosophy and rhetoric from the city.

[3] Dom David Amand (now Emmanuel Amand de Mendieta), *Fatalisme et liberté dans l'antiquité grecque* (Louvain: University of Louvain, 1945), pp. 26–68. Frederick H. Cramer, *Astrology in Roman law and politics* (Philadelphia: American Philosophical Society, 1954), pp. 55–58, passim; see my review in *Speculum 31*, 156–161 (1956).

machos of Carthage (c. 127–110), Philōn of Larissa (110–88),[4] founder of the so-called Fourth Academy, Antiochos of Ascalōn, founder of the Fifth Academy,[5] who had been Philōn's pupil not in Athens but in Rome, Aristos of Ascalōn (c. 68–50), and Theomnēstos of Naucratis (c. 44).

These nine men were all teaching at the Academy at one time or another and had the honor of leading it, but not a single one was an Athenian (one is reminded of the time when the leading professors of the University of Paris were foreigners). Hēgēsinus hailed from Pergamon, Carneadēs from Cyrēnē, Cratēs from Cilicia, Cleitomachos from Carthage (his original name was the great Punic name Hasdrubal), Philōn from Thessaly, Antiochos and Aristos from Palestine, Theomnēstos from Egypt. One could not have a better selection if one had tried to make it as cosmopolitan as possible, and yet it was accidental.

One might add to them Ainesidēmos of Cnossos, a skeptic, who influenced Philōn. None of these ten men was very important, except Carneadēs of Cyrēnē, but they were upholding as well as they could the Platonic tradition.

The Lyceum was not more brilliant. We should remember that the story of every school is very much the same. It is created by a great man and lives for a time on his prestige until another great man appears, sooner or later; in between there are periods of flatness and mediocrity that even the best management cannot lift up. The masters of the Lyceum were Critolaos of Phasēlis (he who accompanied Carneadēs to Rome in 156), Diodōros of Tyros, Erymneus (c. 100), Andronicos of Rhodos (I-1 B.C.), Cratippos of Pergamon, and Xenarchos of Seleuceia. They came from Lycia, Palestine, Rhodos, Pergamon, and Cilicia, all of them from the shores of Asia. Had Greece ceased to be the cradle of genius? Critolaos defended Aristotle against the Stoics and the rhetoricians. Andronicos was ordered by Sulla to prepare, c. 70, an edition of Aristotle's works, the first edition to reach outsiders; he was called the tenth (or eleventh?) successor of the great master. He certainly deserves to be remembered in that way because of his edition, but the living Aristotelian tradition did not begin until almost three centuries later with Alexander of Aphrodisias (III-1), the exegete (*exēgētēs*). Andronicos' edition covered not only the works of Aristotle but also those of Theophrastos, all of which he arranged according to subjects; it may be that those works have come to us in their relative integrity because of his care and if that is true, he cannot be praised too much.

[4] Philōn fled to Rome at the time of the first Mithridatic war in 88; it is not known whether he returned to Athens.

[5] These terms — Third Academy, Fourth Academy, Fifth Academy — suggest greater differences and discontinuities than existed. They were meant to accentuate changes of orientation; those changes were more rhetorical or dialectical than positive. The fundamental scientific knowledge was unchanged.

The Porch was directed by Zēnōn of Tarsos, Diogenēs the Babylonian (II–1 B.C.), Antipatros of Tarsos, Panaitios of Rhodos (II–2 B.C.), Mnēsarchos and Dardanos, Apollodōros of Seleuceia on the Tigris (c. 100), one Dionysios, Antipatros of Tyros who died c. 45. All of them, as far as we know, were Asiatics. Zēnōn was a great teacher, immortalized by his disciples rather than by any writings. Diogenēs was primarily a grammarian and logician. Antipatros of Tarsos wrote on the gods and on divination and had some controversies with Carneadēs of Cyrēnē. Panaitios was by far the leading Stoic; he and his disciple Poseidōnios will be dealt with more elaborately in a moment. Apollodōros wrote treatises on logic, ethics, and physics (all lost); many more treatises are ascribed to Antipatros of Tyros.

The Garden of Epicuros was cultivated by one Dionysios (c. 200), by Basileidēs, by Prōtarchos of Bargylia in Caria, by Apollodōros curiously nicknamed *cēpotyrannos* (tyrant of the garden) – perhaps he was too much of a martinet? – by Zēnōn of Sidōn whom Cicero called "*coryphaeus Epicureorum*," by Phaidros of Athens (?), by Patrōn (c. 70–51).[6] Of course, the greatest Epicurean of that age and perhaps of all ages has been left out, because he lived in Rome, not in Athens. We shall come back to him with pleasure at the end of this chapter.

These four schools flourished together in Athens and they sometimes engaged in polemics, but it would be wrong to consider them as necessarily hostile to one another. Whatever hostility might exist was due to personal jealousies and antipathies. The separation between them was not as rigid as one might think. The Academicians were eclectic and prone to moderate scepticism. I imagine that members of the different schools might attend meetings or festivities arranged by their rivals. One might be a Stoic with Epicurean tendencies or vice versa. The most scientific school was the Lyceum, and yet the best scientific work was done by Epicureans and even by Stoics. More than 1850 years after their death, Montaigne might hesitate between the teachings of Zēnōn of Cypros and those of Epicuros; we may be sure that such hesitations were already felt in antiquity.

THE GROWTH OF STOICISM. PANAITIOS OF RHODOS

Though the four schools had followers in every center of the ancient world, there is no doubt that the Stoa became gradually the most influential. The Academy and the Lyceum were too academic and often too eclectic. Stoicism was the philosophy of the best men, not only of professional philosophers but of civil servants, statesmen, and men of business. If such men were good enough to bother about philosophic questions, they

[6] This is a strange name for a Greek. It has a Roman sound, *patronus* (French, *patron*; Dutch, *baas*; English, boss).

were likely to be Stoics. It was for them more than a philosophy; it was a religion; this explains its relative popularity and also its aberrations.

The fundamental doctrines had been established by Zēnōn of Cition (IV–2 B.C.) and Cleanthēs of Assos (III–1 B.C.). Other disciples had rapidly increased its diffusion; Aristōn of Chios, who flourished in Athens c. 260, taught Eratosthenēs; Persaios of Cition went to the court of Antigonos Gonatas at Pella, was the tutor of Antigonos' son, Halcyoneus, and became powerful in Macedonia; Sphairos of Borysthenēs advised the political reforms of Cleomenēs III (king of Sparta 236–222), Chrysippos of Soloi (III–2 B.C.) completed the Stoic doctrine. Note that the masters of the Early Stoa had already acquired political as well as philosophical influence. Their success was largely due to that combination. The Stoics were not idle rhetoricians; their purpose, from the first, was to strengthen the political conscience; this was urgent and they did it well. Their main ideas, to wit, that virtue is based on knowledge and that the aim of any good man must be to live in harmony with nature (*homologumenōs physei zēn*) and with reason, were principles of individual and political conduct. All this was done before the end of the third century.

The main leaders of the second century were Cratēs (II–1 B.C.) in Pergamon and Panaitios (II–2 B.C.) in Rhodos, and both in Rome. Cratēs, a man of science as well as of letters, was director of the Pergamene library and when he went to Rome in 168 he brought with him the principles of Alexandrian-Pergamene scholarship and helped to organize the Roman libraries.

Panaitios (c. 185–109) of Rhodos was Cratēs' disciple in Pergamon, and then continued his study of Stoic philosophy in Athens under Diogenēs the Babylonian and the latter's successor Antipratos of Tarsos. He returned to Rhodos about the middle of the century and was in Rome c. 144. He was a familiar of P. Scipio Aemilianus [7] and of the historian Polybios (II–1 B.C.). In 141, he traveled in the East with Scipio and then returned to Rome. He succeeded Antipatros as leader of the Stoa and held that office in Athens until his death in 209. Only fragments of his works have come down to us,[8] but his treatise *Peri tu cathēcontos* (*On duty*) is reflected in the *De officiis* of Cicero. He was a man of science as well as a philosopher; he tried to reject astrology and divination, but that was an impossible task and he was bound to fail.

As Cratēs and Panaitios spent many years in Rome and were in touch

[7] Scipio Aemilianus Africanus Numantinus (185–129) was a famous soldier and statesman, the destroyer of Carthage in 146. He was a Stoic, and a highly educated man, who gathered around him the leading men of letters and thinkers (the Scipionic Circle). His friendship with Laelius is immortalized in Cicero's *De amicitia*, and the *Somnium Scipionis* (in Book VI of Cicero's *De republica*) is another reference to him.

[8] Modestus van Straaten, *Panétius; sa vie, ses écrits et sa doctrine avec une édition des fragments* (416 pp.; Amsterdam: H. J. Paris, 1946).

with leading people, the extraordinary success of Stoicism in the Roman world was largely due to them. That philosophy which was spreading from Athens, Pergamon, and Rome was cosmopolitan; it appealed to the Romans at a time when Rome was preparing to become the center of an international empire. It became before Christianity the ethical gospel of the more civilized people.

The Middle Stoa, that is, the teachings and climate of Stoicism during the period extending from the middle of the second century to 30 B.C., was largely the creation of Panaitios and of his most illustrious disciple, Poseidōnios. The latter is so important that a special chapter must be devoted to him, but, before doing so, a few remarks must still be made that concern Stoic doctrine throughout the ages.

The teachers of Stoicism strengthened the individual and political conscience, the sense of duty (*to cathēcon*), the feeling of international brotherhood and universal fellowship (*sympatheia*). Those were their chief merits and they were considerable in evil times. Their weaknesses were, in the first place, their failure to realize that justice must be tempered with mercy,[9] and, in the second place, their leanings to astrology and other superstitions. Their astrological ideas stemmed from the conviction that the universe is an organic whole, of which each part is dependent upon the others, and from their fatalism. They did not believe, as the Babylonians did, in a blind and terrible destiny, but rather in a moral providence. That providence was inscrutable except by means of divination and thus other superstitions were introduced.

It is true that Panaitios resisted astrology and divination and so did his pupils for a time, but the general tendencies were unfortunately in the opposite direction.

POSEIDŌNIOS OF APAMEIA

The most famous disciple of Panaitios was Poseidōnio… rn at Apameia, on the Orontēs, c. 135. After having lived many yea… Athens, studying under his master's direction, Poseidōnios travele… Mediterranean world and finally settled in Rhodos, where he sp… reat part of his life. In 51, he went to Rome and died there soon a… he age of eighty … He was a man of encyclopedic curiosity a… have bee… man of science like Aristotle and … ut th… tegrity was vitiated by Platonic te… in the Stoa. Panaitios, it would see… and less popular. Our judgment … uncertain, because none of hi…

[9] Stoics were trained to be … indifferent concerning mo… that was wise to a large … can detachment be re…

only fragments of them which have been transmitted to us chiefly by Latin writers, such as Cicero and Lucretius, Manilius (I–1), Seneca (I–2), and Pliny the Elder (I–2) and by later compilers, such as Athēnaios of Naucratis (III–1).[10]

He was primarily an expositor of Stoicism and a metaphysical historian (this will be discussed in Chapter XXIV), but he dabbled in many sciences. He was a great teacher, a spellbinder. Cicero attended his school in 78, and Pompey the Great visited him twice. His influence was due to his rhetorical ease rather than to scientific acumen or philosophical depth. It was due also to his spiritualism or rather to the strange combination of spirituality with science. Such a combination has always appealed to people because of its ambivalence; it satisfied their opposite needs of idealism and realism, hope and truth (compare the success obtained later by Galen, Paracelsus, and Swedenborg).

He might be called a Hellenistic Aristotle, and that is correct, if we give to the word Hellenistic the pejorative meaning that generally clings to it. His importance lies in his being one of the main transmitters of Greek science and wisdom to the Roman world. We realize once more that the road from Athens to Rome passed through Rhodos and Alexandria, and that Oriental roads passed through the same places.

CICERO

We may safely assume that Cicero is so well known to the readers of this book that it will suffice to refresh their memories and to recall the main facts of his life.

Marcus Tullius Cicero was born at Arpinum [11] in 106; he was educated in Rome where he attended lectures of the Epicurean Phaidros, c. 90, and the Academician Philōn, c. 88, but the best teacher of his youth was the Stoic Diodotos, w[illegible] was a guest in his father's house from c. 85 on; Diodotos became blind a[illegible] died in Cicero's own house in 59. Cicero was an outstanding lawyer, the gr[illegible] Roman orator, and one of the greatest Latin writers. In 79–78, his healt[illegible]ged him to travel and he attended in Athens the lectures of the Aca[illegible]n Antiochos of Ascalōn and of the Epicurean Zēnōn of Sidōn; he also [illegible] to Poseidōnios in Rhodos, though his main relation with the latte[illegible]ed in Rome much later, c. 51. To complete the list of his teachers, [illegible] same year, 51, and in the same place, Rome, he attended lectures [illegible]demician Aristos of Ascalōn and of the Epi-

[10] Ludwig Edelstein has [illegible] more than twenty years a[illegible] [illegible]donian fragments and [illegible] [illegible]per, "The philosophical [illegible] [illegible]ius," American Journal [illegible] 286–325 (1936) [Isis [illegible] [illegible]ows the many ambig[illegible] thought, which the [illegible] fragments are insufficient to resolve.

[11] Arpinum in the Latium; modern Arpino, not far from Frosinone. This little town (Arpinum) was the birthplace not only of Cicero but before him of the [illegible]us soldier, C. Marius (156–86) and [illegible] him of the statesman, Marcus Vip[illegible] Agrippa (63–12 B.C.).

curean Patrōn. He was deeply influenced by earlier men, whose writings were adapted by him to his own purposes, to wit, Plato in the *De re publica*, Aristotle, whose *Protrepticos* inspired his own *Hortensius*,[12] the Academician Carneadēs of Cyrēnē, one of whose treatises was the model of his *De re publica*, the Stoic Panaitios (d. 109), from whose work he borrowed the substance of his *De officiis*, Hecatōn of Rhodos, Panaitios' disciple. The *Somnium Scipionis* was derived from Poseidōnios.

Cicero was a lawyer and statesman who held many public offices and was mixed up in all the social vicissitudes of his time. It would be impossible to describe his political life without explaining in detail the wars and insurrections that he witnessed, the intrigues and polemics into which he was drawn. Readers who desire to know those facts will find them in the manuals of political history. In spite of his abundant correspondence, it is almost impossible to assess Cicero's character objectively; some historians blame him as much as others praise him. We should remember that he was primarily an author, rather than a politician or a statesman. According to Plutarch (in his life of him), Cicero was generally disliked because of his vanity and continual boasting. He became very rich, yet I believe in his honesty; that is, he was more honest than most of his successful contemporaries. When he was proconsul in Cilicia in 52, he did not plunder the people committed to his charge as was the custom but was considerate of them; he was vainglorious as usual with him but generous; his weakness was remembered, his exceptional virtue forgotten. The noblest moment of his political life was the very end of it; on 7 December 43, at Formiae in the lovely bay of Caieta, he was murdered by order of the Second Triumvirate. He could have saved his life if he had been a coward, but he accepted death. His head and right hand were cut off and taken to the Roman forum to be nailed to the rostrum. It was believed for a long time that his body (or the ashes) had been taken to Greece and buried in the island of Zacynthos (Zante).[13] Who knows?

His philosophy was not original, but a very clear exposition of Greek ideas to which he gave a newer emphasis. Original ideas are exceedingly rare and the most that philosophers have done in the course of time is to erect a new combination of them. What Cicero did was to select wh
considered to be the best ideas of Greek philosophy, chiefly tho
taught in the new Academy and the Stoa.

His main achievement in the tradition of Stoi
nonsense and superstition. This required luc
tious age,[14] and to the many detractors w

[12] The tradition of the *Protrepticos* from Aristotle to St. Augustine, via Cicero, is told in Volume 1, p. 474.

[13] For details see my article, "The death and burial of Vesalius, and, incidentally
(1954).

[14] To
the very
divinatione

MARCI TVLLII CICERONIS OFFI
CIORVM AD MARCVM
FILIVM LIBER PRIMVS INCIPIT

Quanq̃ te marce fili annum
iam audiẽtem cratippũ id q
athenis abundare oportet p
ceptis institutisq; phie ꝓpt
summã & doctoris auctatẽ et
urbis quorũ alterũ te sciẽtia
augere potest altera exẽplis.
tamẽ ut ipe ad meã utilitatẽ semp cũ grecis latina
coniunxi neq; id in phia solũ sʒ etiã in dicẽdi exercita
tione feci id tibi censeo faciendũ ut par sis in utri
usq; orõnis facultate. Quã quidẽ ad rẽ nos ut vi
demur magnũ adiumentũ attulimus hoibus nris.
ut non modo grecarũ lrarũ rudes sʒ etiam docti ali
quantũ se arbitrentur adeptos et ad discendũ et ad
iudicadũ. Quãobrẽ disces tu quidẽ a pricipe huius
etatis philosophorũ et disces quãdiu voles. tãdiu autẽ
velle debebis quoad te quãtũ pficias nõ penitebit.
Sed tamẽ nra legẽs nõ multũ a peripateticis dissi
dentia quoniam utriq; socratici & platonici volumus esse.
de rebus ipis utere tuo iudicio. Nihil enim impe
dio. Orõnem autẽ latinã efficies pfecto legendis
nris pleniorẽ. Nec vero arroganter hoc dictũ exti
mari velim. Nã phandi cedẽs sciẽtiã multis qd
est oratoris ꝓpriũ apte distincte ornate dicere qm

…4. Princeps of Cicero's book on duties, *De officiis* (Mainz: Fust and Schoeffer, …The same volume includes also Cicero's *Paradoxa Stoicorum*. Some of the Greek …ich Cicero could not help using because of the lack of Latin equivalents are …h Greek type. This was the first treatise of classical philosophy to appear in …sy of Pierpont Morgan Library.]

one might say that his struggle against superstitions was a departure as original as it was healthy.

Whatever may have been the weaknesses and mistakes due to the ambition, vanity, and greed of his earlier years, his writings on philosophy and religion after Pharsalos [15] prove him to have been a great man, as were Caesar and Brutus. Pompey and Anthony were not as great as those three, nor was Augustus who obtained the fruits of their efforts.

Let us now consider Cicero's philosophic works in themselves. If we count among them his treatises on political philosophy, he began to write them after the age of fifty.

(1) The six books of *De re publica* (*Republic*), a dialogue based on Plato, were ready in 51 but were lost until the nineteenth century, except the *Somnium Scipionis* (*Scipio's dream*), which was transmitted by the commentary of Macrobius (V-1).[16] In 1820, Angelo Mai discovered a substantial part of the text in a Vatican palimpsest.

(2) The *De legibus* (*On laws*) was begun in 52 but was not published until after the author's death. Three books out of five have come down to us.

His philosophic writings proper were not begun until many years later, when he was discouraged by the bankruptcy of political freedom and by the death of his beloved daughter Tullia (February 45). The books that will now be enumerated were all composed between the time of her death and his own (December 43).

In the following list items 3 to 7 might be classified under ethics, items 8 to 13 under philosophy in a more general way, items 14 to 16 under religion or philosophy of religion. The classification is not emphasized otherwise, because it is not exclusive.

(3) *De officiis* (*On duties*) written in 44 for his son, Marcus, who was then studying at the Lyceum or amusing himself in Athens. It is divided into three books, the first two derived from Panaitios, the third from Hecatōn; examples are taken from Roman history.

(4) *Cato major sive De senectute* (*On old age*), begun in 44 for his friend Atticus.

(5) *Laelius sive de amicitia* (*On friendship*) (c. 44). C. Laelius minor was a learned Stoic, a great friend of Scipio.

(6) *De gloria* (44). Now lost, but Petrarca had a manuscript of it.

(7) *De consolatione sive de luctu diminuendo*, written soon after Tullia's death in February 45 (lost).

---

tached to his edition of it (2 vols.; Urbana: University of Illinois Press, 1920–1923).

[15] Pharsalos in Thessaly, where Caesar defeated Pompey in 48 and thus became the master of the Roman world — not for very long, however, because he was murdered in 44.

[16] Recent English translation by William Harris Stahl, *Macrobius. Commentary on the dream of Scipio* (New York: Columbia University Press, 1952) [*Isis* 43, 267–268 (1952)].

¶ M. Tul. Ciceronis in dialogú de natura deorum ad Brutum Prefatio.

QVm multe sepe res in Philosophia neq̄q̄ satis ad huc explicate sint: tum perdifficilis Brute quod tu minime ignoras: & pobscura questio est de natura deoꝝ: que & ad agnitionem aími pulcherrima est. & ad moderandam religionē necessaria. de qua ꝙ tā uarie sint doctissimoꝝ hominum tamq; discrepantes sententie magno argumento cognoscitur. Nanq; de figuris deorum & de locis atq; sedibus: & actione uite: multa dicuntur. deq; iis súma philosophoꝝ dissensione certatur. Quod uero maxime rem causamq; continet: est utrum nihil agant: nihil moliantur: omni curatione: & ammi/nistratione rerum uacent: an contra ab iis & a ꝑncipio oīa facta & constituta sint: & ad infinitum tempus regantur atq; moueanť. In ꝑmisq; magna dissensio est: eaq; nisi diiudicať in summo errore necesse est homines: atq; maximarum rerum ignoratione uersari. Sunt enim philosophi: & fuerút: qui omnino nullam habere cen/serēt rerum humanaꝝ ꝓcurationem deos. quorum si uera sentētia est: que pōt esse pietas: que sanctitas: que religio? Hec enim oīa pure atq; caste tribuenda deorum numini ita sunt: si animaduer/tuntur ab iis. Et si est aliquid a diis immortalibus hon[illegible]num generi tributum. Sin aút dii neq; possunt nos iuuare: neq; uolunt: neq; omnino curant: neq; qd agamus animaduertunt: neq; est quod ab iis ad hominum uitam permanare possit: quid est: ꝙ ullos dii[illegible] im/mortalibus cultus honores preces adhibeamus? In specie autē ficte simulationis sicut relique uirtutes: ita pietas inesse non potest: cum qua simul sanctitatem & religionem tolli necesse est. Quibus sub/latis perturbatio uite sequitur: & magna confusio. Atq; haud scio an pietate aduersus deos sublata: fides etiā & societas generis hūani & una excellentissima uirtus iustitia tollatur. Sunt aút alii philo/sophi & ii quidem magni atq; nobiles: qui deorum mente atq; rōne omnem mundum amministrari & regi censeant. Neq; uero id solú sed etiam ad hisolem hominum uite consuli & prouideri. Nam & reliqua que terra pariat & fruges: & tempestates ac temporú ua/rietates celiq; mutationes: quibus oīa que terra gignat maturata pubescant: a diis īmortalibus tribui generi humano putāt. multaq;

(8) *Academica* (c. 45). Upon the philosophy of the New Academy as defended by Carneadēs.

(9) *De finibus bonorum et malorum*, written in 45, dedicated to M. Brutus the tyrannicide (d. 42). Discussion of the supreme good and supreme evil in refutation of the Epicureans and the Stoics.

(10) *Tusculanae disputationes* (c. 45–44). Five dialogues on practical problems held at his "Tusculanum" (his villa at Tusculum, near Frascati), also dedicated to Brutus. I. Fear of death; II. Is pain an evil?; III–IV. Distress and its alleviations, pain and its remedies; V. Virtue is sufficient for happiness.

(11) *Paradoxa* (45). Six paradoxes of the Stoics.

(12) *Hortensius*, adaptation of Aristotle's *Protrepticos*. It was written after Caesar's defeat of Pompey's sons at Munda (southern Spain) on 17 March 45. Only fragments of it remain.

(13) *Timaeus*. Translations of Plato's *Timaios*. Only fragments.

(14) *De natura deorum* in three books (c. 45). Dedicated to Brutus. On the nature and qualities of gods according to the Academy, the Porch, and the Garden. It is in this book that Cicero confirms the basis of astrology. The motions of the planets must be voluntary and thus the existence of gods is so obvious that it cannot be denied by a person of sound mind (II:16). This fantastic paralogism, derived from the *Epinomis*, has already been discussed (Volume 1, p. 453). Cicero reconciled his skepticism with acceptance of the Roman state religion, just as many Englishmen reconcile theirs with fellowship in the Anglican Church.

(15) *De divinatione*, written in 44, a continuation of the preceding item, dealing with many forms of divination. He was careful to dissociate religion from superstition.

(16) *De fato*. Dedicated to Aulus Hirtius, one of Caesar's officers and friends, Epicurean, man of letters. Distinction between fatalism and determinism. Only fragments.

It is almost incredible that Cicero could write those fourteen books (3 to 16) in thirty-three months, even if we take into account not only the pre-

---

Fig. 45. Volume 1 of the so-called *Scripta philosophica* of Cicero (Rome, 1471). Conrad Sweynheym and Arnold Pannartz moved their press from Subiaco to Rome in 1467. In 1469 they printed together four Ciceronian treatises, *De officiis, Paradoxa Stoicorum, Cato major sive De senectute*, and *Laelius sive De amicitia*. In 1471 they printed a larger collection of philosophical treatises which bears no general title but is generally called *Scripta* (*sive Opera*) *philosophica* (2 vols., folio). Volume 1 (168 leaves), published on 27 April 1471, included the four treatises already named plus *De natura deorum* and *De divinatione*. Volume 2 (205 leaves), appearing on 20 Sept. 1471, contained the *Tusculanae Quaestiones, De finibus, De fato, De petitione consulatus, De philosophia, De essentia mundi, Academica, De legibus*. These two volumes collected practically the whole of Cicero's philosophy, and their publication in 1471 was a landmark. Many of the treatises issued together in those *Scripta* were reprinted separately or in various combinations, by innumerable printers. Our facsimile reproduces the beginning of the *De natura deorum*, dedicated to M. Brutus. [Courtesy of the Huntington Library, San Marino, California.]

M. T. Ciceronis de somno Scipionis libellus ex vi. de rep. libro exceptus incipit.

Um in Africam venissem. A. Manilio cōsule ad quartam legionem tribunus vt scitis militum. nihil mihi poti⁹ fuit q̄ vt Masinissam cōuenirem regem familiæ nostræ multis de causis amicissimū. Ad quem cum veni. complexus me senex collachrymauit. aliquātoq; post suspexit in cœlum. ⁊ grates tibi ago inquit summe sol. vobisq; reliqui cœlites. q̄ anteq̄ ex hac vita migro conspicio in regno meo. ⁊ his tectis. P. Cornelium Scipionē. cuius ego nomine ipso recreor. Itaq; nunq̄ ex animo meo discedit illi⁹ optimi atq; inuictissimi viri memoria Deinde ego illum de suo regno. ille me de nostra repu. percōctatus est. deinde multis verbis vltro citroq; habitis ille dies nobis cōsumptus est. Post autem regio apparatu suscepti sermonem in multam noctem ꝓduximus. cum senex nihil nisi de Africano loqueretur. omniaq; eius non facta solum. sed etiam dicta meminisset. Deinde vt cubitū discessim⁹. me et de via. ⁊ qui ad multam noctem vigilassem. arctior quam solebat somnus complexus est Hic mihi(Credo equidem quod eramus locuti. Fit enim fere vt cogitationes sermonesq; nostri pariāt aliquid in somno tale quale de Homero scripsit Ennius. de quo sæpissime vigilans solebat cogitare atq; loqui) African⁹ se ostēdit ea forma quæ mihi ex imagine eius quam ex ipso erat notior. Quem vbi agnoui equidem cohorrui. Sed ille ades inquit animo. ⁊ omitte timorem Scipio. ⁊ quæ dicam trade memoriæ Vides ne vrbem illam quæ parere populo romano coacta per me renouat pristina bella. nec potest quiescere. Ostēdebat autem carthaginem de excelso ⁊ pleno stellarum

liminary studies of a lifetime but also his exclusive devotion to the writing of them. Thus did Cicero spend the last thirty-three months of his busy and restless life. Do you know of any famous politician who was able to end his days with the same grace and dignity?

LUCRETIUS

Titus Lucretius Carus is the best, if not the only, example of an author who is known by a single composition. He spent the best part of his life preparing, and at least the last ten years writing, a single poem, *De rerum natura*,[17] which was still unfinished at the time of his death in 55. About himself we know next to nothing, but his poem has come down to us in its integrity and is recognized as one of the greatest in world literature. We shall discuss the man and the poem presently. Let us simply remark at present that the *De rerum natura* is not only an important poem but a very long one; it is made up of 7415 dactylic hexameters; [18] it is of the same order of magnitude as the Western epics, but (and this is its outstanding and almost unbelievable characteristic) it is an epic of scientific philosophy, an epic not of deeds but of thoughts.

Lucretius died in 55 at the age of 44. If these data are correct, he was born in 99. He was a Roman of good family and had been very well educated; he may have been married and he loved children. The most curious statement concerning him was made by St. Jerome under the year 95: "The poet Lucretius was born, he who was driven to madness by a love philter and during lulls of his insanity wrote a few books [19] which Cicero corrected (*emendavit*). He committed suicide in his forty-fourth year." St. Jerome did not like Lucretius, but he would not have invented that cruel account; he probably repeated, not without malice, ancient gossip.[20] Some of it is plausible; love philters (*amoris pocula*) and other charms were used so much in Rome that it was found necessary to forbid them in a law promul-

[17] This Latin title is exactly equivalent to the Greek one, *Peri physeōs*, often used by the early "physiologists."

[18] The *Aeneid* is somewhat longer, 9,895 lines, and the *Iliad* more than twice as long, 15,693 lines. For the length of other poems, see Volume 1, p. 134, and add the Finnish *Kalevala* edited by Elias Lönnrot (1802–1884). The first edition, 1835, extends to 12,000 lines; the second, 1849, to 22,793 lines.

[19] The *De rerum natura* is divided into six books.

[20] St. Jerome died in A.D. 420, 475 years after Lucretius. Such gossip is very plausible; it would be natural enough to invent such calumnies (madness, suicide) in order to punish a freethinker.

Fig. 46. First edition of Cicero, *Somnium Scipionis* (Deventer: anon. printer, 18 July 1489). There were altogether five incunabula editions of it (Klebs, 275.1–5). The *Somnium Scipionis*, the most sublime of Cicero's writings, was his ending of the *De republica*, which he published in 51 B.C. It was the only part of it to be known until 1820 when Cardinal Angelo Mai discovered almost a third of the *De republica* in a Vatican palimpsest of the fourth or fifth century. The *De republica* (as much of it as survived) was first edited by Cardinal Mai (Rome, 1822). [Courtesy of Cambridge University Library.]

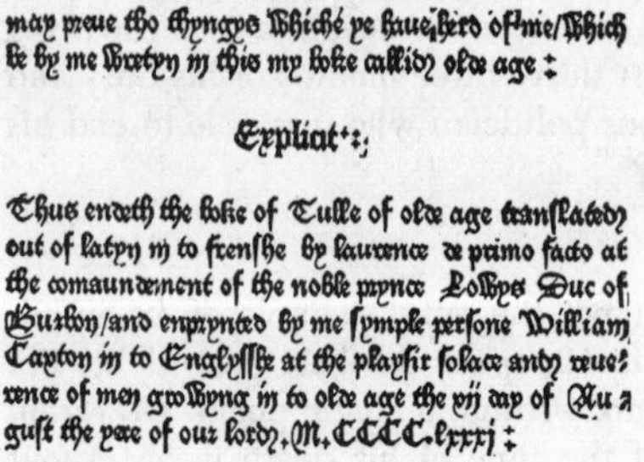
may preue tho thyngys whiche ye haue herd of me/which be by me wretyn in this my boke callid olde age :

Explicit :

Thus endeth the boke of Tulle of olde age translated out of latyn in to frensshe by laurence de primo facto at the comaundement of the noble prynce Lowys Duc of Burbon/and enprynted by me symple persone William Caxton in to Englysshe at the playsir solace and reuerence of men growyng in to olde age the vij day of Augusst the yere of our lord.M.CCCC.lxxxj :

Fig. 47. Colophon (ending) of the first Ciceronian text to be printed in English. The title reads *Tulle of olde age. Tullius de Amicicia*. Printed by William Caxton (Westminster, 7 August 1481). Only twenty-six copies of it have been traced by Seymour de Ricci, *A census of Caxtons* (Oxford, 1909), no. 31, pp. 39–42. It is the translation of Cicero's *De senectute* and *De amicitia*, both of which were dedicated to Atticus in 44/43, that is, during the last year of his own life. Caxton's book was the earliest English translation of a Latin classic. The "explicit," being modernized, reads: "Thus ends the Book of Tullius on old age translated out of Latin into French by Laurent de Premierfait at the command of the noble prince Louis Duke of Bourbon and printed by me simple person William Caxton in English at the pleasure, solace and reverence of men growing into old age, the seventh day of August the year of our lord 1481." Caxton's translation was thus made from the French one made in 1405 by Laurent de Premierfait (d. 1418), for whom see my *Introduction*, vol. 3, pp. 1294, 1313, 1804, 1809, 1811.

gated in 81.[21] Of course, such abuse could not be stopped by any law. Love philters were dangerous poisons; they might eventually kill a man, but they would not cause permanent insanity. It is difficult to conceive that Lucretius' poem was written during intervals of sanity. Cicero may have corrected it or not; it is certain that he and his brother Quintus had read it and approved of it. They read it in 54,[22] and this confirms that it was not quite finished at the time of Lucretius' death and was published posthumously. It may be safely conjectured that Lucretius' poem was "published" and survived because of Cicero's interest in it.

As Lucretius avoided public affairs and was deeply immersed in his meditation and composition, we may assume that he was a very lonely man. This may serve to explain, in the first place, the lack of information about him, and secondly, his suicide. There is nothing to prove that he committed suicide, however, except Jerome's statement, but the idea is plausible and, from the Roman point of view, there was nothing wrong or discreditable in putting an end to one's own life. Many distinguished men committed suicide and they were not blamed for doing so.[23]

[21] By the dictator Sulla (138–78); it is the *Lex Cornelia de sicariis et veneficis*.

[22] Cicero, *Epistulae ad Quintum fratrem*, 2, 11 (9). Cicero's brother, Quintus, to whom the letter is adressed had also read the poem.

[23] When life became unbearable, the anticipation of death was deemed justified. Death was preferred to disgrace. At a time when public executions were arbitrary and frequent, it was right to frustrate them by suicide. Cato of Utica took his own life in 46, Cassius and Brutus, the tyrannicides, in 42, Seneca and his wife Paulina, in A.D.

What does it matter that an author is forgotten if his work is remembered, and what better immortality could he dream of than that of his spiritual offspring?

Let us consider the *De rerum natura* and describe it. It is dedicated to a patrician, Memmius, whom we know much better than Lucretius himself. C. Memmius, who had married Sulla's daughter Fausta Cornelia, was propraetor in Bithynia in 57 and went thither with Catullus in his train; he died after 49 B.C. Lucretius addressed him as a friend rather than a patron, and this confirms our impression that he was a man of substance.

The *De rerum natura* is a defense of Epicurean philosophy, and especially of atomic physics. It is probable that Lucretius had devoted a good part of his life to the study of Greek philosophy, but that he had not always been an Epicurean. Indeed, the poem suggests that he had been recently converted; his enthusiasm and missionary zeal are those of a neophyte. He praised Epicuros like a god and savior. He was well acquainted also with Empedoclēs and it is certain that he had read writings of both which are unknown to us. This makes it more difficult to measure his originality.

He was well acquainted with the writings of other Epicureans, such as Hermarchos of Mytilēnē, Metrodōros of Lampsacos, and perhaps his own contemporary, Philodēmos of Gadara,[24] who died possibly at Herculaneum (c. 40–35).

Though his master was Epicuros who inspired the poem, he also very warmly praised Empedoclēs (I, 715–733) and referred to Anaxagoras (I, 830) and others.

Let us examine the poem and read as much of it as we can. It is divided into six books of which the first three (a little less than half) develop the main argument, atomic physics and cosmology. Books 4 to 6 are additions dealing with a number of secondary topics, yet the whole poem is as well ordered as a systematic treatise and everything hangs together in the best manner.[25] It would be absurd to consider it as the work of a madman, or as having been written during intervals of lucidity. The only madness of Lucretius was his genius; the poem was planned as a whole; the poet's steady inspiration created the unity of the whole work, while the enthusiasm, which could not be sustained throughout, erupted from time to time and gave birth to lyrical passages of great loftiness and beauty.

---

65. Atticus, Cicero's friend, let himself die of hunger in 32 B.C., and Silius Italicus did the same in A.D. 100.

[24] Gadara or Kedar, about 6 miles southeast of the Sea of Galilee. Remember the Gadarenes or Gergesenes in the New Testament — Matthew 8:28, Mark 5:1, Luke 8:26. Philodēmos' writings are preserved in papyrus rolls which were discovered in the ruins of Herculaneum.

[25] Except that the poem was not finished. This does not mean only that it ends abruptly, but there are many lacunae throughout, even in the first book, lines or words missing, and so on. It is clear that considerable work remained to be done at the time of Lucretius' death. Did he despair of perfecting his task?

After an invocation to Venus, goddess of creation, Lucretius expresses his main purpose. He wishes to reveal "the nature of things," their genesis, evolution and dissolution, and to explain the universe in physical terms. This implies a rejection of the religious or mythological accounts. There are thus two sides to his purpose – the defense of science and the attack on superstitions. Religion has been the origin of many crimes (I, 101, *Tantum religio potuit suadere malum*). The fundamental principle is the permanence of matter; nothing can come out of nothing; on the other hand, nothing can vanish. Matter exists in the form of particles separated from one another by empty space. Neither the particles nor the vacuum can be seen, yet they exist. There is naught else. Time is subjective (I, 459, *tempus per se non est*). The particles are solid, indestructible, indivisible (*atomos*). Other theories are refuted, the monism of Hēracleitos, the pluralism of Empedoclēs, the *homoiomereiai* (homogeneous parts) of Anaxagoras.[26] The vacuum is boundless, the universe is infinite, the atoms are innumerable. Many of those statements are "proved," as much as they could be; Lucretius used examples and images to justify them. As the universe is infinite, it can have no center (I, 1070 ff.). Book I ends with a much-needed encouragement to Memmius; the subject is difficult and dark but will be gradually enlightened.

My description is too brief to give a full idea of the richness of the argument. I shall proceed in the same way for the following books, indicating only the main topics as they occur in the reading, leaving out various digressions.

The praise of philosophy and of science introduces Book II, devoted to the study of atomic motions. These motions are not providential. Atoms do not move upward but rather downward; their motions are somewhat irregular and arbitrary, and their "swerving" establishes possibilities of chance and freedom [27] (II, 216–293). The sum of the matter is eternally constant. The whole universe is seemingly immobile. There is a great variety of atomic shapes; that variety is not infinite but its results are, because there is an infinite number of atoms of each shape and the possible combinations are

[26] The ideas of those philosophers have been discussed in Volume 1. Hēracleitos flourished at the beginning of the fifth century; Empedoclēs died c. 435 and Anaxagoras c. 428. The two founders of the atomic philosophy were Leucippos, who flourished like the two last named about the middle of the fifth century, and Dēmocritos, who flourished toward the end of the century and died c. 370. Lucretius was no doubt acquainted with their writings (much better than we can be) but his main inspiration came from their successor and consummator, Epicuros, who died at Athens in 270. All of them came from the Asiatic coast, except Empedoclēs, who was of Sicilian origin, and Dēmocritos, who was a Thracian. The evolution of the ancient atomic theory from Leucippos to Lucretius required four centuries.

[27] This swerving (*prosneusis, inclinatio*) was discussed in Volume 1, p. 591. It is not possible without writing a whole treatise on Lucretius to discuss all of his ideas in themselves, or in relation to their origin. As the writings of Leucippos, Dēmocritos, Epicuros, and others are not known to us in their integrity (we have only a few fragments), it is impossible to say who initiated this or that.

endless. No body is made up of atoms of a single kind. Atoms have no qualities such as color, temperature, sound, taste, or odor. Bodies endowed with life and sensibility are made of atoms just as the lifeless ones are. There is a plurality of worlds in the infinite universe, and each world passes through various phases: birth, growth, decrepitude, and death.

The magnificent invocation to his teacher and father Epicuros which opens Book III is the most moving and the best-known part of the whole poem. I cannot forbear from quoting a few lines of it (III, 1–4, 9–13, 28–30):

E tenebris tantis tam clarum extollere lumen qui primus potuisti inlustrans commoda vitae, te sequor, o Gaiae gentis decus, inque tuis nunc ficta pedum pono pressis vestigia signis.

Tu, pater, es rerum inventor, tu patria nobis suppeditas praecepta, tuisque ex, inclute, chartis, floriferis ut apes in saltibus omnia libant, omnia nos itidem depascimur aurea dicta, aurea, perpetua semper dignissima vita.

His ibi me rebus quaedam divina voluptas percipit atque horror, quod sic natura tua vi tam manifesta patens ex omni parte retecta est.[28]

Never did a disciple speak of his revered teacher with such piety and such pride. Book III, says the author after this ceremonial preamble, will explain the nature of the soul and destroy the fear of death. Mind and soul are parts of the body, they are very closely united, and their substance is material. Their atoms, however, are extremely subtle. Body and soul hang together. The soul is subject to death as well as the body. Is it not subject to disease and to healing? If so, it is mortal. The body's agony is also the soul's. Body and soul cannot exist except together; they die together. The soul is made up of particles and hence cannot be immortal as those particles are. If the soul were immortal, it would be conscious of previous lives; metempsychosis is unthinkable.[29] Could one conceive immortal souls disputing the possession of a mortal body? The soul could not exist outside of a body, hence it is mortal like the latter, and death is not a cause of suffering but a deliverance. Infernal punishments are not veritable but legendary and symbolic. The fear of death is the fruit of ignorance and life is as nothing compared with eternity. The soul being mortal, fear of death is utterly foolish.

[28] This is not easy to translate exactly and beautifully, but the general sense is as follows:

"Out of deep darkness, thou who wast the first to strike out such a clear light and to illuminate the good things of life, I follow thee, O glory of the Greek people, and put now my feet upon the traces left by thy own.

"It is thou, father, who are the inventor of the truth [about things], thou who suppliest us with paternal advice, and it is in thy books, illustrious one, that like bees visiting the flowering meadows we hunt for golden sayings the most worthy of eternal life.

"[When I listen to them] I am seized by a kind of divine volupty and awe, realizing that nature, discovered by thy genius, has been unveiled to be revealed to us."

[29] The concept of metempsychosis (transmigration of souls from one body into another, human or animal) had been favored by Orientals, Pythagoreans, and Orphics; many Greeks, such as Hērodotos, were familiar with it even if they did not accept it (Volume 1, pp. 201, 249, 309).

Book IV deals with the *simulacra*, meaning visions and phantoms and the terrors that they cause. This is a psychologic study of sensations and thoughts. The *simulacra* include the many things that we cannot see clearly, or the illusions (such as optical illusions), or spontaneous images, or emanations of bodies. (Reading this one realizes how observation, not to mention experimentation, was difficult for the ancients, not only because they lacked objective instruments but chiefly because of the abundance of appearances which had not yet been analyzed and which it was impossible to define and classify.) Every body produces emanations, such as sounds, odors, and visions (according to Lucretius, we see objects because atoms emanating from them reach our eyes; he explained vision as we explain smell). Good examples of images are those seen behind a mirror. Various optical illusions are discussed. Sensations are infallible; it is easy to misinterpret them, but, if they are correctly interpreted, they are the very basis of knowledge. Similar views are expressed concerning the other senses (hearing, taste, smell) and the "images" that they produce (such as echoes in the case of hearing). Spiritual visions, dreams. Digression against [Aristotelian] teleology (IV, 822–857), no organ of our body has been created for our use; on the contrary, it is the organ that creates the use of it. Defense of materialism against vitalism.[30] Vision did not exist before the eyes, nor language before the tongue . . . Hunger and thirst, walk and motion, sleep, dreams, puberty and love. Dangers of love, the lovers' illusions and pains. Heredity. Fertility and sterility.

Lucretius had stated the theory of heredity in Book I (lines 149–173, chiefly 167–168) but in Book IV (lines 1218–1222) he expressed ideas which might be called the essence of Mendelism and others (lines 834–835) refuting the doctrine of the inheritance of acquired characters and of pangenesis.[31] Book V is the longest (1457 lines, the five others averaging 1191 lines) and is even more complex than the preceding one. It begins with a new eulogy of Epicuros and proceeds with the discussion of a whole series of phenomena (we might say that Books I to III explain the general theory; Books IV to VI treat various applications of it). The gods are foreign to the human world; they did not create it and do not care about it. It is mortal like all its parts; it had a beginning and will have an end; it is relatively new and is progressing (V, 332–335). Lucretius was the first to express that idea of progress; most of the ancients [32] favored the opposite idea of a "golden age" at the beginning followed by a gradual decadence.

[30] That is how one would put it in modern language. Lucretius rejected Aristotle's special energies (*entelecheia*) determining the growth and form of organisms, just as Jacques Loeb rejected those of Hans Driesch. This type of discussion will never end.

[31] That is the interpretation, perhaps too generous, of C. D. Darlington, "Purpose and particles in the study of heredity," in *Science, medicine and history*, Charles Singer's Festschrift (London: Oxford University Press, 1953), vol. 2, pp. 472–481.

[32] For example, the Sumerians (Volume 1, p. 96) and Hēsiodos (*ibid.*, p. 148). That singular concept of regress (instead of progress) was generally ac-

Lucretius had not rejected the Empedoclean theory of four elements; in a digression (v, 380 ff.) he envisages a cosmic struggle between two of them, fire and water. He then discusses the birth and growth of various parts of the world, movements of celestial bodies, immobility of the Earth, sizes of Sun and Moon, origin of solar light and heat, theories of planetary motion, origin of the inequalities of day and night, Moon phases, eclipses.

This astronomical summary (v, 416–782) is followed by the study of organic evolution, first the plants, then the animals, finally man. Some animals have disappeared or are fabulous (like the centaurs). Prehistoric men were ignorant and inefficient but they acquired knowledge and invented techniques. Beginnings of social life. Origin of language, discovery of fire, of kingdoms and property. The kings are finally dethroned and justice established. Evils caused by belief in gods. The first metals: gold, silver, bronze, lead. Discovery of iron. Development of the art of war (next to religion, there was nothing that Lucretius hated more than war). Origin of clothing, weaving. Horticulture: sowing and grafting. Music. Writing, poetry, and so on. Thus did man advance throughout the ages; his progress was steady but very slow (*pedetemptim*, v, 1453).

The last third of Book v (lines 925–1457) is thus a history of man from primitive times to Lucretius' own sophisticated days; the prehistoric part is especially remarkable, as, for example, the following lines (v, 1283–1287):

Arma antiqua manus ungues dentesque fuerunt
et lapides et item sylvarum fragmina rami,
et flamma atque ignes, postquam sunt cognita primum.
Posterius ferri vis est aerisque reperta.
Et prior aeris erat quam ferri cognitus usus.[33]

This may be considered an adumbration of the discovery made in 1836 by Christian Jürgensen Thomsen of Copenhagen, the first clear enunciation of the "law of the three ages," stone, bronze, iron. Lucretius was almost alone [34] in forestalling Thomsen, nineteen centuries ahead of him. How could he make such a prophecy? He was probably helped by the fact that remains

cepted not only in antiquity but until the birth of modern science in the seventeenth century. For example, Stevin shared it. The idea of progress adumbrated by Lucretius was developed by Seneca (I-2), see *Introduction*, vol. 2, p. 484. It must be added that Aristotle's teleology implied progressive evolution. See Volume 1, p. 498.

[33] "Man's earliest weapons were his nails and teeth, stones and branches torn from the woods. Later, the power of iron and bronze was discovered, bronze first and iron next he learned to use." Note the succession: stones, bronze, iron (*lapides, aes, ferrum*).

[34] To be accurate, the discovery of the wild men of the New World caused a few other anticipations: Michele Mercati (1541–1593), published only in 1717; Aldrovandi (d. 1605), published in 1648; Robert Plott (1686); Joseph François Lafitau, S.J. (1670–1746), in 1724. The succession of stone, bronze, iron was indicated by Johann Georg von Eckhart (Eccardus), *De origine Germanorum* (Göttingen, 1750) and by Antoine Yves Goguet, *De l'origine des lois, des arts et des sciences* (3 vols.; Paris, 1758).

of the stone and bronze cultures could still be observed in his time. The distant past was not as completely obliterated then as it is now around us.

As we have just seen, Book v deals mainly with astronomy, organic evolution, anthropology, and history of culture. In the same way, Book vi deals with meteorology, geography and medicine. The first lines (vi, 1–42) praise Athens and Epicuros. Main topics: thunder, lightning, hurricanes (*prēstēr*), clouds, rain and rainbow, earthquakes, volcanoes (Etna). Many explanations of these phenomena are possible, though only one can be true; in any case, there always is a physical explanation. Infected lakes (Avernus near Cumae), springs. The magnet. Diseases and epidemics. The poem ends rather abruptly with an elaborate account of the plague of Athens (vi, 1139–1285)[35] derived mainly, if not exclusively, from Thucydidēs. The pessimistic author may have wished to complete his cycle with that exceptional calamity, a cycle comparable to every other, from birth to destruction, but, even so, one would have expected a kind of conclusion or a final paean to Venus or to Epicuros.

My analysis was meant to give an idea of the encyclopedic nature of the *De rerum natura*. It must seem very dry and may have taxed the reader's patience, but a more adequate description would have taxed it considerably more. One cannot summarize an encyclopedia. The poem itself, it must be admitted, is very dry and hard to read in any language. Few scholars have read the whole of it, except perhaps at the very beginning when Lucretian knowledge was the freshest available in Latin circles. The general drynss is relieved by lofty invocations, many concrete examples, and a few lyrical outbursts. It is not so much a didactic poem as a philosophic scientific one enlivened with poetic intermezzi. It is a poetic vision of the universe comparable to the visions of Dante and Milton, though extremely different from them, not only in scientific contents but in inspiration. Such as it is, it is unique of its kind in universal literature.

Lucretius was an Epicurean, defending Epicurean doctrines with the ardent conviction of a missionary. His main benefactor was Epicuros, but he was acquainted with other writings of the school, some earlier than Epicuros, others later. It is impossible to determine his indebtedness to each of them, but that hardly matters. Epicuros had been his true initiator and Lucretius' gratitude to him was boundless. It is very warmly expressed in four long passages (i, 62–83, iii, 1–40, v, 1–58; vi, 1–47). Fragments of the second passage have been quoted and translated above. It is in the third that there occur the astonishing lines:

[35] The plague of 430–29 described by Thucydidēs (Volume 1, pp. 323–325). Lucretius' account was in its turn the source of many Latin descriptions, such as Virgil's (*Georgica* 3, 478–566), Ovid's (*Metamorphoses* 7, 517–613), those of Lucanus of Cordova (39–65) and Silius Italicus (25–100).

. . . deus ille fuit, deus, inclyte Memmi, qui princeps vitae rationem invenit eam quae nunc appellatur sapientia, quique per artem fluctibus e tantis vitam tantisque tenebris in tam tranquillo et tam clara luce locavit.[36]

This would shock us if we did not remember the Greek habit of making demigods (*hērōes*) of their great men and of passing easily from hero worship to apotheosis. We do not know the poet's early life, but it is probable that he had suffered much from his passions and his irresoluteness before his "conversion." Epicuros had not been only his teacher but his savior.

The atomic theory enabled him to explain external and internal realities in rational terms and to drive out of his conscience miracles and superstitions. To him that theory was undoubtedly true; to us who have come two thousand years later, it does not appear in the same light. From our point of view indeed his atomism was not an irresponsible theory, but it was not truly scientific, for its experimental basis was too small and too weak. It is therefore quite wrong to compare ancient atomism, a successful guess, with modern atomism, which was a sound hypothesis from the beginning, imperfect at first but capable of endless improvements.

It was nevertheless Lucretius' purpose to explain nature on the basis of facts. Facts were of many kinds; a fruit or a stone was one, but so were feelings. Sensory impressions might be direct or indirect, but all our knowledge was derived from them, and our knowledge would be pure if we were able to interpret those impressions correctly. All this is admirable, but Lucretius was a poet as well as a philosopher and he could not help overreaching himself, as when he spoke of Epicuros' "having advanced far beyond the flaming barriers of our universe and having gone through the whole immensity and come back victorious to teach us the genesis of everything" (I, 73–75). He also referred at least thrice to "nature who creates all things" (*rerum natura creatrix*). He felt the immensity of the universe and the smallness of the earth and of man as intensely as did Pascal. This was lyrical poetry of the highest voltage applied to life and science.

His main conceptions were primarily atomic: the world is made of an infinity of atoms of various shapes, always in motion, but he had drawn from this many bold conclusions. The infinity of space and time, the universal and unavoidable laws of nature, the infinite variety of things, the universal theory of evolution, the oneness of the whole universe and its equilibrium (*isonomia*), the plurality and mutability of separate worlds, heredity, and so forth.

His overreaching was partly conscious. Atoms were not tangible and yet it was necessary to postulate their existence. Sensory impressions are fundamental, yet they must be transcended. In this respect he took no more liberties than the physicists of our own age.

[36] "He was a god, yea, a god, illustrious Memmius, who first discovered the rule of life now called wisdom, who was able by his knowledge to tear us away from such deep darkness and to bring us into a place full of peace and light."

It is not profitable to discuss his views on more definite problems, such as natural selection, magnetism, or the rainbow, because his experimental knowledge was utterly insufficient. Whenever he managed to anticipate modern ideas, that was accidental. Here are two examples. He remarked that the diamond's hardness is due to extreme packing of its atoms but that nothing is absolutely solid except the atoms themselves, and that the embryo owes its creation to the mixture of two kinds of seed (IV, 1229–1232). These were intelligent guesses, not discoveries.

He was so deeply concerned with physical questions that he paid less attention to ethics. His main ethical principle was the need of avoiding superstitions, but this could only be done by the study of physics; hence, ethics brought him back to science, physical science. He denounced the evils and dangers of ambition, glory, and wealth; the struggles required to obtain such illusory goods were worthless; he loved simplicity and aloofness; happiness is the result of an inner equilibrium, self-sufficiency the greatest source of riches.

These were excellent precepts, yet Epicureanism was bound to be defeated because of its own hostility to Stoicism on the one hand and to religion on the other. Let us consider both.

Epicurean views on science could not interest more than a few people, while its moral and social aspects concerned a great many more. Science would develop in its own way, with Epicurean good will or without. The success of Epicureanism itself depended upon whether its rules of conduct would be acceptable to the Roman public or not.

The trouble with the Epicureans was not so much that they were hedonists, but rather that they were escapists. They kept aloof from politics and social commitments. The Stoics did the opposite; they insisted upon the importance of civic duties; according to them, morality was not only of personal but also of social concern. The state needed civil servants; it would find better ones in the Porch than in the Garden.

One may wonder that a book as revolutionary as *De rerum natura* could be published in Rome in 55 or 54, when political freedom was dying. The reason may have been simply Lucretius' political ineptitude. The poet was not interested in the government of Rome but rather in the constitution of the universe. The freedom of writing what he pleased might be granted to him, as it was to Catullus at the same time.

Lucretius was an enemy of superstition in all its forms; he was not only anticlerical but also antireligious. His emotional drive was so strong that he exaggerated the terrors of superstition and the perils of religion. His attack on religion was not directed against the Roman state cult, however, but rather against Platonic tendencies and popular rites. By temperament as well as by doctrine he was a rationalist and positivist; religion meant nothing to him. Once again one would like to know the vicissitudes of his life. His anticlericalism was so intense that one cannot help wondering whether he had been exploited or punished by clerics in his youth. He did not deny the existence of the gods but considered them as indifferent to us. The world is not divine; nature is aimless, the atoms came together by chance.

Now the Hellenistic world, and the Roman world dovetailed with it, were giving more and more scope to superstitions and to irrationality. The social circumstances were so hard and cruel, the miseries caused by wars and revolutions so abundant, that life became unendurable and the people were hungering for some kind of salvation in the hereafter. It is clear enough that Epicureanism could not possibly compete in popularity with the redeeming and messianic creeds that were sprouting everywhere.

At the end of Book III of *De rerum natura*, Lucretius tried to show that the fear of death was foolish and to destroy it. His argument is puzzling. When the body dies, the soul dies too, because body and soul are made of atoms which are equally dissipated when the bondage of life is broken. Confusing the fear of death with the fear of immortality, he argued that the certitude of mortality made it unreasonable to fear death.

He was convinced of man's mortality and had the courage to say so, but seemed to believe that, as soon as man understood the finality of death, he would be tranquilized and happy. But were men really afraid of the afterlife? That is very doubtful. Pleasant visions of the afterlife were evoked by the Eleusinian and other mysteries and also in a subtler way by the Platonic fancies. The ancients had no concept of divine retribution after death; [37] good and evil men alike would share a shadowy and cheerless existence; the best men, however, would be taken to the Isles of the Blest (*Elysium*), placed by Homer at the western margin of the Earth and later by other poets in the lower world (*Inferi*). Why should good men fear the Elysian Fields? And would the common people not prefer survival, any kind of survival, to extinction?

Nevertheless, Lucretius' argument was not offered as a paradox but as being obvious to any clear-minded person. This shows how deeply every man is influenced by the "common sense" of his own contemporaries. He was resigned, as we are, but his resignation was that of a pessimistic man of science who is tired of being duped and has trained himself to be as indifferent as nature itself.

It is as hard for us to understand his fear of immortality as it would have been for him to appreciate the verse, "O death, where is thy sting? O grave, where is thy victory?" (1 Corinthians 15:55).

[37] It would be more correct to say that the idea of eternal sufferings caused by the gods was beginning to shape itself. Lucretius' contemporary, Philodēmos of Gadara, was the first to mention infernal fire as punishment. F. Cumont, *Lux perpetua* (Paris: Geuthner, 1949), p. 226.

Was he not helped at all by Eleusinian or Elysean hopes? I imagine that he had rejected those fancies once and for all with all the other superstitions of which man must free himself to enjoy wisdom and happiness, but perhaps he had not been able to reject the ghostly vision of *manes* and *lemures* drifting aimlessly in the *Inferi*. He may have attended on the ninth, eleventh, and thirteenth days of May the *Lemuria*, that is, the popular ritual of feeding the ghosts and getting rid of them. Did he perhaps confuse the fear of death with the fear of ghosts?

Lucretius' poem is impressive and moving because, in spite of its scientific content and of its objectivity, it contains many personal touches, which help us to evoke the author and to remember that he was a poet. These may be simple words (*vidi*, *videmus*, *mea dicta*, *opinor*), apostrophes to his friend, Memmius, a line "when we ascend high mountains" (VI, 469), a hymn to Venus, or a prosopopoeia of Epicuros. His anxiety to be reasonable did not prevent him from being sensitive. Moreover, his conviction is at once naïve and touching. He is "the one missionary poet of antiquity, the one humanitarian zealot among the philosophers of antiquity."[38] Much of his poem was necessarily prosaic and was made more so by his geometric kind of argument (with such words as *primum*, *deinde*, *huc accedit*, *ergo*) and his wish to be as clear and compelling as possible, yet it was charged with emotion and combativeness and its heavy prose was suddenly leavened with unforgettable lines. It is for that reason that I do not like to call the *De rerum natura* a didactic poem. Lucretius' purpose was not simply to teach but also to convert.

Though the substance of his poem was almost exclusively Greek, the form was Latin and Roman. He continued the tradition of Ennius (II–1 B.C.), rather than that of the Alexandrian poets. In spite of the esoteric nature of his subject, he was as simple as one could be, while they indulged in every kind of sophistication. It was partly because of his simplicity that he was able to influence other Roman poets very different from himself, like Virgil, Horace, and Ovid.

It is a pity that we do not have any inkling of the mistakes, deceptions, and disillusions of his early life, because one has sometimes the feeling that his poem is a defense, rebellion, and vengeance. Because of its queer mixture of humanity, misanthropy, scientific ardor, and Epicurean zeal, it is at once dry and passionate. Cicero, in the letter to Quintus above-mentioned, referred to the "*lumina ingenii*." There is no room here for more than two examples, in addition to those already given.

> The totality of things remains obviously the same . . . but it renews itself repeatedly and mortals are continually borrowing from one another. Some people wax, others wane, in a brief interval the generations replace each other

[38] William Ellery Leonard in his edition of Lucretius (Madison: University of Wisconsin Press, 1942), p. 22.

and like the racers transmit from hand to hand the torch of life (II, 75–79).[39]

True piety does not consist in taking attitudes and observing conventional rites but rather "in being able to view all things with a peaceful mind" (V, 1198–1203).[40]

At its very best, Lucretius' thought is as serene and high as Pascal's, yet these two men were straining their minds in opposite directions.

The Jews were violently opposed to Epicuros, but few of them read Latin[41] and they did not bother about Lucretius. The situation was different for Western Christians. At first view they might agree with him, because the religion that he had attacked was paganism, their own enemy, but there was too much in him that they could not possibly swallow, and they soon denounced not only his atheism but also his self-indulgence and immorality. He might be consulted and some of his arguments against the gods might be handy but he should be handled with great prudence. The Latin fathers could never consider him as an ally, as Tertullian of Carthage (160–225) did Seneca. Christians could make peace with the Porch but never with the Garden, not even in Gassendi's time.

The Lucretian tradition. This tradition is extraordinarily interesting because it is so discontinuous that it makes us think of a river which dives underground, reappears at a great distance, then disappears again, and so on. A Roman elite was prepared to listen to him because Epicureanism had been discussed in the Scipionic Circle after 146 and within Lucretius' lifetime it had been explained by men such as Siro, Virgil's teacher, and Philodēmos of Gadara. Caesar and Atticus were Epicureans, and so were many others (though Stoicism was more acceptable to Roman gentlemen). Lucretius was definitely anti-Stoic, but there is no record of Stoic animus against him.

Cicero's reaction is instructive, because it must have been shared with other thinking men whose leanings were to the New Academy. He was neither a Stoic nor an Epicurean, but had more sympathy with the Porch than with the Garden. One might almost say that he disliked the Epicureans, yet he admired Lucretius and it is possible that, as St. Jerome suggests, he helped to preserve the text of the *De rerum natura.* What is more significant, he interceded with the Athenian authorities for the protection of Epicuros' house against some enterprise of Memmius, the very man to whom Lucretius had dedicated his poem. At the time of the poet's death, in 55, Cicero had not yet begun his philosophic writings, and he was much more interested in ethics and politics than in science. For all that, he did realize Lucretius' singular greatness.

Three Latin poets mentioned Lucretius, to wit, Ovid of Sulmo (43 B.C.–

[39] The final line is especially beautiful: "et quasi cursores vitai lampada tradunt."

[40] "Sed mage pacata posse omnia mente tueri."

[41] At least before the fourteenth century. For early translations from Latin into Hebrew, see *Introduction*, vol. 3, pp. 64, 1073. Lucretius was never translated into Hebrew.

A.D. 17), Statius of Naples (61–95), and Sidonius Apollinaris of Lyon (431–482), and they spoke of him as a poet. Virgil (I–2 B.C.) did not name him but referred to him in the *Georgica* (II, 490–492). Vitruvius (I–2 B.C.) had studied him (*Architectura*, preface to IX).

A critical edition of his text was prepared by Valerius Probus of Beirūt, a grammarian who flourished in Rome under Nero (emperor 54–68).

The first Christian to discharge his fury against him was Lactantius (c. 250–317), "the Christian Cicero." St. Jerome (IV–2) knew of him and so did Servius [42] (who wrote a commentary on Virgil), Bishop Sidonius, Isidore of Seville (VII–1). All in all, knowledge of his work was very restricted and received little acknowledgment. He was probably read *sub rosa* as Ovid was, but with less pleasure. The *De rerum natura* was not as attractive as the *Ars amatoria*.[43]

There was no Islamic tradition because the Arabic writers did not read Latin and obtained their knowledge of Epicureanism and atomism straight from the Greek sources.[44]

To return to the Catholic world, the manuscript edited by Valerius Probus a century after Lucretius' death must have been copied, otherwise the tradition would have been broken and lost, but we suspect that the manuscripts were rare. While Cicero's were copied by the dozens and produced wholesale, those of Lucretius were issued in few copies, or copied one by one. Two excellent manuscripts of the ninth century have come down to us and are both preserved in the library of the University of Leiden.[45] The earlier of these, the *Codex oblongus*, was copied at the beginning of the ninth century at Tours or by a scribe from Tours and was used by Hrabanus Maurus (IX–1) at the abbey of Fulda.[46] The second codex, called *Quadratus*, was preserved for centuries in the abbey of St. Bertin, at Saint Omer.[47] This proves that there was a medieval tradition of Lucretius, quiet but tenacious, in such places as York, Tours, Fulda, St. Bertin. We should never underestimate Christian acquaintance with non-Christian books; most of the Christian clerks were fanatics but some were scholars.

[42] Servius is mentioned here in chronologic order, because he flourished in the fourth century, but he was not a Christian.

[43] The comparison is not as artificial as the reader may think. Lines 1030–1287 of Book IV deal "*de rebus veneriis*" and Lucretius explains at great length the dangers of sexual love.

[44] Salomon Pines, *Beiträge zur islamischen Atomenlehre* (150 pp.; Berlin, 1936) [*Isis 26*, 557 (1936–37)].

[45] Both manuscripts are available in complete facsimile editions prepared by Emile Chatelain (Leiden, 1908–1913). They are generally called Codex Vossianus oblongus and Codex Vossianus quadratus. The adjective Vossianus refers to two famous Dutch philologists, Gerard John Vossius (1577–1649) who collected the two manuscripts, and his son, Isaac Vossius (1618–1689), who sold them to the University of Leiden.

[46] Fulda in Hesse-Nassau, 54 miles northeast of Frankfurt am Main. The abbey of Fulda was founded in the eighth century and in the tenth century its abbot was primate of Germany. It played an important part in the development of German culture.

[47] Saint Omer near Corbie, which is 10 miles northeast of Amiens (Somme; Picardie).

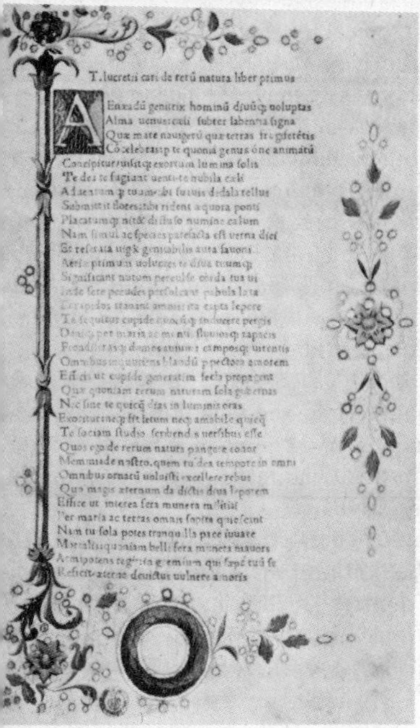
T. lucretii cari de rerũ natura liber primus

Fig. 48. First page of the princeps of the *De rerum natura* printed by Ferrandus at Brescia in 1473 (Klebs, 623.1). This first page contains lines 1 to 34 of the first book, the invocation to Venus beginning: "*Aenæadum genitrix, hominum divumque voluptas, Alma Venus* . . ." [Courtesy of the Bibliotheca Medicea-Laurenziana in Florence.]

TITI
LVCRETII CARI
DE RERVM NATVRA
LIBRI SEX.
A DIONYSIO LAMBINO
Monstroliensi litterarum Græcarum in vrbe Lutetia doctore Regio, locis innumerabilibus ex auctoritate quinque codicum manu scriptorum emendati, atque in antiquum ac natiuum statum ferè restituti, & praeterea breuibus, & perquàm vtilibus commentariis illustrati.
PARISIIS,
Et Lugduni habentur.
In Gulielmi Rouillij,
Et Philippi G. Rouillij Nep.
ædibus, via Iacobæa sub Concordia.
CVM PRIVILEGIO REGIS.

Fig. 49. First edition of Lucretius based upon one of the ancient manuscripts, the Quadratus. It was edited by Denis Lambin and printed by Guillaume and Philippe Rouille (Paris and Lyons, 1563). Many other editions had appeared between 1473 and 1563, but this one was the most important of the Renaissance. [Courtesy of Harvard College Library.]

Lucretian traces appear in the works of such men as Guillaume de Conches (XII-1) and Jean de Meung (XIII-2), but the poem itself disappeared from view, to reappear only a few centuries later. Thus, in addition to the two ninth-century manuscripts, we have some thirty-five written six centuries later.

The tradition was renewed as follows: Poggio Bracciolini, being apostolic secretary at the Council of Constance (1414–1418), took advantage of his position to hunt for classical manuscripts in the monastic libraries; in 1418, he discovered (at Murbach?, Alsatia) a manuscript of Lucretius and sent a transcript of it to his friend Niccolo de' Niccoli. The latter's copy, made in 1418–1434, is still available in the Laurentian library of Florence. It was probably the model of all the other fifteenth-century manuscripts and it was certainly the source of all early editions prior to 1563.

The success of Poggio's discovery is typical of Renaissance paganism; Lucretius' ideas were so libertine that he has often been compared with such rebels as Omar Khayyam and Voltaire, but they did not frighten the early humanists. The first printed edition was published at Brescia in 1473 (Fig. 48), and was followed by four other incunabula (Klebs, 623), all Italian. The last of these, prepared by Hieronymus Avancius of Verona (Venice: Manutius, December 1500), was the best, but that Aldine edition was improved by the Juntine, prepared by Pier Candido Decembrio (Florence: Junta, 1512) and these editions were reprinted many times.

A great step forward was made in the new edition prepared by Denis Lambin with the assistance of Adrien Turnèbe and Jean Dorat (Auratus), printed by Rouillius (Paris and Lyons, 1563) (Fig. 49). This was the first edition based upon one of the ancient manuscripts, the Quadratus. Lambin's task was completed three centuries later by Karl Lachmann (Berlin, 1850), whose edition was based not only on the Quadratus but also on the *Oblongus* and upon the labors of many scholars.

Of later editions it will suffice to mention the Latin-English by H. A. J. Munro (2 vols.; Cambridge, 1864), the Latin-French by Alfred Ernout (Collection Budé, 2 vols.; Paris: Belles Lettres, 1920), and the Latin edition with elaborate commentary by William Ellery Leonard and Stanley Barney Smith (Madison: University of Wisconsin Press, 1942) [*Isis* 34, 514 (1943)].

A few more remarks about the modern tradition. It is difficult to separate the Epicurean tradition from the purely Lucretian. For example, Gassendi's [48] defense of atomism was based on Epicuros: *De vita et moribus Epicuri libri octo* (Lyons, 1647), *De vita moribus et placitis Epicuri seu Animadversiones in decimum librum Diogenis Laertii* (Lyons, 1649), *Syntagma philosophiae Epicuri* (The Hague, 1659). In Jewish eyes Epicuros had been a master of impiety; for Latin Christians he was too far away to be blamed and Lucretius was the real imp (*suppôt de Satan*).

The French cardinal de Polignac [49] wrote a long poem against Lucretius, the *Anti-Lucretius, sive de Deo et Natura, libri novem*, which was issued posthumously (2 vols.; Paris, 1747) (Fig. 50). It is said to be one of the outstanding scientific poems in modern Latin.[50] I have not read it.

The illustrious French poet, Sully-Prudhomme (1839–1907), published a French translation in verse of the first book of *De rerum natura* (Paris, 1869).

Other items of the same kind might be quoted, because Lucretius' single

[48] Pierre Gassendi, born in Provence, in 1592, died at Paris in 1655.

[49] Melchior de Polignac was born in Puy-en-Velay in 1661 and died in Paris in 1742. His *Anti Lucretius* was directed also against Pierre Bayle (1647–1706).

[50] It was very successful, being almost immediately translated into French by Jean Pierre de Bougainville (Paris, 1749), four times reprinted in eighteen years; it was translated into English in 1757.

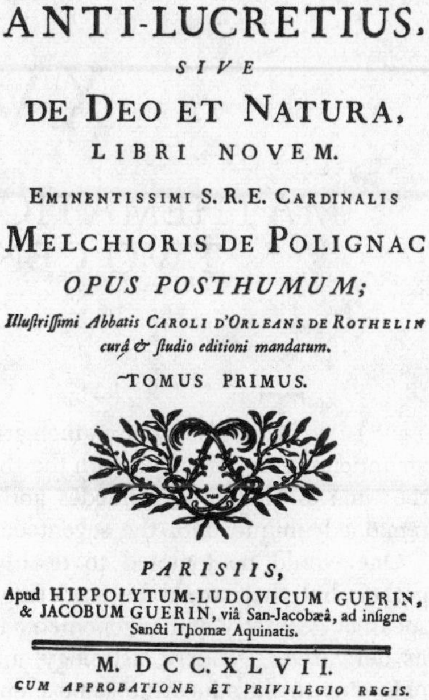
ANTI-LUCRETIUS,
SIVE
DE DEO ET NATURA,
LIBRI NOVEM.
EMINENTISSIMI S. R. E. CARDINALIS
MELCHIORIS DE POLIGNAC
OPUS POSTHUMUM;
Illustrissimi Abbatis CAROLI D'ORLEANS DE ROTHELIN
curâ & studio editioni mandatum.
TOMUS PRIMUS.
PARISIIS,
Apud HIPPOLYTUM-LUDOVICUM GUERIN,
& JACOBUM GUERIN, viâ San-Jacobæâ, ad insigne
Sancti Thomæ Aquinatis.
M. DCC. XLVII.
CUM APPROBATIONE ET PRIVILEGIO REGIS.

Fig. 50. First edition of the *Anti-Lucretius, sive de Deo et Natura, libri novem* by Melchior de Polignac (1661–1742), cardinal and diplomat. It is one of the most famous Latin poems of modern times. It was published posthumously (2 vols., 22 cm; Paris: Guérin, 1747), and was often reprinted in Latin, French, Dutch, Italian, and English. The first edition includes an engraved portrait of the author. [Courtesy of Harvard College Library.]

achievement fascinated the imagination of poets and philosophers all over Christendom. Some admired him, others hated him, but all were attracted and excited.

FREEDOM OF CONSCIENCE

After Pharsalos (48), power was concentrated in Caesar's hands, the Republic was dead, the Roman empire was beginning to shape itself, democracy was eclipsed, and political freedom was dying. Fortunately, some of the leading men of Rome had been trained by Greek philosophers. It remained possible to discuss philosophy and even religion, if one were careful not to attack the state ritual.

Freedom of conscience was upheld by such men as Lucretius and Cicero, who wrote in Latin and whose books have continued to inspire mankind to this very day. Neither of them was a man of science in the technical sense, yet both helped powerfully to salvage the Greek heritage of science and wisdom. Both were defenders of reason in the face of growing irrationalism. On that ground alone they deserve the attention of historians of science and the gratitude of every lover of freedom.

XVIII

MATHEMATICS IN THE LAST TWO CENTURIES[1]

The history of mathematics during the last two centuries seems an anticlimax as compared with the third century, but that is simply because the time of Euclid, Archimēdēs and Apollōnios was a golden age which remained unique until the seventeenth century, two millennia later.

One would be tempted to deal with mathematics and astronomy together, but that would be confusing instead of clarifying. It is better to speak of geometry and trigonometry in one chapter, and of astronomy with its derivatives, geodesy, astrology, and the calendar, in another. This will oblige us to introduce the same men twice but that does not matter very much.

HYPSICLĒS OF ALEXANDRIA

The outstanding name in geometry was that of Hypsiclēs (II–1 B.C.), and it is also the one that is best known. He flourished in Alexandria in the second century and is the author of the so-called XIVth book included in a good many editions of Euclid's *Elements*. That is, many early editions of Euclid were restricted to plane geometry (Books I to VI); the editions that included Books VII to XIII were likely to include XIV (and XV) as well.[2] That was logical enough because XIV (and XV) dealt with regular solids and were thus appendixes to XIII.

Book XIV contains eight propositions relative to the more complex polyhedra, those of twelve and of twenty faces. The author gives credit for some of the propositions to Aristaios the Elder (IV–2 B.C.) and to Apollōnios (III–2 B.C.), but we must be grateful to him for having preserved and proved (or re-proved) some very astonishing results, which may be summarized as follows.

Aristaios the Elder had shown that "the same circle circumscribes both the pentagon of the dodecahedron and the triangle of the icosahedron in-

[1] For mathematics in the third century B.C., see chapters III, V, VI.

[2] Book XV is an inferior and much later work. Its author was a pupil of Isidōros of Milētos, the architect of Hagia Sophia in Constantinople, c. 532.

Fig. 51. Division of the line AB in extreme and mean ratio by the point C: AB/AC = AC/CB

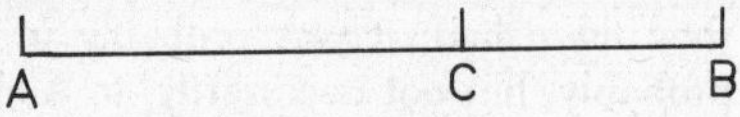

scribed in the same sphere."[3] This is proposition 2 of Book xiv. On that basis Hypsiclēs established the other theorems.

Let us assume the existence of a straight line *AB* and of a cube, a dodecahedron, and an icosahedron, all regular and inscribed in the same sphere. Divide the line *AB* in extreme and mean ratio by the point *C*, *AC* being the larger segment (Fig. 51). Then the side of the cube is to the side of the icosahedron as $(AB^2 + AC^2)^{\frac{1}{2}}$ is to $(AB^2 + BC^2)^{\frac{1}{2}}$, also as the area and volume of the dodecahedron are to the area and volume of the icosahedron. These are three different propositions which are equally remarkable and which one did not expect to find enmeshed. The fundamental cause of these beautiful propositions is the unexpected fact that the perpendiculars from the center of the sphere to the faces of both polyhedra are equal. This was a worthy addition to the *Elements*.

According to Diophantos (III–2), Hypsiclēs gave a general definition of polygonal numbers:[4] they are the sums of sequences of numbers in arithmetic progressions. If the common difference is 1, the sums are triangular numbers; if it is 2, the sums are square numbers; if it is 3, the sums are pentagonal numbers; if it is 4, the sums are hexagonal numbers; and so on. In each case the number of angles is equal to the common difference plus 2. The following examples will make this clearer:

triangular numbers	0	1	3	6	10	15	21	28
difference 1	1	2	3	4	5	6	7	
square numbers	0	1	4	9	16	25	36	49
difference 2	1	3	5	7	9	11	13	
pentagonal numbers	0	1	5	12	22	35	51	70
difference 3	1	4	7	10	13	16	19	
hexagonal numbers	0	1	6	15	28	45	66	91
difference 4	1	5	9	13	17	21	25	

Hypsiclēs was also an astronomer and his astronomical work will astonish us as much as his mathematics, though in a different way. He was an ingenious man, though not comparable with his illustrious predecessors.

A FEW OTHER GREEK MATHEMATICIANS

The date of Hypsiclēs cannot be determined with any accuracy. The same can be said of the five following men; we do not know when and where

[3] Volume 1, p. 505.

[4] The first conception of such numbers, ascribed to Pythagoras, was of geometric origin (Volume 1, p. 205). Diophantos' statement occurs in his treatise on polygonal numbers. See Thomas L. Heath, *Diophantus* (Cambridge, ed. 2, 1910), p. 252.

they flourished; it was probably in the second century or soon after and probably, but not necessarily, in Alexandria. They might have flourished in any one of many Greek cities because of the emulation obtaining between those cities and of frequent communications.

These five men, Zēnodōros, Perseus, Nicomēdēs, Dionysodōros and Dioclēs, are ghostlike personalities, but each of them did something definite which will be recorded presently.

Zēnodōros (II–1 B.C.) is famous because of his treatment of isoperimetric plane surfaces in a treatise entitled *Peri isometrōn schēmatōn.* He stated that, of all regular polygons having an equal perimeter, that one is the greatest in area which has the greatest number of angles (or sides); that a circle is greater in area than any polygon of equal perimeter; that regular polygons are greater in area than irregular ones of equal perimeter and equal numbers of sides. He also showed that, of all solid figures the surfaces of which are equal, the sphere is the greatest in volume.

The original text of Zēnodōros is unfortunately lost but the substance of it was incorporated by Pappos (III–2) in his *Synagōgē* (Book v) and there are fragments of it in a commentary by Theōn of Alexandria (IV–2). This was a brilliant anticipation of a new branch of mathematics, too precocious to be exploited until much later.

There are no glimpses of such problems in Arabic literature, except in the *Rasā'il ikhwān al-ṣafā* (X–2) and very few in Latin: Leonardo of Pisa (XIII–1), Thomas Bradwardine (XIV–1), Albert of Saxony (XIV–2), Regiomontanus (d. 1476).[5]

To appreciate the originality of Zēnodōros' thinking one has but to reflect that even now there are a number of people, yea educated people, who misunderstand the relation between perimeter and area.

A full treatment of such problems became possible only with the calculus of variations (Johann Bernoulli, 1696; Euler, 1744; Lagrange, 1760); the physical realization of minimal surfaces was accomplished with soap bubbles by Joseph Plateau of Brussels a century ago (1843 to 1873).[6] Zēnodōros could not imagine such things but his first adumbration of them is admirable.[7]

Perseus. If Heath's claim that Perseus may be anterior to Apollōnios is correct, our ignorance concerning his time is even greater than in the other cases. What we know of him is derived from Proclos (V–2), a very late witness. In his commentary on Euclid I, Proclos wrote, "Apollōnios derived the properties of each of the three conics, Nicomēdēs, of the conchoid, Hippias, of the quadratrix, and Perseus, of the spirics."

[5] Sarton, "The tradition of Zēnodōros," *Isis 28*, 461–462 (1938).

[6] Sarton, "The years 'forty three'," *Isis 34*, 195 (1942–43).

[7] It was only in 1884 that the isoperimetric properties of circle and sphere were rigorously proved by Hermann Amandus Schwarz, using Weierstrass' method. B. L. Van der Waerden, *Science awakening*, trans. Arnold Dresden (Gro-

The spirics are plane sections of the surfaces generated by the revolution of a circle around an axis in its plane but not passing through its center.[8] There were three kinds of spirics. The simplest kind was obtained when the axis was outside the circle, in which case the surface was a real *speira* (tore or anchor ring). Archytas of Tarentum had already made use of such a simple spiric to find two mean proportionals between two given lines.[9] If the axis was tangent to the circle, one obtained a tore without a hole in the middle. The third kind was obtained when the axis intersected the circle; the surface was then falling back into itself (*epiphaneia empeplēgmenē*).

There is thus a great variety of spirics, the *hippopedē* and Bernoulli's lemniscate being special cases. It is not probable that Perseus was able to exhaust that variety, but it is astounding that he was able to investigate some of those curves without any kind of algebraic apparatus.

Nicomēdēs. Another member of the same group was Nicomēdēs, referred to by Proclos (see above); he lived probably at the turn of the third century, in Pergamon (?). Van der Waerden tells us that he flourished between Eratosthenēs and Apollōnios, but that is not easy to conceive.[10] Nicomēdēs had invented the conchoid to solve the same problem as Archytas, to wit, to find two mean proportionals between two given lines, and he applied it also to the solution of another classical problem, the trisection of a given angle. According to Pappos (III–2), Nicomēdēs had devised an instrument to draw the curve, which he called *cochliōdēs* (in the shape of a shell, slug or snail); the name *conchoeidēs* (in the shape of a shell) was given later by Proclos (V–2).[11]

It is said that Nicomēdēs applied another curve, invented by Hippias of Elis,[12] to the squaring of the circle, but that had already been done by Dinostratos (IV–2 B.C.). It is for that application that the curve was called the *quadratrix* (*tetragōnizusa*).

Dionysodōros of Amisos,[13] who flourished probably in the second century, solved the Archimedean problem concerning the division of a sphere by

ningen: P. Noordhoff, 1954) [*Isis 46*, 368 (1955)], p. 269.

[8] For the relation of spiric lines to other special curves, see R. C. Archibald, "Curves," in *Encyclopaedia Britannica* (ed. 14, 1929), vol. 6, pp. 887-899, nos. 11, 58.

[9] In order to solve the problem of the duplication of the cube Volume 1, p. 440.

[10] The approximate dates of the first are 273–194, and of the second, 262–190. They were almost exact contemporaries.

[11] The corresponding English names are cochloid and conchoid. Later mathematicians called Nicomēdēs' curve "conchoid of a line" to distinguish it from the "*limaçon de Pascal*," "conchoid of a circle." See Archibald, nos. 13, 14, 57. "Conchoid" and cochloid" are derived from two Greek words, *conchos* and *cochlos*, meaning shell. "Cochlos" also means snail or slug (*limaçon*).

[12] Hippias (V B.C.) was a much older mathematician, immortalized by two Platonic dialogues. See Volume 1, p. 281.

[13] Amisos in Pontos, the modern Samsun, on the south shore of the Black Sea.

a plane in a given ratio, by means of the intersection of a parabola and a rectangular hyperbola. He wrote a book on tores (*Peri tēs speiras*). According to Hērōn of Alexandria, this contained an adumbration of Guldin's theorem (really discovered or rediscovered by Pappos).[14]

Dioclēs, who flourished at about the same time, solved the same Archimedean problem, invented the curve called the cissoid,[15] and applied it to the duplication of the cube. He wrote a book on burning mirrors (*Peri pyreiōn*) and may have invented the parabolic burning mirror.

These six mathematicians may be divided into three groups: first, Zēnodōros, who is entirely original; second, Hypsiclēs, who continued Euclid; and third, four followers of Archimēdēs, who investigated the properties and applications of special curves, Perseus, Nicomēdēs, Dionysodōros, and Dioclēs. Note that these men were still obsessed by the three classical problems of the fifth century; these problems continued to obsess mathematical thought until the sixteenth century.[16]

HIPPARCHOS OF NICAIA

Hipparchos (II–2 B.C.) was one of the greatest astronomers of all times, and we shall deal with him at greater length in the following chapter, but we must speak of him here and now because he was also an outstanding mathematician. This is sometimes forgotten, because his mathematical efforts were auxiliary to his astronomical ones, means to an end, yet they were fundamental. He was not simply a mathematician, but the founder of a new branch of mathematics, trigonometry, without which astronomical computations were almost impossible. The subordination of trigonometry to astronomy was so deep that for a long time the former was considered a part of the latter. Even today, spherical trigonometry is an integral part of a course on astronomy (or navigation) and is hardly studied otherwise.

Ancient astronomers were not concerned about the distances of stars, which they assumed to be located on a single sphere. As long as they believed that all the stars turn around the Earth at the same speed, the coexistence of those stars on a single sphere was almost a logical necessity. When they studied the relations of, say, three stars, they had to consider their angular distances (from the point of view of the observer), or, to put it otherwise, the segments of great circles joining these stars two by two. The segments

[14] Sarton, *Ancient science and modern civilization* (Lincoln: University of Nebraska Press, 1954), p. 80.

[15] *Cissoeidēs*, resembling ivy (*cissos*). See Archibald, "Curves," nos. 3, 49, 51, 53, 55.

[16] The three problems were the quadrature of the circle, the trisection of the angle, and the duplication of the cube (Volume 1, 278). For a late example, see Mollā Luṭfī'l-maqtūl, *La duplication de l'autel* (Paris: Bocard, 1940) [*Isis* 34, 47 (1942–43)]. Luṭfī'l-maqtūl was librarian to Muḥammad the Conqueror (ruled 1451–1481).

joining three stars [17] constitute a spherical triangle, and all of the problems of mathematical astronomy were problems of spherical trigonometry.

Trigonometry was studied because of its applications, but it is as much a branch of pure mathematics as geometry. The student of trigonometry learned to solve spherical triangles just as the student of geometry solved plane triangles. The sides of spherical triangles, being arcs, were measured by angles; hence, the spherical triangle was a combination of six angles; A, B, C, the three vertices, and a, b, c, the three sides. The solution of spherical triangles was comparable to that of plane triangles, except that it was more complex: given some of the six elements of the triangle, determine the others.

Now it was realized by Hipparchos that those problems would be simplified if the consideration of arcs were replaced by the consideration of the chords [18] subtending those arcs. But in order to make this possible two things were necessary: (1) to establish a number of propositions concerning the relations of various arcs or chords of a given sphere; (2) to compile a table of chords for the sake of computation.

These two achievements were accomplished by Hipparchos, but our knowledge of them is indirect and incomplete.

Who was Hipparchos? He was born at Nicaia,[19] and the time of his life is determined by Ptolemy's references to astronomical observations extending from 161 to 127. The earlier ones, made in Alexandria from 161 to 146, may not be his own, but there is no doubt with regard to the period 146 to 127; he was in Rhodos at least during the years 128–127. We are safe in saying that he flourished in the third quarter of the second century. One does not know when and where he died.

What is much worse, all his writings are lost, except for a youthful commentary on the *Phainomena* of Eudoxos of Cnidos (Plato's younger contemporary) and on the astronomical poem that Aratos of Soloi (fl. 275) had derived from it. It is a description of the constellations, apparently made with the help of a celestial globe. This is definitely a minor work, the existence of which is a very poor consolation for the loss of all the others.

Our knowledge of Hipparchos is derived from Strabōn (I–2 B.C.) and later writers, above all the great astronomer, Ptolemy (fl. 127–151). The latter's *Almagest*, the astronomical bible until the times of Copernicus and Kepler, refers very frequently to him and sometimes quotes him verbatim.

[17] One or more of those stars might be the projection of a planet upon the stellar sphere, or one of the segments might be the spherical distance to a great circle (meridian, equator, or ecliptic).

[18] "Chord" is the Greek word *chordē*, meaning a string of gut, or the string (chord) of a lyre.

[19] East of the Propontis (Sea of Marmara). Nicaia, or Nicaea, was one of the main cities of Bithynia; it is best known because of the council held there in A.D. 325, the first general council of the Christian church. The seventh council was gathered there in 787. The modern Turkish name is Iznik; the English sometimes use the form Nice.

Ptolemy praised his predecessor very highly; he called him a man loving work and loving the truth, or the greatest truth lover; he was anxious to give due credit to him, yet it is not always possible to draw the line between the two men and to give each his due exactly. It would seem that Hipparchos wrote many astronomical monographs but no general treatise. The encyclopedic nature of the *Almagest*, its superior value, and its formal perfection were probably the main causes of the loss of Hipparchos' original writings. The early copyists must have felt that the *Almagest* rendered previous writings obsolete and superfluous. The essential of Ptolemy's task (mathematical as well as astronomical) was already done by Hipparchos, yet Ptolemy completed it, worked out the necessary details, compiled new tables, and so on. The case of the *Almagest* is closely parallel to that of Euclid's *Elements*; both authors superseded and almost obliterated their predecessors, because they combined an astounding power of synthesis and exposition with original genius.

Though Ptolemy mentions the dates of Hipparchos' observations, he speaks of him as one would speak of an older contemporary. Nothing can give one a more striking impression of the slowness of scientific progress in antiquity than the spectacle of those two "collaborators" separated by almost three centuries.[20]

On that account one might be tempted not to say much of Hipparchos, except retrospectively when discussing Ptolemy, but that would not be fair to the former and would distort the historical picture. Our purpose now is to show the mathematical height that a Greek astronomer had reached more than 125 years before Christ. We shall therefore explain Hipparchos' work briefly in this chapter and the following, because a complete statement of his work can be made only by considering simultaneously that of Ptolemy (second century).

The reader must bear in mind that whatever is ascribed to Hipparchos, in this and the next chapter, is necessarily tentative, because we have no text to support our assertions. We are on safer ground in the few cases when Ptolemy ascribes an invention to his predecessor, but even then we cannot evaluate the original invention and Ptolemy's addition to it, or modification of it.

According to Theōn of Alexandria, Hipparchos wrote a treatise in twelve books on straight lines (chords) in a circle. So large a treatise must have contained a general theory of trigonometry plus tables, perhaps as much as is given us in the trigonometric chapters of the *Almagest*. Using the trigonometric method which he had invented, Hipparchos was the first to determine exactly the times of rising and setting of the zodiacal signs.

This implies that he had compiled for himself a table of chords. How did he do that? First, he must have given himself means of measuring the

[20] The time interval between Hipparchos and Ptolemy (c. 285 years) is much greater than that between Newton and Einstein (c. 220 years).

circle and the chords. Hypsiclēs had had the idea of dividing the ecliptic into 360°; Hipparchos was the first to generalize that idea and to divide every circle into 360° (as we still do now). He divided the diameter into 120 units or "parts." Smaller quantities than the degree and the part were probably expressed by means of sexagesimal fractions. The problem was then to express the length of the chord of any arc of a circle in terms of those parts. The Euclidian knowledge of regular polygons made this easy in some special cases. Thus the chord of 60°, which is the side of the hexagon inscribed in the circle, is equal to the radius of the circle, or 60^p (60 parts); the chord of 90° (side of a square) equals $\sqrt{(2r^2)}$[21] or 84^p 51′ 10″ (that is, 84 + 51/60 + 10/3600 parts); the chord of 120° (side of a triangle) equals 103^p 55′ 23″; and so on. If one knew the chord of x°, one could quickly deduce that of $(180 - x)$°, because the sum of the squares of supplementary chords equals the square of the diameter.

By such means Hipparchos was able to measure a number of fundamental chords. How did he measure the many other chords in between them? If he did it at all, he must have been acquainted with the so-called "Ptolemy's theorem"[22] or with an equivalent proposition. This enabled him to find the chords of $(a \pm b)$ in terms of the chords of a and b, and thus to calculate as many additional chords as he pleased. If his table was as developed as Ptolemy's eventually was, it gave the length of chords for every half degree from 0° to 180°, each chord being expressed in parts of the radius $(r/60)$, minutes, and seconds.[23]

The reader familiar with modern trigonometry may be puzzled by the chords, for he uses sines (and other ratios). The sines were invented much later (say in the fifth century) by Hindu astronomers, adopted by al-Khwārizmī (IX-1) and other Arabic astronomers, and transferred to the Latin West in the fourteenth century. It is not difficult to pass from chords to sines, though it took considerable genius to think of it.

Consider the angle α subtended by the chord AB and draw the perpendicular OC (Fig. 52). If the radius is 1, AC is the sine of the angle $\frac{1}{2}\alpha$ and one sees immediately that chord $\alpha = 2 \sin \frac{1}{2}\alpha$.

Why was the change made? Simply because the handling of sines (and of other trigonometric ratios) is much simpler than that of chords. Our trigonometric formulas (both plane and spherical) are relatively simple and elegant because of their symmetry. Similar formulas based on chords

[21] The radius r equals 60 parts; then $\sqrt{(2r^2)} = \sqrt{7200} = 84^p 51' 10''$.

[22] In any quadrilateral with sides a, b, c, d inscribed in a circle, the product of the two diagonals st equals the sum of the products of opposite sides: $st = ac + bd$.

[23] In the Latin translation, the sexagesimal fractions of the first order were called *partes minutae primae*, those of the second order, *partes minutae secundae*. Our words "minutes" and "seconds" are stupidly derived from the first adjective in the first case and from the second adjective in the other.

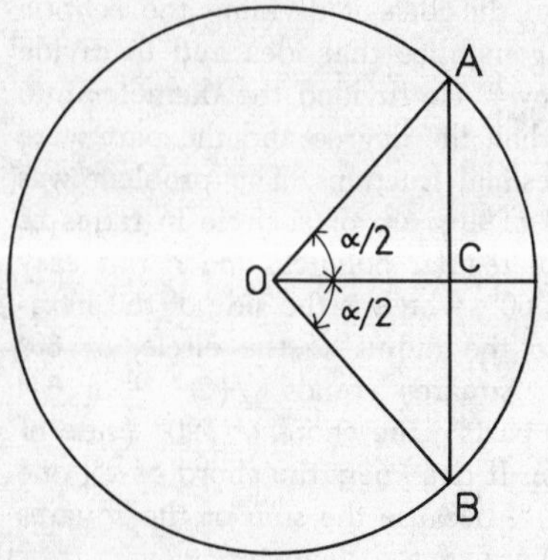

Fig. 52. Diagram of the relation between chords and sines: AC = sin ½α

would be more complicated and less elegant. Nobody has had the courage, however, to deduce them, because the sines drove the chords away forever.

Ptolemy's table of chords (and presumably Hipparchos' table) for every half degree from 0° to 180° could be easily changed into an equivalent table of sines for every quarter degree from 0° to 90°.

Hipparchos' mathematical intuition appears also in his awareness that the method of epicycles, invented by Apollōnios, and the method of eccentrics, probably invented by the same man, were kinematically equivalent, that is, one could choose the one or the other. This proved that those methods were simply expedients, which have not necessarily a natural basis.

We are not surprised to hear that Hipparchos' curiosity extended to mathematical problems independent of astronomy. According to Plutarch he was interested in combinatorial analysis (permutations and combinations). According to Arabic mathematicians he studied algebraic problems. We shall come back to this in the section of our next chapter on Babylonian influences.

These are side matters, however. If one should ask to give the proof of his mathematical genius, it would suffice to answer: He invented trigonometry. That is quite enough, is it not?

THEODOSIOS OF BITHYNIA

Hipparchos' new mathematical tool was not immediately accepted because of the old tradition of studying spherical problems in a geometric way and because its value was not apparent except when one tried to solve definite astronomical problems and to find numerical solutions of them. It is thus not surprising that the old tradition was continued by the mathematician Theodosios (I–1 B.C.), who flourished probably after Hipparchos and before Strabōn (that is, at the end of the second century or the beginning of the first).

He is generally called Theodosios of Bithynia after his native country, south of the Black Sea; he is also called in the manuscripts Theodosios of Tripolis,[24] because he resided in that city(?). He may be identical with a namesake who invented a sundial for all latitudes (*pro pan clima*; Vitruvius, *De architectura*, IX, 8, 1). He is the author of three treatises which have been preserved in the "Little Astronomy"; [25] three other books of his are lost, one of them being a commentary on Archimēdēs' method (*ephodion*). One would give much to have that one! We shall mention Theodosios again in Chapter XIX.

The most important of his extant works is the *Spherics* (*Sphairica*); it is the earliest of its kind to reach us but it was partly derived from a lost treatise anterior to Autolycos of Pitanē (IV–2 B.C.).[26] The *Spherics* is divided into three books. Book I and part of Book II (prop. 1–10) explain the properties of great and small circles on the sphere, tangent planes, distances of various circular sections, circles that touch or cut one another, parallel circles. The rest of the work is devoted to a variety of astronomical applications.

The *Spherics* is written in Euclidean style and is a necessary continuation of the *Elements*. There are no propositions in Euclid concerning the sphere except those proving that spheres are to one another as the cubes of their diameters (XII, 16–18), and a few references to spheres apropos of regular polyhedra. It is strange indeed that Euclid paid so much attention to the regular solids and so little to the sphere, which was their womb and their limit.

The purely geometric part of the *Spherics* is very much like Book III of the *Elements*. One finds in it such propositions as that any plane section of a sphere is a circle (prop. I), how to find the center of a given sphere (prop. II), and so on.

Theodosios tried to do geometrically what Hipparchos did by means of trigonometry; the geometric method was good and it threw much light on the subject, but as it implied no quantitative measurements it was not practical.

In spite of Hipparchos and Ptolemy, the *Spherics* of Theodosios and two other treatises of his were preserved because of their inclusion in the "Little

[24] Unfortunately, we do not know which Tripolis is meant. The western Tripolis (Ṭarābulus al-gharb) may be rejected, but there are at least three cities called Tripolis in the East. The name, meaning "three cities," was given to various agglomerations of three urban centers. The best known Oriental Tripolis was on the Phoenician coast (modern Ṭarābulus al-Sham in Lebanon), but it does not follow that Theodosios resided there. It is not even certain that he resided in any city of that name. He may have been called Theodosios of Tripolis by confusion with another man. It is thus better to call him Theodosios of Bithynia.

[25] *Ho micros astronomoumenos* (*topos*), a collection of astronomical writings, a part of which has come down to us through Arabic translation, *Kitāb al-mutawassiṭāt*. See *Introduction*, vol. 1, pp. 142, 211, 759; vol. 2, pp. 624, 1001; vol. 3, p. 633.

[26] Volume 1, p. 511.

Astronomy," and, moreover, the *Spherics* was part and parcel of the Euclidean tradition, which appealed very strongly to the Arabic mathematicians.

The *Spherics* was translated twice into Arabic by Thābit ibn Qurra (IX–2) and by Qusṭā ibn Lūqā (IX–2); the Arabic text was twice translated into Latin, first by Plato of Tivoli (XII–1) and second by Gerard of Cremona (XII–2); the Arabic version by Qusṭā was put into Hebrew by Zerahiah Ḥen (XIII–2). The Arabic versions were reëdited by Naṣīr al-dīn al-Ṭūsī (XIII–2) and by Muḥyī al-dīn al-Maghribī (XIII–2).

To return to the Western tradition of the *Spherics*, Gerard's Latin translation was overlooked but the earlier one by Plato of Tivoli was first printed with various other treatises by Octavianus Scotus (Venice, January 1518), and again by the Junta (Venice, June 1518); the Latin text was edited a second time by Johann Voegelein of Heilbronn (Vienna, Joannes Singrenius, 1529) and a third time by Francesco Maurolico (Messina, 1558).

The first edition of the Greek text by Jean Pena appeared in the same year (Paris: André Wechel, 1558) (Fig. 53), together with a new Latin version. The same Greek text was reprinted with relatively small modifications by Joseph Hunt (Oxford, 1707) and by the Danish Hellenist, Ernest Nizze (Berlin, Reimer, 1852). A critical edition of the Greek text was finally prepared by Johan Ludvig Heiberg (Berlin, 1927) [*Isis 11* 158, 409 (1928)].

The popularity of Theodosios' *Spherics* is proved by the existence of various adaptations by Conrad Dasypodius (Strassburg, 1572), Christophe Clavius (Rome, 1586), Denis Henrion (Paris, 1615), Pierre Hérigone (Paris, 1634),[27] Jean-Baptiste Du Hamel (Paris, 1643), Marin Mersenne (Paris, 1644), Camillo Guarino Guarini (Turin, 1671), Claude François Milliet de Chales (Lyons, 1674), Isaac Barrow (London, 1675).

The first complete translation into French we owe to Paul Ver Eecke, *Les Sphériques de Théodose de Tripoli* (175 pp.; Bruges, Desclée De Brouwer, 1927). This translation was prepared with great care (as every translation by Ver Eecke is) but, unfortunately, upon Nizze's imperfect text, as Heiberg's was not yet available to him.

The tradition of this text has been described in some detail because of its importance in the history of Greek mathematics. It deserves to be remembered together with Euclid's *Elements*, which it completed.

MATHEMATICAL PHILOSOPHERS

Most of the men already discussed in this chapter were primarily mathematicians or else astronomers obliged to solve mathematical problems in order to complete their own task. The representatives of Greek culture, however, the literary elite, were deeply concerned with philosophy and philol-

[27] Henrion and Hérigone wrote in French, all the others in Latin. Note the international distribution of those men. Dasypodius and Clavius were German; Guarini, Italian; Barrow (Newton's teacher), English; Henrion, Hérigone, DuHamel, Mersenne, and de Chales, French.

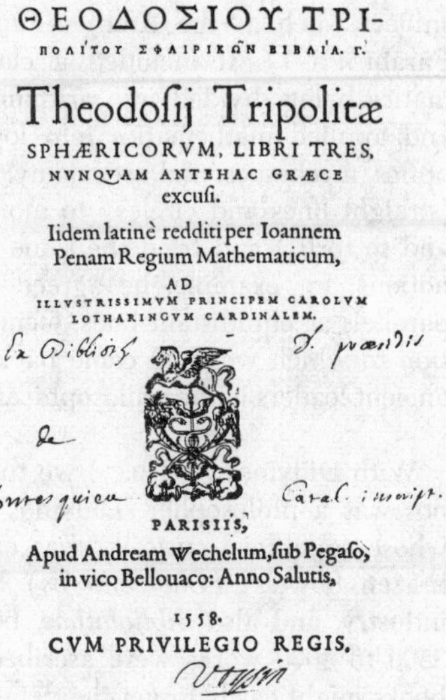
ΘΕΟΔΟΣΙΟΥ ΤΡΙ-
ΠΟΛΙΤΟΥ ΣΦΑΙΡΙΚΩΝ ΒΙΒΛΙΑ Γ.

Theodoſij Tripolitæ
SPHÆRICORVM, LIBRI TRES,
NVNQVAM ANTEHAC GRÆCE
excuſi.

Iidem latinè redditi per Ioannem
Penam Regium Mathematicum,

AD
ILLVSTRISSIMVM PRINCIPEM CAROLVM
LOTHARINGVM CARDINALEM.

de Montesquieu

PARISIIS,
Apud Andream Wechelum, ſub Pegaſo,
in vico Bellouaco: Anno Salutis,

1558.
CVM PRIVILEGIO REGIS.

Fig. 53. Princeps of the treatise on spherical geometry by Theodosios of Bithynia (I-1 B.C.). It was edited and translated by Jean Pena, who taught mathematics at the Collège de France from 1555 to his death in 1558 (Paris: Andreas Wechelus, 1558). [Courtesy of Harvard College Library.]

ogy, both of which included science (that was the chief characteristic of Hellenic and even Hellenistic humanism). Consider the following men: Zēnōn of Sidōn, Poseidōnios, Geminos and Didymos – the first an Epicurean, the second and third Stoics, the last-named a philologist and man of letters.

Zēnōn of Sidōn was probably the head of the Garden before Phaidros; Cicero came to listen to him in Athens in 79–78 [28] and Cicero's contemporary, Philodēmos of Herculaneum, borrowed from him. He discussed the preliminaries of Euclid's *Elements*, claiming that they implied unproved assumptions. The Epicureans (and the Skeptics) were impatient with mathematical abstractions; their criticism may have been vexing, but it was not useless. It was bound to draw the fire of the Stoics. Poseidōnios wrote a treatise to refute Zēnōn's arguments, but he was more concerned with mathematical astronomy and with geodesy than with mathematics pure and simple.

His pupil, Geminos of Rhodos (I-1 B.C.), who flourished c. 70 B.C., wrote an introduction to mathematics of which only fragments remain. It was probably entitled *On the arrangement* (or *theory*) *of mathematics; Peri tēs tōn mathēmatōn taxeōs* (or *theōrias*). It was the main source of Proclos' [29] commentary on the first book of Euclid, and of later writers on the

[28] Cicero, *De natura deorum*, 1, 59, stating that "our friend Philōn used to call him the leader of the Epicurean choir" (*coryphaeus Epicureorum*). Philōn of Larissa (c. 160–80) was the founder of the Fourth Academy.

[29] Proclos the Successor (V-2), one of the last Greek mathematicians, who

subject, such as the Arab al-Faḍl ibn Ḥātim al-Nairizī (IX–2) and al-Fārābī (X–1). It included a classification of mathematics, pure mathematics being divided into arithmetic (theory of numbers) and geometry, and applied mathematics into logistics (reckoning), geodesy, harmonics, optics, mechanics, and astronomy. He also classified lines, from simple ones (straight lines and circles) to more complex ones (conics, cissoids, spirics, and so forth) and tried the same for surfaces. He insisted on fundamental notions; for example, he agreed with his master Poseidōnios in defining parallels as equidistant lines. Geminos also wrote an astronomical introduction to which we shall come back in the next chapter. He was one of the ancient leaders in the philosophical development of mathematical knowledge.

With Didymos (I–1 B.C.) we turn in the opposite direction. While Geminos was a philosopher, Didymos was an incontinent author, a polymath, whose superficial curiosity was endless. He was nicknamed "The man of brazen bowels" (*Chalcenteros*) because of his prodigious and heartless industry, and also *Bibliolathas*, because he forgot his own writings. Some 3500 to 4000 works were ascribed to him,[30] but an author of but a few books might easily forget the details of his earlier ones. Most of these works dealt with literary matters, lexicography, grammar, archaeology, mythology. In one of them he criticized Cicero's *De re publica*, and this helps to fix his date. He flourished in the first century (c. 80–10) in Alexandria and belonged to the philological school that had been founded a century before by Aristarchos of Samothracē (II–1 B.C.). He wrote a little treatise on the measurement of wood.[31] This is more interesting from the metrological than the mathematical point of view. He used throughout Egyptian fractions (fractions with numerator one, and also 2/3),[32] for example, 1 1/2 1/5 1/10, meaning 1 4/5, or 5 1/2 1/5 1/10 1/50 1/125 1/250, meaning 5 104/125.

Didymos' presence at the end of this chapter is perhaps out of place, and it is a kind of anticlimax, because he was neither a philosopher nor a mathematician, but he could not be left out, and there was no other place to put him in. Other authors might occasionally quote measurements and make computations. He represents them all and helps to prove the persistence of Egyptian arithmetic.

His little treatise was first edited by Angelo Mai, *Iliadis fragmenta* (Milan, 1819), then by the great master of ancient metrology, Friedrich Hultsch, *Heronis geometricorum . . . reliquiae* (Berlin, 1864), finally by Johan Ludvig Heiberg, *Mathematici graeci minores* (Copenhagen, 1927) [*Isis 11*, 217 (1928)].

French translation by Paul Ver Eecke,

---

flourished almost eight centuries after Euclid.

[30] This is not as terrible as it seems, because a "work" might be as short as a newspaper or magazine story. Many newspaper men have written more than 4000 "pieces."

[31] Cubic measurement is meant, as when we speak of a cord of wood. His explanations are poor and confusing.

[32] For those Egyptian fractions, see Volume 1, p. 37. The Egyptians used also,

"Le traité du métrage des divers bois de Didyme d'Alexandrie," *Annales de la Société scientifique de Bruxelles* *56*(A), 6–16 (Louvain, 1936).

The references to sexagesimal fractions and to Egyptian ones suggest the query: To what extent were the dozen mathematicians dealt with influenced by Oriental methods? The query is very important, but we reserve the discussion of it for the next chapter, because those influences concerned astronomy as well as mathematics.

THE GREEK MATHEMATICAL PAPYRUS OF VIENNA

It has often been remarked (by myself and others) that the Greeks were interested in the properties of numbers (what we call the theory of numbers) but hardly in computation. That remark would be very misleading if one did not hasten to qualify it. The philosophers and mathematicians were primarily interested in the theory of numbers, to which the Pythagoreans and the Platonists had given a cosmologic significance, but how many were they? The average Greek did not care so much about that theory but, being a practical man and, in many cases, a money-lover, he could not help being deeply interested in computations of every kind. The business of life, however simple, obliged everybody to cast accounts, and merchants, bankers, craftsmen had to cast a great many of them. It was necessary to take measurements, to figure prices, to arrange for part payments. Much of that could be done and was done with an abacus or with stones (hence our word "calculate"), but there developed also unavoidably an art of calculation (*hē logisticē technē*).

What is true is that our word "arithmetic" (derived from *arithmos*, number) was reserved for higher purposes. An account was called *logismos*; an accountant, *logistēs* (that name was also given in Athens to the auditors of public accounts); the methods of computation would be called *logistica* (hence the English word "logistics," now obsolete in this sense). In the Hellenistic world and almost certainly also in the Greek world, there was an art of computation comparable to ours but having no academic status; it belonged to the field of domestic economy and was a part of craftsmanship. The ability to cast accounts and to cast them fast and correctly[33] was an essential part of the endowment of every craftsman and merchant, and to a lesser extent of every intelligent person.

There is in the Vienna library a Greek papyrus (No. 19,996) containing a stereometric treatise which experts ascribe to the second half of the first century B.C. That document is of great interest from three points of view. First, it gives some idea of the state of geometric knowledge in Egypt

---

though very rarely, the fraction ¾. There is no example of it in Didymos.

[33] *Logismos* means calculation, *paralogismos*, a false calculation; *paralogizomai*, to make an error, to make it deliberately, or, in the passive, to be the victim of an error, to be cheated. There are many other words derived from *logismos*, which are good witnesses of computing habits.

toward the end of the Hellenistic age; second, the computations included in it are good specimens of contemporary logistics; and third, it shows that the Greeks who were then living in Egypt were submitted to Babylonian influences as well as to Egyptian ones.

It is a collection of 37 problems of stereometry which are solved without geometric demonstrations but are solved correctly. They are the kind of problems that surveyors or architects would have to face and the proper formulas are given them for their solution. The first item is simply the definition of a unit of volume, the cubic foot. Then follow the 37 problems, leading to the climax, the determination of the volume of a pyramidal frustum. This was not a novelty, of course; the same problem was already solved in the Golenishchev papyrus of Moscow, dating from the Thirteenth Dynasty (say the eighteenth century B.C.).[34] Strangely enough, geometry of a rather advanced kind is mixed up with the old Babylonian assumption that is equivalent to putting $\pi = 3$.

The calculations are of the old Egyptian type, with a notable difference. The old Egyptians used only unit fractions (fractions the numerator of which is 1), plus the fractions 2/3 and (very rarely) 3/4. In this papyrus the fractions are generally of the unit type, for example, the scribe writes 52 1/2 1/8 1/16 1/32 1/64 for 52 47/64 (note the geometric progression of the denominators as in the old Egyptian texts), but there are a few examples of more general fractions, such as 2/5, 4/5, 7/15, 3/20. The Greek texts of that time also preferred unit fractions (plus 2/3, 3/4) in the Egyptian style, yet Archimēdēs had already noted that $3\ 1/7 > \pi > 3\ 10/71$.[35] The scribe of the Vienna papyrus was possibly a man of genius and his use of general fractions remained exceptional; not only Egyptians but also Greeks and Romans preferred unit fractions (except sexagesimals in astronomy) and those unit fractions continued to crop up in medieval texts.

For more details see Hans Gerstinger and Kurt Vogel, "Eine stereometrische Aufgabensammlung im Papyrus Graecus Vindobonensis 19996," *Mitteilungen aus der Papyrussammlung der Nationalbibliothek im Wien* (*Papyrus Herzog Rainer*) [Neue Serie] *1*, 11–76 (1932); Kurt Vogel, "Beiträge zur griechischen Logistik," *Sitzber. bayer. Akad. Wiss., Math. Abt.*, 357–472 (München, 1936) [*Isis* 28, 228 (1938)].

---

[34] Volume 1, pp. 36, 38.

[35] *Cyclu metrēsis* (measurement of a circle), proposition 3. Before reaching that relatively simple result Archimēdēs had used far more complicated fractions, such as

$$\frac{1351}{780} > \sqrt{3} > \frac{265}{153},$$

$$\frac{96 \times 153}{4673\frac{1}{2}} > \pi > \frac{96 \times 66}{2017\frac{1}{4}}.$$

XIX

ASTRONOMY IN THE LAST TWO CENTURIES HIPPARCHOS OF NICAIA[1]

SELEUCOS THE BABYLONIAN

The main hero of this chapter is Hipparchos, but before speaking of him, it is well to say a few words about Seleucos (II–1 B.C.), who lived about a century after Aristarchos and was the last defender of the latter's ideas until the time of Copernicus.

Unfortunately, we know almost nothing about him. He was born or flourished in Seleuceia[2] on the Tigris. We are not even sure of that, but, if he really flourished in that city, he may have obtained there some knowledge of Greek astronomy. Indeed, Seleuceia had been founded in or after 312 by Seleucos I Nicatōr and became the capital of his empire; it had replaced Babylon as the main center of trade between East and West. It had a mixed Greek, Babylonian, and Jewish population. Seleucos may have visited Alexandria, but that was not absolutely necessary; he might have heard of Aristarchos' work in Seleuceia itself, or it may well be that he hailed from the West.

No matter how the theory concerning the diurnal rotation of the Earth and its annual revolution around the Sun reached him, he recognized its value and was even more affirmative about it than Aristarchos. The latter had introduced that theory only as a hypothesis (*hypotithemenos monon*), but Seleucos declared it to be true (*apophainomenos*).[3] This story is very plausible; the astonishing thing is not that the Aristarchian theory was accepted, but rather that its acceptance was of such short duration; we shall explain that when we speak of Hipparchos presently.

[1] For astronomy in the third century B.C., see Chapter IV.

[2] Strabōn refers thrice to him, saying "Seleucos of Seleuceia is a Chaldean" (XVI, 1, 6), and calling him Seleucos the Babylonian (I, 1, 9), and Seleucos of the region of the Erythraian Sea (III, 5, 9).

[3] Plutarch, *Platonicae quaestiones*, VIII, 2.

Mediterranean tides are so small that they had escaped observation; larger tides could not remain unnoticed, however; some were noticed by Pytheas (IV–2 B.C.) in the Atlantic, others by Nearchos (IV–2 B.C.) in the Indian Ocean, and the influence of the Moon was not difficult to detect. Dicaiarchos of Messina (IV–2 B.C.) observed that the Sun had some power over them too. Poseidōnios (c. 135–50) was the first to complete that theory and to account for the tides by the joint action of Sun and Moon; this enabled him to explain those that were abnormally high or low (spring and neap tides). Now, if Seleucos flourished a century after Aristarchos, he was a generation ahead of Poseidōnios, and if he flourished in the Tigris valley, he may have been acquainted with the Persian Gulf, the Indian Ocean, and perhaps even the Red Sea. According to Strabōn (III, 5, 9), he had observed periodic inequalities in the tides of the Red Sea, which he connected with the stations of the Moon in the zodiac. He tried to account for them by the resistance opposed to the Moon by the diurnal rotation of the Earth's atmosphere. His conclusions were erroneous, but they showed the independence and originality of his mind.

HIPPARCHOS OF NICAIA

The reader has already made Hipparchos' acquaintance as a great mathematician, but we must now review his astronomical work, which was at least of equal importance. That work is known indirectly through the *Almagest*, published almost three centuries later, and, as we have already explained, it is next to impossible to know how much Ptolemy modified Hipparchos' ideas. It is generally assumed, however, that the essential work was done by Hipparchos, except for a general theory of planetary motions which he had no time to complete. Another fundamental question, "How much did Hipparchos receive from his predecessors?" will be answered by and by and discussed more fully later.

Instruments. To make astronomical observations, instruments are needed, and the value of the observations is very largely dependent on the goodness of the instruments that made them possible. Hipparchos certainly used a celestial sphere to study the constellations. This enabled him to make remarks concerning their shape and the alignment of stars, without computation. In his commentary on Aratos he mentioned many more stars than were to be included later in his catalogue; his knowledge of those stars was at first graphical (drawings on the sphere), not arithmetic. Though Ptolemy refers to Hipparchos as the inventor only in the case of an improved diopter [4] (*Almagest*, v, 14), we may assume that Hipparchos' tools were not essentially different from those of his successor. The parallactic instrument

[4] He used a ruler 4 yards long provided with sights. That device is so simple, however, that it would be strange if it had not been used before him, say by Eratosthenēs, or by earlier astronomers still.

(*Almagest*, v, 12) and the mural quadrant (I, 10) were possibly Ptolemaic improvements; on the other hand, it is highly probable that Hipparchos was already using a meridian circle (I, 10) and a universal instrument, *astrolabon organon* (v, 1). He made a great many observations that were remarkably accurate within the scope of his instrumental possibilities. He was the first to divide instrumental circles into 360 degrees, though Hypsiclēs, who flourished in Alexandria just before him, had divided the ecliptic in the same way.

Planetary theories. An admirable chapter of the *Almagest* (IX, 2) explains the difficulties that Hipparchos had to overcome in order to rationalize his observations. Much work had already been done before his time by Eudoxos of Cnidos (IV–1 B.C.) and by Apollōnios of Pergē (III–2 B.C.) to account for the changing magnitudes of the planets, for the irregular motions of Sun and Moon, and for the greater irregularities of the planets, especially their mysterious retrogressions. Apollōnios had originated the method of epicycles and perhaps also that of eccentrics,[5] and Hipparchos was first to use both methods. He was thus able to reduce solar and lunar trajectories to combinations of circular motions, but he could only begin the analysis of planetary trajectories which was to be completed three centuries later by Ptolemy. Here again, it is impossible to say just how much was done by each man.

The "Hipparchian system." Hipparchos was resolved "to save the phenomena" (*sōzein ta phainomena*), that is, to account as exactly as possible for the accumulated observations with a minimum of hypotheses systematically worked out. His scientific prudence was carried so far that he rejected the heliocentric theory that had been so boldly advocated by Aristarchos of Samos and had been reaffirmed by his older contemporary, Seleucos the Babylonian. Hipparchos is responsible for that repudiation and for the formulation of what is often called the "Ptolemaic system" as against the "Copernican system." We should not blame him for that but rather praise him, because the Aristarchian theory of c. 280, and even the Copernican one of eighteen centuries later, did not solve the main difficulties. These were due to Pythagorean prejudices according to which celestial movements had to be circular; these prejudices were set aside only by Kepler in 1609. It is strange to think that the so-called Copernican system was defended (roughly) before the Ptolemaic one, but so it was. The progress of science is not as simple as some people imagine; it involves "retrogressions" like those of the planets. A good idea may be premature and therefore ineffective; that was the case for the Aristarchian idea in 280 B.C. and, to a smaller extent, for the Copernican one in 1543.

[5] I must refer again to Otto Neugebauer's paper, "Apollonius' planetary theory," *Communications on pure and applied mathematics* 8, 641–648 (1955), too technical to be summarized here.

Precession of the equinoxes. This introductory section is inserted for the sake of readers whose memory needs to be refreshed. The equinoxes (of spring and autumn) are the points of intersection on the celestial sphere of two great circles, the equator and the ecliptic. The latter may be assumed to be fixed but the former is not; it glides slowly and hence the equinoxes move; their motion is retrograde (clockwise for an observer north of the ecliptic) and amounts to about 50″.2 a year. That is, the vernal equinox advances on the ecliptic about 50″.2 a year; it precedes the Sun by that amount (hence the name precession). For the same reason the angle of the intersecting circles (the obliquity of the ecliptic) diminishes by about half a second (0″.48) each year.

The equator moves or "slides" as it does because it is always (by definition) perpendicular to the axis of the Earth (Fig. 54); the direction of that axis *OE* is not fixed, it describes a cone around the perpendicular *OP* to the ecliptic; the angle α of the cone (that is, half the angle at its summit) is equal to the obliquity of the ecliptic. Inasmuch as a given generatrix (say *OE*) of the cone moves clockwise 50″.2 a year, it will return to its original position in $360°/50''.2 = 26{,}000$ years.

This secular increase could not be realized at first; it was thought on the contrary that precession would not continue indefinitely in the same direction but would be reversed after a time, then the equinox would return, and so on, oscillating around a normal position. That was the idea of trepidation, which was entertained for an incredibly long time.[6] Arabic astronomers of the ninth century and later rejected it, but Copernicus (1543) did not. The final rejection was due to Tycho Brahe in 1576,[7] yet Brahe, and Kepler after him, had doubts concerning the regularity and continuity of the precession. Their doubts were natural enough, because by 1600 precession had been observed for less than twenty centuries, while the observation of a complete cycle would require 260 centuries, thirteen times as long!

Their doubts would vanish only when the phenomenon was completely explained. This became possible after the discovery of universal gravitation. Newton explained the precession of the equinoxes in his *Principia* (1687). The axis of the Earth rotates and the Earth behaves like a top because of the attraction of Sun and Moon upon its equatorial belt.[8] The theory was reëxplained by Euler in 1736 and generalized by him in 1765. Precession was first discovered in the case of the Earth, but it is a mechanical

[6] The history of the theory of trepidation was told in Volume 1, pp. 445–446.

[7] J. L. E. Dreyer, *Tycho Brahe* (Edinburgh, 1890), pp. 262, 354–355. According to Dreyer, the first to realize the steady continuity of the precession was Girolamo Fracastoro (1538) and the same was reasserted by Padre Egnazio Danti (1578). Such assertions were arbitrary, however.

[8] The Earth is flattened at the poles and heavier at the equator. The equatorial radius is 22 km longer than the polar one. The "flattening of the Earth," that is, the ratio of that difference to the equatorial radius, equals 1/292.

phenomenon of frequent occurrence. For example, some atomic nuclei are comparable to small magnets and when they are submitted to a strong magnetic field they begin to spin like tops and to "precess"; the stronger the field, the more rapid the precession.

Hipparchos discovered the precession and measured it, but he could not understand nor even guess its cause. Yet the discovery of the precession may be considered his outstanding achievement; it proves the relative precision of his stellar observations, his confidence in them, and the masterly freedom of his mind. He had written a treatise on the subject bearing those very words "precession of the equinoxes" or their equivalent (*metaptōsis*) in the title (*Almagest*, VII, 2). Comparing his observations of stars with those of Timocharis of Alexandria at the beginning of the third century, he found that their longitudes had decreased. For example, according to Timocharis (III–1 B.C.), the longitude of Spica (in Virgo) was 8° in 283 (or in 295); the longitude determined by himself in 129 was only 6°. Hence, 2° had been lost in 154 (or 166) years, corresponding to a loss (or precession) of 46″.8 (or 43″.4). The true value, as we have seen, is about 50″.2. We shall come back to this presently.

Year and month. Hipparchos' knowledge of the precession enabled him to distinguish the sidereal from the (shorter) tropical year, that is, the interval between returns of the Sun to a given star from that between its returns to the precessing equinox. He compared two observations of the summer solstice, one made by himself and another made by Aristarchos of Samos 145 years earlier, and found that the tropical year was not 365¼ days but 1/300 day less, which amounts to 365 days 5 hours 55 minutes 12 seconds (his value was less than 6½ minutes in excess of the real value).[9]

From Hipparchos' estimate of the Great Year (304 years with 112 intercalary months), the length of the mean lunar month [10] is 29.531 days or 29 days 12 hours 44 minutes 3⅓ seconds (instead of 2.9 seconds, the error being less than 1 second!). This precision was possible because of the availability of ancient Babylonian observations; in its turn it facilitated a more correct prediction of eclipses.

Distances and sizes of Sun and Moon. Hipparchos made a new investigation of these problems and revised the results obtained by Aristarchos. If the diameter of the Earth is *d*, then the diameters of Sun and Moon are respectively 12⅓*d* and ⅓*d*, and their distances from the Earth are 1245*d* and 33⅔*d*. These results are so far away from the truth that they hardly count. The only thing that counts for Hipparchos, as well as for Aristarchos, is

[9] The real value is 48 minutes 46 seconds. To put it in the form of regular fractions, 1 tropical year = 365.242 days, 1 sidereal year = 365.256 days. The former is a little shorter and the latter a little longer than the old Callippos approximation, 365.250 days.

[10] The synodic month, at the end of which Sun and Moon are in the same position relative to the Earth.

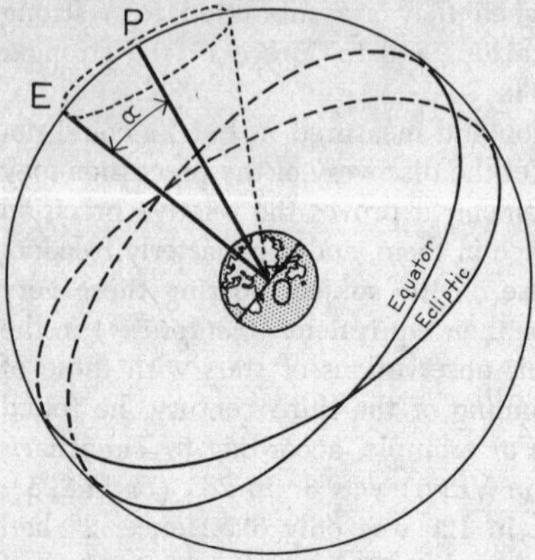

Fig. 54. Diagram to illustrate the precession of the equinoxes.

that both realized the possibility of making such measurements; their own method of making them was utterly insufficient.

Catalogue of stars. Hipparchos' earliest work (the only one extant) is a commentary on the *Phainomena* of Aratos, a Greek poem describing the constellations (see end of Chapter IV). The scientific value of that work is negligible, but its practical importance is considerable. It helped to popularize the names of stars and constellations which are preserved in our own nomenclature and the general outlines with which we are familiar. The stars might have been grouped in other ways, as was done sometimes by Egyptian and Babylonian astronomers, but the tradition from Eudoxos to Aratos to Hipparchos . . . fixed the *sphaera graecanica* (the celestial sphere of the Greeks) as opposed to the *sphaera barbarica.*

It is possible that Hipparchos' study of Aratos was his astronomical initiation, but he soon realized the need of doing something better. He began to make astronomical observations, in great numbers and with increasing precision. The determination of star longitudes and their comparison with earlier longitudes of the same stars led him to the discovery of the precession of the equinoxes; it may be also that he was thus induced to compile a catalogue of the main stars. His interest may have been increased by his discovery of a new star; according to Chinese reports, a nova appeared in Scorpio in 134.[11] His discovery of it is described by Pliny in a passage which is so naïve and yet so prophetic that we reproduce it *in extenso.*

[11] The Chinese nova (and Hipparchos') was more probably a comet. According to the Latin historian, Justinus (third century A.D.?), the greatness of Mithridatēs the Great was predicted by the appearance "of the star Cometes" at the time of his conception and of his enthronement (120). There are Chinese records of comets in 134 and 120; Mithridatēs was perhaps born in 133 (con-

Hipparchus . . . who can never be sufficiently praised, no one having done more to prove that man is related to the stars and that our souls are a part of heaven, detected a new star that came into existence during his lifetime; the movement of this star in its line of radiance led him to wonder whether this was a frequent occurrence, whether the stars that we think to be fixed are also in motion; and consequently he did a bold thing, that would be reprehensible even for God – he dared to schedule the stars for posterity, and tick off the heavenly bodies by name in a list, devising machinery by means of which to indicate their several positions and magnitudes, in order that from that time onward it might be possible easily to discern not only whether stars perish and are born, but whether some are in transit and in motion, and also whether they increase and decrease in magnitude – thus bequeathing the heavens as a legacy to all mankind, supposing anybody had been found to claim that inheritance! [12]

The naïveté of Pliny's statement will be underlined when we discuss astrology; its prophetic aspect concerns his understanding of the ultimate value of a catalogue of stars.

It might be argued that Hipparchos' discoveries of precession and of the nova of 134 were less the causes of his catalogue than fruits thereof, to which it will suffice to answer that his catalogue was not compiled in a single year and that many of its data were available to him long before its completion. Moreover, another catalogue had already been compiled by Eratosthenēs, and there may have been others still.[13]

Hipparchos' catalogue did not include more than 850 stars but for each of them he gave (and was apparently the first to do so) the ecliptic coördinates (latitude and longitude) and the magnitude. Unfortunately, it has not come to us in its integrity, and we know it only from the larger catalogue compiled by Ptolemy three centuries later and including 1028 stars.

Babylonian influences. It is clear that Hipparchos could not have discovered the precession (or be certain of it) nor have measured year and month

ceived in 134); these comets may have been observed in the West, and the one of 134 was probably Hipparchos' nova. J. K. Fotheringham, "The new star of Hipparchus and the dates of birth and accession of Mithridates," *Monthly Notices of the Royal Astronomical Society* (January 1919), pp. 162–167.

[12] Pliny, *Naturalis historia,* II, 24, 95, translation by Harris Rackham in the Loeb Classical Library edition (1938).

[13] My friend, Solomon Gandz, drew my attention (in a letter addressed on 5 July 1953 from Atlantic City to Jerusalem) to what he called "the oldest reference or allusion to a catalog of stars." Quoth Isaiah (40:26), "Lift up your eyes on high and behold who hath created these things, that bringeth out their host by number: he calleth them all by names . . ." This is the Second Isaiah (fl. 550–540), the "Milton of Hebrew poetry." This suggestion is very interesting, but confusing. Early people gave names to stars, at least to the most brilliant ones, because they were deeply concerned with them, and it was impossible to refer to them (as to other things, such as minerals, plants, and animals) without identifying names. As soon as enough stars had been named, it was natural to compile lists of them. Yet a list of star names is essentially different from a catalog of stars, such as that of Hipparchos.

with as much precision as he did on the basis of the Greek observations alone, for reliable (Greek) observations did not go back more than a century or two. The Babylonians had not only accumulated a large number of observations, but they had started the habit of referring stars to the ecliptic, that is, of measuring longitudes instead of right ascensions. This facilitated the discovery of the precession. It was believed for a time that that discovery had actually been made c. 315 by a Babylonian named Ki-din-nu (Cidēnas of the Greeks, not mentioned by Ptolemy);[14] we must give the credit for it to Hipparchos, but he could not have made it if the Babylonian longitudes had not been available to him. As soon as it was possible to compare the longitudes of stars at different epochs, sufficiently distant, the discovery of the precession was inevitable. The crudity of the observations was compensated for by their temporal distance. The difference of about 1° after a century might be unobservable or overlooked, but the difference of a little less than 4° after four centuries called for an explanation.

Hipparchos' lunar and planetary theories were partly derived from Babylonian (or "Chaldean") observations. Ptolemy says so definitely (*Almagest*, IV, 2; IV, 10; IX, 7; XI, 7) and Father Kugler has shown that Hipparchos' determinations of the length of the month (average, synodic, sidereal, anomalistic, draconic) correspond exactly with those found in contemporary Chaldean tablets.

Hipparchos needed Babylonian data to discover the precession and to increase the precision of his own results. On the other hand, the Alexandrian conquests (334–323) and the Wars of the Diadochoi (322–275) had started a tremendous welter of peoples and ideas in the Near East. Some Chaldean astronomers may have influenced Greek ones, and vice versa. At the beginning Babylonian and Greek methods had been very different (the fundamental points of view being respectively arithmetic and trigonometric), but now each nation borrowed from the other, or, even when there was no actual borrowing, each stimulated the other in various ways; the result at first was ambiguity and chaos, and the final synthesis begun by Hipparchos was completed only by Ptolemy. The most curious episode of that period of confusion is the final defense of heliocentrism by Hipparchos' older contemporary, Seleucos the Babylonian!

We shall see in the final section of this chapter that some Chaldeans disregarded the new Greek astronomy and remained faithful to their own traditions.

A FEW OTHER GREEK ASTRONOMERS

Hipparchos dominated the whole period, even as Ptolemy would dominate the twilight of antiquity and the Middle Ages from beginning to end.

[14] The proponent of that theory was Paul Schnabel from 1923 on; he thought he had completely proved it in his memoir "Kidenas, Hipparch und die Entdeckung

There were other astronomers, however, whose varied activities illustrate the fermentation of astronomical ideas that was occurring in many places of the Greek world, especially in Alexandria and Rhodos. It will suffice to say a few words about each of them.

Hypsiclēs. The mathematician Hypsiclēs (II–1 B.C.), being anterior to Hipparchos, had no knowledge of trigonometry. He wrote a treatise (*anaphoricos*) on the rising (and setting) of the zodiacal signs,[15] in which the times of rising and setting were determined arbitrarily in the Babylonian manner. The times of rising for the signs from Aries to Virgo form an ascending arithmetic progression, those of the signs from Libra to Pisces a descending one. He was the first Greek to divide the zodiac circle into 360 degrees, and he made a distinction between the degree of space (*moira topicē*) and the degree of time (*moira chronicē*).[16]

Arrianos (II–1 B.C.) *ho meteōrologos* (the meteorologist) was so called because he wrote treatises on meteorology and on comets; he flourished probably in the second century B.C.

The Papyrus of Eudoxos. Alexandria was perhaps the main center of astronomical studies. Hypsiclēs was probably working in that city. There has come down to us a Greek papyrus (now in the Louvre) called the Papyrus of Eudoxos, because an acrostic at the beginning reads *Eudoxu technē* (*Eudoxi ars*). It deals with astronomy and the calendar and has the appearance of a student's notebook. The astronomical data correspond to the latitude of Alexandria and the years 193–190. Such a notebook is not of great importance in itself, but it is a witness of astronomical thinking and teaching.

Theodosios of Bithynia (I–1 B.C.) is of deep interest to us as a mathematician completing Euclid's *Elements* (see Chapter XVIII), but his own

---

der Praezession," *Zeitschrift fur Assyriologie 37*, 1–60 (1927) [*Isis 10*, 107 (1928)]. Otto Neugebauer, "The alleged Babylonian discovery of the precession of the equinoxes," *Journal of the American Oriental Society 70*, 1–8 (1950).

[15] The zodiac circle (*zōdiacos cyclos*) or circle of animals is a belt in the heavens about 16° wide on both sides of the Sun's path, the ecliptic; the Moon, planets, and many stars travel in that belt which is divided into twelve "mansions" or "signs": 1, Aries (ram); 2, Taurus (bull); 3, Gemini (twins); 4, Cancer (crab); 5, Leo (lion); 6, Virgo (virgin); 7, Libra (balance); 8, Scorpio (scorpion); 9, Sagittarius (archer); 10, Capricornus (goat); 11, Aquarius (waterbearer); 12, Pisces (fishes). The Sun enters a new sign each month, for example, Aries on 20 March, Libra on 22 September, Aquarius on 20 January. The zodiac is made so conspicuous by the Sun's and Moon's paths that it drew the attention of primitive peoples as well as of professional astronomers everywhere.

[16] The degree of space is the 360th part of the zodiac circle; the degree of time is the 360th part of the time it takes for any part of the zodiac to return to a given position.

curiosity was primarily astronomical and we have two treatises of his, *On days and nights* (*Peri hēmerōn cai nyctōn*) and *On habitations* (*Peri oicēseōn*) giving the positions of the stars at various times of the year as seen from various parts of the Earth. Two other astronomical treatises are lost, entitled *Diagrams of the houses* (*Diagraphai oiciōn*) and *Astrology* (*Astrologica*).

Poseidōnios. The Stoic philosopher Poseidōnios of Apameia (I–1 B.C.) made a new measurement of the size of the Earth, which was inferior to that of Eratosthenēs; his estimates of the diameter and distance of the Sun were much better than those of Hipparchos (and Ptolemy), yet very far from the truth. He was the first to explain the tides by the joint action of Sun and Moon and to call attention to spring and neap tides.

According to Cleomēdēs, Poseidōnios' measurement of the size of the Earth was based on the following assumptions: (1) Rhodos and Alexandria lie on the same meridian; (2) their linear distance is 5000 stades; (3) their angular distance is 1/48 of the circle. Hence, the circumference of the Earth is $5000 \times 48 = 240{,}000$ stades. According to Strabōn, Poseidōnios' result was lower, 180,000 stades.[17]

Cleomēdēs (I–1 B.C.) and Geminos were both disciples of Poseidōnios, but that does not necessarily imply that they were his contemporaries;[18] we may assume that they flourished in the first century B.C. Cleomēdēs [19] wrote a treatise on the cyclical motion of celestial bodies (*cyclicē theōria meteōrōn*), which is a good summary of Stoic astronomy; no author posterior to Poseidōnios is mentioned. He did not accept the latter's opinion that the equatorial region is inhabited. The *Theōria* is divided into two books. Book I explains that the world is finite but surrounded by an infinite vacuum, defines the celestial circles and the five terrestrial zones, discusses

[17] Eratosthenēs' method was the same but applied to other data. He assumed that Syēnē (Aswān) was on the same meridian as Alexandria, that their linear distance was 5000 stades, and that their angular distance was 1/50 of the circle. Hence, the circumference of the Earth measures $5000 \times 50 = 250{,}000$ stades, a result corrected later by him to 252,000 stades. The correctness of those estimates depends upon the values of the stade. For discussion, see Aubrey Diller, "The ancient measurements of the Earth," *Isis 40*, 6–9 (1949). Cleomēdēs referred to Lysimachia at the northeast end of the Hellespont (Dardanelles) as being on the same meridian as Alexandria. The cöordinates of those four localities, supposed to be on the same meridian, are as follows:

	Long. E	Lat. N	Differences: Long.	Differences: Lat.
Lysimachia	c. 27°	c. 40°30′	c. 1°16′	c. 4° 3′
Rhodos	28°16′	36°27′	1°37′	5°15′
Alexandria	29°53′	31°12′	3° 4′	7° 7′
Syēnē	32°57′	24° 5′		

[18] Some scholars, such as Albert Rehm, Pauly-Wissowa, vol. 21 (1921), 679, would place Cleomēdēs as late as the second century after Christ and even later. The only certainty is that he is later than Poseidōnios; he was almost certainly earlier than Ptolemy.

[19] Cleomēdēs may have been born or may have lived for a time at Lysimachia at the northeast end of the Hellespont, for he refers many times to that place. Otto Neugebauer, "Cleomedes and the meridian of Lysimachia," *American Journal of Philology* 62, 344–347 (1941).

the inclination of the zodiac upon the equator and its consequences. Our information concerning the measurements of the Earth by Eratosthenēs and Poseidōnios is derived exclusively from that book. The Earth is but a dot as compared with the heavens.

Book II begins with an anti-Epicurean diatribe apropos of the size of the Sun, which is probably borrowed from Poseidōnios. It includes explanations of the Moon's phases and of eclipses and a few data concerning the planets.

Cleomēdēs made various remarks on refraction (*cataclasis*), even on atmospheric refraction, for example, that the refraction of light may cause the Sun to be still visible when it is below the horizon.

His book remained unknown to ancient and Arabic astronomers, but it was known to a few Byzantine scholars, such as Michaēl Psellos (XI–2) and Jōannēs Pediasimos (XIV–1), and attracted the attention of the early printers.

The Latin text edited by Giorgio Valla was published in his *Collectio* as early as 1488 and again in 1498 (Venice: Bevilaqua); it was also printed separately (Brescia: Misinta, 1497) (Fig. 55); there were thus no fewer than three incunabula.[20] The Greek text was first published by C. Neobarius (Paris, 1539) (Fig. 56).

Modern edition with Latin version by Hermann Ziegler, *Cleomedis de motu circulari corporum caelestium libri duo* (264 pp.; Leipzig: Teubner, 1891).

Geminos. The life of Geminos of Rhodos (I–1 B.C.) is practically unknown; he was a disciple of Poseidōnios and was anterior to Alexander of Aphrodisias (III–1), who quoted him. Such limits are a little too distant to be useful, but his *floruit* may be determined more closely. He said that in his time the feast of Isis was a month away from the winter solstice; this gives the date c. 70 B.C., which is plausible on other grounds. He would thus be not only Poseidōnios' disciple but also his contemporary.

In mathematics he followed Euclid and in astronomy Hipparchos and the Babylonians. At any rate, in his astronomical introduction he used a Babylonian method to calculate the Moon's speed in the zodiac. His treatise on mathematics has been discussed in the preceding chapter; his astronomical introduction was entitled *Eisagōgē eis ta phainomena*. While his mathematical work is known only through later commentators, such as Proclos (V–2), Simplicios (VI–1), or the Arab al-Nairizī (IX–2), his astronomical introduction is extant. It covers the whole field of astronomy in an elementary way and is a valuable source for the history of Greek astronomy.

Geminos' *Introduction* was translated into Arabic; the Arabic text was translated into Latin by Gerard of Cremona (XII–2), *Liber introductorius ad artem sphaericam*, and into Hebrew by Moses ibn Tibbon (XIII–2). The Hebrew version, *Ḥokmat ha-kokabim* or *Hokmat ha-tekunah*, was made

[20] Klebs, Nos. 280, 1012. He missed the first edition, 1488.

CLEOMEDIS DE CONTEMPLATIONE ORBIVM EXCELSORVM. CAROLO VALGVLIO BRIXIANO. Interprete liber primus.

Diffinitio Müdi

Vm uaria mundus ſignificatione dicatur : de eo nobis ſuſcepta oratio eſt:qui ab ornamento:atq; ordine nuncupat̃.cuius diffinitio eſt huiuſmodi. Müdus eſt conſtitutio cæli & terre:& naturæ rerũ:q̃ intra cõcludunt̃.Hic oĩa continet corpora nullo extra penitus relicto:ut alio loco probatur.haud infinitus qdem:ſed terminatus:argumento q̃ a natura gubernatur. nullius.n. infiniti natura eſſe poteſt: cum id continere:& ei dominari naturam oporteat:cuius ſit natura.Quod autem adminiſtrantẽ heãt naturam ſatis arbitror eſſe cognitum:Primum qdem ex ordine partiũ ipſius : deinde earum rerum:q̃ oriuntur inde.tertium a ſocietate atq; conſenſu inter ſe ſe partium. Quartum q̃ ſingula alicuius rei gratia facta ſunt ad quam referunt̃.Poſtremo q̃ maximas afferunt utilitates. Quæ quidem particulari quoq; naturæ ꝓpria ſunt & conſentanea.Quare cum naturam habeat regentem ipſe qdem eſt neceſſario finitus

Quod Mundus ſit finitus

Quod uero eſt extra:uacuum eſt.& inane ab omni parte permeãs infinitum huius autẽ inanis infiniti. Quod a corpore occupat̃ locus appellatur.quod uero non occupatur:uacuum eſt. Quod autem ſit uacuum paucis exponemus.Omne corpus in aliquo ſit neceſſe eſt.Id autem in quo eſt.ab eo quod occupat:& implet: aliud eſſe oportet:incorporeum uidelicet & uelut tactui non obnoxiũ. Huiuſmodi igitur ſubſiſtentiam:quæ corpus ſuſcipere:continereq; poſſit:uacuum eſſe dicimus. Quod autem in huiuſmodi aliqua re corpora ſint:in aqua potiſſimum:& omni humida ſubſtantia licet intueri.Nam cum eſt uaſe in quo humor:& aliquod ſolidum corpus inſit:ipſum ex trahimus ſolidum : in aius exempti locum aqua

Quod Vacuum ſit extra Mundum

Fig. 55. First Latin edition of Cleomēdēs, *De contemplatione orbium excelsorum*, by Carolus Valgulius of Brescia, secretary of Cardinal Caesar Borgia to whom the book is dedicated (21 cm; Brescia: printed by Bernardinus Misinta, 3 April 1497). [Courtesy of Harvard College Library.]

Chardon de la Rochette

ΚΛΕΟΜΗΔΟΥΣ ΚΥΚΛΙΚΗ ΘΕΩΡΙΑ ΕΙΣ ΒΙΒΛΙΑ Β'

Nunc primùm typis excuſa prodit, cum Regio Priuilegio in quinquennium.

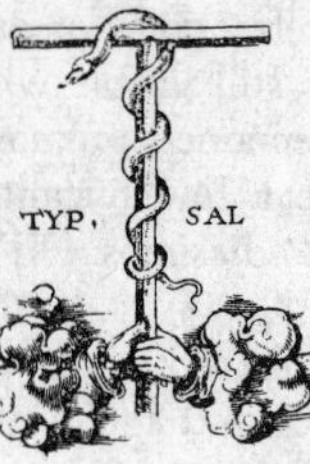

PARISIIS

PER CONRADVM NEOBARIVM, REGIVM IN GRAECIS TYPOGRAPHVM.

M. D. XXXIX.

Fig. 56. Princeps of Cleomēdēs, *Cyclicē theōria* (21 cm, 44 leaves; Paris: Conradus Neobarius, 1539). This copy bears the signature of Chardon de la Rochette (1753–1814), French Hellenist, great friend of Coraēs. *Lettres inédites de Coray à Chardon de la Rochette, 1790–96* (Paris, 1877). [Courtesy of Harvard College Library.]

in Naples in 1246. A part of Geminos' *Introduction* was printed as early as 1499 at the end of the *Astronomici veteres* edited by Aldus in Venice under the title *Proclu sphaira* (*The sphere of Proclos*). This was really a medieval compilation of extracts from Geminos' treatise. By 1620 more than 20 editions of it had been issued. Geminos' *Eisagōgē* was edited for the first time in Greek by Edo Hildericus (Altdorf, 1590) (Fig. 57), reprinted, Leiden, 1603.

Modern edition with German translation by Karl Manitius (413 pp.; Leipzig: Teubner, 1898). See also Otto Neugebauer, *Quellen und Studien zur Geschichte der Mathematik Astronomie und Physik*, vol. 3, *Mathematische Keilschrift-Texte* (Berlin: Springer, 1937), p. 77.

The Keskinto inscription. The importance of Rhodos, next to Alexandria, as an astronomical nursery, is proved by the activities of Hipparchos, Poseidōnios, Cleomēdēs and Geminos. In addition, we have an astronomical inscription dating from the period 150–50 B.C. which was found in Keskinto, ancient Lyndos, in that island.

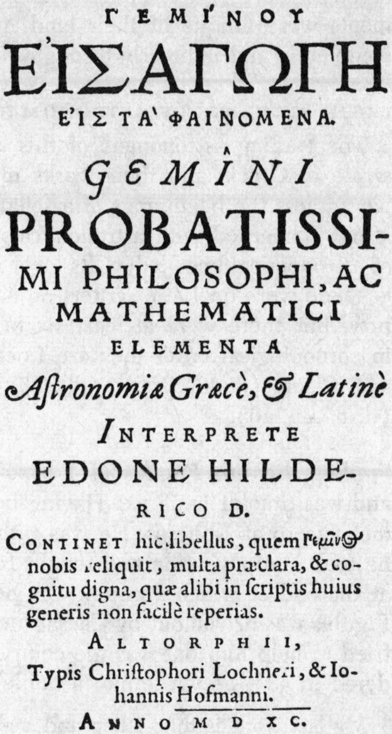
ΓΕΜΙ'ΝΟΥ
ΕἸΣΑΓΩΓΗ
Ε'ΙΣ ΤΑ' ΦΑΙΝΟΜΕΝΑ.
GEMÎNI
PROBATISSI-
MI PHILOSOPHI, AC
MATHEMATICI
ELEMENTA
Astronomiæ Græcè, & Latinè
INTERPRETE
EDONE HILDE-
RICO D.
CONTINET hic libellus, quem Γεμῖνος nobis reliquit, multa præclara, & cognitu digna, quæ alibi in scriptis huius generis non facilè reperias.
ALTORPHII,
Typis Christophori Lochneri, & Iohannis Hofmanni.
ANNO MDXC.

Fig. 57. Princeps of Geminos, *Astronomical introduction*, edited in Greek and Latin by Edo Hildericus (14.5 cm, 18 pp.; Altdorf, 1590). Fragments of it had appeared in Latin translation in the *Astronomici veteres* (Venice: Aldus, 1499), under the title *Proclu sphaira* (*The sphere of Proclos*), and had been frequently reprinted. [Courtesy of Harvard College Library.]

Xenarchos of Seleuceia[21] (I-2 B.C.) in Cilicia flourished at the end of the first century B.C. in Alexandria, Athens, and Rome. While in Rome, he was befriended by Augustus. He was a Peripatetic philosopher and grammarian; Strabōn was one of his disciples. He wrote a treatise against the quintessence[22] (*Pros tēn pemptēn usian*) wherein he had the boldness to criticize the principles of Aristotle's astronomy: the natural motions of celestial bodies are not exclusively circular, uniform, and homocentric. Such state-

[21] This was Seleuceia Tracheōtis. Many cities had been called Seleuceia in honor of Seleucos Nicatōr, founder of the Seleucid dynasty. One was on the Tigris (Seleuceia Babylonia); another was Seleuceia Pieria, a fortress overhanging the sea, north of the Orontēs, west of Antioch. Seleuceia Tracheōtis in Cilicia Aspera had an oracle to Apollōn and annual games in honor of Zeus Olympios. The other cities bearing the name were less important.

[22] Plato had postulated a quintessence or fifth element in order to be able to establish a parallel between the five regular solids and the elements. In the *Timaios*, the fifth solid was equated with the whole universe. In the *Epinomis*, the fifth element, beyond fire, was called aether (Volume 1, 452). For Aristotle the aether was the supreme element; and this remained the Peripatetic doctrine, but the Stoics reverted to four elements. The quintessence came back with the revival of Platonism. Philōn (I-1) made no distinction between the aether, the celestial fire in the astral region, and the substance of souls. Xenarchos' treatise (*Pros tēn pemptēn usian*) was a criticism of the Aristotelian aether.

ments were unique of their kind; unfortunately, we know them only incompletely and indirectly through Simplicios' commentary on the *De coelo.*

LATIN STUDENTS OF ASTRONOMY

The leading astronomers of this age, Seleucos, Hipparchos, and others, wrote in Greek, and their works might have been studied in Rome, but we witness the beginnings of a scientific literature in Latin. Such literature is not concerned with astronomical research but rather with the diffusion of astronomical knowledge. Its level is not high, but what could one expect?

There were no Latin writers on astronomical subjects in the second century, but there were at least six in the following, the last before Christ; in chronological order they are Lucretius (d. 55), P. Nigidius Figulus (d. 44), Cicero (d. 43), Varro (d. 27), Virgil (d. 19 B.C.), and finally Hyginus (d. c. A.D. 10).

Publius Nigidius Figulus (I–1 B.C.) was a politician who became senator and was praetor in 58 B.C. Having been sent on an embassy to the East, he met Cicero at Ephesos. He was rather conservative; during the Civil War he took Pompey's side and fought for him at the battle of Pharsalos (48), at the end of which Pompey was beaten and Caesar master of the world. Figulus was driven out by Caesar and died in exile in 44. His friend, Cicero, tried to help him but was too compromised himself to do so (he was murdered in 43); he wrote him a noble letter, ending,

> My last word is this; I beg and beseech you to be of good courage and to bethink you not only of the discoveries for which you are indebted to other great men of science, but also of those you yourself made by your own genius and research. If you make a list of them, it will give you every hope.[23]

This letter shows that Cicero had a great respect for him. Nigidius Figulus was a very learned man and deeply concerned with philosophy and with astronomy; his interest in astronomy was natural enough because he shared the cosmic views of the Stoics, as explained by Poseidōnios, and was the center of a new so-called "Pythagorean" school in Rome. He, and shortly afterward Varro, were the first Latin champions of astrology.

Nigidius Figulus was a defender not only of astrology but also of other forms of divination and magic. He wrote many books, of which only fragments remain,[24] dealing with mythology, divination, astrology, meteorology, geography, zoölogy. In his book *On the gods* (*De diis*), he combined "Zoroastrian"[25] with Stoic astrology, and discussed the Stoic ideas of

[23] Quoted from Frederick H. Cramer, *Astrology in Roman law and politics* (Philadelphia: American Philosophical Society, 1954) [*Speculum 31*, 156–161 (1956)], p. 64.

[24] The whole of those fragments is small. Latest edition by Anton Swoboda, *Nigidii operum reliquae* (143 pp.; Prague, 1889).

[25] The word "Zoroastrian" is put between quotation marks because the Greek tradition of Zoroastrianism was very different from the real thing. Zoroastrianism was mixed with Babylonian and Chaldean

ecpyrōsis and palingenesis.[26] His most important contributions to posterity were his studies on the stars, the *sphaera graecanica* (as described by Aratos) and the *sphaera barbarica* (derived from Oriental sources); he was the first to give the Latin names of the constellations and stars, and that was especially useful in the case of the "barbaric" sphere. For him astrology was a valid application of astronomical knowledge and his influence as an astrologer was great. He cast the horoscope of Octavius, born on 23 September 63 (the future Octavianus and Augustus); it is said that he had warned the father, Octavius senior, that his child would become master of the world according to the stars.

Lucretius and Cicero. Much as Cicero admired Nigidius Figulus for his learning, he did not share his astrological beliefs. Cicero had been influenced by the Epicureans, chiefly his friend Lucretius, and also by Carneadēs and by the Stoic Panaitios. His *De divinatione* (written in 44,, after Caesar's death) was a very strong attack against divination in general and astrology in particular. One cannot be too grateful to Lucretius and Cicero for their defense of rationalism in a very critical age; it demanded much lucidity to do so in that age because of the growing popularity of astrological nonsense, much courage also because of the gradual decline of freedom.

M. Terentius Varro. The astrological as well as the encyclopedic tendencies of Nigidius Figulus were continued by his older contemporary, Marcus Terentius Varro (116–27 B.C.).

Varro was born at Reate[27] in the Sabine country in 116; he was educated in Rome, being a pupil of the Stoic grammarian, L. Aelius Stilo, and later in Athens where he sat at the feet of the Academician, Antiochos of Ascalōn. The greatest part of his life was largely devoted to public affairs, which meant politics and war. He was in Pompey's service and obtained under him the offices of tribune, curule aedile, and praetor. In 76, he was Pompey's proquaestor in Spain, and in 67 he was taking part in Pompey's war against the pirates of the Eastern Mediterranean Sea; he fought in Pompey's war against Mithridatēs and in 49 he was fighting for him in Spain, and again in Greece. Caesar pardoned him twice, the second time after Pharsalos (48), and entrusted to him the arrangement of his Greco-Latin library.[28] Antony was less generous and persecuted him twice, the

---

ideas, astrology, and much else. For example, Zoroaster himself was often called an astrologer. Joseph Bidez and Franz Cumont, *Les Mages hellénisés. Zoroastre, Ostanès et Hystaspe d'après la tradition grecque* (2 vols.; Paris: Belles Lettres, 1938) [*Isis 31*, 458–462 (1939–40)].

[26] For *ecpyrōsis* and palingenesis, see Volume 1, p. 602.

[27] Reate, in the Latium, ancient capital of the Sabines, became a Roman *municipium* (a city with a certain amount of local independence). Modern name, Rieti; 42 miles north-northwest of Rome.

[28] This illustrates once more Caesar's generosity and his appreciation of literary merit. Caesar could be generous, while Antony could not, because Caesar was great while Antony was petty.

second time when the Second Triumvirate was established in 43. Varro was outlawed and much of his property, including his library, was taken away from him, but his life was saved, probably thanks to Octavianus' intervention. When Octavianus became emperor, Varro could resume the task begun under Caesar and was put in charge of the Augustan library.

In 43 he was an old soldier, aged 73, yet he still had 16 years to live and he devoted them to intense study and writing. He began his real life at a time when most people end theirs and his fame is almost exclusively based upon the work done between the ages of 73 and 90.

His literary activity was immense and most of it, certainly the best of it, was accomplished in his old age. Quintilian (I–2) called him, justly, "the most learned of the Romans." [29] We shall be obliged to speak of him, again and again, in other chapters of this book. At present we must restrict ourselves to a general survey of his books, a discussion of his astrological views and of his scientific encyclopedia.

His main works are seven which I shall list as far as possible in chronologic order: (1) *Saturarum Menippearum libri CV*, written c. 81–67, the "Menippean satire," a mixture of prose and verse; (2) *Antiquitatum rerum humanarum et divinarum libri XLI*, the "human and divine antiquities," written in 47; (3) *Logistoricon libri LXXVI*, a collection of dialogues on various subjects, begun in 44; (4) *De lingua latina libri XXV*, published before Cicero's death (7 December 43), probably in that fatidic year; (5) *Hebdomades vel de imaginibus libri XV*, 700 biographies of famous Greeks and Romans, written in 39; (6) *Rerum rusticarum libri III*, in 37; (7) *Disciplinarum libri IX*, date unknown, presumably toward the end of his life.

Of those books and of many others not listed only two survive, his treatise on agriculture (*Res rusticae*), and Books v to x of his treatise on the Latin language; the first of these will be dealt with in Chapter XXI, the second in chapter XXVI, but this seems to be the best place to discuss his *Disciplinae*, an encyclopedia, one of the earliest works of its kind, certainly the earliest in Latin.

The *Disciplinae* was divided into nine books: (i) *Grammatica*, (ii), *Dialectica*, (iii) *Rhetorica*, (iv) *Geometrica*, (v) *Arithmetica*, (vi) *Astrologia*, (vii) *Musica*, (viii) *Medicina*, (ix) *Architectura*.

The nine books have been divided by me into three groups to help the reader realize that he is facing the traditional "seven liberal arts," the trivium and the quadrivium, which can be traced back to Greek antiquity, almost to the time of Archytas of Tarentum (IV–1 B.C.), that is, to Plato's time.[30] The two main groups represented grammar plus the art of speech and discussion (the basis of any kind of knowledge), on one hand, and mathematics (conceived as science) on the other. The two final books were

[29] *Vir Romanorum eruditissimus*. Quintilian, *Institutio oratoria*, x, 1, 95.

[30] For the origins of the seven liberal arts, see Volume 1, pp. 434, 440.

devoted to applications, medicine and architecture, which to this day are not taught in the college of liberal arts but are relegated to special professional schools.

The trivium and the quadrivium together constituted the "seven liberal arts," which were the basic education of late antiquity, the Middle Ages, and the Renaissance; the relics of them still exist in our colleges of arts, and in the degrees of bachelor and master of arts.

We are most interested in the quadrivium, of course, and may well ponder on its fourfold division: geometry, arithmetic, astrology, and music. Notice that the main division, into quadrivium and trivium, was not a division between science and the humanities. For one thing, is not music a part of the humanities? You will probably answer: full-fledged music, yes, but not solfeggio and the beginnings of instrumental technique. Everybody will agree that there is no humanity in that; it is a torture for the students and for their neighbors. Solfeggio and the rest are to music what grammar is to language. Therefore, I have long maintained that the main cleavage in education is not vertical, between humanities on the right and science plus technology on the left, but rather horizontal, between grammar at the bottom and the humanities above it.[31] There were humanities as well as grammar in the trivium, there were humanities as well as science in the quadrivium; it all depended on the personalities of teachers and students.

The first and second parts of the quadrivium obliged Varro to discuss geometry and arithmetic and he also wrote separate (lost) treatises on those subjects, for example, a treatise on mensuration (*Mensuralia*), another on geometry in which he stated that the Earth is egg-shaped, another still on arithmetic (*Atticus sive de numeris*). The third part of the quadrivium was called *astrologia*, a word that might mean astronomy as well as what we call astrology.[32] As a matter of fact, Varro had not been an astrologer to begin with, for in his youth he had shared the skepticism of the New Academy; as he grew older, he was more and more influenced by Nigidius Figulus and by other Roman stoics and "Pythagoreans"; he became more conservative and mystical. After Caesar's and Cicero's deaths, astrologers like Nigidius and advocates of astrology like Varro had an open field. Varro was astrologically minded, but he was unable to draw a horoscope. On the other hand, he loved to cogitate on astrological fatalism, on numerology, and on similar fantasies. He wrote a treatise *De principiis numerorum*. His *Hebdomades* (a collection of biographies) was so entitled because he liked to indulge in mystical ideas concerning the number seven; he created

[31] G. Sarton, *The appreciation of ancient and medieval science during the Renaissance* (Philadelphia: University of Pennsylvania Press, 1955), p. x.

[32] The ambiguity of the terms has been discussed in *Introduction*, vol. 3, p. 112. The two terms are equally adequate to designate genuine science, for *logos* is as good as *nomos*. For "astronomy," compare "agronomy," "taxonomy," "bionomy;" for "astrology," compare "geology," "biology," "meteorology." The word "nomology" is made of both endings!

or transmitted the fear of climacteric years (multiples of 7-year periods).[33] He also played with the Pythagorean idea of a cycle of 440 years [34] for each individual and with the Stoic concept of palingenesis. He was so enamored of such fancies that his final wish was to be buried with Pythagorean ritual.[35] He died in 27 B.C.

Varro's sources were Greek, yet he was deeply Roman. His *Romanitas* was even more intense than Cicero's, and it was applied to learning rather than to morals and politics. Lucretius, Cicero, Virgil, and Varro were the first great teachers of Greek philosophy and science in the Latin language. Of the four he was perhaps the greatest teacher; he was not a poet like Lucretius and Virgil nor a stylist like Cicero; he was more interested in erudition than in literary conceits and his main purpose was always to teach. Hence, the *Disciplinae*, which became one of the patterns of ancient and medieval thought.

As Varro's historical writings are full of astrological references, he played as great a part as Nigidius Figulus in the diffusion of astrology (he was less of an astrologer but more popular as a writer); they both helped powerfully to prepare the astrological climate of the Roman empire. However, his influence as a friend of astrology was indirect. The Varronian tradition concerns primarily his *Res rusticae*; it will be described when we speak of this, his greatest work.

Virgil, Vitruvius, Hyginus, Ovid. In the second half of the first century B.C. and during the Augustan age (27 B.C.–A.D. 14), astronomy, or at least astral mythology, was an essential part of the education of a Roman gentleman. We may thus expect the leading authors to have some knowledge of astronomy. This expectation has already been verified in the cases of Cicero and Varro; let us consider a few more examples: Virgil, who died in 19 B.C.; Vitruvius, who flourished under Augustus; Hyginus, who was still at the head of the Palatine library in A.D. 10; and Ovid, who lived at least until A.D. 17.

The main source of their knowledge was the poem of Aratos (III–1 B.C.), which they could read in the original or in Cicero's translation. An improved translation was made by the general, Germanicus Caesar (15 B.C.–A.D. 19), but that was too late to be used, except perhaps by Vitruvius and Ovid. As witnessed by Book IX of his *Architectura*, Vitruvius was well acquainted with Greek astronomy and even with Chaldean "astrology." He

[33] The word *climactēr* meant a round of a ladder, hence a difficult or decisive step in life. Our word "climacteric" stems from the adjective *climactēricos*; the Greeks also used the verb *climactērizomai*, to be in a climacteric age. In French lore, the grand *climatérique* ($7 \times 9 = 63$) was especially fatidic. When Viète (1540–1603) died, he was 63 and the matter was gravely commented upon. All that nonsense began, as far as I know, with Varro.

[34] $440 = 2^3 \times 5 \times 11$. I do not understand the importance given to that number.

[35] Pliny, *Natural history*, XXXV, 46.

was convinced that astrology was a Chaldean specialty. His statement on the subject is so remarkable that we must quote it verbatim:

> As for the remainder of *astrologia*, to wit, the effects produced by the twelve signs on the human course of life, and also by the five planets, the Sun, and the Moon, we must accept the calculation of the Chaldeans, because the casting of nativities (*ratio genethlialogiae*)[36] is their specialty, so that they can explain the past and the future from astronomical calculations. Those who have sprung from the Chaldean nation have handed on their discoveries about matters in which they have approved themselves of great skills and subtlety.[37]

Among the prisoners of war brought home by the great Sulla was a Greek from Milētos or from Caria who became a famous teacher in Rome and wrote so many books that he was given the nickname of Alexander Polyhistor. His most promising pupil was another prisoner or slave, Gaius Julius Hyginus (I–2 B.C.), who had been brought by Caesar from Alexandria.[38] Alexander recognized Hyginus' exceptional merit and called Augustus' attention to him. Not only did the emperor emancipate him, but he put him in charge of the Palatine library. Hyginus imitated the polygraphic and encyclopedic tendencies of his patron and wrote considerably on a great variety of subjects. Astronomy was naturally one of those subjects; he not only exploited Aratos' *Phainomena* as the others did but used a celestial sphere. He may have been one of Virgil's teachers. He was still director of the Palatine library in A.D. 10. It may be that his fame as a scholar was partly due to his directorship, because innocent people are generally led to believe that a librarian must be *ipso facto* an exceptionally learned man.

He almost sank into oblivion until Isidore of Seville (VII–1) rescued him; thanks to Isidore, his fame revived in the Middle Ages and some of his writings were saved. There are fragments of those on agriculture and apiculture, and we have almost the whole text of his astronomical treatise (*De astrologia*, or *De signis caelestibus*). It describes 42 constellations and the myths relative to them, and is divided into four parts: (1) explanation of the cosmos, the celestial sphere and its sections, (2) stories of the constellations, (3) their shapes, (4) the planets and their motions (the end of the book is lost).

[36] Prognostications by astronomical means were divided into two main groups; the first, called *catholicos*, concerned races, countries, nations, cities; the second, called *genethlialogicos*, concerning individuals (*Tetrabiblos*, II, 1). When one speaks of astrology, one generally refers to the second group. *Genethlē* is the birth, origin, birthplace; *genethlios* refers to the birthday (*natalis*); *genethlialogia* is the horoscope.

[37] Vitruvius, *De architectura*, IX, 6, 2. Vitruvius' views on Chaldean astrology remained traditional for a considerable time, as is proved by the bad fame that the Chaldeans enjoyed. It is only since 1880 that Chaldean "astronomy" has been gradually revealed. See the final section of this chapter.

[38] According to another story, Hyginus was of Spanish origin.

This work enjoyed considerable popularity, as is proved by many manuscripts and no less than five incunabula, four of them in Latin – Ferrara: Carnerius, 1475 (Fig. 58); Venice: Ratdolt, 1482; Venice: T. Blavis, 1485 and 1488. When Erhard Ratdolt moved to Augsburg, he published a German translation, *Von den zwölf Zeichen*, in 1491 (Fig. 59; Klebs, 527.1–4, 528.1). There are many sixteenth-century editions. A new Latin text was edited by Johann Soter (Cologne, 1534).

Modern editions by Bernhard Bunte, *Hygini astronomica* (130 pp.; Leipzig, 1857) and by Emile Chatelain and Paul Legendre (Paris: Champion, 1909).

Virgil (I–2 B.C.) was influenced by Nigidius Figulus and Hyginus, and also by an Epicurean tutor and Stoic models. This explains a certain amount of ambivalence; he accepted astrological ideas as everybody did, but with moderation. He had studied medicine and mathematics (which included astrology). One of his eclogues has a definite astrological flavor in the Stoic manner or, call it, a messianic intimation; it is *Ecloga IV*, dedicated to Pollio after the peace of Brundisium (40 B.C.).[39] Virgil announced the beginning of a new age; a child would be born who would bring back the golden days; prosperity would grow with him and would be completed at the time of his maturity. Who was the child? Pollio's (a child was born to him in 40 B.C.)? It is more likely that Virgil had no living child in mind. The whole *Ecloga* (the fourth in date and number of the *Bucolica*) is more political than bucolic, in spite of the pastoral background. Political and prophetic, it brings to mind the predictions of the Sibylla of Cumae[40] as well as Orphic and Etruscan prophecies, according to which the life of the world was divided in periods or "years," each of which, announced by Apollōn and inaugurated by Saturn and the virgin Astraia, was a complete renewal. The tone of the prophecy is so religious that the ancients, beginning with Constantine (emperor 306–337) and St. Augustine (V–1), thought that the child was to be the very Messiah already announced in the Old Testament! That interpretation is not favored, and yet a great Jewish archaeologist, Salomon Reinach (1858–1932), could declare, "That poem written in 40 B.C., entirely religious, is the earliest Christian work."[41]

As will be explained in another corner of this book, Virgil's knowledge of astronomy was not simply of the Aratean type; he was a farmer and

[39] Brundisium (Brindisi) had been besieged by Antony when Octavian tried to prevent his landing in Italy. The negotiations between the two triumvirs led by Pollio evoked great hopes and popular joy. After his return, Pollio was consul. Gaius Asinius Pollio (I–2 B.C.) had fought on Caesar's side and later on Antony's in the Civil War. He founded the first Roman public library in the Atrium Libertatis; he was a critic and patron of letters, a friend of Virgil, Horace, and others.

[40] Cumae (modern Cuma) in the Phlegraean fields, just west of Naples. Its fame was due chiefly to its being the residence of the earliest Sibyl. See Chapter XX.

[41] This is quoted from Henri Goelzer's Latin-French edition of the *Bucolica* (Paris, 1925?), p. 41.

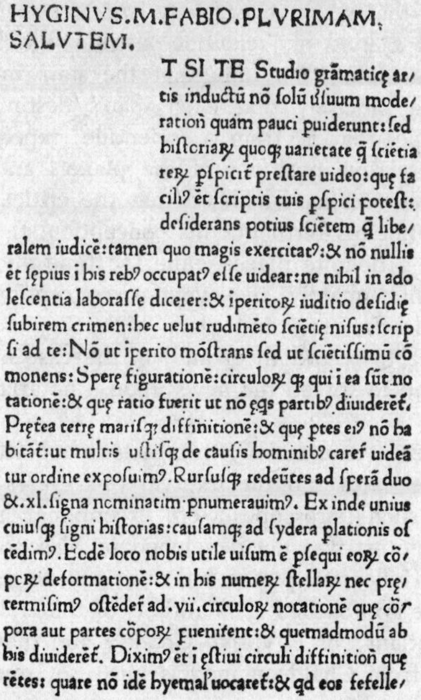

HYGINVS.M.FABIO.PLVRIMAM.
SALVTEM.
T SI TE Studio grāmaticę ar/
tis inductū nō solū usuum mode/
ratiō quam pauci puiderunt: sed
historiaꝝ quoq; uarietate q̄ sciētia
reꝝ pspicit prestare uideo: quę fa
cilꝰ ēt scriptis tuis pspici potest:
desiderans potius sciētem q̄ libe/
ralem iudicē: tamen quo magis exercitatꝰ: & nō nullis
ēt sępius i his rebꝰ occupatꝰ esse uidear: ne nihil in ado
lescentia laborasse dicerer: & iperitoꝝ iuditio desidię
subirem crimen: hec uelut rudimēto sciētię nisus: scrip
si ad te: Nō ut iperito mōstrans sed ut sciētissimū cō
monens: Sperę figuratione: circuloꝝ q̄ qui i ea sūt no
tationē: & quę ratio fuerit ut nō ęqs partibꝰ diuiderēt.
Prętea terrę marisq; diffinitionē: & quę ptes eiꝰ nō ha
bitāt: ut multis usitisq; de causis hominibꝰ carēt uideā
tur ordine exposuimꝰ. Rursusq; redeūtes ad sperā duo
&.xl.signa nominatim pnumerauimꝰ. Ex inde unius
cuiusq; signi historias: causamq; ad sydera plationis os
tēdimꝰ. Eodē loco nobis utile uisum ē psequi eoꝝ cō/
pcꝝ deformationē: & in his numeꝝ stellaꝝ nec pre/
termisimꝰ ostēdēt ad.vii.circuloꝝ notationē quę cōr
pora aut partes cōpoꝝ ꝑuenirent: & quemadmodū ab
his diuiderēt. Diximꝰ ēt i ęstiui circuli diffinitiō quę
rētes: quare nō idē hyemal' uocarēt: & qd eos fefelle/

Fig. 58. Princeps of Hyginus' Latin astronomical poem, *Poeticon astronomicon* (Ferrara: Augustinus Carnerius, 1475). [Courtesy of Huntington Library, San Marino, California.]

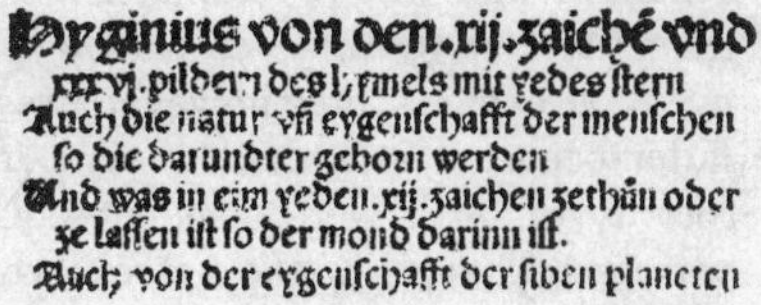

Hyginus von den.xij.zaichē vnd
xxxvj.pildern des hymels mit yedes stern
Auch die natur vñ eygenschafft der menschen
so die darunder geborn werden
Und was in eim yeden.xij.zaichen zethūn oder
ze lassen ist so der mond darinn ist.
Auch von der eygenschafft der siben planeten

Fig. 59. German translation of Hyginus' astronomical poem, *Von den zwölf Zeichen* (Augsburg: Erhard Ratdolt, 1491). [Courtesy of the Armed Forces Medical Library, Cleveland, Ohio.]

loved to talk with other farmers; he was familiar with meteorology and astronomical folklore.[42]

My last example is Ovid, who was well acquainted with astrology, yet remained somewhat skeptical. He was probably typical of the most civilized Romans, who accepted astrological fancies but did so without enthusiasm. It would not do to reject ideas that enjoyed so much popularity in the highest circles, but one should keep one's own counsel. Poetical images were always permitted, of course, and practical hints might be taken, but that did not affect one's inner belief.

ASTROLOGY

The origin of astrology was told in Chapter XI. The astrological ideas emerging from Persia and Babylonia were early combined with Pythag-

[42] P. d'Hérouville, *L'astronomie de Virgile* (35 pp.; Paris: Belles Lettres, 1940), includes a list of all the stars mentioned by Virgil. Virgil's selection was arbitrary; for example, he named but six of the zodiacal signs. A map illustrates the Virgilian constellations and stars. See also William Ernest Gillespie, *Vergil, Aratus*

orean and Platonic fancies. Much of that was not astrology, *stricto sensu*, but astrolatry and astral mythology. A kind of "scientific" astrology was established when somebody had the notion, not only that the stars influenced human destinies, but more precisely that each man's destiny could be deduced from his horoscope, that is, from a "scientific" representation and interpretation of the relative positions of the planets and main stars at the time of his birth. It was soon realized that the critical event in a man's life, however, was not his birth but his conception; the latter occurred at a definite time in a definite place, while the time and locality of birth were accidental. Unfortunately, the event of conception was secret, even to the actors, while the time and place of birth were tangible and obvious; it was possible to record them and, in the case of important people, to witness them and to spread the news upon a notarial document.

The first known horoscopes are recorded in cuneiform tablets dated 410 and 263. Note the earliness of those dates and the interval (147 years) between them, which suggest that such horoscopes were exceptional in Chaldea.[43] At any rate, the horoscopic art was not developed in Mesopotamia, but in Egypt, throughout the Hellenistic period but more rapidly as the end of it and the Roman age were approached. The authors of the Hellenistic horoscopes were Greek Egyptians (or Egyptian Greeks) and they borrowed their knowledge not only from Chaldean models but also from Pharaonic ones.[44]

Greek-Egyptian astrology seems to have reached a climax during the Augustan age and its popularity was very much increased by Stoic philosophy and pantheism, and also by imperial favor. Astrology conquered the Roman world, survived it, traversed the Middle Ages and the Renaissance, and is still popular today.

The early documents were sometimes modified and amplified, but in many cases they were simply copied or literally translated. A great mass of astrological texts has been published under the direction of Franz Cumont, in a collection entitled *Catalogus codicum astrologorum Graecorum* (12 vols.; Académie Royale de Belgique, 1898–1953) [*Introduction*, vol. 3, p. 1877; *Isis 45*, 388 (1944)], abbreviated CCAG. The majority of those texts are late, sometimes very late, but that does not matter so much because of their innate conservativeness and the lack of astrological evolu-

---

and others; the weather-sign as a literary subject (80 pp.; doctoral dissertation, Princeton University, 1938), for Virgil's weather lore as exemplified in *Georgica*, I.

[43] For details about those early horoscopes, see Frederick H. Cramer, *Astrology in Roman law and politics* (quarto, 292 pp.; Memoirs of the American Philosophical Society, vol. 37, Philadelphia, 1954), [*Speculum 31*, 156–161 (1956)] p. 5–7.

[44] Excellent study by Franz Cumont with the collaboration of Claire Préaux, *L'Egypte des astrologues* (254 pp.; Brussels: Fondation égyptologique Reine Elizabeth, 1937) [*Isis 29*, 511 (1938)]. This book deals chiefly with the social environment of the Egyptian astrologers: Ptolemaic kings and government officers, life in the towns and country, games, trades, arts and crafts, religion and ethics.

tion. One might safely repeat about them the old slogan, "Plus ça change, plus c'est la même chose."

The most famous astrological work of the second century B.C. bears the double name "Nechepsō-Petosiris," which it is almost impossible to elucidate. It is in that text that one finds the earliest account of the astrological meaning of the signs of the zodiac, as well as other novelties. The text is lost but many fragments were collected by Ernest Riess, "Nechepsonis et Petosiridis fragmenta magica," *Philologus, Suppt.* vol. 6, 325–394 (1894), and many more have been revealed in the CCAG.

During the last century B.C., there were many more Greek astrologers, such as Timaios, but the best astrological information was now available in Latin, rather than in Greek writings, notably in those of Cicero, P. Nigidius Figulus, and M. Terrentius Varro which have already been dealt with.

The technical elements of that literature are less interesting, however, than the social ones. The astrological fancies were popular, less because of their intrinsic value (which amounted to almost nothing) than of the value given to them by human needs and religion and by Stoic approval. The political vicissitudes and the social miseries created a climate that was favorable to spurious consolations. Many Greeks and Romans accepted their fate more readily in the same spirit as the Muslim who says "*maktūb*" (it is written) and resigns himself to the unavoidable. During the Augustan age there was more security, but no freedom and no spiritual peace.[45]

While the writings of the professional astrologers may be neglected, we must pay attention to the opinions of the greatest astronomer of that time, Hipparchos. Those opinions are not well known, but they are reflected in the *Tetrabiblos* of Ptolemy (II-1) just as Hipparchos' astronomical knowledge is reflected in the *Almagest*. It is not true, as Tarn claims,[46] that Hipparchos' rejection of heliocentrism assured the success of astrology, but his acceptance of the astral religion implied astrological possibilities. Granted his belief that there is some connection between souls and stars and his belief in divination (which he shared with almost every one of his contemporaries), the step down to astrology was almost inevitable.

How could that happen? As a pure astronomer, Hipparchos was isolated from the people around him and yet he craved for sympathy. He must share the religion of his neighbors and the astral religion was then the highest, the purest religion. He accepted it and astrology came with it; when we join a religious circle we cannot help sharing to some extent its superstitions. Moreover, Epicureanism was discredited, while Stoicism was the most respected philosophy; Epicureans rejected astrology, while Stoics

[45] For the social background of astrology the best accounts are those of Cumont, *L'Egypte des astrologues* and of Cramer, *Astrology in Roman law and politics*. Cramer's book extends to the murder of Severus Alexander in A.D. 235 and even beyond that.

[46] Tarn and Griffith, *Hellenistic civilisation*, p. 348.

Fig. 60. Monumental horoscope of Antiochos Epiphanēs, king of Commagēnē, referring probably to his coronation, under Pompey's auspices, in 62 B.C. It is a bas-relief, 1.75 × 2.40 m, which was found at Nimrud-Dagh, not far from Samosata (in Commagēnē, north Syria). It represents the meeting of three planets in the sign of the Lion, the Sun itself is symbolized by the Lion and the Moon by a crescent. The inscription at the top reads: "Pyroeis Hēracleus (Mars), Stilbōn Apollōnos (Mercury, sometimes dedicated to Apollōn), and Phaethōn Dios (Jupiter)." [A. Bouché-Leclerc, *L'astrologie grecque* (Paris, 1899), pp. 373, 439.]

favored it. Thus, the noblest feelings, the highest religion as well as the highest philosophy, all the good things in his environment, conspired to make him share the astrological delusion. How could a man resist such a convergence of social obligations? This is mere guessing, of course; we have no means of penetrating Hipparchos' mind, much less his soul, but is it not plausible enough? How else could we explain his betrayal, a betrayal to be repeated three centuries later by his disciple and superseder, Ptolemy?

There is much discussion nowadays concerning "science and society," meaning the impact of society upon science and the opposite impact of science upon society. The latter is of necessity very slow,[47] because men of science are rare and seldom powerful, but the former is immediate and overwhelming. The "case of Hipparchos and Ptolemy" is the best example of that; the two greatest astronomers of classical antiquity were so completely overweighed, overturned, and defeated by their environment that, instead of attacking astrology, they proceeded to give scientific arms to it.

We may be sure that they were careful to distinguish between the pure astrological doctrine (as it was eventually formulated in the *Tetrabiblos*) on the one side and the stupidities and impostures of astrological soothsayers on the other. Yet they had betrayed the scientific stronghold. The people did not make that distinction and did not care. The great Hipparchos had approved astrology; every charlatan could take refuge behind him, and did so.

Moreover, after Hipparchos had given astrology a certificate of scientific respectability, the Stoic philosophers were fortified in their beliefs and increased their own propaganda. This was especially true of Poseidōnios, who flourished in Rhodos not long after Hipparchos and was the head of the Stoic school in that island. He had opportunities of preaching his Stoic-astrological doctrines not only in Rhodos but also in Rome (he was there

[47] The impact of technology may be rapid, because the invention of new tools or machines creates new needs; in ancient times, however, the new tools were not mighty enough to disturb the rhythm of life.

in 87 and again toward the end of his life in 51) and in many other places, for he was a great traveler. Thanks to Hipparchos and Poseidōnios, astrology had all the highbrow credentials it might need, and its success was almost complete. Instead of saying with Pliny (see quotation above) that Hipparchos can never be sufficiently praised for his defense of the astral religion, I would say, all honor to the relatively few men, such as Cicero,[48] who had the insight and courage to resist the astrological avalanche.

Hipparchos was the greatest astronomer; Cicero, a layman. It is interesting to note that in this case the layman was right and the "specialist" wrong; such a case is not unique in the history of science.

In Rome and the ever-extending Roman world astrological beliefs became almost official, though subordinated to the state religion. The story of the relations of astrology with the state is a part of political history and does not concern us here. The Senate made a distinction, however, between theoretical astrology, which was never interfered with, and practical astrology, which had to be repressed because it was abused by charlatans and other rascals. The majority of astrologers preying upon Roman citizens were Greek exiles, some of whom were good men, while others were unscrupulous adventurers.

In 139 B.C., a *senatus consultum* decreed the expulsion of every astrologer from Rome. Other *consulta* of the same kind were decreed from time to time, the last one in A.D. 175. Such decrees were difficult to apply and went too far. In the year A.D. 11, Augustus issued an imperial edict against special kinds of astrological activities; it outlawed "consultations *à deux* and curbed the scope of topics on which astrologers were free to speak and clients allowed to consult them." [49] Those conversations *à deux* were outlawed because they might be seditious; astrological judgments against the government could easily be obtained for a consideration and were then used as political weapons.

The best illustrations of the political importance attached to astral religion and indirectly to astrology are given by a number of Hellenistic and Roman coins representing the Sun and various constellations, crescents and stars, the zodiac and separate signs.[50]

In our judgment of ancient astrology, we should always remember that, while pure astrology was innocent and innocuous, astrological predictions could be used (and were used) in the same way as black magic. One consulted an astrologer as one would a necromancer for the gratification of sexual desire, hatred, ambition, greed, or any other evil. The astrological

[48] Cicero had been one of Poseidōnios' auditors.

[49] Cramer, *Astrology in Roman law and politics*, p. 232. Abundant details are included in Cramer's book, primarily concerned with the political aspects of astrology.

[50] Cramer, Plate 12, plate of coins; pp. 29–44, catalog of 142 coins.

fancies of a Stoic philosopher did nobody any harm, and his equanimity immunized him against soothsayers. The situation was very different, however, in the underworld,[51] which organized the astrological racket and was in its turn the main victim of it.

We should not be too severe, however, because those errors are not yet uprooted from the public mind and the infamous racket is still flourishing. To illustrate, an excellent astronomical journal, published not for astronomers but for educated readers,[52] had to be discontinued for lack of support, while journals devoted to astrology are enriching their owners. Astrological columns are run in many newspapers; the columnists earn probably more money than honest astronomers. The same situation obtains more or less in other countries. It is so disgraceful that we have no right to throw stones at the ancient astrologers, nor at the social organization that permitted their existence.

THE CALENDAR

It is not part of our purpose to tell the whole history of the calendar, because that is an endless subject, which is to a large extent not scientific but rather political and religious; the determination of any calendar implies scientific considerations, but those considerations are often peripheral rather than central. Moreover, the origins of many calendars are obscure, being part of the anonymous folklore, rather than datable creations. This is definitely the case for the early Roman calendar, about which we know very little with any certainty.[53]

The earliest Roman calendar was probably lunar and the priests were charged to proclaim or call[54] the new moon. Solar considerations were necessarily introduced because of the seasons; the farmer's calendar always tended to be solar as well as lunar. In 303, the *aedilis curulis*[55] Cn. Flavius drew a list of *dies fasti* and *dies nefasti*[56] (holy days and others), and it

[51] I do not mean the physical underworld, that is, the poor and downtrodden, but the spiritual underworld, which included rich people as well as poor, and imperial families as well as beggars and prostitutes.

[52] The reference is to *Popular Astronomy*, published in Northfield, Minnesota (59 vols., 1893–1951).

[53] There is an immense literature on the subject, full of contradictions and polemics. The latest book received by me is John Phelps, *The prehistoric solar calendar* (107 pp.; Baltimore: Furst, 1955). It discusses the old Celtic calendar represented by the Coligny inscription of A.D. 71, the old Roman calendar, the Etruscan calendar, and the Sumerian one.

[54] *Calare;* hence *calendae* (the calends, first day of each month), and also *intercalaris*, intercalary (month), *intercalaris annus*, bissextile year.

[55] The *aedilis curulis* was a Roman magistrate entitled to use a special chair (*sella curulis*) and a bordered robe (*toga praetexta*). According to Cicero (*De legibus*, III, 3, 7), they were "*curatores urbis, annonae, ludorumque sollemnium*" in charge of markets, taxes, and solemn games.

[56] *Fastus* is anything agreeing with *fas* (divine law). The *fasti* were the lawful days. They included Calendae, Nonae, Idus, Nundinae, and various festivals.

was he who established a calendar of twelve months (the Flavian calendar), totaling 355 days, intercalary months of 22 or 23 days being added every two years (the average year was thus too long, 366 days). The intercalary month was inserted after February 23.

The Romans were not very proficient in such matters; their scientific incompetence is amusingly revealed in another astronomical matter. The first sundial established in the Forum in 263 came from Catana,[57] which is 4°23′ south of Rome, yet the Romans were satisfied with it for a century, being unable to correct it and perhaps unaware of the necessity for correction. The first sundial adapted to Roman astronomical needs was set up in 164 B.C. by Q. Marcius Philippus during his censorship. Yet astronomical insouciance continued to be the rule rather than the exception. Ovid could still say that the ancient Romans knew arms better than the stars:

> Scilicet arma magis quam sidera, Romule, noras.[58]

He explained that the legendary founder of Rome, Romulus, estimated the length of the year to be ten months because that was the duration of a woman's pregnancy! [59]

The errors of the calendar (its chronic getting out of step with seasonal events) was corrected from time to time by new intercalations. In 191 B.C., the Lex Acilia empowered the pontiffs to deal with intercalations at their discretion; this shows that the calendar was a religious matter.[60] It is probable that some of those pontiffs were negligent and did not pay much attention to small differences. Those differences accumulated, however, and in Caesar's time the *Floralia*,[61] a spring festival, was celebrated in summer.

As Caesar's establishment of the Julian calendar occurred in Egypt, we must return there for a moment. Calendrical difficulties were extreme in that country, because it was necessary to harmonize Greek dates with Egyptian and Chaldean ones. The translation from one system into another was always difficult and sometimes hopeless.

The Egyptians themselves had first tried the use of a lunar year but had rejected it very early (from the First Dynasty on) in favor of a solar calendar. They had the wisdom to keep clear of a mixed lunisolar calendar. They

[57] Catana (Catanē in Greek; now Catania) on the east coast of Sicily, at the foot of Mount Aetna, was a Greek city. It was conquered by Rome during the first Punic War (264–241) but remained essentially Greek for a long time afterward.

[58] Ovid, *Fasti*, 1, 27–34. The *Fasti* was written toward the end of his life; he died c. A.D. 18; it is a kind of poetic yearbook.

[59] That duration was generally estimated to be ten lunar months (or about 9/12 of a year). *Introduction*, vol. 3, pp. 252, 268, 1230, 1698.

[60] The pontiffs (*pontifices*) constituted a college of priests established very early (under the legendary Numa Pompilius, second king of Rome). Their chief was the *pontifex maximus*, a title used to this day by the popes.

[61] *Floralia* or *Florales ludi*, a festival of peasant origin instituted in 238 B.C. in honor of Flora, goddess of flowers and spring. When botanists speak of the flora of a country, they refer unwittingly to her.

divided the year into twelve months of three decades each (corresponding to the 36 decani), but they soon added a holiday season of five days.[62] Hence, their year measured $(30 \times 12) + 5 = 365$ days, which was a little too short. In the decree of Canōpos (238),[63] pronounced by an assembly of priests during the rule of Ptolemaios Evergetēs (king 247–222), it was decided to add one day every fourth year. This was all right, but Hellenistic astronomers had spoiled the Egyptian calendar by introducing lunar considerations. Apparently, the decree of Canōpos was not enforced, because the divergencies continued to the extent that Julius Caesar felt obliged to do something.

After the battle of Pharsalos (48) which made him master of the world, Caesar sojourned for some time in Egypt, and it was there that he began to think of the calendrical difficulties which were an increasing nuisance to the Roman government. Caesar was thinking in terms of Roman imperialism and unity, and, as he was interested in astronomy, it was natural enough for him to consider the necessity of a reformed calendar which would become the official calendar of the Roman commonwealth.

He secured the collaboration of the Peripatetic philosopher and astronomer Sōsigenēs of Alexandria,[64] employed a scribe named M. Flavius, and was probably advised by his colleagues in the college of pontiffs; he had been a pontiff since c. 75, and *pontifex maximus* since 63.

The victory of Thapsos (46), ending the Civil War, gave him the opportunity of promulgating the much-needed reform. In order to restore the balance he inserted between November and December 46 two intercalary months of 67 days, the month of February having already received an intercalation of 23 days; thus, that year 46 (*annus confusionis*) totaled $355 + 23 + 67 = 455$ days. The new calendar (Julian calendar) began on 1 January 45;[65] it extended to 365 days, and a single day, intercalated after 23 February[66] was added each fourth year; this day was called *bissextum* and the year to which it was added was an *annus bissextilis* (or *intercalaris*). The year was still divided into 12 months: Januarius, Februarius, Mars, Aprilis, Majus, Junius, Quinctilis (later called Julius in honor of Julius Caesar), Sextilis (later called Augustus after the first emperor), September, October, November, December. At first the year began in March, and this explains the names of the last four months (called seventh, eighth, ninth,

[62] More details in Volume 1, p. 29.

[63] Canōbos or Canōpos, near the westernmost mouth of the Nile, just east of Alexandria. The inscription recording the decree of Canōpos was discovered in 1881 and is preserved in the Museum of Cairo. The decree is written in hieroglyphics, demotic, and Greek.

[64] He was said to be Egyptian. His name is a good Greek one; many Greek names begin with *Sōsi* or end with *genēs*. This means nothing, however, for Egyptians and Jews often assumed Greek names.

[65] The year 45 B.C. was equal to OL. 183–4, and to A.U.C. 709.

[66] It was intercalated after 23 February because a month had been intercalated after that day every second year in the Flavian calendar (see above). Such is the force of habit, or call it tradition.

and tenth); the beginning of the year had been moved to 1 January in 153 B.C.[67]

There were three principal days in each month: *calendae*,[68] the first day; *nonae*, the fifth (or seventh) day; *idus*, the thirteenth (or fifteenth) day.[69]

The other days were counted backward from those principal days, being spoken of as the *n*th day before the calends, the nones, or the ides; thus,

2 January = *quarto [die] ante nonas Januarias*,
6 January = *octavo ante idus Januarias*,
14 January = *undevicesimo ante calendas Februarias*,
31 January = *pridie calendas Februarias*.

Note that the day preceding the *calendae* is called *pridie* (the first day before) and is preceded by *tertio*, *quarto*, . . . to *undevicesimo*. There is no day *secundo*, because the *calendae* is itself considered the first day before the *calendae*! In the same way the first days before *nonae* and *idus* are considered the second days. This was shockingly illogical.

When it was necessary to add a day in each fourth year that day was inserted between 23 February and the normal 24 February (becoming 25). The normal 24 February was called *sexto ante calendas Martias*, and the day inserted just before it was called *bis sexto ante calendas Martias*.[70] Hence, the word "bissextile."

It may be explained in passing that the English term for bissextile year is leap year. Why? A normal year of 365 days is equal to $(7 \times 52) + 1$ days. When two years of 365 days follow each other, there is thus a shifting of one day for each date; when a bissextile year intervenes, it is necessary to leap two days (after 29 February).

For example, in 1942, 4 July was a Saturday; in 1943, a Sunday; but in 1944, a Tuesday; then in 1945, a Wednesday.

Much of this will be clearer in graphical form (Fig. 61). In the line representing the first four months of the year, the divisions are drawn to scale; *K*, *N*, *I* represent the 1st, 5th, and 13th of January, February, and April, and the 1st, 7th, and 15th of March. The letter *B* shows the position of the bissextile day, the second 23rd (or 24th) day of February. The days included in the spaces, *a*, *b*, *c* of each month are counted backwards, respectively,

[67] Thus, our New Year's Day originated in 153 B.C., but it was not used continuously from that time on.

[68] The word *calendae* was generally written *kalendae*, the letter *k* being simply an ancient form of *c* which was kept for the sake of religious archaism. Note that the words *calendae, nonae, idus* designating single days are plural. Those fixed days were of lunar origin, the *calendae* corresponding (at first) to the first crescent, *nonae* to the first quarter, *idus* to the full moon. In the course of time the Roman calendar became more solar and the fixed days had less and less connection with lunar phases.

[69] Seventh or fifteenth in March, May, July, and October; *nonis Martiis* = 7 March; *Idibus Octobris* = 15 October.

[70] The Church preserved the insertion of the extra day between 23 and 24 February. Thus, the feast of St. Matthew, 24 February, is celebrated in bissextile years on 25 February; E. Cavaignac, *Chronologie* (Paris, 1925), p. 20.

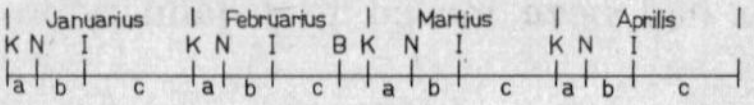

Fig. 61. First third of a Julian bissextile year.

from the following *N*, *I*, or *K*. Thus, the days of the second half of each month are counted backward from the first day of the next month. Out of the 366 days of a bissextile year, there are only 36 that have a direct name; the other days, that is, the vast majority of them (330), are counted backward from the next of the 36 principal days.

It was proper to explain the Roman calendar with some detail (though not by any means completely), because this reveals a new aspect of Roman life and of the Roman mind. The Romans are generally believed to have been matter-of-fact and practical, yet their method of counting the days was really backward and was as awkward as possible. Why did they behave in that strange way? The answer is simply that the determination of the calendar was a religious business; the pontiffs in charge of it preferred to keep it as esoteric as they could. The more obscure it was, the more sacred would the calendar appear.

The priests were deliberately abstruse, but, in my eagerness to be quickly understood by my readers, I go perhaps to the other extreme. For example, I quote all my dates in B.C. style, and obviously there could not be any such dates before Christianity. The use of *Anno Domini* dates was first suggested by Dionysius Exiguus (VI–1) and did not obtain any currency until the tenth century; the use of dates B.C. was a much later innovation.

Many writers on ancient history affect to quote dates *ab urbe condita* (U.C.), from the foundation of Rome, but that era remained for very long uncertain. It was established somewhat arbitrarily, seven centuries after the event, by Varro, as equivalent to 753 B.C.[71] The Romans hardly used that style of dating, however. Their common practice was to designate the years by the names of the ruling consuls. We could not do that without burdening the reader with unnecessary ambiguities. The years B.C. are by all means the simplest,[72] and it is for that reason that I am using them exclusively.

According to the Julian calendar, the average length of the year was 365¼ days, which was a little too much. The excess was small, only 11 minutes and 14 seconds, or 0.0078 day, yet that added up to 1 day in 128 years. In 1000 years the Julian calendar fell behind nearly 8 days. A new reform was long felt to be necessary and was finally realized on 4 October 1582 by Pope Gregory XIII. Such as it was, the Julian calendar had been in current use for more than sixteen centuries (1627 years).

The Roman count of days from the calends, the nones, or the ides con-

[71] Hence, x U.C. = $(753 - x + 1)$ B.C.; 753 U.C. = 1 B.C.; 754 U.C. = A.D. 1.

[72] Tables for the translation of consular years into U.C. or B.C. dates were compiled by Dante Vaglieri and published in Ettore de Ruggiero, *Dizionario epigrafico di antichità romane* (Spoleto, 1910), vol. 2, pp. 1143–1181; they range from 509

tinued throughout the Renaissance and even later. Erasmus' letters to his friends or the letters he received from them were generally dated in the Roman way.[73] There are still living so-called humanists who, when they have to indite a letter in Latin, instead of dating it, say, *Vicesimo quinto Augusti* 1955, prefer to put it *Octavo ante kalendas Septembres* 1955. That is tradition with a vengeance.

Julius Caesar had taken so great a part in the reform of the calendar, not only because that was his duty as *pontifex maximus*, but also because he had a genuine interest in astronomy. He composed a treatise, *De astris*, which was a kind of "farmer's almanac" in which data concerning the stars, the seasons, and the weather were combined. As far as the stars and weather signs were concerned, this continued the tradition of Aratos; other Hellenistic data were made available to him by Sōsigenēs; Caesar and his secretary were naturally familiar with Roman weather lore. It is probable that the Julian calendar and the *De astris* were introduced together. The calendar lasted until 1582; the *De astris* could not last as long, yet its popularity was remarkable; it extended almost to the end of antiquity. Iōannēs Lydos,[74] who flourished in the sixth century, was still making use of it. The *De astris* might be called astro-meteorological, as our farmer's almanacs still are, but it was not astrological *stricto sensu*. Caesar was ready to accept farmer's signs and predictions but not to put up with horoscopic nonsense. He shared to some extent the healthy skepticism of Lucretius and Cicero, and was one of the last Romans to uphold their rationalism.

Shakespeare helps us to remember that a soothsayer[75] had warned Caesar to beware of the thirty days preceding and including the ides of March. In spite of that warning and of the fearful entreaties of his wife, Calpurnia, Caesar went to the Senate on that fatal day and was murdered (15 March 44 B.C.).

My discussion was restricted to the Roman calendar, which became the supreme calendar in the empire. The Hellenistic (Greek) calendars were not discussed, because the matter is too complicated. Here again, the contrast obtaining between Roman unity and Greek anarchy is striking. There

---

B.C. to A.D. 631. Shorter tables by Willy Liebenam, *Fasti consulares imperii Romani vom 30 v. Chr. bis 565 n. Chr.* (128 pp.; Bonn, 1910); these tables begin only with Julius Caesar.

[73] Erasmus is singled out as an example, because it is so easy to consult his *Opus epistolarium* as edited by Percy Stafford Allen (1869–1933) and his successors (11 vols.; Oxford, 1906–1947). Some of Erasmus' letters are dated in our way, most of them in the Roman way.

[74] Iōannēs Lydos was born in 490 in Philadelphia, Lydia. He wrote treatises on the months (*Peri mēnōn syngraphē, De mensibus liber*) discussing the Roman calendar, on wonders (*Peri diosēmeiōn, De ostentis*), and on Roman magistrates (*Peri archōn, De magistratibus reipublicae Romanae*). The best edition of the whole, as far as it is preserved, is that by Immanuel Bekker (Bonn, 1837), Greek and Latin.

[75] The haruspex (inspector of entrails) Vestritius Spurinna. Shakespeare, *Julius Caesar* (Act 1, scene 2; Act 3, scene 1).

were separate calendars in every Hellenistic state, with little agreement except with regard to the great games (*ta*) *Olympia, Isthmia, Nemea, Pythia.*

The Olympic games were celebrated at Olympia, Elis (northwest Peloponnēsos) every fourth year (corresponding to B.C. years divisible by four). The Pythic games, near Delphoi in Phōcis (north of the middle of the Gulf of Corinth) were also celebrated every fourth year, but two years after the Olympic. The Isthmic games and Nemeic games were celebrated every second year, the Isthmic at the Isthmus of Corinth, the Nemeic at Nemea in Argolis (northeast Peloponnēsos). Thus, at least one or another of the games occurred every year; for example: 480, *Olympic* and Isthmic; 479, Nemeic; 478, *Pythic* and Isthmic; 477, Nemeic; 476, *Olympic* and Isthmic; 475, Nemeic; 474, *Pythic* and Isthmic. (The names of the four-year games are italicized).

The names of victors were recorded and the games duly listed. As those games, especially the Olympic, which were by far the most important, interested every Greek, the lists of them provided a chronological background. This was described when I dealt with Timaios of Taormina, in Chapter XII.

Aside from this athletic dating, the most successful and long-lived of Hellenistic chronologies was the Seleucid in Syria and Mesopotamia, the era of which was the entry of Seleucos Nicatōr in Babylon in 312/311. That chronology is of great importance not only for political historians but also for historians of science, as it was extensively used in cuneiform tablets, some of which record mathematical, astronomical, and other scientific facts. The success of a chronological system may be said to be established when it is adopted by other nations. The Seleucid was adopted by the Arsacid or Parthian dynasty.[76] The acta of the first Oecumenical Council of Nicaia were dated SE 636 (= A.D. 325). What is more, the Arabs adopted it, at least for astronomical purposes, under the name of Dh'ūl-qarnain (the two-horned one, Alexander the Great). That name is somewhat justified in that the Seleucid calendar was a belated fruit of the Alexandrian revolution.

A few words must suffice concerning the Jewish calendar. Its era is 3761 B.C., but that is a late invention of Jewish rabbis who wanted to begin with the assumed date of the Creation. The Jewish calendar, purely lunar and religious, was begun only at the end of the second century after Christ, and hence an account of it does not find its proper place in this volume.

THE WEEK

The year, the month, and the day were astronomical units of time, but they did not suffice for the arrangements of civil and religious life. The

[76] The Seleucid dynasty lasted from 323 or 312 to c. 64 B.C.; the Arsacid from 250 B.C. to A.D. 226. The Arsacid had their own chronology, but generally added the Seleucid to the Arsacid date.

month was too long and the day too short; some intermediary between them was needed. It is true that the four phases of the moon (new, first quarter, full, last quarter) suggested a quaternary division of the month, but the exact lengths of those phases were not easy to determine. Those phases are probably the origin of the unit that we call the week, yet a long evolution was required before that additional unit could be sufficiently standardized.

Among ancient nations the Babylonians and later the Jews were the first to think of a seven-day week. In the case of the Babylonians, the seven days were of planetary origin (they knew seven planets, Sun and Moon included); in the case of the early Jews, there is no evidence of planetary influence and the days were numbered, as in Genesis 1 or Exodus 20:11, their first day corresponding to our Sunday and the seventh being the day of rest or Sabbath.[77]

The Egyptians used a longer unit, the decan or decade. Each of their months was divided into three decans, the year into thirty-six. We find something analogous in the Attic calendar. The full months (of 30 days) were divided into three decades; the hollow months (of 29 days) were also divided into three periods but the third lacked one day. It is remarkable that the days of the third period (not those of the first two) were numbered backward (as in the Roman calendar), the first day of that period being called *decatē* (*hēmera*) *mēnos phthinontos* (tenth day of the declining month). In the case of hollow months either the tenth or the second day (that is, the first or the ninth day) of the third period was canceled.

The Romans had a week of eight days and the eighth day was called *nundinae* (short for *novem dies*). Why nine? The days were marked in calendars by letters:

A B C D E F G H,

and the last day, a market day, was the ninth counting inclusively from the preceding market day! That is, from one H to the next, one counted nine, if the first H was called one. It is clear that an eight-day week could have no planetary meaning. Periodic market days were needed, and they had been spaced that way by the buyers and sellers for the sake of convenience without any religious afterthought.

In Babylonia, each day was consecrated to a planet, and the same usage established itself in Hellenistic times, the names of the planets being translated into Greek, or being given Egyptian equivalents in Greek Egypt. The story is very long and complex and we must restrict ourselves to the main pattern, which may be represented briefly in a synoptic table.[78]

[77] The choice of seven days in Genesis may have been partly inspired by the existence of seven planets, but it is impossible to prove that.

[78] Franz Cumont, "Les noms des planètes et l'astrolatrie chez les Grecs," *Antiquité classique 4*, 5–43 (1935).

Names of the planets [79]

Modern	Babylonian	Greek	Egyptian	Latin
Moon	Sin	Selēnē	Thoth	Luna
Mercury	Nabu	Hermēs		Mercurius
Venus	Ishtar	Aphroditē	Isis	Venus
Sun	Shamash	Hēlios	Rē [80]	Sol
Mars	Nergal	Arēs	Ertōsi	Mars
Jupiter	Marduk	Zeus	Osiris	Jupiter
Saturn	Ninib	Cronos	Horus	Saturnus

Strictly speaking, many of those divine names were not really names but short cuts for expressions like *ho astēr tu Hermu, tēs Aphroditēs, tu Dios*, or, in Latin, *stella* (or *sidus*) *Mercurii, Veneris, Jovis* (the star of Mercury, of Venus, of Jupiter). It was only about the end of the Hellenistic age that attempts were made to give Greek names to the planets,[81] and their use was poetic, pedantic, or esoteric, never popular. A good example of it occurs in the monumental horoscope of Antiochos I Epiphanēs, king of Commagēnē, representing the conjunction of Mars, Mercury, and Jupiter at the time of his coronation in 62 B.C.[82]

The synoptic table above illustrates the fact that the association of the seven planets with seven deities was universal. In the course of time, the association became a real identification; the star of Venus became Venus herself. It is impossible for us to recapture that illusion, but there can be no doubt concerning its existence almost everywhere.

The seven-day week, the planetary week, was accepted all over the Roman world toward the end of the first century B.C. This is in itself very remarkable, but what is more remarkable is that the acceptance of the seven-day week was as implicit and casual as any bit of folklore.

How could that happen? The seven-day concept was favored by many tendencies. Seven days is the closest approximation to the length of one of the moon phases; [83] from that point of view, the seven-day period was natural. Belief in hebdomadism (the sacredness of the number seven, p. 165) was widespread. The Jewish account of creation in Genesis specifies seven days. The seven-day week is physiologically adequate; six days of labor and one of rest is a good rhythm.[84]

[79] In the order of increasing distances from the Earth.

[80] Atum and Horus-Harakhte were also considered as sungods.

[81] Mercury, Stilbōn (scintillating); Venus, Phōsphoros, Lucifer (light carrier); Mars, Pyroeis (fiery); Jupiter, Phaethōn (radiant); Saturn, Phainōn (luminous). Compare also the association of the Sun with Apollōn Phoibos, Latin Phoebus, the radiant one.

[82] He had been reinstated by Pompey but was deposed again by Antony in 38. The kingdom of Commagēnē, detached from the Seleucid territory in 162 B.C., suffered various vicissitudes and was finally annexed to Rome by Vespasian in A.D. 72.

[83] The first phase (from new moon to first crescent) lasts about 7.5 days, the second 6.75, the third 7.75, the fourth 7.5; total, 29.5 days, the synodic month (more exactly 29.52 days).

[84] The decade was a little too long;

It was the extraordinary convergence of those tendencies that insured the success of our week. It was established automatically; at any rate, we have no document or monument evidencing any kind of governmental or religious sanction.

The acceptance and universal diffusion of the week is comparable to the acceptance and universal diffusion of the basis ten in number systems (that is, as far as integers were concerned). In both cases, unanimity was obtained with relative ease, because it was casual and instinctive. If some administrative busybodies had organized conferences for the discussion of the seven-day week (or the decimal basis, or both), there would have been dissenters explaining the superiority of a shorter or longer week (or of one of the bases 2, 8, 12, 60),[85] and they would have created disagreements and disunions; in the course of time there would have been dissenting minorities, heresies, rebellions, and so on.

Humanity was spared infinite troubles by the anonymous inventors and early sponsors of the hebdomadal week and of the decimal basis of numeration.

The religious origin of the week is proved by the existence in any week of a religious day which is either its beginning (in Christianity) or its climax (the Jewish Sabbath). The astrological origin is even more evident, at least in the majority of the calendars, because of the names given to the days. Consider for example, the English and Italian [86] names of the days and the corresponding planets:

Sunday	Domenica	Sun	Thursday	*Giovedi*	Jupiter
Monday	*Lunedi*	Moon	Friday	*Venerdi*	Venus
Tuesday	*Martedi*	Mars	*Saturday*	Sabato	Saturn
Wednesday	*Mercoledi*	Mercury			

The connection with our planets is obvious in the words italicized. It is hidden in the English names of the third to sixth days, because they are derived from those of Anglo-Saxon or Scandinavian gods corresponding to the classical ones, Tiw, Woden, Thor, and the goddess Frig.

nine days of work instead of six was a little too much. The calendar established during the French Revolution included decades instead of seven-day weeks. It lasted only fifteen years (1792–1806). I have often wondered whether its early death was not partly due to a physiologic reason; one day of rest or recreation out of ten is too little for human comfort.

[85] For discussion of decimal and non-decimal bases, see G. Sarton, "Decimal systems early and late" *Osiris* 9, 581–601 (1950), 2 fig. It is interesting that the binary system is now used in the electronic computing machines, but the results are translated into the decimal system. The binary would be unbearable in everyday life, because the numbers, even very small ones, include so many digits; for example, $64 = 2^6 =$ binary 1,000,000. The resurrection of the binary system, at least in the machines, is a good example of the unpredictability of human affairs.

[86] Strictly speaking, there were no Latin names, any more than for the planets themselves. The planets were called *Mercurii stella, Veneris stella,* and the days *Mercurii dies, Veneris dies,* and so on. Only the gods had names.

The Italian names of the first and last days of the week are, respectively, Christian (the day of the Lord) and Jewish. The names occurring in other Romance and Germanic languages are of the same origin as those used in Italian and English. It is astonishing that the Catholic Church was never able to extricate itself from the astrological terminology.[87]

The Orthodox Church was more vigilant. For example, the Greek names of the days are *cyriacē, deutera, tritē, tetartē, pempē, parascevē, sabbaton*, that is, the day of the Lord, second, third, fourth, fifth, preparation, sabbath. The only one of those names that requires explanation is the sixth. Preparation means the Jewish preparation for the sabbath; the Hebrew word for it, *netot*, was translated into Greek in the New Testament (Mark 15:42). Good Friday is called in Greek *hē megalē* (*hagia*) *parascevē*. None of the Orthodox day names is astrological.

The numbering of the days, beginning with Sunday, the first day, is the rule not only for the Orthodox Christians but also for the Jews and Muslims and they all call the last day sabbath. The Muslims call the sixth day *yawm al-jum'a*, for it is their day of religious meeting.

The year, month, day are incommensurable between themselves; that is, none can be expressed correctly in terms of the two others. Hence all the calendrical troubles. The weeks did not introduce similar difficulties, because they continued across the months and the years without reference to them.

The only exception to this was the Babylonian week, which was part of their month. The Babylonians attached a special importance to the 7th, 14th, 21st, and 28th days of each month, which was thus divided into four periods of seven days plus a remainder. Those days were to some extent sacred days, but the weeks were not real weeks, because they were not continuous. The first day of each month was always the first day of a week.

On the contrary, the octonary weeks of the Romans were continuous. There was a restriction to that, however. The *nundinae* was a market day and the farmers who had created its periodic recurrence did not want it to coincide with the *nonae* or with the *calendae Januariae*. This was simply a taboo which could not be overcome except by inserting from time to time a day between two weeks. These insertions were finally established in a cycle of 32 years, because 32 Julian years = 11,688 days, including 1461 *nundinae*.

Hence, the Babylonian and the Roman weeks were different from ours,

[87] It is less astonishing, however, if we remember the extraordinary mixture of Christianity and paganism that obtained during the Renaissance, even in high ecclesiastical or academic circles. The Latin liturgy followed Jewish usage, at least since Tertullian's time (c. 160–230), calling the days *feria prima* (Sunday), *feria secunda, tertia,* and so forth, but those terms were never used otherwise and remained unknown to the laity.

the first because it was not continuous, and the second (if we disregard the little discontinuities just referred to) because it was octonary.

Our week, the astrological *hebdomas*, is strictly continuous, suffering no interruption from month or year. Any day may be the first of a year, or the first of a month.

THE HOURS

There is still an important feature of the astrological week to be explained. The seven planets known to the ancients were, in order of decreasing distances from the Earth, Saturn, Jupiter, Mars, Sun, Venus, Mercury, Moon. One would expect to find them in that order (or the reverse one), while the calendrical order is very different.

To account for that it is necessary to speak of still another division of time, a fraction of the day, the hour.

The Egyptians divided the day into 12 hours, and the night also into 12 hours, but as the day increased (or decreased) the length of the day hours increased (or decreased), while the length of the night hours decreased (or increased).[88] The Sumerians divided the days into three watches and the night into three watches (and those watches increased or decreased during the night or day). The Jews did the same (*ashmoreh*, Exodus 14:24; *phylacē*, Matthew 14:25). The mathematical genius of the Sumerians revealed itself somewhat later, and they realized the impracticality of unequal watches for astronomical purposes; they then divided the whole day (day plus night, *nychthēmeron*) into 12 equal hours of 30 *gesh* each. There were thus 360 *gesh* in each full day, just as there were 360 days in each year.

We have inherited the division of the full day into 24 hours from the Egyptians and the very important concept of equal hours from the Babylonians.

That concept was so advanced, however, that it was not understood by the ancient people, except the astronomers. Hipparchos divided the *nychthēmeron* into 24 equinoctial [89] hours.[90] For all other people (not only the

[88] When speaking of unequal hours, we mean unequal from day to day; but all the day hours of a single day were equal, and so were all the night hours.

[89] Equal hours are called equinoctial because the unequal hours of day and night become equal at the time of the equinoxes.

[90] The term *hōra* in our sense of hour is relatively recent; at the beginning, the 12 parts of the day or night were simply called parts (*merē*). The word *hōra* meant any kind of period (year, month, season); the technical meaning later was that of hour of the day (unequal or equal). The semantics of the English word hour is similar to that of the Greek *hōra*. The poetical genius of the Greeks created the *Hōrai* (*Horae, Hours*) who were goddesses of the order of nature, the seasons, rain-giving, and so on. The *Hōrai* were three, Thallō, Carpō, and Auxō, a divine group symmetric with two others, the three *Moirai* (*Parcae*, Fates), Clōthō, Lachesis, and Atropos, and the three *Charites* (Graces), Euphrosynē, Aglaia, and Thalia. Statues of those goddesses generally appeared in groups of three, six, or nine.

common people but the most educated ones as well) the day was divided into 24 unequal or seasonal hours (*hōrai cairicai*), 12 day hours of one length plus 12 night hours of another length. Some sundials or clepsydras were arranged to show the correct hours throughout the year (*hōrologia hēliaca, sciothērica*).

The Romans used unequal or seasonal hours. At the equinoxes, those hours were equal and the day from our 6 A.M. to our 6 P.M. was divided into 12 hours called *prima hora*, . . . , *duodecima hora*. The *septima* (seventh hour) began throughout the year at noon (*meridies*). The day was also divided into four watches: *mane*, from sunrise to the end of the second hour; *ad meridiem*, from the third hour to the end of the sixth; *de meridie*, from noon to the end of the ninth hour; *suprema*, from the tenth hour to sunset. The night was divided into four watches (*vigiliae*) of unequal length throughout the year, but the third always began at midnight (*media nox*, *noctis meridies*).

The division of the whole day into unequal hours continued in Europe in some places as late as the eighteenth century.

We may now return to the astrological week and justify the succession of its days. The astrologers, being astronomers to begin with, divided the nychthēmeron into 24 equal hours. Each hour was dedicated to one of the seven planetary gods, and each day was called after the god of its first hour.

Let us begin with the day of Saturn (*Saturni dies*), so called because its first hour was dedicated to Saturn; the second hour was the hour of Jupiter; the third, of Mars; the fourth, of the Sun; the fifth, of Venus; the sixth, of Mercury; the seventh, of the Moon.

Not only the first hour, but also the 8th, the 15th and the 22nd were dedicated to Saturn. The 23rd and 24th hours were dedicated to Jupiter and Mars, and therefore the first hour of the following day belonged to the Sun, and that day was called *Solis dies.* Therefore, the astronomical order of planets:

Saturn, Jupiter, Mars, Sun, Venus, Mercury, Moon,

was replaced by a new order obtained by jumping two items after each item of the first series. One thus obtains

Saturn, Sun, Moon, Mars, Mercury, Jupiter, Venus,

which is the order of our days,

Saturday, Sunday, Monday, Tuesday, Wednesday, Thursday, Friday.

This can be illustrated more clearly by means of diagrams (Figs. 62 and 63).

Note that the planetary week proves two things. First, astrologic beliefs were so strong in antiquity that our weekdays, which constitute an outstanding part of our vocabulary, still bear the stamp of that superstition.

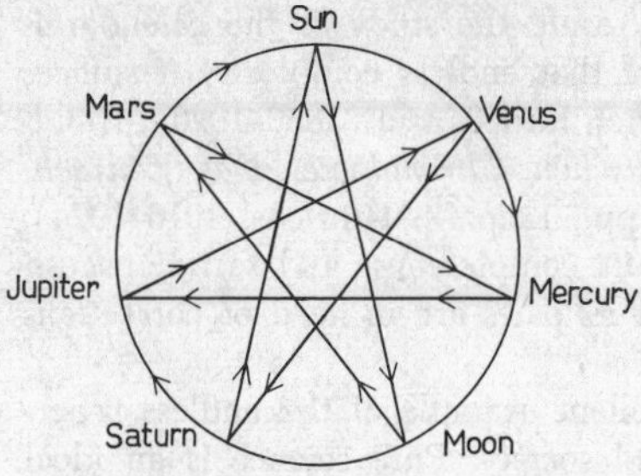

Fig. 62. Diagram for deriving the order of the weekdays from that of the planets; start with the Sun and, skipping two, follow the diagonals from Sunday to Monday, Tuesday (Martedi) . . . Saturday. The planets are shown in their ancient order clockwise around the circle from Saturn, the most distant, to the Moon, the least distant.

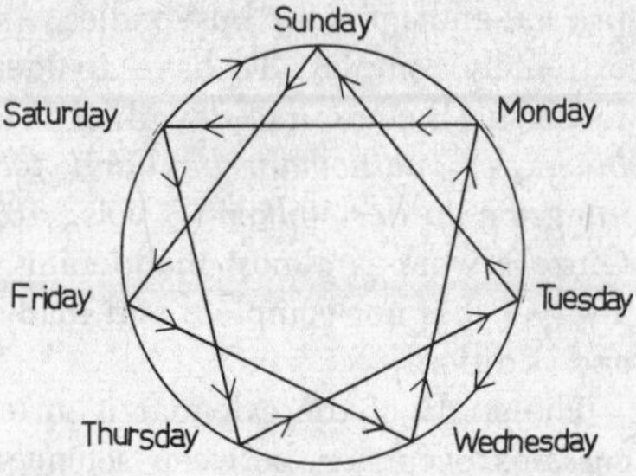

Fig. 63. Diagram for deriving the order of the planets from that of the weekdays; start with Saturday and, skipping one, follow the diagonals from Saturday to Thursday, and so on, passing from Saturn to Jupiter, Mars, Sun, Venus, Mercury, Moon, that is, reviewing the planets in the order of decreasing distances from the Earth, according to ancient ideas. The weekdays are shown in their natural order clockwise around the circle.

We are using astrological terms willy-nilly many times each day. Second, it proves that the division of the day into 24 hours was accepted by the astrologers, if not yet by the common people.

I hope that the reader will forgive me for having devoted so much space to the calendar from eras and years to months, weeks, days, and hours. This may seem very remote from science, yet the establishment and regulation of each of those periods implied astronomical knowledge, and in their turn they influenced astronomy very deeply. To say that they influenced astronomy is an understatement; no astronomy would be possible without determination of time. Even to this day it is necessary to continue that determination with greater and greater precision, and that is one of the major tasks of the observatories and of some physical laboratories.

That is only one side of the picture, however. Chronology is not only one of the basic requirements of the astronomer; it is also a fundamental tool of the historian and, as it expresses the many rhythms of our lives, it concerns every man. Rational men have helped to build up the science of chronology but irrational men, far more numerous, have not been inactive. Therefore, the calendar is not simply a scientific achievement, or that achievement is far from pure but is mixed with an incredible amount of irregularity and impurity. The historian of chronology is obliged to deal not only with science but also with folklore (the folklore of every nation), astrological and other superstitions, and the dogmatic arbitrariness of magistrates, priests, and ignorant busybodies. As a result, the study of the calendar is extremely complex. To have an idea of that endless complexity it suffices to consult the admirable work of Friedrich Karl Ginzel (1850–1926), *Handbuch der mathematischen und technischen Chronologie. Das Zeitrechnungswesen der Völker* (3 vols., 1652 pp.; Leipzig: Hinrichs, 1916–1924). Ginzel's work is almost maddening in its completeness and fastidiousness, and yet it is not complete and many of its parts are in need of corrections and additions.

The study of the calendar is an excellent example of the endless repercussions occurring between science and society. Pure science is an ideal which could be realized only in a social vacuum, and this is simply a way of saying that it cannot exist, or that it could never exist for very long.

Ginzel is the main source of reference. There are many other books or memoirs. For the weeks, see F. H. Colson, *The week* (134 pp.; Cambridge: University Press, 1926). Solomon Gandz, "The origin of the planetary week or the planetary week in Hebrew literature," *Proceedings of the American Academy for Jewish Research 18*, 213–254 (1949).

EGYPTIAN ASTRONOMY. THE ZODIAC OF DENDERA

If one sails up the Nile from Cairo to Luxor, one passes in latitude 26°+ the city of Qena (the Greek Cainēpolis = Newton!) in the neighborhood of which, on the western side of the valley, is Dendera,[91] one of the most ancient cities of Egypt. Dendera was dedicated to Hathor, goddess of joy and love (identified by the Greeks with Aphroditē), and it boasts a temple to her. The temple, which exists today, was built very late, at the end of the Ptolemaic age and during Augustus' rule, upon the site of a much older one going back to the Ancient Empire. On the ceiling of one of the rooms on the temple roof is a representation of all the constellations, generally called the zodiac of Dendera. It is a bas-relief framed in a circle, the diameter of which measures 1.55 m. The original is now in the Bibliothèque Nationale in Paris and is replaced *in situ* by a plaster cast.

The zodiac of Dendera was discovered in 1798 by General Louis Desaix de Veygoux, whom Bonaparte had sent on an expedition to Upper Egypt, and its existence was first announced, together with that of five other monuments of Egyptian astronomy, in the *Description de l'Egypte.*[92] It attracted considerable attention[93] because it was first believed to be extremely an-

[91] Dendera (or Dandara), is a corruption of the Greek *ta Tentyra;* its distances from Cairo and Luxor along the river are about 400 and 60 miles.

[92] *Description de l'Egypte, ou recueil des observations et des recherches qui ont été faites en Egypte pendant l'expédition de l'armée française* (19 vols.; Paris, 1809–1828).

[93] The literature devoted to it is considerable, much of it published in 1822 and following years. There is not yet a good and full account of it. A list of the 48 constellations represented (21 boreal, 12 zodiacal, 15 austral) is given by E. M. Antoniadi, *L'astronomie égyptienne* (Paris, 1934) [*Isis* 22, 581 (1934–35)], pp. 60–74. For Dendera literature, see Ida A. Pratt, *Ancient Egypt* (New York), vol. 1 (1925), pp. 124–125; vol. 2 (1942), p. 95.

cient. J. B. J. Fourier (who had accompanied Bonaparte to Egypt), writing in 1830, considered it to be 40 centuries old. Fourier was a mathematician of considerable genius, but not an Egyptologist.[94]

Scholars are now agreed that the zodiac of Dendera is very late and the only disagreement is whether it is late Ptolemaic or Augustan. According to François Daumas, the most probable date would be 100 ± 20 B.C.[95] The exact dating of that monument does not matter very much, if we recognize it as late Ptolemaic; even if it had been completed only in Roman times, that would hardly affect its nature; it is definitely an Egyptian monument, preserving ancient traditions.

We may call it the last astronomical monument of Egypt. It is the only monument of its kind framed in a circle.[96] One might even say, it is the only example of Egyptian decorative art of circular shape; that in itself is a sufficient proof of low or late antiquity.

BABYLONIAN ASTRONOMY

It was very necessary in my Volume 1 to explain the Babylonian (or, more exactly, Sumerian) mathematics which was very much – say a thousand years – earlier than the Greek and helped to account for certain oddities of Greek mathematics. We now realize that the Greeks were standing on the shoulders of Oriental giants, some of whom were Egyptian, others "Babylonian" – some of whom had dwelt along the Nile, the others along the two rivers Euphratēs and Tigris and in the space extending between them (*hē mesē tōn potamōn*, Mesopotamia).

The ancient Babylonian knowledge of mathematics and astronomy percolated into the Greek world at least as early as the time of Pythagoras, and much more actively in the post-Alexandrian age when Babylonian, Egyptian, and Greek astronomers had opportunities of meeting one another in the Aegean islands, in Egypt, and in Western Asia.

The best proof of mathematical percolation is given by the survival of the sexagesimal fractions; the best proof of astronomical percolation is the discovery of the precession of the equinoxes by Hipparchos, partly on the

[94] Jean Baptiste Joseph Fourier (1768–1830). The Fourier formula, series, and theorem are named after him.

[95] In a letter to me dated Castelnau-le-Lez (Hérault), 20 February 1954. The epigraphy in the part of the temple containing the zodiac is not of the Roman kind conspicuous in other parts.

[96] According to Richard A. Parker (letter dated Providence, R. I., 23 September 1955); there are several other zodiacs as yet unpublished in the tombs of Sohāg on the Nile (about 50 miles southeast of Asyūṭ). They are circular, but rather crude as compared with the zodiac of Dendera. They are probably Roman of the first century after Christ. Professor Parker does not remember any Egyptian monument of circular shape, whether astronomical or not, earlier than the Dendera zodiac. However, consider the solar symbols, a bas-relief in the wall of the subterranean tomb of Seti II (c. 1205 B.C.) in Thebes. The symbols, a scarab and Ammōn, are inserted in a circle which is itself a solar symbol, the solar disk Atōn. José Pijoán, *Summa artis* (Madrid, vol. 3, 1932), Fig. 560.

basis of Babylonian observations. There were other Babylonian elements in Hipparchos, which were transmitted to his successors and appeared in the *Almagest*.

Still another proof of mutual influence, though this time in the opposite direction, was the defense of the heliocentric system by Seleucos the Babylonian, one of Hipparchos' contemporaries.

It would be extremely interesting to know more exactly how Babylonian knowledge was transmitted to the Greeks or vice versa, but such information is lacking. It is highly probable that the exchange of data and even of methods was very largely personal and oral; it was a secret transmission which left but few traces and can be inferred only from the results, sometimes very distant ones, like the *Almagest*. Oral transmission is still important with us, as in the scientific meetings and international congresses, but it was infinitely more important in ancient times. Even when information has been transmitted to us orally, we are not satisfied with it until we have read the explanations with our own eyes. The ancients depended upon oral information because in the majority of cases no written explanations were available.

The Seleucid empire was weak and chaotic; some vassals were always plotting against their king. It had far less cohesion than the Lagid (or Ptolemaic) kingdom of Egypt. The Seleucid rulers were not very distinguished (much less so than the early Ptolemaioi) and their main virtue, perhaps, was their defense of Hellenism in Asia. The Greeks were a very small minority, however, and we can easily imagine, from our own experience, that there was considerable native hostility against them, something like the anticolonialism, nationalism, and xenophobia of our own days. Religion provides the best focus and furnace for such feelings. It was so in the Seleucid empire. The native priests had the power of denouncing their rulers in the most confidential and effective manner, of rallying the natives around approved leaders, and of exciting the passions of the populace.

As the Chaldean calendar was purely lunar (like the contemporary Hebrew one), the determination of the first crescent (and of other lunar times) was one of the main responsibilities of the priests. The latter were or became astronomers and, under the influence of old Babylonian traditions and of new circumstances, they developed a very original astronomy which will be briefly explained in the following section.

The originality of their efforts is astonishing; not only were those efforts independent of the Greek (we can easily understand that, if only on account of national prejudice), but they were also curiously independent of ancient Babylonian astronomy. The Chaldean astronomy is as original as the old Chinese astronomy and the Maya astronomy, which were developed in parts of the world as distant from the eastern Mediterranean as it is possible to reach. China was then inaccessible and Central America unthinkable.

CHALDEAN ASTRONOMY [97]

At about the time when Hipparchos was working in Alexandria and Rhodos, and when Seleucos was still championing the heliocentric system of Aristarchos, Chaldean priests were computing ephemerides of the Moon and planets in Mesopotamian temples. They did not develop a coherent astronomical system but an empirical method of recording and even of anticipating the positions of the Moon and planets. Their tables of the Moon were of special importance to them, because their calendar was purely lunar (like the contemporary Hebrew one); their main duty was to determine the first visibility of the new crescent. The tables indicated a little in advance the time when the crescent could be expected and thus facilitated the task of the observers.

A corpus of all the known Chaldean tablets and fragments, 300 in number, has been edited with commentary by Otto Neugebauer.[98] Those tablets were written in cuneiform script, one third of them in Uruk [99] and the rest probably in Babylon. Most of them were written during the Seleucid period (312–64 B.C.), some are later, down to A.D. 49. Many are dated, the dating being according to the Seleucid era (S.E. 1 = 311 B.C.).

The astronomers and scribes were priests in the service of Chaldean temples; the various scribes of the Uruk temples signed their tablets in the colophons and thus we know that they belonged to two families, Ekur-zākir and Sin-legē-unninnī; as their names are recorded in the usual Semitic style, "A son of B son of C . . . ," it is possible to reconstruct the genealogies of those two families.[100]

In spite of the fact that most of those tables belong to the Seleucid period, I prefer to call them Chaldean, because the term Seleucid evokes the Hellenistic government, while the priests-astronomers-scribes were natives. If the Seleucid rulers had wished to promote astronomy, they would have preferred to patronize the disciples of Aristarchos or Hipparchos, rather than the Chaldean priests. Moreover, it is extremely unfair to take away from the Chaldeans their best scientific work (and call it Seleucid) while giving full credit to them for their abundant superstitions. It is easy enough to damn any nation, if we credit all their bad deeds to them and

[97] This heading, "Chaldean astronomy," is specific, as will appear below, as opposed to the generic title, "Babylonian astronomy," of the whole section. "Babylonian" is far more general in its several acceptations than "Chaldean."

[98] Otto Neugebauer, *Astronomical cuneiform texts. Babylonian ephemerides of the Seleucid period for the motion of the Sun, the Moon and the planets* (quarto, 2 vols. of text, 528 pp., 1 vol. of 255 pl.; published for the Institute for Advanced Study in Princeton, New Jersey, by Lund Humphries, London, June 1955) [*Journal of the American Oriental Society*, 75, 166–173 (1955)].

[99] Uruk, also called Erech (Genesis 10:10) and Warka, is on the lower Euphrates, much below Babylon.

[100] Neugebauer, *Astronomical cuneiform texts*, p. 14.

their good deeds to others (that has often been done by politicians; it should not be done by historians of science).

The historical term, Chaldean, is short for late-Babylonian or Neo-Babylonian; it is used for the Neo-Babylonian empire (625–538) and the same Semitic people, the Chaldeans of Babylonia, were later ruled by the Persians (538–332), by Alexander (332–323), by the Seleucid diadochoi (312–64), by the Parthians (Arsacid dynasty 171 B.C.–A.D. 226), and by the Persians again (Sassanian dynasty) from A.D. 226 to the Muslim conquest in 641.

The geographic term Chaldea refers to the southern part of Babylonia, extending along the Euphratēs river from Babylon to the Persian Gulf. As far as their origin is known, all the tablets published by Neugebauer come from that very region.

One might call those tablets Babylonian, but it is better to use the term Chaldean (or Neo-Babylonian), as Babylonian evokes in most minds hoary antiquity, while the Chaldean tablets are relatively recent, some of them as recent as Jesus Christ, who is closer to us than to the earliest Babylonian mathematicians.

While the Greeks were concerned with trajectories and invented various geometric theories to account for them, the Chaldeans' purpose was a much humbler one; they tried to determine in advance, on the basis of anterior observations, the times of conjunction and opposition (syzygies), of first and last visibilities, and of eclipses. Their method was arithmetic rather than geometric. Following the old Babylonians, they used arithmetic progressions to describe periodic events; they had also inherited from their Babylonian ancestors the concept of the zodiac as a frame of reference for solar, lunar, and planetary motions, the sequences characteristic of those motions, variable durations of days and nights, and their extraordinary arithmetic skill. Their results were remarkably good, except in the case of solar eclipses, where they neglected an essential element, the parallax of Sun and Moon.[101]

The lunar ephemerides were necessary for religious purposes; we do not know the use of the planetary ephemerides, though they were very probably for divination. It is astonishing that the Chaldeans were much more interested in Jupiter than in the other planets; Jupiter is more brilliant than Sirius, the brightest star, but it is less brilliant than Venus can be when it is nearest to us.[102]

The Chaldean astronomers were very conscientious and tried various arithmetic methods for the computation of their tables. The two main

[101] In the Neugebauer corpus there are only three tables (out of 300) concerned with eclipses (two lunar, one solar). There are 41 texts and fragments on Jupiter, and only 40 on the other four planets together.

[102] I call the planet Jupiter to be understood by my readers. For the Babylonians, old and new, it was the star of Marduk, the chief god of their pantheon. The Greeks replaced Marduk by their own chief god Zeus, and the Romans by Jupiter. But why did they associate a planet

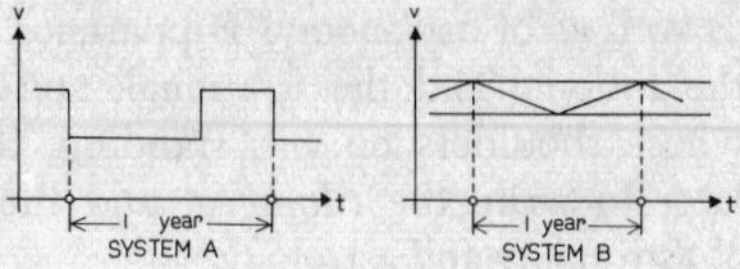

Fig. 64. Diagram illustrating two methods ("system *A*" and "system *B*") used by Chaldean astronomers to compute their ephemerides of the Moon. [Borrowed from O. Neugebauer, *Astronomical cuneiform texts* (London: Lund Humphries, 1955), vol. 1, p. 41.]

methods are called "system *A*" and "system *B*" (Fig. 64); the assumption of *A* is that the Sun moves with (different) constant velocities on two different arcs of the ecliptic; the assumption of *B* is that the solar speed varies gradually throughout the year. The second assumption is more refined than the first and yet it is not certain that it is posterior to it. At any rate, we must face the following facts. The Uruk tablets extend from 231 to 151 B.C., while the Babylonian ones extend from 181 B.C. to A.D. 49; that is, the Babylonian tablets are very largely later than the Uruk ones and yet most of them are of the *A* type; the Uruk tablets, more ancient, are almost exclusively of the *B* type.

It has been explained in another section of this chapter that the Chaldeans invented horoscopes, but that the practice was elaborated chiefly in Ptolemaic Egypt and the rest of the Greco-Roman world. The tablets edited by Neugebauer contain no trace of astrology, but more astrology is included in other tablets and there were very probably more Chaldean horoscopes than is realized today.[103]

In addition to the tablets investigated by Neugebauer there are others, which Father Kugler called "second-class ephemerides,"[104] giving the dates of entrance of planets into zodiacal signs. That was just the kind of information that astrologers needed to draw their horoscopes.

In spite of their rationalism, the Greeks were well prepared for the acceptance of the astrological aberration because of their belief in the astral religion, which seemed more "rational" and had become more acceptable to them than their fantastic mythology. From the astral religion to astrology was an easy step, and the political and economic miseries of their age pushed them to take it.

As far as theory is concerned, the Greeks were the creators of astrology

---

that was not the brightest with the chief god?

[103] A corpus of all Greek horoscopes is being prepared by Otto Neugebauer and Henry Bartlett Van Hoesen. Dr. Van der Waerden (in a letter to me dated Zürich, 11 January 1956) recalls that during the Seleucid period many legal and commercial documents were no longer written in clay; this was probably the case for the horoscopes and accounts for their rarity. The only Chaldean horoscopes to survive were the few written on clay tablets.

[104] The information of this paragraph and the following I owe to Professor Van der Waerden (letter, 11 January 1956). For the "second-class ephemerides" see F. X. Kugler, *Sternkunde und Sterndienste in Babel* (Münster in Westfalen, 1926), vol 2, pp. 470–513.

as well as of astronomy; Hipparchos worked powerfully in both directions, the rational and the irrational, and it was because of Hipparchos, upon whose shoulders he was standing, that Ptolemy was able three centuries later to write the *Almagest* and the *Tetrabiblos*, the bibles, respectively, of astronomy and astrology.[105]

Nevertheless, astrological fancies continued to be diffused by the Chaldeans themselves, as is witnessed by their reputation. Their influence upon posterity was double. The healthier kind was caused by the astronomical knowledge obtained from them by Hipparchos (for example, with regard to lunar motions), transmitted to Ptolemy, and integrated into Western astronomy. Van der Waerden has shown that planetary tables from the time of Augustus to Hadrian were computed by Chaldean methods. There was some progression in this, for the Hadrian tables were better than the older ones. Chaldean elements can be traced also in writings of Hypsiclēs (II–1 B.C.), Cleomēdēs (I–1 B.C.), Geminos (I–1 B.C.), Manilius (I–1),[106] not to mention the *Tetrabiblos* and the *Anthology* of Vettius Valens.[107] They all used Chaldean methods to compute the rising and setting of the Moon, its velocity, the rising times of zodiacal signs, and so forth. Manilius, Ptolemy, and Vettius have brought us back into astrology. The other kind of Chaldean influence, less healthy but more pervasive, was astrological. One might say that Chaldean computational methods were carried East and West by the makers of horoscopes, or by the astrologers who published tables or manuals for the guidance of practical horoscopists. Traces of Chaldean astrology can be detected in Sanskrit and Tamil literature,[108] and from India they percolated into Persian and Arabic writings. When Arabic writings were translated into Latin, those traces reached Western authors, such as Pietro d'Abano (XIV–1) and Western art, for example, the frescoes, dated c. 1470, in the Shifanoja Museum in Ferrara.[109] All this, however, did not count as far as the progress of astronomy was concerned; the only Chaldean elements that reached modern astronomers are those which came along the Hipparchian-Ptolemaic channel, being as it were incorporated and lost in the Greek tradition.

[105] G. Sarton, *Ancient science and modern civilization* (Lincoln: University of Nebraska Press, 1954), pp. 37–73.

[106] The list might be continued, but later Greek (or Roman) writers borrowed from those already named, for example, Pliny (I–2), Firmicus Maternus (IV–1), the Michigan papyrus (*Introduction*, vol. 1, p. 354), the *Geōponica* (*Introduction*, vol. 1, p. 370), Martianus Capella (V–2), Gerbert (X–2).

[107] Both the *Tetrabiblos* and the *Anthology* date from the middle of the second century after Christ. The *Tetrabiblos* is a formal treatise, the *Anthology* — as its name indicates — is a collection of astrological cases, horoscopes. Otto Neugebauer, "The chronology of Vettius Valens' Anthologiae," *Harvard Theological Review 47*, 65–67 (1954) [*Isis 46*, 151 (1955)].

[108] Otto Neugebauer, "Tamil astronomy," *Osiris 10*, 252–276 (1952).

[109] Otto Neugebauer in his elaborate review of Louis Renou and Jean Filliozat (editors), *L'Inde classique, manuel des études indiennes* (Hanoi: Ecole française d'Extrême-Orient, 1953) in the *Archives internationales d'histoire des sciences* No. 31 (April 1955), pp. 166–173.

The Chaldean virtuosity in astrology and other forms of divination is sufficiently proved by their fame which began early enough. The Greek word *Chaldaios* had already acquired the meaning of astrologer. Lucretius [110] referred to the *Babylonica Chaldaeum doctrina*, the Babylonian doctrine of the Chaldeans (a good way of combining the two adjectives), as opposed to the Greek doctrine. The Chaldeans were spoken of in the Old Testament as astrologers and magicians, people who were a little too clever. The Babylonians were not treated much better in the Old Testament and were mercilessly denounced in the New (Revelation 17:5). The labels stuck and throughout the ages Chaldean suggested not only astrology but also magic, occultism, and imposture, while Babylonian came to mean an astrologer and a papist! The word "Chaldean" was often used to designate a soothsayer or a fortune teller; it was considered a better insult than "Babylonian," except for religious purposes.[111]

The Chaldeans deserved their bad reputation, for they created superstitions in abundance. A good many of those superstitions survive in the folklore of the Mandaeans, a tribe of Gnostic Christians; the Mandaeans of today live in the same territory as the Chaldeans of old and are perhaps, to some extent, their physical as well as their spiritual descendants.[112]

It is a strange turn of fate that while their bad fame continued steadily throughout the ages their more serious achievements remained almost unknown until 1881. Since that time, they have been discovered, edited, and commented upon by three Jesuit pioneers, Joseph Epping (1835–1894), Johann Nepomuk Strassmaier (1846–1920), and Franz Xaver Kugler (1862–1929). We owe the most important studies to the last named, especially *Die babylonische Mondrechnung* (Freiburg im Breisgau: Herder, 1900) and *Sternkunde und Sterndienst in Babel* (Münster in Westfalen: Aschendorf, 2 vols., 1907, 1909–1924, and 3 supplements, 1913, 1914, 1935) [113] [*Isis* 25, 473–476 (1936)]. Kugler's activities are continued in great style by Otto Neugebauer,[114] Abraham Sachs, and B. L. Van der Waerden. Their resurrection of Chaldean astronomy is extremely interesting but cannot affect the astronomical thought of today. Except for the Chaldean ingredients transmitted to us by Hipparchos and Ptolemy, the growth of astron-

[110] *De rerum natura*, v. 727.

[111] Note that the adjective "Egyptian" was also given evil meanings, connected with astrology, occultism, or gypsy lore!

[112] The living Mandaean folklore was carefully investigated by Mrs. Ethel Stefana Drower (E. S. Stevens). She edited the Mandaean Book of the zodiac, *Sfar Malwasia* (London: Royal Asiatic Society, 1949). My review of it in *Isis 41*, 374 (1950) aroused an excellent rebuttal by Otto Neugebauer, "The study of wretched subjects," *Isis 42*, 111 (1951).

[113] A fourth supplement, to contain chronology and indices, was announced in 1935 but did not appear and probably never will.

[114] For a summary of his views, see Neugebauer's article on "Ancient mathematics and astronomy" in Charles Singer, *History of Technology* (Oxford: Clarendon Press, vol. 1, 1954) [*Isis 46*, 294 (1955)], pp. 785–803. This will be useful for students who wish a full survey of Babylonian astronomy, old and new, from the Sumerian to the Christian age and beyond.

omy would have been essentially the same if those ingenious priest-astrologers of Chaldea had not intervened.[115]

The following book was received too late to be used in the writing of this chapter: *Late Babylonian astronomical and related texts,* copied by Theophilus Goldridge Pinches and Johann Nepomuk Strassmaier, prepared for publication by Abraham J. Sachs and J. Schaumberger (Brown University Studies, vol. 18, 327 pp.; Providence: Brown University Press, 1955). This contains more than 1300 previously unpublished texts, excavated at Babylon about 75 years ago and now in the British Museum; the majority are astronomical texts of the last few centuries before Christ.

---

[115] B. L. Van der Waerden is preparing a memoir on "the widespread influence of Chaldean astronomy" (letter, 11 January 1956).

XX

PHYSICS AND TECHNOLOGY IN THE LAST TWO CENTURIES CTĒSIBIOS, PHILŌN OF BYZANTION, VITRUVIUS[1]

CTĒSIBIOS

The story of Hellenistic physics and technology used to be summarized with three illustrious names, Ctēsibios of Alexandria, Philōn of Byzantion, and Hērōn of Alexandria, the dating of whom was uncertain except that they appeared in that order. In my *Introduction*, volume 1, I placed them tentatively in the following periods: II–1 B.C., II–2 B.C., I–1 B.C. I was certainly wrong with regard to the last, whom it is better to place in I–2 after Christ.[2] Hērōn thus belongs to a later, post-Christian age, and my account of Hellenistic physics will be focused upon two men only, Ctēsibios and Philōn.

An ancient epigram suggests that Ctēsibios constructed a singing cornucopia for the statue of Arsinoē erected by her brother and husband, Ptolemaios II Philadelphos, c. 270 B.C. If that is true, Ctēsibios flourished a century earlier than I first thought. According to Tannery, he lived under Ptolemaios III Evergetēs (247–221). Whether he flourished in the third century or in the second, he was a barber and an engineer. The combination is curious but not implausible; he was a craftsman and inventor, and the trimming of hair and beard was a kind of craftsmanship. He wrote a book describing his inventions and experiments, but that book is lost, and whatever knowledge we have of him is derived primarily from Vitruvius (I–2 B.C.), secondarily from Philōn of Byzantion (II–2 B.C.), Athēnaios

[1] For physics and technology in the third century B.C., see Chapter VII.

[2] It is probable that Hērōn flourished after A.D. 62 and before 150 [*Isis 30,* 140 (1939); *32,* 263 (1947–49); *39,* 243 (1948)]. In 1938, Otto Neugebauer concluded that either Hērōn must be put at the end of the first century after Christ or else all dates from −100 to 200 must be considered equally probable; *Isis 30,* 140 (1939).

the mechanician (II–2 B.C.), Pliny (I–2), Hērōn (I–2), Athēnaios of Naucratis (III–1), and Proclos (V–2).

He invented a force pump, a water organ, and water clocks. When we say that he invented the force pump this means simply that he realized the need of three essential parts of it, the cylinder, the plunger, and the valve. His model was eventually improved by Philōn and others; it was the prototype of the two pumps found in Bolsena (now in the British Museum) and of a third one found near Civitavecchia.[3]

The water organ, which he called *hydraulis*, was the application of pumps to music; the air of a wind instrument instead of being supplied by the player's lungs was supplied by a machine. The nature of Ctēsibios' invention can be imagined from Vitruvius' imperfect description (*Architectura* x, 8, 6) and from various antique terracotta models. The air chamber containing water was needed to keep up the wind, and in order to direct the latter into the one or the other of various pipes, a keyboard was necessary. The essential parts of the organ were the pump, the air chamber, the pipes, and the keyboard. All organs are variations and improvements of the one devised by Ctēsibios.

The invention of the water organ was, as far as we can know, an absolutely new departure. The water clocks, however, were improvements of earlier devices to measure time. We need not refer to sundials, which were of no use except when the Sun was shining; water clocks were invented in Egypt in the second millennium.[4] Most clepsydras were used to measure a certain length of time as a whole rather than its subdivisions and the gradual passage of it; a speaker was allowed the time that it took a clepsydra of definite contents to empty itself; variations in the speed of outflow did not matter.[5] Ctēsibios' invention consisted in regularizing that speed and in making it possible to follow the passage of time. He realized intuitively that the outflow would be steady only if the head of water above it remained the same[6] and if the aperture of the outlet was constant. The aperture might be blocked by dirt or it might be enlarged by erosion; dirt would be avoided by the use of clear water and erosion by fashioning the aperture in gold or in a hard stone.[7] The head of water could be maintained constant only if the clepsydra was constantly replenished; then the

[3] For details, see Aage Gerhardt Drachmann, *Ktesibios, Philon and Heron* (Copenhagen: Munksgaard, 1948 [*Isis* 42, 63 (1951)], p. 4.

[4] Alexander Pogo, "Egyptian water clocks," *Isis* 25, 403–425 (1936), with illustrations.

[5] For the ancient use of clepsydras to measure the time allotted to speakers, see A. Rome, "La vitesse de parole des orateurs attiques," *Bulletin de la classe des lettres, Académie royale de Belgique* 38, 596–609 (1952); 39 (1953). Many years ago, I witnessed the same usage in the churches of Dalecarlia in Sweden. Clepsydras were placed in the pulpit ostensibly to put a limit to the preaching.

[6] This was first expressed clearly by Frontinus (I–2): the speed of outflow is a function of the height of the water above the outlet.

[7] Such as onyx. Arabic writers called the aperture *jaz'*, meaning onyx or agate.

water flowing out would be collected in another vessel. The elapsed time would be measured by the amount of water in it. Figure 65 is a diagram of the device. Water falls from *A* into the vessel *BC*; *B* is an overflow keeping the level constant; the water flows out at *C* and fills the vessel *D*; the amount of water present in *D* can be evaluated at any time by the position of the float *E*. Note that to transform the clepsydra into a water clock it was necessary to add an inflow vessel to the outflow one, as had been done by the Egyptians ages before.

Ctēsibios' inventions were fundamental; he might have applied for "basic" patents if such things had existed in his time. His ideas of the force pump, the water organ, and the water clock were susceptible of endless improvements.

PHILŌN OF BYZANTION

The last of the Hellenistic mechanicians whose name has come to us was Philōn of Byzantion (II–2 B.C.). He flourished after Ctēsibios and before Vitruvius (I–2 B.C.), probably closer to the former than to the latter; he spent much time in Alexandria and some in Rhodos. He was probably a military engineer employed by the state.[8] Fortifications had been built for centuries and war is one of the oldest human activities. In Philōn's time the arts of building fortifications and of besieging them (poliorcetics) were already well developed, nowhere better than in the island of Rhodos. The main city of the Rhodians was besieged in 305 by Dēmētrios, king of Macedonia. This Dēmētrios obtained so much fame as a taker of cities that he was nicknamed Poliorcētēs (the besieger). In spite of his use of gigantic siege engines, he failed to subdue the Rhodians, and made a treaty with them in 304. He so much admired their brave resistance that he gave them the machines he had used against them. The machines were sold and the sale price helped to build the famous Colossus. Rhodos was engaged in many conflicts and the arts of war were cultivated there more assiduously than anywhere else. We may assume that Philōn learned much in Rhodos; on the other hand, it is possible that his books were composed for the technical education of the rulers of that island.

He was the first man who tried to cover the engineering[9] arts of war — offense and defense — in a complete way. He wrote a great mechanical treatise (*Mēchanicē syntaxis*), divided into eight (or nine) books, of which only a third has come down to us. We are not quite sure of the division of the treatise, except for the extant parts, but it was probably as follows:

[8] Military engineering was one of the first technical professions. Think of the tradition represented by . . . Archimēdēs, Ctēsibios, Philōn, Vitruvius, Hērōn . . . Leonardo da Vinci, Vannoccio Biringuccio . . . Vauban . . . the builders of atomic bombs . . .

[9] The engineering arts as opposed to the human arts (training of soldiers or sailors, tactics, strategy). The selection and creation of weapons is a matter of engineering; their use, a matter of training and psychology.

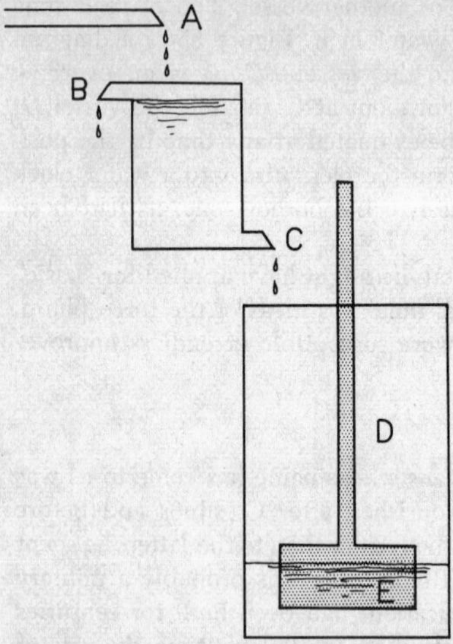

Fig. 65. Ctēsibios' water clock. [After A. G. Drachmann, *Ktesibios, Philon and Heron* (Copenhagen, 1948), p. 18, fig. 2.]

1. Introduction, generalities, mathematical preparation; for example, the duplication of the cube was discussed (lost);
2. *Mochlica*, the use of levers in machines (lost);
3. *Limenopoïca*, construction of harbors (lost);
4. *Belopoiïca*, construction of engines for shooting; first published in Greek and Latin by Melchisédech Thévenot in *Veterum mathematicorum opera* (folio; Paris, 1693), pp. 49–78 (Fig. 66);
5. *Pneumatica*, lost in Greek but preserved in Arabic; a small part is also extant in medieval Latin from the Arabic; the Latin text was edited by Valentin Rose, *Anecdota graeca et graecolatina* (281–314; Berlin, 1870) reprinted by Wilhelm Schmidt, *De ingeniis spiritualibus*, in Latin and German in *Heronis Alexandrini opera omnia* (Leipzig: Teubner, 1899), vol. 1, pp. 458–489; Baron Carra de Vaux, *Le livre des appareils pneumatiques et des machines hydrauliques*, in Arabic and French (Notices et extraits des MSS de la Bibliothèque Nationale, 38, 211 pp., Paris 1902).

6(?) *Teichopoiïca*, building of walls or fortifications (lost);

7. *Parasceuastica*, preparation of equipment and resources; defense of fortifications;
8. *Poliorcētica*, methods of besieging; books 7 and 8 are partly preserved in Greek; part of them was included in Thévenot's edition of 1693; Albert de Rochas d'Aiglun, *Traité de fortification, d'attaque et de défense des places*, French translation in *Mémoires de la Société d'émulation du Doubs* (vol. 6, Besançon, 1872).

The little treatise on the seven marvels of the world (*Peri tōn hepta theamatōn*) ascribed to one Philōn Byzantios is a composition of a later age (fourth or fifth century).

The most interesting of Philōn's genuine writings is the *Pneumatics*, whose influence was considerable. Out of 65 chapters in the Arabic text, only 16 exist in the Latin text,[10] and it has been argued that the Arabic text contained Arabic interpolations. It is difficult to hold that the medieval Latin text is closer to the Greek original, because it was derived from an Arabic translation, as is proved by the *basmala*[11] at the beginning. Arabic interpolations are possible, because the Arabic writers were fascinated by this subject, but the substance was already available in Greek, and we may safely assume that the Arabic version represents essentially the ancient original. Therefore, it is best to describe the contents of the longer text as edited in Arabic by Carra de Vaux. Chapters 1 to 8 constitute a theoretical introduction, which is very promising. Read chapter 1:

Says the writer, " I have been aware, dear Aristōn, of your desire to know elegant apparatus, and therefore I have wished to answer your request by giving to you this book, to serve as a model for your mechanical investigations. I shall first describe the pneumatic apparatus, and I shall mention all the appliances known to earlier men of science.

"The philosophers who have cogitated upon physical matters have recognized that a vessel which common people believe to be empty is not really empty but full of air. That has been ignored as long as one has not been certain that air is a body from among the bodies. I do not wish to recall what has been said on the subject nor repeat the controversies relative to it. That air is one of the elements [*isṭuqish*, *stoicheion*] is not simply a theory but a fact, made obvious by tangible observations. I shall report what is necessary to attain my purpose and prove that air is a body." [12]

That is a beginning in the best Greek style, in spite of a few Arabic idioms.[13] Philōn describes a series of experiments showing that air is a material body which fills out space; a vacuum cannot exist, so that water cannot be poured out of a container unless air is allowed to get in and replace it, and if air is withdrawn from a container, water will follow it, even

[10] The 16 Latin chapters are chapters 1 to 11 and 17 to 21 of the Arabic text. Hence, the most important part, the physical introduction (chaps. 1–8), is available in medieval Latin as well as in Arabic.

[11] Every Arabic Muslim text began with the words Bismi 'llāhi -l-raḥmāni -l-rahīmi (In the name of God, the merciful, the compassionate). The Latin text of this treatise begins with the words "In nomine Dei pii et misericordis."

[12] Carra de Vaux (Arabic, p. 17; French, p. 98). The Aristōn to whom the book is dedicated is otherwise unknown. The name is spelled in Arabic Arisṭūn (or Yārisṭūn). The Greek name Aristōn was not uncommon; there are two philosophers bearing it, the Stoic Aristōn of Chios (c. 260 B.C.) and the Peripatetic Aristōn of Ceōs (fl. 230 B.C.). One of the earliest editors of Aristotle was named Aristōn of Alexandria; he flourished in the second half of the first century B.C. (Volume 1, pp. 495, 604).

[13] For example, the opening sentence, "Qālā innī 'alamtu yā Arisṭūn al-ḥabīb shawqaku . . ." and the words "jasad min al-ajsād" (a body from the bodies), "laisa min al-qawl faqaṭ bal min al-fa'l" (that is not simply a theory but a fact).

ΕΚ ΤΩΝ ΦΙΛΩΝΟΣ ΒΕΛΟΠΟΙΙΚΩΝ ΛΟΓΟΣ Δʹ.

EX OPERE PHILONIS LIBER IV.

DE TELORUM CONSTRUCTIONE.

ΦΙΛΩΝ Ἀρίστωνι χαίρειν. Τὸ μὲν πρότερον ἀποσταλὲν πρὸς σὲ βιβλίον περιεῖχεν ἡμῖν τὰ λιμενοποιϊκά. νῦν δὲ καθήκει λέγειν, καθ' ὅτι τὴν ἐξ ἀρχῆς διάταξιν ἐποιησάμεθα πρὸς σέ, περὶ τῶν βελοποιϊκῶν, ὑπὸ δέ τινων ὀργανοποιϊκῶν καλουμένων. Εἰ μὲν οὖν συνέβαινεν ὁμοίᾳ μεθόδῳ κεχρῆσθαι πάντας τοὺς πρότερον πεπραγματευμένους περὶ τοῦ μέρους τούτου, τάχα ἂν οὐθενὸς ἄλλου προσεδεόμεθα, πλὴν τοῦ τὰς συντάξεις τῶν ὀργάνων ὁμολόγους οὔσας * σαφηνίζειν· ἐπεὶ δὲ διενηνεγμένους ὁρῶμεν οὐ μόνον ἐν ταῖς πρὸς ἄλληλα τῶν μερῶν ἀναλογίαις, ἀλλὰ καὶ ἐν τῷ πρώτῳ καὶ ἡγουμένῳ στοιχείῳ, λέγω δὲ τῷ τὸν τόνον μέλλοντι δέχεσθαι τρήματι, καλῶς ἔχον ἐστὶ * τὸ τὰς μὲν τῶν ἀρχαίων παρεῖναι, τὰς δὲ τῶν ὕστερον παραδεδομένας μεθόδους, * τέχνας δυναμένας ἐπὶ τῶν ἔργων τὰ δέοντα ποιῆσαι, ταύτας ἐμφανίζειν. ὅτι μὲν οὖν συμβαίνει δυσθεώρητόν τι τοῖς πολλοῖς, καὶ ἀτέκμαρτον ἔχειν τὴν τέχνην, ὑπολαμβάνω μὴ ἀγνοεῖν σε· πολλοὶ γοῦν ἐνστησάμενοι κατασκευὴν ὀργάνων ἰσομεγεθῶν, καὶ χρησάμενοι τῇ τε τοιαύτῃ συντάξει, καὶ ξύλοις

* ἐμφανίζειν

* περὶ

* περὶ τῆς καθόλου τέχνης

PHILO Aristoni salutem. Superior quidem liber ad te missus ea complectitur quæ pertinent ad portuum constructionem. Nunc vero dicendum est juxta ordinem quem tibi polliciti sumus, de telorum, seu ut quidam vocant, machinarum fabricatione. Quod si omnes qui ante nos de hoc argumento scripserunt, simili methodo usi essent, nulla alia re fortasse opus haberemus, quam ut instrumentorum constructiones quæ sunt ejusdem rationis ac proportionis explicaremus. Sed quoniam eos reperimus dissentientes, non solum in partium ad se invicem proportionibus, verum etiam in eo quod primum ac præcipuum est elementum, in foramine scilicet quod funem accipere debet: consentaneum est veterum quidem methodos omittere, eas vero proponere, quæ a recentioribus traditæ possunt in machinis perficere id quod intenditur. Et artem quidem ipsam habere aliquid quod difficile comprehendi possit a multis, nec facile conjectura percipi, te ignorare non arbitror. Multi certe qui instrumenta ejusdem magnitudinis instituerant, & eadem compositione, iisdem

upward. That is, Philōn had gone as far as one could go before Torricelli (1643). In one of the experiments (chap. 8), a light is placed in a closed vessel over a body of water; some water is gradually drawn into the vessel. This is because the flame has destroyed the air, and the water goes up to fill the vacuum. In this he was going as far as one could go before Lavoisier (1772).

The other chapters, 13 to 65, describe siphons, various apparatus, means of keeping the water level constant in vessels (necessary for water clocks), a jar containing six fluids which can be poured out separately, more apparatus of many kinds, with water wheels and water pumps, hydraulic toys, water jets. If the Arabic interpreters were tempted to add a few more tricks, that hardly matters. The core of the book is Hellenistic.

It is probable that much of this had already been invented by Ctēsibios, but as Ctēsibios' own book is lost it is impossible to know for certain.

The Ctesibian-Philonian tradition was continued by Hērōn of Alexandria (I–2) and later by the Arabs, as is best proved by the fact that it was only because of Arabic translations that Philōn's most original work was preserved. It is possible that the Arabic translation (edited by Carra de Vaux in 1902) was preceded by Armenian and Persian versions which are lost. The translator is not named and this suggests that he belonged to the early period of Arabic translators, the age of Caliph al-Ma'mūn (IX–1).

One of the most curious items in the Philonian collection of apparatus and gadgets is an octagonal inkpot [14] which has an opening on each side; one can turn it around, put one's pen in any hole and have it inked. This is made possible because the inkpot itself inside of its octagonal house is hung in gimbals. Philōn invented what we now call the Cardan's suspension as applied to the ship's compass and barometer, or anything that must keep the same position in spite of outside motions. Girolamo Cardano (1501–1576) may have reinvented that clever trick, but Philōn had invented it eighteen centuries before. The gimbals were known to the Chinese as early as the Han dynasty [15] and they were described also in the *Mappae clavicula* (VIII–2). The first description of a compass suspended in gimbals occurs in a Spanish book by Martin Chavez, *Breve compendio de la esfera y de la arte de navigar* (Cadiz, 1546, 1551; Seville, 1556).[16]

[14] No. 56; Carra de Vaux's edition (Arabic, p. 82, French 171).

[15] Berthold Laufer, *Cardan's suspension in China* (William Henry Holmes Anniversary volume; Washington, 1916), pp. 288–292, 1 plate; *Introduction*, vol. 3, p. 715.

[16] It occurs in the third edition (1556), dealing with magnetic declination; I do not know whether it is in the first two editions.

Fig. 66. Treatise on the construction of engines for shooting by Philōn of Byzantion (II–2 B.C.). Princeps included in the *Veterum mathematicorum . . . opera graece et latine pleraque nunc primum edita ex manuscriptis codicibus Bibliothecae Regiae,* edited by Melchisédech Thévenot (1620–1692) (splendid folio, 44 cm; Paris: Royal Press, 1693), pp. 49–104. [Courtesy of Harvard College Library.]

These Chinese, medieval, and sixteenth-century rediscoveries may be independent, or objects mounted in gimbals may have passed from hand to hand. The tradition might very well have been (as so many technical traditions are) manual instead of literary. We do not expect the men of Han to have heard of Philōn, but some real gimbals may have reached them, as objects of virtue and curiosity.

VITRUVIUS

In spite of the fact that Greek was the learned language of this period, the outstanding technical book was composed in Latin. It is a treatise on architecture written by Vitruvius (I-2 B.C.) and as it is unique of its kind the briefest reference, *De architectura*, or "Vitruvius," is sufficient to identify it.

In spite of his fame, Vitruvius is practically unknown. We do not even know where and when he was born and died.[17] He flourished for a time at Fanum Fortunae, for he was the architect of its basilica.[18] He lived probably in the second half of the first century B.C.

He was the author of a single work, *De architectura*. According to its first line, it was dedicated to "imperator Caesar," which must mean Octavianus, the adopted son of Julius Caesar. The dedication was written shortly before 27 B.C., when Octavianus was given the title Augustus; if it had been written in 27 or later, that title would certainly have been used. Nevertheless, Vitruvius flourished in the Augustan age and held some office, as architect and engineer, in the rebuilding of Rome; he was in charge of plumbing and of war engines.

The *De architectura* is divided into ten books: I, architectural principles; II, history of architecture, materials; III, Ionic temples; IV, Doric and Corinthian temples; V, public buildings: theaters (and music), baths, harbors; VI, town and country houses; VII, interior decoration; VIII, water supply; IX, dials and clocks; X, mechanical and military engineering.

The scope is encyclopedic and overlaps in many ways architecture *stricto sensu*. The main purpose was to give the young architect a general education, including history, science, music, and many other things.

The first book explains the principles of such an education as well as the principles of architecture itself. In chap. 3 we are told that there are three departments in architecture: the art of building (Books I to VIII), the making of timepieces (Book IX), and the construction of machinery (Book X). The architect was an engineer as well as a builder and an artist. That is

[17] Two places have been suggested as his birthplace. The first is Formiae (Mola di Gaeta), on the Campanian coast, where Cicero had a villa and in the neighborhood of which he was murdered. The other is Verona.

[18] Vitruvius, V. 1, 6. The place was called Fanum Fortunae because of a famous temple to Fortune. Augustus sent to it a colony of veterans, and it was then called Colonia Julia Fanestris. It is located on the Adriatic shore of the Marches (Le Marche); modern name, Fano.

still true today, except that the functions of an architectural firm are generally divided among various men; one is the designer and artist, the master builder, another is the manager and treasurer; others still are put in charge of technical matters, such as plumbing, lighting, ventilation, acoustics. In Vitruvius' time all the work had to be done by a single man.[19] Chapter 4 explains how to choose the site of a city; chap. 5 how to build the city walls; chap. 6 how to draw the streets with due regard to the prevailing winds, and the final chapter 7, how to determine the sides of the public buildings.

In other words, a good part of Book I was devoted to what we would call "city planning," a subject that is relatively new with us, but has an old Greek ancestry.[20]

It would take too long to analyze every book of the *De architectura*, but some items will be mentioned to emphasize its complexity and its importance in the history of art and technology.

Book II tells the history of dwellings from prehistoric times and discusses the use of materials, such as brick, sand, lime, pozzuolana, stone, timber, how to build walls (*opus incertum*; ancient style and the *opus reticulum*, "now used by everybody").[21] Pozzuolana (*pulvis puteolanus*), a volcanic earth found at Puteoli but also at and near Rome, was used together with lime to make a kind of concrete. From the second century B.C., when the Romans realized the strength and durability of concrete, they used it frequently for the building of walls and vaults (concrete floors are discussed in Book VII, 1).

Book III, devoted to the construction of temples, begins very appropriately (from the Greek point of view) with an essay on symmetry; symmetry and proportion (*analogia*) in temples and in the human body. For Vitruvius the proportions of the body were fundamental, those of the temple derivative from them.[22] The Greek *entasis*, a swelling in the middle of columns for better perspective, is explained at the end of chap. 3.

[19] The Latin name *architectus* was a transcription of the Greek name *architectōn*, meaning chief artificer, master builder, director of works. In Athens the title was given also to organizers, such as the manager of the state theater and of the Dionysia.

[20] The founder of city planning was Hippodamos of Milētos, who flourished about the middle of the fifth century (Volume 1, p. 295).

[21] For deeper study of this see, the great work of Esther Boise Van Deman and Marion Elizabeth Blake, *Ancient Roman construction in Italy from the prehistoric period to Augustus* (Washington: Carnegie Institution, 1947) [*Isis 40*, 279 (1949)].

[22] The search for a canon of beauty was not exclusive to the Greeks; it occurred in Egypt and India (*Introduction*, vol. 3, p. 1584). A canon of human beauty was established in Greece by Polycleitos of Argos (fl. 452–412) and modified by Lysippos of Sicyōn (fl. 368–315). Architectural canons or "orders" varied from time to time, and are symbolized by the terms Doric, Ionic, Corinthian. The Greeks understood that a canon should not be fixed forever; what really matters is the search for it. If the canon is too rigid, it loses its virtue; it dies. It would be interesting to compare Greek ideas on the architectural canon of beauty with Indian ideas. Tarapada Bhattacharyya, *A study on vāstuvidyā* (382 pp.; Patna, 1947) [*Isis 42*, 353 (1951)].

In Book IV he discusses the origins and characteristics of the three orders. The most interesting part of this book is perhaps his treatment of Tuscan temples, which are hardly known otherwise; these were the temples built by the Romans before their submission to Greek models.

Book V, devoted to public buildings, such as basilicas, theaters, baths, gymnasiums, and palaestras, includes a very important study of music and acoustics. Vitruvius explains sound as a displacement of air in waves which he compares with the waves that can be observed on the water's surface when a stone is thrown into a pond.[23] What is more remarkable was Vitruvius' application of the wave theory to architectural acoustics. The wave theory of sound was Greek, its application to the acoustics of a hall typically Roman. This achievement was duly admired by Wallace Clement Sabine (1868–1919), the American master in the field of architectural acoustics.[24]

In chap. 8, Vitruvius analyzes the acoustics of a theater and the phenomena that may spoil it, which we call interference, reverberation, echo. Chapter 5 of the same book is entirely dedicated to the sounding vases used in theaters to reinforce human voices; this matter is not clear to me. Vitruvius gave to those resonators the Greek name *ēcheia* (drum, gong); no ancient examples of them have yet been discovered, but there are quite a few medieval examples in Christian Europe.[25]

His description of the basilica at Fano,[26] the building of which he superintended, may have been an addition to his original text. It is very brief and suggests what modern architects call "specifications."

Book VI. Building of town and country houses; the need of adapting them to the climate; proportions of the main rooms and their exposure. He recommended the use of arches in the substructures (chap. 8). This was not a novelty; arches had been used in Egypt, Greece, Etruria, but the Romans were the first to rely extensively on semicircular arches.

Book VII. Interior decoration; preparation of floors and walls, slaking of lime for stucco, stucco coverings of walls, fresco painting, various pigments and colors.

Book VIII. Water supply; how to find water (by rational means only; no divining rod), various kinds of water, rainwater, leveling instruments, aque-

[23] Archytas of Tarentum (IV-1 B.C.) and Aristotle (IV-2 B.C.) were well aware of the fact that sound was due to air vibrations. Aristotle made many other remarks on sound, for example, that it is heard more clearly in winter than in summer and at night than in the daytime. After Lucretius and Vitruvius, no further progress was made until Ptolemy (II-1). The first to prove that sound was caused and transmitted by air waves was the physician Günther Christoph Schelhammer (1649–1716) of Jena in 1684 or 1690. A more spectacular proof was given by Champion and Henri Pellet in 1872; they showed that chemical reactions can be released by sound waves, for example, "nitrogen triiodide" ($NI_3 \cdot NH_3$) can be exploded by certain sounds.

[24] Author of *Architectural acoustics* (Cambridge, 1906), *Collected papers on acoustics* (Cambridge, 1922).

[25] See the article "Acoustic vases" in my *Introduction*, vol. 3, p. 1569, apropos of specimens found embedded in the vaults of St. Mary of Carmel in Famagusta (Cypros), built c. 1360.

[26] See note 18.

ducts, wells and cisterns; reference to lead poisoning caused by lead pipes (chap. 6, 11); use of a lighted lamp to test the purity of air (chap. 6, 13).

Book IX. Dials and clocks. This is an unexpected digression on chronology and horology, including the necessary astronomical introduction; zodiac and planets, phases of the Moon, course of the Sun, constellations, astrology and weather signs, the analēmma, a kind of sundial, and its uses; sundials and water clocks.

Book X. Practical mechanics (this continues the efforts of Ctēsibios and Philōn and is one of our best sources for the study of their own achievements). Vitruvius made a distinction between *mechanica* and *organica*, that is, between the principle of the *mēchanē* and that of the *organon*, the latter being susceptible of greater autonomy or automatism, while the ordinary machines required more human labor. It is interesting to find that distinction in pre-Christian times. Vitruvius describes hoisting machines, engines for raising water, water wheels and water mills, water screws, Ctēsibios' pump, water organs, odometers (taximeters), and he passes from peace engines to war ones, catapults or scorpiones, ballistae, stringing and tuning of catapults, siege engines, tortoises (*testudo*) for filling ditches, Hēgētōr's [27] ram and tortoise; means of defense. His final words are:

> In this book I have fully set forth the mechanical methods which I could furnish, and which I thought most useful in times of peace and war. Now in the previous nine books I have dealt with the other several topics and their subdivisions, so that the whole work, in the ten books, describes every department of architecture.[28]

We are reminded that architecture was far more comprehensive for Vitruvius than it is for us; it included engineering, astronomy and horology, and machines of every kind.

Vitruvius' style was generally clear but poor; he wrote like an engineer who is more familiar with instruments than with the Muses and for whom writing is less a pleasure than an unavoidable necessity. He was either too brief or too florid. His grammar is so poor that some scholars were tempted to assign to the *De architectura* a later date, say the third century after Christ or even later. It could not belong, so they thought, to the golden age of Latin literature. They forget that Vitruvius was not a man of letters. He tried to write well but is generally at his worst when he grows rhetorical. The ending of his book, quoted above, is typical of his labor; the other books were ended in the same tired manner, as if he was glad that the task was done. There were no ghost writers in his time; if there had been, he might have asked one of them to do that unpleasant job for him. He did his best and promised to speak as he was able (*ut potuero, dicam*; II, 1, 7).

[27] Vitruvius calls him Hagetor Byzantius (X, 15, 2). This Hēgētōr of Byzantion is otherwise unknown. He must be different from the surgeon Hēgētōr (II-2 B.C.) for whom see Galen, K. G. Kühn, *Galeni opera omnia* (20 vols.; Leipzig, 1821–1833), vol. 8, p. 955.

[28] Quoted from the Latin-English edi-

L. VICTRVVII POLLIONIS AD CESAREM AVGVSTVM DE ARCHITECTVRA LIBER PRIMVS.

PREFATIO

CVm diuina mens tua: & numen Impator Cæsar imperio potiretur orbis terrarú: inuictaq; uirtute cunctis hostibus stratis triumpho uictoriaq; tua ciues gloriarentur: & gentes oẽs subacte tuũ spectarent nutum. P. Q. R. & Senatus liberatus timore amplissimis tuis cogitatióibus cósiliisq; gubernaretur. Non audebam tantis occupatióibus de Architectura scripta & magnis cogitatióibus explicata ædere. Metuens ne nó apto tpe interpellans subirẽ tui animi offensionẽ. Cum uero attenderem te non solú de uita cói oĩum curam. P. Q. rei constitutionẽ habere. Sed etiam de oportunitate publicorumq; edificiorũ ut ciuitas aperte nó solú prouinciis esset aucta. Verú etiã ut maiestas imperii publicorum edificiorum egregias haberet auctoritates. Non putaui pretermittendum quin primo quoq; tpe de his rebus ea tibi æderẽ. Ideoq; primum parẽti tuo de eo fueram notus & eius uirtutis studiosus. Cum aút cócilium celestium in sedibus ĩmortalitatis eú dedicauisset. & Impium parentis in tuam potestatem transtulisset. Illud idem studium meum in eius memoria permanens in te contulit fauorẽ. Itaq; cum. M. Aurelio & . P. Numidico &. CN. Cornelio ad preparationẽ balistarum & scorpionum reliquorúq; tormentorũ refectióem fui presto: & cum eis cómoda accepi: q̃ cum mihi primo tribuisti recognitionẽ per sororis cómendationem seruasti. Cum ergo eo beneficio essem obligatus ut ad exitú uite non haberẽ inopie timorẽ hec tibi scriberẽ cepi. q̃ animaduerti te multa ædificauisse & nunc ædificarẽ. Reliquo quoq; tpe & publicorum & priuatorum edificiorum pro amplitudine rerum gestarú ut posteris memorie traderent curam habiturum. Conscripsi prescriptióes terminatas ut eas attendens & ante facta & futura qualia sint opera per te nota posses habere. Nãq; his uoluminibus aperui omnes discipline rationes.

The original manuscripts were illustrated, but none of those illustrations have come down to us, except perhaps a diagram of the winds which was not needed.

Vitruvius' sources. Vitruvius knew Greek and occasionally used Greek words or was obliged to coin new Latin ones, for he was the first or almost the first writer in the field. He was acquainted with the writings not only of the Hellenistic mechanicians but of many other authors. There is a long list of them in the preface to Book VII, and others are mentioned *passim.* It is possible that many of them were not known to him directly but only indirectly, for example, through Varro's *Disciplinae.*

His best source of information, however, was not literary but manual and oral. He knew how things were done and could do them himself. He had a technical knowledge of many monuments and helped to build new ones. His knowledge was the practical kind available to a craftsman of genius, obtained from past achievements and enriched with his own experience.

Tradition. The *De architectura* must have been known to Roman architects of the Augustan age, because the author was an Augustan official. He was quoted by Pliny the Elder (I–2) and, with special reference to plumbing, by Frontinus (I–2); he was mentioned much later by Sidonius Apollinaris of Lyon (431–488). The Vitruvian tradition is much simpler than that of the technical treatises written in Greek, because it was restricted to the Latin world which was becoming more and more drowsy. Vitruvius' existence was ignored by the Byzantine and Islamic writers. It is true that the Arabs shared a part of Vitruvius' knowledge, because they used some of his own sources (Ctēsibios, Philōn) and some of his tools. As far as machines are concerned, no capital invention can be ascribed to Vitruvius, but he made Hellenistic inventions available to Latin readers.

One of the first medieval scholars to study Vitruvius was Einhard (IX–1), who was employed by Charlemagne as architect, diplomat, and educator. This facilitated the diffusion of Vitruvius in the Carolingian empire and later in the German countries.

The oldest manuscript of the *De architectura* to reach us, the British Museum Harleianus 2767, was long believed to be of German origin, but it was produced in the Saxon scriptorium of Northumbria, probably in Jarrow or Wearmouth about the eighth century. It was probably copied from a manuscript that had been in the hands of Cassiodorus (VI–1) in

tion by Frank Granger (Loeb Classical Library; Cambridge, 1934), vol. 2, p. 369.

Fig. 67. Princeps of Vitruvius (I–2 B.C.), *De architectura*, the greatest architectural treatise of antiquity. Edited by Joannes Sulpitius (folio, 29 cm, 98 leaves; Rome: Eucharius Silber, 1487). Our facsimile reproduces the first page of Vitruvius' text. [Courtesy of Harvard College Library.]

Squillace (eastern Calabria) or in those of the Benedictine monks of Monte Cassino. There are various other manuscripts anterior to the twelfth century. The most important of these is, strangely enough, another Harleian [29] (3859, eleventh century), written in the Benedictine abbey of St. Peter in Ghent and used by Fra Giocondo.

Boccaccio's (XIV–2) scientific knowledge was partly derived from Vitruvius, and the interest of Renaissance scholars in him was much increased when Poggio Fiorentino [30] found a new manuscript.

There are no fewer than three incunabula editions: Rome: Silber, 1486–87 (Fig. 67); Venice: Christophorus de Pensis, 1495–96; the second text was reprinted by Simon Bevilaqua in his edition princeps of Cleōnidēs,[31] Venice, 1497 (Klebs 1044.1–2, 281.1). These editions were superseded by the one prepared by Fra Giocondo of Verona [32] (Venice: Joannes de Tridino alias Tacuino, 1511) (Fig. 68), the first to be illustrated.[33] This edition was revised and reprinted by the Juntae (Florence, 1513; again, 1522). Fra Giocondo was mainly responsible for the Vitruvian vogue of the Renaissance. There were many other sixteenth-century editions and translations: first Italian (Como, 1521); first French, by Jean Martin (Paris, 1547); first German, by G. H. Rivius (Nuremberg, 1548); first Spanish (Alcalá de Henares, 1602).

It should be stated here that the princeps of Vitruvius was preceded by the first edition of the *De re aedificatoria* by Leone Battista Alberti (1404–1472), published posthumously at the request of the author's brother Barnardo (Florence: Nicolaus Laurentii, 29 December 1485). Alberti was well acquainted with Vitruvius, to whom he referred many times, but his work was partly derived from the new architecture of Filippo Brunelleschi (1377?–1446);[34] strangely enough, though he greatly admired Brunelleschi, he did not mention the latter's famous cupola of Santa Maria del Fiore in Florence. The *De re aedificatoria* obtained considerable success, being translated into Italian (Venice, 1546), then again by Cosimo Bartoli (Florence, 1550), into French by Jean Martin (Paris, 1553), and from Italian into English by Giacomo Leoni (London, 1726). A facsimile reprint of the

[29] The Harleian library was collected by Robert Harley (1661–1724) and his son Edward (1689–1741), first and second earls of Oxford; it was purchased by the British Museum in 1753. *Catalogue* (4 vols.; London, 1808–1812).

[30] Poggio Bracciolini of Florence (1380–1459); *Introduction*, vol. 3, p. 1291.

[31] Cleōnidēs (II–1 B.C.) was a writer on music, a late follower of Aristoxenos (IV–2 B.C.).

[32] Giovanni Monsignori (c. 1435–1515) of Verona; Dominican "Fra Giocondo"; flourished in Paris 1499–1506; died in Rome. He was an archaeologist, an architect, a collector of manuscripts and of inscriptions. He found in Paris Trajan's correspondence with the younger Pliny; he edited Pliny's letters in 1508, and Caesar's works in Venice, 1513.

[33] Those illustrations, the first of their kind, were very important. They were for many readers a revelation. For example, the readers found in Fra Giocondo's edition the first plan of a Roman house. In his edition of Caesar (1513) he reproduced Caesar's bridge across the Rhine.

[34] Frank D. Prager, "Brunelleschi's inventions and the 'renewal of Roman masonry work'," *Osiris* 9, 457–554 (1950).

Fig. 68. Title page of a much better edition of Vitruvius with many figures and index, by Giovanni Monsignori of Verona (c. 1435–1515), better known under his Dominican name, Fra Giocondo. It is a folio of 30 cm, dedicated to Julius II (pope 1503–1513). (Venice: Joannes de Tridino, alias Tacuino, 22 May 1511.) [Courtesy of Harvard College Library.]

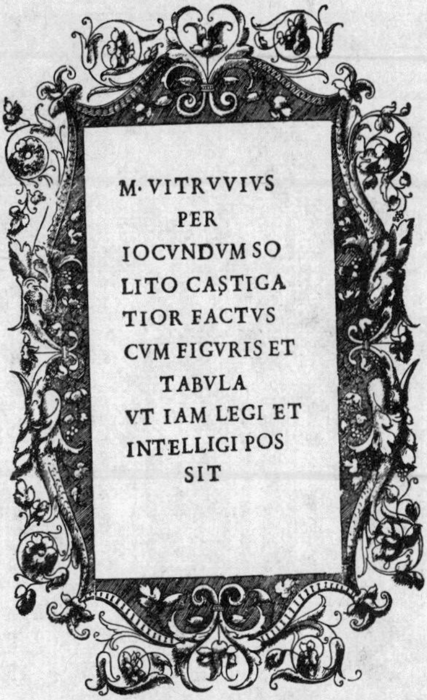

M·VITRVVIVS
PER
IOCVNDVM SO
LITO CASTIGA
TIOR FACTVS
CVM FIGVRIS ET
TABVLA
VT IAM LEGI ET
INTELLIGI POS
SIT

third edition of Leoni's translation (London, 1755) has just been published (London: Tiranti, 1956). Alberti's influence was small, however, as compared with that of Vitruvius, to whom we now return.

One of the minor academies of the Renaissance, the Accademia della Virtù, was concerned mainly with the study of Vitruvius. It was founded by Claudio Tolomei [35] and others under the patronage of Cardinal Ippolito de' Medici (c. 1511–1535), a nephew of Leo X.

Vitruvius' fame was so great at that time that Girolamo Cardano (1501–1576) included him among the twelve leading thinkers of all ages, the only true Roman.[36] His glory was consecrated by the work of Andrea Palladio (1518–1580), which caused the triumph of classical architecture all over Europe and the temporary contempt of Gothic architecture. Palladio's book, *I quattro libri dell' architettura*, was first published at Venice in 1570 (Fig. 70) and was translated into French and into English. To the English translation were added notes by Inigo Jones (1573–1652).[37] Both Palladio and Jones were primarily architects, the creators of many monuments, and the

[35] Claudio Tolomei (1492–1555), of Siena, was bishop of Korčula (an island off the Dalmatian coast) and the founder of the new Tuscan poetry with Latin meter (*la poesia barbara*).

[36] The whole list is given in my *Introduction*, vol. 3, p. 738.

[37] An English translation of Book I appeared in 1668 ("second edition"), the sixth edition in 1700; *The four books* with Inigo Jones' notes (15 parts; London, 1715), Italian, English, and French.

Fig. 69. Frontispiece of the French translation of Vitruvius by Claude Perrault (1613–1688) with abundant commentary and splendid illustrations. It is a large folio (43 cm), dedicated to Louis XIV (Paris, 12 June 1673). Perrault designed the colonnade of the Louvre. He was also a very distinguished anatomist. [Courtesy of Harvard College Library.]

Fig. 70. First edition of Andrea Palladio's *Architecture* (folio, 30 cm; Venice, 1570), which revived the influence of Vitruvius. The central medallion shows Fortuna Audax in a ship steered by Justicia; *Lychnos* (Uppsala, 1954–55), pp. 165–195. [Courtesy of Harvard College Library.]

triumph of Greco-Roman architecture ("Palladianism") was established by their artistic creations as well as by their books.

All considered, Vitruvius was one of the most influential authors of the whole of classical antiquity, and historians of science should pay full attention to him. His book is a kind of encyclopedia, comparable in its field to the lost *Disciplinae* of Varro and to the *Historia naturalis* of Pliny the Elder. Next to the monuments it was the best source for the study of Greco-Roman architecture. He was himself a historian of science and technology; for example, see his notes concerning the development of the architectural orders (Books III–IV), the history of astronomy (Book IX), of geography (Book VIII, 3), of mechanics (Book X). His notes were not always correct (he was not a good historian), and he helped to transmit errors, for instance, that the Niger was a branch of the Nile and that one ought to look for the Nile's sources in the far west,[38] but that was unavoidable.

Recent editions. Critical edition by Valentinus Rose (Leipzig: Teubner, 1867; again, 1899) and by Friedrich Krohn (Leipzig: Teubner, 1912).

English translations by Morris Hicky Morgan (344 pp.; Cambridge: Harvard University Press, 1926) and by Frank Granger in the Loeb Classical Library (2 vols.; Cambridge: Harvard University Press, 1931–1934).

A FEW OTHER GREEK AND ROMAN PHYSICISTS AND TECHNICIANS

We put together in this section the notes concerning a few men whose names have reached posterity, bearing in mind that more important novelties or simple gadgets were discovered by illiterate people or by people who did not bother to write and have long been forgotten.

The investigation of burning mirrors, an Archimedean tradition, is ascribed to one Dioclēs (II–1 B.C.). Arrianos (II–1 B.C.) was called the meteorologist because of his research in that field. Cato the Censor (234–149), of whom more will be said presently, published the earliest recipe for ordinary mortar and the earliest description of a *bain-marie*.[39] Athēnaios (II–2 B.C.) wrote a short treatise on siege engines (Fig. 71), which tells the origin of some of them and which it is interesting to compare with Book X of Vitruvius.[40] Carpos of Antioch (date unknown) was called *ho mēchani-*

[38] Book VIII, 2, 6–7. See my *Introduction*, vol. 3, p. 1772.

[39] A double boiler. This is not simply a piece of scientific apparatus; it has become a common kitchen tool. The housekeeper should give thanks to Cato the Censor when she uses one and not confuse him with his illustrious great-grandson, Cato of Utica (95–46). The origin of the expression bain-marie is unknown. According to Du Cange, *Glossarium mediae et infimae latinitatis*, the phrase *balneum Mariae* occurs in Arnold de Villanova (XIII–2), meaning *fornax philosophicus*.

[40] The Greek text was first edited with Latin translation by Melchisédech Thévenot in his *Veteres mathematici* (Paris, 1693); the Greek text was reëdited by Carl Wescher, *Poliorcétique des Grecs* (Paris, 1867); French translation by Rochas

Fig. 71. Title page of the mechanical treatise of Athēnaios (II–2 B.C.) in M. Thévenot's collection, *Veteres mathematici* (Paris, 1693), pp. 1–11. [Courtesy of Harvard College Library.]

ΑΘΗΝΑΙΟΥ ΠΕΡΙ ΜΗΧΑΝΗΜΑΤΩΝ.

ATHENÆI DE MACHINIS.

ΟΣΟΝ ἐφικτὸν μὲν ἀνθρώπῳ τοὺς ὑπὲρ μηχανικῆς ποιουμένῳ λόγους, ὦ σεμνότατε Μάρκελλε, ἐμνήσθην τοῦ Δελφικοῦ παραγγέλματος, ὡς ἔστι θεῖόν τι τὸ ὑπομιμνῆσκον ἡμᾶς χρόνου φείδεσθαι, ὡς ἔστι σχεδὸν εἰπεῖν ἅπαντα καταχρώμεθα ἀφειδῶς εἰς τὰς κατεπειγούσας τοῦ βίου χρείας, καὶ χρημάτων μὲν καὶ τῶν ἄλλων δοκούντων ἡμῖν εἶναι πολυτελῶν, μὴ τὴν τυχοῦσαν ἐπιστροφὴν καὶ φυλακὴν ποιησώμεθα. [ἀλλὰ τοῖς τῶν ἀρχαίων προσέχωμεν συντάγμασι, καὶ αὐτοί τε μικρὸν ἐπιτείναντες ἑαυτούς, οὐκ ἀσκόπως εὑρήσομεν, καὶ παρ' ἄλλων ῥᾳδίως ἂν μεταλάβωμεν.] τοῦ χρόνου δὲ μεταβλητοῦ γε ὄντος, καὶ ῥευστοῦ, ἀφειδοῦμεν, ᾧ εὐχερὲς τὸ τέλος, καὶ ταῦτα τῆς φύσεως νεμούσης μὲν ἡμέρας δυνάμιν ἱκανὰ εἰωθυίας εἰς τὸ κατεργάσασθαί τι τῶν ἐν τῷ βίῳ χρησίμων, ὕπνον δὲ νυκτός, ἀλλὰ παντελῶς ἀκαριαῖον. ὁ δὲ μόνος καὶ εἷς *δικαίως ποιητής, οὐδὲ τὸν δοθέντα παρὰ τῶν θεῶν εἰς ἀνάπαυσιν ἡμῖν τοῦ σώματος ὕπνον, παννύχιον οὐδὲν ἐᾷ· οὕτω πολλὴν φαίνεται ποιούμενος πρόνοιαν, τοῦ μὴ καταργεῖσθαι τὴν διάνοιαν ἐπὶ πολὺν χρόνον. οἱ δὲ γράφοντές τι, ἢ παραγγέλλοντες ἡμῖν, καὶ τῆς ὠφελείας εἵνεκα δοκοῦντες αὐτὸ πράττειν, οὐκ ἂν εἰ οὕτως πολυγραφοῦντες, εἰς οὐκ ἀναγκαίους λόγους καταναλίσκουσι τὸν χρόνον, ὅπως ἐμφήνωσι τὴν ἑαυτῶν πολυμαθίαν, παρεκβάσεως γοῦν πληρώσαντες, ἀπολείπουσι τὰ βιβλία,

* δίκαιος

QUANTUM quidem licet homini qui de machinis ſcribit, ampliſſime Marcelle, memor fui præcepti Delphici, ut pote divini, quo admonemur, ut tempori parcamus. Illo enim fere univerſo ſine parcimonia abutimur ad urgentes hujus vitæ neceſſitates: ac pecuniarum quidem, & cæterarum rerum, quæ precioſæ eſſe nobis videntur, haud levem curam ac ſollicitudinem gerimus. [Sed Antiquorum incumbamus libris, & noſmetipſos paululum intendentes non infeliciter inveniemus; & ab aliis inventa facile accipiemus.] Tempori vero, quod mutabile & fluxum eſt, nequaquam parcimus, cum tamen finis ejus in promptu ſit. Idque cum natura diei quidem attribuerit virtutem, quo ea quæ ad vitam utilia ſunt operemur; nocti vero ſomnum aſſignaverit, ſed eum omnino brevem. Is enim qui ſolus Poetæ nomen meretur, ſomnum, qui nobis a Diis ad corporis quietem conceſſus eſt, tota nocte capere non ſinit: adeo ei curæ fuit, ne mens longo temporis ſpatio otioſa eſſet. Hi vero qui nunc ſcribunt, aut præcepta nobis tradunt, etſi utilitatis gratia id facere videantur, tamen adeo prolixe ſcribentes, in ſermonibus minime neceſſariis tempus conterunt, ut multiplicem eruditionem ſuam oſtentent. Libros enim exceſſibus plenos nobis relinquunt:

cos (the mechanician); he invented a kind of level, the *chōrobatēs*, which, according to Theōn of Alexandria (IV–2), was similar to the *alpharion* or to the *diabētēs*. Now the *diabētēs* (or *libella*) had been invented by Theodōros of Samos (VI B.C.), and before him by Egyptians of the Twentieth Dynasty (1200–1090), if not before. It is the application of a plumb line to leveling purposes. This is a good example of an indispensable instrument (one cannot build safely without it) that was invented over and over again; the original invention may have been copied, but the instrument is so simple that independent reinventions of it are plausible enough.[41]

Meteorological studies were continued during the first or last century B.C. by Poseidōnios and by two other men who were probably his disciples, first Cleomēdēs, who is more important as an astronomer, and then Asclēpiodotos. Cleomēdēs studied refraction (*cataclasis*), including atmospheric refraction. Asclēpiodotos wrote a short treatise on tactics (*technē tacticē*), which was illustrated by means of figures and diagrams. It was plagiarized by Claudios Ailianos (III–1), yet its own text has come down to us.[42]

All that does not amount to much. The best theoretical work was done, strangely enough, not by Greeks but by Romans writing in Latin, two contemporaries of Virgil, Vitruvius and Varro (in his lost works). And the very best was not written at all, but done. This was a great age of public works, a few of which will be examined presently.

PUBLIC WORKS

Hellenistic Asia. Out of many Hellenistic cities, the best exemplar was Pergamon, which was built on a magnificent site, in Asia Minor, at about the latitude of Lesbos, some fifteen miles inland. Three rivers came together at that point, and close to the lovely valleys rose a steep hill. The masters built themselves an acropolis on the top, whence they could dominate the region. The lower city was built gradually, but its golden age was the second century (a century after Alexandria). It was only after they had defeated their most dangerous rivals, the Gauls (or Galatians)[43] that their

---

d'Aiglun in *Mélanges Charles Graux* (Paris, 1884), pp. 781–801, with 12 illustrations.

[41] For the *diabētēs* see Volume 1, pp. 124, 191. Adolphe Rome, "Un nouveau renseignement sur Carpus" *Annuaire de l'Institut de philologie et d'histoire orientales* 2, 813–818 (Brussels, 1934). The word *chōrobatēs* is used by Vitruvius, VIII, 5, 1.

[42] Greek-English edition by William Abbott Oldfather, *Aeneas Tacticus, Asclepiodotus, Onasander* (Loeb Classical Library; Cambridge, 1923), pp. 229–340, with technical glossary.

[43] These Gauls were real Gauls or Celts who had moved eastward at the invitation of Nicomēdēs I, king of Bithynia (278–250), settled in his kingdom, and expanded south of it, covering the central part of Anatolia. There were three tribes of these Gauls; one of them, the Tectosages in Galatia prima, had for its capital Ancyra, which is now the capital of the Turkish republic (Ankara). It is better to call them Galatians than Gauls, however, for they intermarried with native women or with Greek immigrants and must have become very different from their ancestors of western Europe. For one thing, they spoke Greek and affected Greek manners; the Romans called them *Gallo-Graeci*. They are the people to whom Paul the Apostle addressed one of his epistles.

economic and cultural expansion could begin in earnest. The Galatians were first beaten by the Seleucid king of Syria, Antiochos I Sōtēr, in 276, and then again by Attalos I Sōtēr, c. 235, who was first to assume some time later the title of King of Pergamon, added a big slice of the Seleucid kingdom to his own, and began his dangerous flirtation with Rome. His political instinct was weak, but he proved himself to be a great patron of arts and letters. He wanted Pergamon to emulate Alexandria, and entrusted the building of his capital to a Greek architect who made sure that the public buildings, constructed at different levels on the hillside, would appear in all their glory.

The Pergamene renaissance begun under Attalos I (241–197) reached its climax under his son and successor, Eumenēs II (197–60). The artistic achievements will be described in Chapter XXVII. Pergamon became one of the most beautiful cities of the Hellenistic world. One of its features, less obvious to the visitors than its beautiful monuments but of very great importance, was its elaborate water system. The water came from the Madaras-dag and conduits brought it across valleys almost to the top of the acropolis (332 m above sea level). The water circuits were extremely long and pressure in the pipes must have been as high as 16 to 20 atmospheres. The hollowed stones through which the pipes passed have been found but not the pipes *in situ.* Were these in lead or in bronze? Some ceramic pipes are extant, 6 to 9 cm in diameter and 48 cm long.[44]

The Roman world. The Romans were great builders not only of temples, theaters, stadiums, triumphal arches, and other monuments, but also of roads, aqueducts, bridges, docks. A few examples must suffice.

The Emporium of Rome was built in 194 by the aediles, M. Aemilius Lepidus and L. Aemilius Paulus. It was a kind of market or warehouse for the ships landing in Rome. *Navalia* or dry docks as well as other monuments are credited to Hermodōros of Salamis, who flourished in Rome in the second half of the second century B.C.[45]

The Pontine marshes south of Rome were dried up c. 160 B.C. They extended from Nettuno to Terracina, being about 60 km long and 6 to 15 km wide. Traces of the old drainage system have been discovered. Water was drained out by means of open trenches and also of pipes. Those lands were cultivated and some Romans built themselves villas surrounded by fields and gardens. After the fall of Rome the drainage, which required continuous supervision, was neglected, new swamps were formed, and malaria decimated or discouraged the people. The Campagna was abandoned almost until our own days.

[44] More details in Curt Merckel, *Die Ingenieurtechnik im Altertum* (Berlin, 1899), p. 508. It is difficult, however, to know how much of that is Pergamene and how much is Roman of a later date. The Roman domination began in 133 B.C. but lasted many centuries.

[45] Cicero, *De oratore*, I, 62.

Another feat of drainage was accomplished a little later (c. 109) in the middle Po valley at Placentia.[46]

The earliest Roman aqueducts, the Aqua Appia (312) and the Anio Vetus (272), have been briefly described in Chapter VII. As the city grew, more water was needed and more aqueducts were built. The construction of the Aqua Marcia was ordered in 144 by the praetor, Quintus Marcius Rex, and completed in 140. At this time Greek influences were strongly felt, and this aqueduct was far superior architecturally to the older ones. It was built with new materials and new methods. It included beautiful bridges and lofty arches; since much of it was built above the ground, it was the first of the "high-level" aqueducts of ancient Rome. It was about 59 miles long, but was shortened afterward by the substitution of arches across the valleys for the long channels around them. The dating of such immense works is as illusive as that of cathedrals. One can say it was begun in such a year and "completed" so many years later, but that completion was not final. For example, consider one of the bridges of the Aqua Marcia, the Ponte Lupo, which carries it above another stream, the Aqua Rossa. It is a colossal structure, 365 feet long, 70 feet wide at its base, and having a maximum height of 100 feet. It has been restored so often that Dr. Van Deman described it as "an epitome in stone and concrete of the history of Roman construction for almost nine centuries."

A fourth aqueduct, the Aqua Tepula, was begun in 125, on a higher level than the Marcia but with a smaller capacity. It brought water from the Alban hills; as that water was tepid, it was called "*tepula*"; what is worse, the water was not very wholesome.

In 33 B.C. Marcus Vipsanius Agrippa was commissioned by the Senate to repair and reorganize the older aqueducts. He built a new aqueduct called Aqua Julia in honor of his patron, C. Julius Caesar Octavianus (Augustus in 27 B.C.), and rearranged the Aqua Tepula, combining its course partly with that of the Julia. The Julia was largely built of concrete instead of the expensive cut stone.

In 19 B.C. Agrippa began the building of a new aqueduct to supply the public baths, which he had erected. As the water started to flow on Vesta's feast day, the new aqueduct was called Aqua Virgo. The reference is to the priestesses of the goddess, the Vestal Virgins (*Vestales virgines*), who were the guardians of the sacred water as well as of the sacred fire.[47] The source of the Virgo was only 8 miles out of Rome, but so many detours were needed because of the configuration of the ground that the length of its channel was about 14 miles.

[46] Modern Piacenza. The Padus (Po) valley was in Cisalpine Gaul. Placentia and Cremona were Roman colonies founded in 219 on the right (south) bank of the Po. Placentia was retaken and destroyed by the Gauls in 200, but rebuilt by the Romans. It became an important city.

[47] Vesta (Greek, Hestia), guardian of the hearth, was not represented by a statue but by a fire which the Vestals kept burning.

At the very end of our period, Augustus caused the Aqua Alsietina to be built, so called because its water was taken from Lake Alsietinus. It was 25 miles in length, its cost must have been enormous, and yet its only purpose was to provide an abundance of water for the naumachia, or mimic sea fights. These fights were held in the arena of a circus, flooded for the occasion by a complicated system of pipes and sluices, or sometimes in an artificial lake. The Aqua Alsietina was used for the first time in 2 B.C. at the dedication of the temple to Mars Ultor (Mars the Avenger), when Augustus gave a naumachy in a special basin dug for the occasion, 1200 by 1800 feet, surrounded by extensive gardens. The actors in such entertainments were, like the gladiators, criminals, prisoners, or beggars, whose lives were forfeited and easily expendable.

Like the Appia and the Virgo, the Alsietina had no clearing tanks; its surplus water unfit for drinking was used for irrigation.

Water systems existed not only in Pergamon and Rome but in many cities; we might even say they existed in every Roman city of sufficient size. For example, in Aletrium[48] were found water circuits and drainage pipes established in 100 B.C. by L. Betilienus Varus (as recorded in a local inscription). Siphons were used and the fall was over 100 m; some of the pipes were of lead (10 cm in diameter and 10 to 35 mm thick). Other relics of Roman water circuits have been found in Lyons, Arles, Nîmes, Sens, Lutetia (Paris), Antibes, Vienne, and in Strassburg, Metz, Mainz, Cologne, Vienna.

The aqueduct of Tarragona[49] (Acueducto de las Ferreras) dates from the beginning of the imperial age; it extended to 35 km. It crosses a valley at a height of 30 m over an enormous bridge 211 m long, built in two stories, the lower of which contains 11 arches and the upper 25.

The Pont du Gard (Fig. 72), built in 18 B.C., was a part of the water system of Nîmes, measuring almost 50 km. The architect is unknown, but it was built when Agrippa was in Nîmes as governor of Gaul. It is built in three levels, each consisting of a series of circular arches, six very large ones on the lowest level (maximum diameter 24 m), ten smaller on the middle level, and a great many very small ones on the top level.[50]

These bridges, which were intrinsic parts of aqueducts, suggest other bridges needed to extend ordinary roads across valleys, but an enumeration

[48] Modern Alatri, Frosinone province, in central Italy. It has remains of pre-Roman walls of cyclopean masonry.

[49] Ancient Tarraco, 54 miles west-southwest of Barcelona, on the Mediterranean. It was the capital of Hispania Tarraconensis, the eastern half of the peninsula. The better-known aqueduct of Segovia (40 miles north-northwest of Madrid) is a later work of Trajan's time.

[50] The Pont du Gard is a monument of great beauty. Remember the impression it made upon Jean Jacques Rousseau, c. 1740 (he was then 28): "Que ne suis-je né Romain! Je restai là plusieurs heures dans une contemplation ravissante. Je m'en revins distrait et rêveur, et cette rêverie ne fut pas favorable à madame de Larnage. Elle avait bien songé à me prémunir contre les filles de Montpellier, mais non pas contre le pont du Gard. On ne s'avise jamais de tout." *Confessions*, partie 1, livre VI.

of Roman roads and bridges would be too long. The first stone bridge in Rome was the pons Aemilius built in 179; it was made of stone piles supporting a wooden deck; stone arches were added in 142 B.C. (that bridge is the modern Ponte di S. Maria, or Ponte Rotto).

A temple to Asclēpios had been built in 292 on an island of the Tiber; the island was connected to the banks by two bridges which were eventually rebuilt in stone, the pons Fabricius in 62 B.C. supported by two arches of c. 25 m each, and the pons Cestius under Tiberius.

Another kind of bridge was the temporary wooden one (Fig. 73) thrown across the Rhine in 55 B.C. by Caesar and described by him in *De bello Gallico* (IV, 16–19). It was the first military bridge of its size. The site of it is not exactly known; it was somewhere between Andernach and Cologne. It was a trestle bridge (*pont de chevalets*) and was built by means of pontoons; the legs were rammed down into the river bed, the rigidity of the whole structure was increased by means of other legs, and special precautions had to be taken to protect the bridge against the impact of the stream. Caesar ends his clear description with the words, "The whole work was completed in ten days from that on which the collecting of timber began and the army was taken across."

Considering on the one side the lack of precedent, then the breadth, rapidity, and depth of the Rhine, finally the speed of the workers, this was an astounding achievement. It helps to explain the strength of the Roman armies of the golden age; it was due to the will power of great commanders like Caesar, the energy and discipline of the soldiers, and, last but not least, the availability of clever engineers.

It was necessary also to dig canals. The first of these outside Italy was the Fossa Mariana (so-called after Gaius Marius, c. 155–86) dug in 101 in the Rhône delta in order to ensure its navigation. In general, the existence of a delta tends to close the entrance of a river; access to its lower course is made possible only after the execution of engineering corrections (canals and dikes). Sometimes, similar corrections must be applied to the river itself. This was done by the Romans to improve the navigability of the Rhine, not so much for the sake of business as for military reasons. Naval stations were established in Mainz, Coblenz, and Cologne and the river bed was kept clear and strong by the building of embankments in the years 13 B.C.–A.D. 47.

Roads and bridges were built mainly for military reasons, and this was even more true of harbors many of which were created for the satisfaction of imperial needs. Tarragona is a good example; the engineering needs were materially increased under Augustus when that city became the capital and its harbor the mouth of Eastern Spain. A great harbor was created by Agrippa in 36 B.C. in Baiae, just West of Naples. It was called, in Octavianus' honor, the Portus Julius. Rome's ancient harbor, Ostia, was not materially improved until a little later, but the navigation of the Tiber required con-

Fig. 72. Pont du Gard. Roman aqueduct built in 18 B.C. across the river Gard (tributary of the Rhône) near Nîmes.

Fig. 73. Figure in Fra Giocondo's edition of Caesar's *Commentaries* (Venice, 1513) showing how Caesar's bridge was built across the Rhine c. 55 B.C. [Courtesy of Harvard College Library.]

tinuous supervision; it was entrusted to special magistrates, the *Curatores riparum et alvei Tiberis* (curators of the Tiber's banks and bed). A harbor was created in Caesarea [51] by Hērōdēs the Great and greatly increased the commercial importance of that city, which he had founded (in 22) and named in Augustus' honor. The building of the harbor lasted ten years and was inaugurated by Hērōdēs in 9 B.C., the twenty-eighth year of his rule. Gigantic blocks of stone were used, and strong embankments and very high walls were built around it. The harbor was adorned with beautiful buildings and with heroic statues of Augustus and Roma.

The Phlegraean Fields. Let us now return to Italy, to a part of it where many wonders of nature could be witnessed, as well as lesser wonders of Roman engineering. I am referring to the Phlegraean Fields [52] at the sea-

[51] Caesarea Palaestinae, Caisareia in Greek, Qaiṣāriya in Arabic. In A.D. 70 the city became the capital of Roman Palestine and the residence of the Roman procurators. It is now in Israel. Abundant ruins remain, which it was my privilege to visit in August 1953.

[52] Derived from the Greek *ta phlegraia*, meaning "the burning" (fields). The Latin term is Phlegraei campi.

shore, west of Naples, the volcanic plain of Campania, where Mother Earth was playing all kinds of tricks: burning lava could be witnessed at times, hot and mineral springs, fumaroles, sulfur mines, earthquakes, and, less obviously, the slow subsidence of the bay. It was the strangest kind of landscape one could dream of — infernal and sinister oddities mixed up with incredible loveliness, the blue sky and the laughing sea, luxuriant vegetation, gay flowers and luscious fruits, figs, olives, and grapevines in abundance. The place was sacred in at least three ways. It was there, in Cumae,[53] that the Greeks had established c. 750 their first Italian colony. Cumae was for centuries a living center of Greek culture for the Etruscans as well as for the Romans.

It was in Cumae that the oldest and most famous sibyl of Italy prophesied, and the very impressive grotto where she officiated can be visited to this day.[54] The *sibylla* of Cumae was a priestess of Apollōn; she and other sibyls were mediators between men and the unknown, comparable to the oracle of Dōdōnē or the Pythia of Delphoi.[55] They acted as mediums do when they are in a frenzy and their ecstatic and obscure oracles might be interpreted for the public good or their own advantage by clever politicians. The sibyl of Cumae was said to write some of her oracles on oak (or palm) leaves which were abandoned to the winds; this does not agree with the fact that her oracles were preserved in special books.

It is said that Tarquinius Superbus (d. c. 510) acquired the Sibylline Books, which were placed for safekeeping in the temple of Jupiter Capitolinus under the guardianship of special priests (the *duo viri sacris faciundis*, whose number was increased to ten and finally to fifteen).[56] It is probable that that early collection included oracles not only of the sibyl of Cumae but also of the older sibyl of Erythrai (in Iōnia, opposite Chios) and perhaps of other sibyls. The collection was gradually increased and in cases of dire emergency the Senate ordered the *Decemviri* to inspect (*adire, inspicere*) the sacred books (*libri fatales*) and to interpret them. The result of the interpretation was generally the adoption of a ritual designed to expiate evil or avert calamity. In 83 B.C., the whole collection was lost in the conflagration that destroyed the temple of Jupiter.

Those *chrēsmoi sibylliacoi* (sibylline oracles) were written in Greek verse (hexameters); they symbolized Greek religion and Greek worship in the Roman world.

[53] The Greek name was Cymē, Latinized Cumae, Italianized Cuma. Tarquinius Superbus, seventh and last of the legendary kings, died there (c. 510) in wretched exile.

[54] Its entrance, which had been hidden for centuries, was excavated in May 1932 by Amedeo Maiuri. Before its discovery the "grotto of the Sibyl" described by Virgil, *Aeneid*, VI, 11, was wrongly identified with a grotto close to Lake Avernus (see below).

[55] For those Greek oracles see Volume 1, p. 196.

[56] This college of fifteen continued to function until the time of Flavius Stilicho in A.D. 405. It represented the Greek rite, as against the Pontifices, who represented the Roman rite.

After 83 B.C. a new collection was hastily formed from many temples all over the Greek world, so many, indeed, that it was necessary to make a selection; Augustus ordered some 2000 spurious items to be burnt.

There gradually came to be two kinds of sibylline oracles, those uttered by real Sibyls and those, far more numerous, that were ascribed to sibyls as imaginary as the Muses. Those of the second kind constitute a definite literary genre, cultivated until the end of antiquity and even in the Middle Ages. The genre was Greek but with occasional Latin imitations, of which the *Ecloga quarta* of Virgil, composed in 40 B.C., is the most famous example. Virgil's eclogue was concerned with the end of the world (or a new golden age).

The *Sibylline oracles*, which have come to us in many sixteenth-century editions, are definitely apocryphal, though composed in the traditional manner: Greek hexameters, archaic vocabulary, decent obscurity. They continued to be written until the sixth century of our era and even later. Their purpose was political or eschatological, or both combined; they expected to win the heathen world to the Jewish or Christian faith. Thanks to the popularity of those apocryphal writings, the ancient sibyls to whom they were invariably ascribed were given an importance comparable to that of the prophets of the Old Testament. Their influence upon arts and letters was considerable, especially during the Renaissance. In many decorative compositions the sibyls were represented together with the prophets. The best examples of such compositions (among many) were given by Michelangelo in his frescoes of the Cappella Sistina in the Vatican (1508–1510) and by Raphael in Santa Maria della Pace (1514).[57]

The earliest edition of the *Oracula sibyllana* was published by Oporinus (Basel, 1545); Latin translation, Basel, 1545–46; Greek-Latin edition, Basel, 1555. Many other editions in the sixteenth century and later. Modern editions of the Greek text by Aloisius Rzach (Vienna, 1891), and by Johann Geffcken (Leipzig, 1902). Rzach's text was Englished by M. S. Terry (292 pp.; New York, 1899).

The best general discussion is still, I believe, the one by Auguste Bouché-Leclerq, *Histoire de la divination dans l'antiquité* (4 vols.; Paris: Leroux, 1879–1882), vol. 2 (1880), pp. 93–226. For Virgil's Eclogue IV see Henri Jeanmaire, *La Sibylle et le retour de l'âge d'or* (150 pp.; Paris: Leroux, 1935). For the Jewish oracles, Alberto Pincherle, *Gli oracoli sibillini giudaici; Orac. Sibyll., LL. III–IV–V* (178 pp.; Rome: Libreria di cultura, 1922), Italian translation with notes.

The awfulness of the Phlegraean Fields was vastly increased in the popular mind by the ominous presence of the Solfatara (Forum Vulcani), which was the crater of a sleeping volcano, and of Lake Avernus, a deep lake in gloomy surroundings; the mephitic smells arising from it inspired the belief that it communicated with the underworld (in a sense it does, for it is a volcanic lake with outlets of sulfurous gases at the bottom).[58]

[57] Emile Mâle (1862–1954), *Quomodo Sibyllas recentiores artifices repraesentaverint* (80 pp.; Paris, 1899).

[58] Lake Avernus is an old crater, 3 km

The main cause of sacredness of the region is the fact that Virgil spent there a part of his life. In Book VI of the *Aeneid* he sang of the sibyl, of Lake Avernus, and of the underworld. When we travel across the Phlegraean Fields, as I did a few years ago, we travel with him, and we are conscious all the time of his presence. After his death at Brindisi in 19 B.C. his ashes were collected from the funeral pyre and translated to Naples; they were placed in a tomb on the Via Puteolana between the first and second milestone. "Virgil's Tomb" is still shown and it thrills many visitors to think that it is genuine; it has also given pleasure to many scholars who tried their best to prove or disprove its genuineness.[59]

Cumae was the oldest and in early times the most important city of the region. It is located at the west end of it. At the southeast end of a promontory just below it was the little bay of Misenum (Misēnon), of which Augustus made a military harbor, the main station of the Roman fleet in the Tyrrhenian Sea. The lovely bay of Puteoli is an irregular half circle extending from Misenum on the west to Mons Pausilypus (Posillipo) on the east. When walking from Misenum around the bay, one passes Baiae where there are famous springs and which was the favorite spa and bathing resort of the Roman elite; a number of palaces and villas were built in the hills above the shore.[60] A little farther, in the middle of the bay, was Puteoli, a colony founded by the Cumaeans as early as 521 and colonized by the Romans in 194. It was an excellent harbor which acquired considerable importance under Roman management.[61] By 125 B.C. it was already the main commercial center for business with Alexandria and with Spain; as a warehouse it was second only to Dēlos. It became an opulent city famous for its lighthouse, amphitheater, trade guilds, fire brigade, outgoing roads, post station, and other amenities. Its richness was the cause of its ruin, for as soon as Roman arms were no longer strong enough to defend it, it was repeatedly sacked by barbarians.[62]

Before leaving the Phlegraean Fields, I would like to add two remarks. First, this region just west of Naples is very different from the Vesuvian, southeast of it. Vesuvius destroyed Pompeii and Herculaneum in A.D. 79

---

in circumference, 60.50 m deep; it is surrounded by precipitous banks, which in ancient times were covered with thick woods. Agrippa cut the trees down, built a tunnel connecting Avernus with Cumae, and canals leading from Avernus to Lake Lucrinus, and thence to the sea. Lake Avernus then became a safely hidden harbor, the Portus Julius (see below).

[59] See the excellent little guide by the best authority, Amedeo Maiuri, *The Phlegraean Fields* (146 pp., ill.; Rome: Libreria dello Stato, 1947).

[60] Among the famous men who owned villas at Baiae were Licinius Crassus the Orator (d. 91 B.C.), Caius Marius (d. 86 B.C.), Caesar and Pompey, Varro, Cicero, Hortensius the Orator (d. 50 B.C.).

[61] Its original Greek name was Dicaiarcheia; Puteoli is the Roman name (now changed to Pozzuoli). Its importance diminished when a new harbor was built at the mouth of the Tiber, Ostia, which was very close to Rome. This was begun by Claudius in A.D. 42 and finished by Nero in 54.

[62] Such as the Visigoth, Alaric, in 410, the Vandal, Genseric, in 455, and the Ostrogoth, Totila, in 545. After that there was not much left to plunder.

and is still active; in the Phlegraean region, on the contrary, in spite of its volcanic nature, no such catastrophe has ever occurred and life has continued without great interruption as it was and is today. The only great change is caused by the steady subsidence of Baiae, much of which is now under water.[63]

Another remark which I like to make is that a school of archaeology, the Villa Vergiliana, was established a few years ago in Cuma, thanks to the enthusiasm and the solicitude of Signora Mary A. Raiola. It was my privilege to visit it in July 1953 when Rev. Raymond V. Scholer, S.J., was director. The school is not meant for research but simply to accommodate American students and teachers for a month of archaeological promenades in what is perhaps the most evocative part of the Roman world. This is very important, because American students lack the opportunities of Italian, French, Spanish, or other Mediterranean students and do not grow up, like them, in a kind of living past, in the familiarity of Roman and Greek relics. The classical world may seem a little unreal to those young Americans, but a full month spent in the Phlegraean Fields, in Naples, and in the Vesuvian region is sure to give them a deeper understanding of the ancient world; this may be more instructive to them than years of literary studies.

The students established in Cuma may observe in their immediate neighborhood a whole series of "travaux d'art": the old acropolis, the Greek and Roman walls, temples, thermae, the amphitheater, cisterns, drains, tunnels and other hypogea. Some of those buildings are very ancient, but many are Roman, and more exactly from the time of Octavianus – Augustus. For example, the architect Cocceius, who was in Octavianus' service, dug and built the so-called *Crypta neapolitana*, a tunnel across the hills separating Naples from the western fields. It is 700 m long but very narrow (3.20 m; height varying from 2.80 to 5.60 m) and is badly lighted by a series of vertical or oblique loopholes. The Grotta di Seiano in the neighborhood is another tunnel which must probably be credited to the same architect. Those diggings were facilitated by the nature of the rock, tufa, composed of volcanic detritus in various stages of consolidation but generally easy to cut.

Lake Lucrinus close to the sea in the bay of Baiae was famous for its oysters and shellfish. Agrippa ordered the building of moles to protect it from storms. In the early part of the first century B.C., an oyster farm was established there by one Sergius Orata;[64] this was a very profitable undertaking. Orata's method, breeding the oysters on a large number of pales emerging from the waters, has been followed until now. In the hills around

[63] One may add, though this belongs to a much later time, the eruption of a new volcano, Monte Nuovo near Lake Avernus on 30 September 1538. It is 139 m high with a deep crater at the top. See my *Six wings* (Bloomington: Indiana University Press, 1956).

[64] Varro, III, 3, 9; Columella, VIII, 16, 5. Orata (or Aurata) was a nickname given to him because of his fondness for the gilt bream (*Abramis brama*).

Lucrinus were hot springs, and this induced Romans to build many thermae and villas. One of those villas was Cicero's, which he called Academia (or Cumanum); after his death it fell into other hands and was finally part of Hadrian's demesne; it was there that Hadrian was buried in A.D. 138.

At Misenum where Agrippa and Augustus established the naval base necessary for the domination of the Tyrrhenian Sea, a subterranean reservoir of fresh water was built for the provisioning of sailors and marines. This was the "*Piscina mirabilis,*" an immense rectangular tank (70 × 25.5 m; 15 m high) supported by 48 pilasters arranged in four rows down the length and twelve rows across, making five long aisles in one direction and 13 shorter ones in the perpendicular direction. The *Piscina* had a capacity of 12,600 m^3. The sight is very impressive; it suggests a temple rather than a reservoir.[65]

Lake Avernus having become a naval port and dockyard, a large and beautiful tunnel was built through Monte Grillo to connect the lake with Cumae. It was 1 km long and wide enough to permit carts to run in both directions; it received daylight at all points by means of six vertical or oblique light wells or ventilation shafts; it included an aqueduct (a gallery within a gallery) with its own niches, ventilation shafts, and wells for descent. The aqueduct was needed to convey drinking water to the fleet. This was another of the public works ordered by Agrippa and executed by Cocceius; "it stands as the greatest piece of civil and military engineering achieved by the Romans in the field of subterranean roads."[66] A military dockyard (Navale di Agrippa) was constructed close to the lake as a necessary addition to the naval harbor, Portus Julius. There was another tunnel near by, some 200 m long, 3.75 m wide, and 4 m high without light shafts. It is wrongly called the Grotto of the Sibyl. It was built in the Augustan age to provide a secret passage from Avernus to Lucrinus.

The most impressive monument in the district is the real "grotto of the Sibyl," which had been hidden by landslides and stone falls and was discovered only in recent years (1932). It was built by Greeks as early as the fifth century B.C., if not before, but was modified by them in the fourth and third centuries. This does not belong to the period covered by this volume but is so important that it is impossible to overlook it when we happen to be so close to it. Indeed, it is one of the most impressive monuments in the whole Mediterranean world. The main feature of it is a long trapezoidal dromos or gallery, 131.5 m long, 2.4 m wide on the floor but getting narrower above, and 5 m high; its dimensions surpass those of other dromoi in Mycenaean and Etruscan architecture. It was lighted by six lateral galleries

[65] It evokes comparisons with the Byzantine cisterns of Constantinople, though some of these are much larger. The Basilica Cistern (Yere batan serai), still in use, is 140 x 70 m in area and is supported by 336 columns each 8 m high; the Philoxenos Cistern (Bin bir direk, Thousand and One Columns) is 64 x 56 m and is supported not by 1001 columns but by 224.

[66] Maiuri, *The Phlegraean Fields,* p. 127.

opening westward over the sea. As one walked along that underground gallery, one was gradually brought into the proper mood to face the Sibyl in her inner room (*oicos endotatos*; the adyton). The very unbelievers would be duly impressed; as to the true believers, they were overwhelmed, became enthusiastic, and lost every power of criticism. The hysterical Sibyl would not rave in vain; every word of hers would be treasured as a divine message. Virgil who was with us during our visit must have felt many times that awful experience; he helped us to share it and to understand it with charity.

Marcus Vipsanius Agrippa. Two names have obtruded themselves repeatedly in the preceding pages – Agrippa and Cocceius. Who were they?

Marcus Vipsanius Agrippa (63–12 B.C.) was one of the leading personalities of the Augustan age. In spite of his being the scion of an obscure family, he was sent to complete his studies in Appollōnia,[67] where Octavius (later Octavianus, Augustus) was his classmate. After Caesar's murder, he returned to Rome with Octavianus. He took an active part in the Civil Wars; in 41, he commanded a part of Octavianus' army around Perugia; in 38, he was governor in Gaul, suppressing a rebellion of the Aquitani and directing a punitive expedition across the Rhine. In the following year he was shifted to naval service; he supervised the organization of Octavianus' fleet and obtained the naval crown, *corona navalis* (it was then that he built Portus Julius). In 36, he won naval victories at Mylai and Naulochos (both on the northeast coast of Sicily) and defeated Pompey's fleet; in 35–33 he was engaged in the Illyrian campaign; in September 31 at Actium his naval victory was the main cause of Antony's defeat. Soon after, he was sent on a political mission to the East, his headquarters being at the island of Mytilēnē, then was recalled home and did some more fighting in Gaul and Spain. He was associated with Augustus, sharing a part of his power for ten years (18–8) and was appointed one of the Fifteen Men, the college *sacris faciundis*. During the years 16–13, a second mission to the East was intrusted to him; it was then that he established Polemōn as king of Pontos and Bosporos (15 B.C.), organized Roman colonies in Syria, at Hēliupolis (Ba'albek) and Bērytos (Beirūt),[68] and cultivated Hērōdēs' friendship. His last mission was to Pannonia (south and west of the Danube) to prevent a rebellion. He then returned for the last time to Italy, where he died in the following year (12 B.C.). He bequeathed his property to Augustus and was buried in the imperial mausoleum.

In 21, he had married Augustus' daughter Julia; through his children

[67] Apollōnia in Illyria, not far from the Adriatic shore. This was a prosperous Greek colony, where young Romans were sent to be properly Hellenized.

[68] Hēliupolis and Bērytos were ancient cities, the former a sacred place for the worship of Ba'al. Bērytos was an old Phoenician harbor which had been destroyed by the Syrian usurper, Tryphōn Diodotos, in 140 B.C. From c. 15 B.C. it was a Roman colony; Agrippa settled two legions in its territory.

and grandchildren of three marriages he had many ramifications in the imperial families. He was praetor in 40 and consul thrice, in 37, 28, and 27. He wrote an autobiography which is unfortunately lost; it might have been a revealing document.

This long enumeration, incomplete as it is, serves to illustrate what it might mean to be a Roman general and statesman in those days.

The most useful of Agrippa's accomplishments, however, were the many public works erected during his aedileship in 33 and following years, some of which have already been mentioned. Among them were the restoration of aqueducts and the building of two new ones (Julia and Virgo), the construction of sewers, roads, and tunnels, of the Porticus Neptuni (a stoa or colonnade), of the Thermae Agrippae, of the Pantheon (in 27), of the Portus Julius in Lake Avernus, the aqueduct of Nîmes or Pont du Gard (in 18). He organized a survey of the empire, of which we shall speak in another chapter.

Agrippa was a Roman, a practical man, whose primary interest was the building of utilitarian works (aqueducts and sewers, harbors, roads, and tunnels), yet some of his works were artistic creations of a very high order. The Pont du Gard was a marvel; the Pantheon was another;[69] a rotunda of such boldness and beauty had never been erected before (Fig. 74). It may be finally said to Agrippa's artistic credit that he ordered the translation from Lampsacos (on the Asiatic side of the Dardanelles) to Rome of the Fallen Lion of Lysippos.[70]

We may assume that most of Agrippa's activities were purely administrative; he planned and ordered this or that, yet the mass and variety of his accomplishments are very impressive. It is never enough to plan and order, one must plan wisely and be sure that the orders are implemented; this requires many diverse qualities, primarily the ability to obtain the collaboration of competent assistants. One of his assistants was L. Cocceius Auctus who built tunnels in the Phlegraean Fields. Another was Valerius of Ostia, architect of the Pantheon. Still another was Vitruvius, though we do not know exactly their relation.[71] Many of Agrippa's great works ought to be credited to those assistants, but it is perhaps a little nearer to the truth to credit them to him than to Augustus, as is often done.

[69] The Pantheon was dedicated to all the gods after the victory of Actium; it was completed by Agrippa during his third consulate (27 B.C.). It is a circular temple with a cupola; the diameter of the rotunda and the height of the cupola are equal, 43.20 m. The Pantheon was badly damaged by a fire in A.D. 80 and rebuilt. It is now a church, Santa Maria Rotonda (or ad Martyres).

[70] The story is told by Strabōn, XIII, 1, 19. Lysippos of Sicyōn was one of the most famous artists of Greece; he was the official sculptor of Alexander the Great. The number of monuments ascribed to him is immense, but most of them, including the Fallen Lion, are lost.

[71] Vitruvius does not mention Agrippa, Cocceius, or Valerius. Cocceius is mentioned by Strabōn, v, 4, 5, and Agrippa repeatedly, but not Valerius.

Fig. 74. Interior of the Pantheon, as painted by Giovanni Paolo Panini c. 1740. According to its own inscription the Pantheon was completed in 27 B.C. by Agrippa; it was dedicated to all the gods (hence its name, pantheon), but more especially to Mars and Venus, patrons of the Julian family to which Caesar and Augustus belonged. It was burned twice, in 80 and 110, and was repaired or rebuilt in 127 by Hadrian (emperor 117–138), who generously replaced the original inscription. In 609 it was transformed into a Christian church and dedicated to Mary and all saints or martyrs (hence its present name, Sancta Maria ad Martyres; more commonly Santa Maria Rotonda). It is the burial place of kings and queens of Italy and of artists, notably Raphael. Its cupola was by far the largest and most beautiful of antiquity. Inasmuch as the whole building is but a single room its most significant feature is its general proportions, which are the same now as they ever were. [Samuel H. Kress Collection, National Gallery of Art, Washington, D. C.]

MINING AND METALLURGY

"The Seleucid kingdom, at the time of its greatest extent, had been the first civilized state in western history able to supply all its requirements in metals. But in many cases the enormous cost of land-transport would have made it cheaper to import from abroad than to draw upon internal sources; for instance it must have been easier to buy tin from Bohemia or the Atlantic coastlands than obtain the products of Drangiana." This astonishing statement is the opening of Davies' book on Roman mining.[72] He then proceeds to show that the Roman situation was very different, for the Romans were spread out in so many countries that they were almost self-sufficient in metals, and their domination of sea roads enabled them to bring materials from a long distance at a relatively low cost. Not only had the Romans enough metals for their own use, but they could supply them to outsiders and thus obtain a political hold upon them. They exported gold to India, silver and copper to Germany, but in the republican period the Senate tried to regulate the outflow of gold, and in the later empire the export of iron was forbidden, lest it be made into weapons by the barbarians.[73] They might be obliged, however, to import from the distant non-Roman world some materials of superior quality, but the cost of transportation of such expensive materials did not matter. The best example of that kind was the so-called Seric iron, which probably came from India [74] (not China) and could be brought by sea to Roman harbors.

Mining was the main industry of Hellenistic times and its practice then (and before) [75] was cruel in the extreme, absolutely inhuman. The work was a punishment meted out to slaves, criminals, prisoners of war; the mines were labor camps of the worst kind with no respect whatsoever for life, no mercy. The best (or worst) examples of such cruelty were given in the Nubian gold mines, exploited by the Ptolemies; labor in those mines was so terrible that the workers welcomed death when it finally came to deliver them.[76]

[72] Oliver Davies, *Roman mines in Europe* (302 pp., 10 maps, 49 ill.; Oxford, 1935) [*Isis* 25, 251 (1936)]. Drangiana (Drangianē) was south of the Oxus river and west of the Indus (modern Sijistān, now divided between western Afghānistān and eastern Īrān).

[73] Davies, p. 1, 1935.

[74] Ferrum Sericum; Pliny, *Natural history*, xxxiv, 14, 41. It was probably the famous Indian steel, strangely called "wootz" (Henry Yule and A. C. Burnell, *Hobson-Jobson: A glossary of colloquial Anglo-Indian words and phrases, and of kindred terms, etymological, historical, geographical and discursive*, ed. William Crooke [London: John Murray, 1903], p. 972).

[75] For the silver mines of Laurion in Attica, see Volume 1, pp. 296–297, 229. Equally scandalous conditions obtained in the Cappadocian quicksilver mines. W. W. Tarn and G. T. Griffith, *Hellenistic civilisation* (London: Arnold, 1952), p. 254. On ancient mining in general, see Oliver Davies in *Oxford Classical Dictionary*, p. 573.

[76] As stated by a contemporary, the geographer Agatharchidēs of Cnidos (II–1 B.C.). Karl Müller, ed., *Geographi graeci minores* (Paris, ed. 2, 1892), vol. 1, pp. 123–127. This Karl Müller is often called

This suggests to me the following reflection. It has been said that agriculture and mining were the two fundamental industries. That may be true, yet they are utterly different. Agriculture, or husbandry, is eminently social; it is very deeply connected with the most natural human group, the family; its very name is characteristic; it is the business of a husbandman, a husband and father. Mining, on the contrary, was eminently antisocial; miners were slaves and prisoners, and their labor was so hard and painful that it hardened and brutalized them.

The most flourishing period of Roman mining seems to have been the late republic and the early empire, but it is difficult to be more precise. The history of ancient mining is almost dateless; slag heaps are the best evidence of mining, but it is impossible to say when they were built; the only chronological landmarks are provided by labor troubles and repressions, but these are not always available.[77]

It is thus impossible to explain the development of mining techniques. The Roman technique was derivative from the Egyptian, the Greek, and the Etruscan. As Roman surveyors obtained considerable experience in many countries, East and West, their intuitive ability in prospecting grew keener. They developed new ways of flushing, pitting, driving galleries, sinking shafts, lighting and ventilating, draining, propping, hauling, and surveying. They had better iron tools, as well as stone picks, stone wedges, and stone hammers. Their metallurgical technique was also improved, giving better methods of crushing ores, washing, roasting, better furnaces of many kinds, better smelting, liquation,[78] cupellation, and so forth. It is not certain that cast iron was already known in Rome (it was in China); it may have been obtained occasionally, if not in Rome itself, at least in the barbaric countries of Central Europe. Steel had been known for centuries and better steel may have been obtained in certain places, for example, in Como, where the goodness of the steel was ascribed to the virtues of the lake water.

To give two more specific examples, it is possible that some Romans were already able to refine gold by the salt or stibnite[79] process, and that they knew a method comparable to the Pattinson process for the separation of silver from lead-silver ores.[80] It cannot be said, however, when that knowledge was obtained, except that it was before Pliny's time.[81]

Karl (Charles) Müller of Paris to distinguish him from the immense tribe of Müllers. In spite of his outstanding merit, he is practically unknown; we do not even know his date of birth and death. His works from 1841 to 1868 were published in Paris, later until 1883 in Göttingen. Was he perhaps obliged to leave Paris at the time of the War of 1870?

[77] For example, history of the silver mines of Laurion, Volume 1, p. 296.

[78] Liquation is the separation of a fusible substance from one less fusible by heating. According to Davies, *Roman mines in Europe*, p. 17, the Etruscans had already made use of that process.

[79] Stibnite is the native antimony trisulfide, used mainly as a cosmetic.

[80] Hugh Lee Pattinson patented his process for desilvering lead in 1833.

[81] See R. J. Forbes, *Metallurgy in antiquity* (Leiden: Brill, 1950), p. 205 [*Isis*

There is no doubt about the extension of mining of many kinds all over the Roman world nor about the relative complexity of metallurgical techniques, but chronological precision is unobtainable.

43, 283–285 (1952)]. Forbes gives many technical details, but without chronological precision; that is not his fault.

XXI

NATURAL HISTORY, CHIEFLY AGRICULTURE

This chapter is divided into three sections, Carthaginian, Hellenistic, and Roman or, more exactly, Latin. The first will astonish many readers, for it is to some extent Oriental, and they did not expect a new Oriental intrusion in the Western Mediterranean world.

CARTHAGINIAN AGRICULTURE

Carthage was founded c. 814 B.C. on the North African shore, just southwest of Sicily, by a swarm of Tyrians,[1] that is, Phoenicians. It became the leading Phoenician colony of the Mediterranean and because of its intrinsic power and of its location, south of the Tyrrhenian Sea, was the main rival and enemy of Rome. The Punic Wars (I, 264–241; II, 218–201; III, 149–146) were the bitter fruits of that rivalry; the final defeat of Carthage in 146 opened the way for Roman imperialism. We know little about Punic culture. The only names emerging from the past are those of the two Carthaginian explorers, Hannōn (V B.C.) and Himilcōn (V B.C.), both of whom flourished in the fifth century. Their language, Phoenician or Punic, very close to Hebrew, was secret to the Romans and the more so because the script was different from theirs, yet the Romans had heard of a treatise on agriculture, written at an unknown time by a Carthaginian called Mago [2] (II–2 B.C.). After the destruction of Carthage (146 B.C.), the Roman Senate ordered the translation of Mago's book into Latin.

We know practically nothing of the author, but his name was that of an illustrious Punic family. The eponym who flourished c. 520 was the founder of the military power of Carthage; four others of the name distinguished

[1] Inhabitants of Tyre (Hebrew Ẓor, Greek Tyros, Arabic Ṣūr), on the coast of S. Lebanon. Tyre was itself a colony (c. fifteenth century) of Sidon (Ṣaidā'). It was a city of great importance, political and commercial, the capital of Phoenicia from the eleventh century to 774 B.C. It is frequently mentioned in both the Old Testament and the New. For Phoenician history, see Volume 1, pp. 102 (map), 108, 222.

[2] Magōn in Greek. In the same way, the names Hanno and Himilco in Greek are spelled Hannōn and Himilcōn.

themselves in the military service of their country. One of them was an admiral (fl. 396) in the war against the great Dionysios of Syracuse; another was commander of a Carthaginian army in Sicily in 344; another was the youngest brother of Hannibal, under whom he served in the Second Punic War; he was defeated by the Romans in the Po Valley in 203, managed to reëmbark his army for Africa, but died of wounds on the voyage; still another was commander of New Carthage (Carchēdōn hē nea; Cartagena on the southeast coast of Spain) when that city was taken by Scipio Africanus in 209 and he was sent prisoner to Rome.

Mago the agriculturist may have been a member of that family; at any rate, his name was well known in Rome and that helps to explain Roman interest in a Punic treatise and the wish to have it put in Latin.

In his treatise on agriculture, Varro recites the long list of his Greek authorities (I, 1, 8), more than fifty, and ends by saying, "All these are surpassed in reputation by Mago of Carthage, who gathered into twenty-eight books, written in the Punic tongue, the subjects they had dealt with separately."[3] Columella (I-2) called Mago the father of agriculture (*pater rusticationis*). This label, like all other labels of the same kind, is misleading. After all, if Mago had put together the knowledge accumulated by so many authors, he would hardly be called the father of the subject, but let that be.

The loss of the Punic original is not surprising, but it is odd that nothing remains of the Latin translation, and that the little we know of his work we owe to a second translation made into Greek c. 88 B.C. by Cassios Dionysios. Did Cassios make it from the Latin or from the Punic? The second hypothesis is not impossible, because he flourished in Utica,[4] the greatest city of North Africa after Carthage; it was, like the latter, a Phoenician colony, but in the Third Punic War sided with Rome. In other words, Cassios might have known Punic, or he might have been familiar with Punic scholars ready to help him. According to Varro (I, 1, 10), Mago's twenty-eight books were translated by Cassios Dionysios and "published in twenty books dedicated to the praetor Sextilius." In these volumes he added not a little from the Greek writers whom I have named, taking from Mago's writings an amount equivalent to eight books. Diophanēs, in Bithynia, further abridged these in convenient form into six books dedicated to King Dēiotaros. I shall attempt to be even briefer and treat the subject in three books." There is no proof that Varro knew the Latin translation of Mago or that he knew the latter otherwise than through Cassios'

[3] This quotation and others in this chapter are taken from the Latin-English edition of Varro by William Davis Hooper (Loeb Classical Library; Cambridge: Harvard University Press, 1934).

[4] Utica is famous because it was there that the Pompeian party made its last stand against Caesar and because of the suicide of the virtuous "Cato of Utica," who preferred to die rather than to fall into Caesar's hands (46 B.C.).

Greek version. When he refers to Mago he refers as well to Cassios and his references are not important.

The history of that Punic text is curious. Translated into Latin after 146, then abbreviated in Greek c. 88 B.C. by Cassios Dionysios, then abbreviated a second time by Diophanēs of Nicaia about the middle of the same century and dedicated to Dēiotaros, one of the tetrarchs of Galatia,[5] it is a good witness of international confusion – a text written in Punic, translated into Latin presumably in Rome, then Hellenized twice, in the West and in the East.

HELLENISTIC BOTANY

Cassios Dionysios (I–1 B.C.), dealt with in the previous section as Mago's translator into Greek, was a botanist in his own right. Not only had he added to his translation many extracts from Greek authors, but two treatises were ascribed to him, one on roots (*rhizotomica*), the other on materia medica. This second treatise was illustrated.

Two other botanists, if we may give them that noble title, were kings, Attalos III of Pergamon and Mithridatēs VI of Pontos. The first, Attalos Philomētōr (king from 138 to 133), he who bequeathed Pergamon to the Romans,[6] is quoted by Varro as one of the authorities of his treatise on husbandry[7] and was referred to also by Columella (I–2) and by Pliny. It is significant that he was especially interested in poisonous plants; he prepared poisons and experimented with them. Mithridatēs Eupator was also making toxicological experiments.[8]

A better claim to botanical fame is held by Mithridatēs' physician, Cratevas. He wrote a materia medica wherein he showed some knowledge of the action of metals upon the body (this may have been a part of the toxicological investigations that interested the king so deeply), but, what is more note-

[5] A strange character, this Dēiotaros. He had sided with the Romans in their war against Mithridatēs VI Eupator, the Great, king of Pontos (120–63), and been rewarded by them with the title of king and the addition of Armenia Minor to his territory. In the Civil War he took sides with Pompey, but submitted later to Caesar. Being accused of complicity in a plot against Caesar, he was defended by Cicero, whose oration is extant. In 42, he joined the party of Brutus and Cassius, and died shortly afterward in old age. He was an Asiatic chieftain whose ambition led him to be embroiled in all the Roman quarrels.

[6] The will was contested by his half brother, Aristonicos, who succeeded in delaying the annexation a few years but was made prisoner in 130, carried to Rome, and put to death. In 130, the state of Pergamon became the Roman province of Asia, the city of Pergamon being its capital. Pergamon enjoyed much prosperity under the Roman protection because of its commercial opportunities and of its famous *asclēpieion* or medical temple, but its famous library was taken away in 40 B.C. by Mark Antony to be presented to Queen Cleopatra. (Did that really happen?) In the first century after Christ, Pergamon was evangelized and became one of the Seven Churches of which St. John the Divine spoke in *Revelation* 2:12–17.

[7] In his list of authorities (I, 1, 8) Varro brackets Attalos with Hierōn of Sicily, who may be Hierōn, king of Syracuse (270–216).

[8] We shall come back to them in the medical chapter, below.

worthy, he composed a treatise on roots (*rhizotomicon*) divided into at least five books and illustrated. He may have been himself a rhizotomist, that is, a "culler of simples." He omitted the description of plants but determined the form of the "herbal." He is the "father of botanical illustration." [9] But is he really?

This reminds us that the book of Cassios Dionysios was also illustrated and that drawings and diagrams were included in some of the technical writings of the same age. This is not surprising. As the men of science of the Hellenistic period were interested in special investigations and concrete analyses they would feel the urge of adding drawings without which the description of an instrument, and a fortiori of a natural object, can never be complete. Those tendencies have been forgotten because the drawings have almost always dropped out of the manuscripts; it was easy enough to copy the text but difficult to copy the images, or, if these were copied, the intention of the original drawing might easily be falsified and traduced. Consider the words *asphodelos* and *acantha* (or *acanthos*); those words were easy enough to recognize, even if the spelling was modified (as it is in English, asphodel, acanthus), but what about the drawings of these plants? Those drawings were infinitely more instructive than the names but difficult to reproduce.

It is claimed that a few of Cratevas' illustrations were actually transmitted and preserved for posterity in the incomparable Greek manuscript of Damascus (Nicolaos ho Damascēnos I–2 B.C.), born in that city c. 64, Juliana; this is possible, but how can it be proved? [10]

The last botanist deserving to be mentioned in this section is Nicholas of Damascus, (Nicolaos ho Damascēnos, I–2 B.C.), born in that city c. 64, and friend of Hērōdēs I (Herod the Great), king of the Jews from 40 to 4 B.C., whom he survived.[11] This Nicholas was primarily a historian, but the Aristotelian work *De plantis* was ascribed to him.[12]

Nicholas' work is a real botanical treatise, not a herbal; it is written not in the style of Dioscoridēs but in that of Theophrastos and of Aristotle himself; no wonder that the latter was believed to be the author. It is divided into two books and deals with the generalities of plant life.[13]

[9] Quoted from Charles Singer, "The herbal in antiquity," *Journal of Hellenic Studies 47* (52 pp., 10 pl., 46 fig.; London, 1927) [*Isis 10*, 519–521 (1928)].

[10] That manuscript was found in Constantinople by Ogier Ghiselin de Busbecq, c. 1562, and is now in the State Library of Vienna. G. Sarton, "Brave Busbecq," *Isis 33*, 557–575 (1942), p. 566, figs. 4–7. There is a complete facsimile of it (2 enormous folio vols.; Leiden, 1906). The value of botanic illustrations is discussed in my little book, *The appreciation of ancient and medieval science during the Renaissance* (Philadelphia: University of Pennsylvania Press, 1955), pp. 86–95.

[11] Herod died at about the time of Christ's birth. If Jesus was martyred in A.D. 28 at the age of 33, then he was born in 5 B.C. E. Cavaignac, *Chronologie* (Paris, 1925), pp. 197–210.

[12] Sarton, *Appreciation*, pp. 63–64. This text is lost in Greek and known to us only in a Latin translation from the Arabic by Alfred of Sareshel (XIII-1). English translation by Edward Seymour Forster in the English *Aristotle* (Oxford, 1913), vol. 6.

[13] For contents, see Volume 1, p. 546.

The four Hellenistic botanists dealt with were all Asiatics, Attalos of Pergamon, Mithridatēs and Cratevas of Pontos, and Nicholas of Damascus.

Knowledge of animals was derived from the practice of farming, from the chase, and from the organization of zoölogical parks and menageries. Such menageries were very ancient institutions, for it was tempting to keep ferocious animals in cages as a means of showing the power of the king who owned them. For example, see references to the zoölogical parks of Astyagēs (king of Media from 594 to 559) in Xenophōn's *Cyropaedia*.[14]

W. W. Tarn has this to say about Hellenistic menageries: "Seleucus sent an Indian tiger to Athens and Ptolemy II had a zoölogical garden containing, besides 24 great lions, leopards, lynxes and other cats, Indian and African buffaloes, wild asses from Moab, a python 45 feet long, a giraffe, a rhinoceros and a polar bear, together with parrots, peacocks, guinea-fowl, pheasants and many African birds."[15]

LATIN WRITERS ON AGRICULTURE

The most important technical work of this age was written not in Greek but in Latin, by Vitruvius; so also, the most important agricultural writings were in Latin, those of Cato the Censor, Varro, Virgil, and Hyginus. The first wrote before the middle of the second century; the three others in the second half of the first century B.C.

CATO THE CENSOR

Cato the Censor (II–1 B.C.), or Cato Major,[16] was born in Tusculum in 234 and died in Rome in 149. He was educated on his father's farm near Reate, and that practical training was so deep that it molded the whole of his life, as is witnessed by the work of his old age, the *De rustica*, an account of which will be given presently. His military career, begun at 17, continued for a good many years, interrupted and extended by a long series of political offices. It is typical of Roman expansion that this young farmer, after distinguished service in the Second Punic War (218–201), took part in military expeditions in Thrace, Greece, and eastern Spain, was a civil servant in Sicily, Africa, Sardinia, and Spain, and was censor in 184. He delivered a good many political and judicial orations,[17] and did

[14] *Cyropaedia*, I, 3, 14; I, 4, 5. For more information on menageries, see *Introduction*, vol. 3, pp. 1189, 1470, 1859.

[15] W. W. Tarn and G. T. Griffith, *Hellenistic civilisation* (London: Arnold, 1952), p. 307.

[16] So called to distinguish him from his great-grandson, Cato of Utica (95–46), Stoic, conservative, honest but unamiable defender of Pompey to the last, who committed suicide at Utica in 46 B.C.; one of the noblest figures of the Roman Republic. Cicero wrote a pamphlet, *Cato*, in his praise and Caesar replied with the *Anticato*. His best praise is a line of the Spanish poet Lucanus (Cordova, A.D. 39–65): "Victrix causa Deis placuit, victa Catoni" (The winning cause was pleasing to the gods, the lost one to Cato).

[17] Not only did he deliver his speeches, but he was one of the first Romans to write them out and to publish them. Cicero had

his duty with so much severity that he was nicknamed Censorius.[18] Having been sent on a political mission to Carthage in 175 (26 years after the second war) he was so infuriated by the Carthaginians' renaissance, truculence, and perfidy that he nourished against them an intense hatred. He realized the need of destroying the city for Roman safety, and ended every one of his speeches in the Senate with the assertion, however irrelevant it might appear, "Carthage must be destroyed." [19] His obsession was finally shared by the Senate, and the Third Punic War began in the year of his own death, 149, at the age of 85. I rejoice that Providence did not grant him the savage joy of contemplating three years later the destruction of Carthage, which he had called for so often.

In spite of his military and political activities and of his lack of literary genius, he wrote considerably. All of his books were primarily teaching books; he wrote not to please the Romans but to persuade and to instruct them. He did that so well that such men as Cicero and later Pliny and Quintilian could not help admiring him. He hated every form of sophistication and luxury and was therefore a constant adversary of everything Greek. He was tough, hard, cruel, narrow-minded, mean-souled, bigoted, yet he was the first educator of his people and the greatness of Rome was due in a good measure to his single-mindedness and to his stubborn efforts. He meant what he said and repeated it endlessly.

The only work of his that has survived in its integrity is one which he wrote in his old age, entitled *De agri cultura* in the manuscripts or *De re rustica* in the earliest editions. The composition of that book was his last duty to Rome, for he felt that good farming was the necessary basis of a sound republic. It is hard to believe that that book, written by him in the second quarter of the second century B.C., is actually the earliest extant specimen of a treatise in Latin prose. Think of Greek masterpieces like the histories of Hērodotos and Thucydidēs, written before the end of the fifth century, and of the appearance more than two centuries later of the first substantial writings in Latin prose, those of Cato. It is not that Rome was so new – the conventional date of its birth is 753 – but that Latin culture was very slow and very late. However, it is not the lateness of those writings which astonishes us but their mediocrity, their commonplaceness, their lack of elevation.

Let us examine Cato's masterpiece. I was wrong to call it a treatise, for it is not sufficiently organized for that; it is simply a collection of warnings, admonitions, and recipes put together without much order. The *De agri cultura* is a small book of less than 80 pages; it is divided into 162 chapters averaging 17 lines, but extending from 2 lines each to as many as 140.

---

read more than one hundred fifty. All are lost, except for occasional excerpts.

[18] This was a play on words. *Censorius* refers to the office of censor which he was holding, but it also meant rigid and severe.

[19] "Delenda est Carthago." "Ceterum censeo Carthaginem esse delendam."

The tone of the book is heard at the very beginning. Here is the opening paragraph, complete and verbatim, serving as principium or introduction:

It is true that to obtain money by trade is sometimes more profitable, were it not so hazardous; and likewise money-lending, if it were as honourable. Our ancestors held this view and embodied it in their laws, which required that the thief be mulcted double and the usurer fourfold; how much less desirable a citizen they considered the usurer than the thief, one may judge from this. And when they would praise a worthy man their praise took this form: "good husbandman," "good farmer"; one so praised was thought to have received the greatest commendation. The trader I consider to be an energetic man, and one bent on making money; but, as I said above, it is a dangerous career and one subject to disaster. On the other hand, it is from the farming class that the bravest men and the sturdiest soldiers come, their calling is most highly respected, their livelihood is most assured and is looked on with the least hostility, and those who are engaged in that pursuit are least inclined to be disaffected. And now, to come back to my subject, the above will serve as an introduction to what I have undertaken.[20]

This introduction, wherein farmers are compared with moneylenders and traders, is ungracious but not as irrelevant as it may seem. Money is one of the keynotes of the book. It should be noted that the subject of it is much broader than the earliest title, *De agri cultura*, would suggest; the title chosen by Renaissance editors is better, *De re rustica*; the subject is not simply the cultivation of land but the much larger one signified by the good English word "husbandry." Cato, earthbound, shrewd, and mean, realized strongly that the farmer who is not a good businessman to begin with cannot be a good farmer. When Roman gentlemen who owned estates and tried to operate them opened this book, they knew from the first lines that the author was not a man of letters but a real farmer, a "dirt farmer," who knew his business, was not sentimental about it, would not allow other people to fool him, and in his turn would not try to fool his readers with pretty words.

The best way to give an idea of the book's contents and structure is to indicate roughly the main subjects dealt with and to indicate the chapters dealing with them. The reader will then see at a glance that sometimes a number of connected chapters hang together; but that at other times they are widely separated. A few remarks are added here and there.

In order to operate a farm one must have one. How to acquire one, what to look for, precautions to be taken (1). A youth should plant trees, and when he is older, say 36, build himself a farm but be cautious (3). Cato might have been called "caution," he was a suspicious old man, always on the defensive.

Value of a plantation (*arbustum*) in the suburbs, so that firewood and faggots may be sold or used in the master's city house (7).

Building of a farm (*villa*) from the

[20] Quoted from the translation by William Davis Hooper and Harrison Boyd Ash in the Loeb Classical Library (Cambridge: Harvard University Press, 1934), p. 3.

ground up (14). Walls (15). Pressing room (12, 13, 18), wine press (19), press rope (63). Mill (20–22). Threshing floor (91, 129). Plastering of the walls (128). Making of brooms (152). Lime kiln (38). Burning lime on shares (16).

"What is good cultivation? Good ploughing. What next? Ploughing. What third? Manuring" (61).[21] Manuring is often discussed (29, 36–37, 39, 50). Ditches and drainage (43, 155).

What to plant, when, and where (6, 8, 9, 34–35). Transplanting trees (28, 49). Things harmful to crops (37). Work to be done in the spring (40). Plant nurseries (46–48). Fruit trees (48, 51). Hay (53). Firewood (7, 55, 130). Wood props (17). Fig trees (42, 94, 99). Olive yard, oil (10, 31, 42, 64–65, 68–69, 93, 100, 117–119, 153). Vineyard (11, 33, 41, 44–45, 47, 49). Cypresses (151). Pruning and trimming trees (32). Layers (52, 133). Grafting (40–42).

Sundry vegetables: asparagus (161), cabbage (156–57). The two chapters on cabbage are largely medical; the second one on Pythagoras' cabbage (*de brassica Pythagorea*) is by far the longest in the whole book (c. 140 lines).

The vintage (23, 25–26, 68), wine in general (107–11), special wines, "Greek wine" (24, 105), Coan wine (112),[22] grape juice (120).

Best markets for buying clothes, shoes, pots and pans, etc. (135). Cato names not only towns but in many cases individual merchants.

Building of good stalls, stout pens and latticed feed-racks for animals (4). Forage for cattle (27, 30, 54). Yearly ration for a yoke of steers (60). "You should have as many carts as you have teams, either of oxen, mules or donkeys" (62).[23] There is only one chapter on dogs (124) which we quote in its entirety: "Dogs should be chained up during the day, so that they may be keener and more watchful at night." No mention of cats; Cato would have considered them as useless animals but he probably had no knowledge of them.[24]

There are a good many domestic recipes, for food (74–82, 84–87, 121), for bleaching salt (88); how to cram fowls (89–90), to fight pests (92, 95, 98), to grease axles, belts, shoes (97), to preserve food (101, 116, 120), for sea water (106), for a sweet aroma (113), for curing hams (162). This is the last chapter. The book ends in the most abrupt manner: "No moths or worms will touch them."[25]

It is clear that Cato did not care about an elegant conclusion.

We now come to the most important parts of the book from the point of view of economic and social history. Cato explains the master's duties (2), the duties of overseers (5, 142), those of the female housekeeper (143), the *vilica*, who often was the overseer's wife; finally, the duties and treatment of slaves, how to feed and clothe them (56–59, 104), duties of the watchman and the ladler (66–67).[26] Cato was truly ferocious (the word is

[21] Let us give the Latin of this: "Quid est agrum bene colere? bene arare. Quid secundum? arare. Quid tertium? stercorare."

[22] Coan wine was famous (Volume 1, p. 384); in this case, however, Coan refers to a method not to a provenience, as when we speak of "Russian leather."

[23] "Quot iuga boverum, mulorum, asinorum habebis, totidem plostra esse oportet."

[24] Cats were very well known in ancient Egypt but almost disappeared from the Western world until medieval times; the Greeks and Romans fought mice not with cats but with some kind of weasel (*galē*, *mustela*); *Introduction*, vol. 3, pp. 1422, 1863.

[25] "Nec tinia nec vermes tangent."

[26] The overseer (*vilicus*) was the top man. The watchman or custodian (*custos*) was in charge of the stores and pressing

not an exaggeration); slaves should be fed just enough so they would not starve. We read incidentally (56–57) that some of the slaves working in the fields were chained together and we know from Columella (I, 8, 16) that at night they were kept in an underground prison, the *ergastulum*.

Many chapters emphasize the fact (fact No. 1 in Cato's mind) that farming is business (136–137, 144–150). They explain how to draw contracts for letting the land to a tenant farmer, for letting a vineyard to him, for letting the gathering of olives, for milling the olives, for the sale of olives on the tree, of grapes on the vine, of wine in jars, for the lease of winter pasturage, for the sale of the increase of the flock; how to measure wine for sale (154).

The many chapters dealing with the treatment of ill persons or animals, and various superstitions *ad hoc*, will be discussed in our medical chapter below.

Cato mentions no literary authorities but he names a few people from whom the farmer may obtain the things he needs: Lucius Tunnius of Casinum and Gaius Mennius of Venafrum for press ropes, Minius Percennius of Nola for the cultivation of cypress, Rufrius of Nola for oil mills.

It is worth while to compare Cato's book on estate management (for that is really what it is) with one written two centuries earlier, the *Oiconomicos* (*oeconomicus*) by Xenophōn (IV–1 B.C.). The comparison is entirely to Cato's disadvantage. The Greek writer was a man of letters and a humanist, and his book is beautifully written, gentle and attractive. As compared with him, Cato was perhaps a man of more experience but a boor and a brute. The difference between Xenophōn and Cato is an excellent illustration of the difference between Greek and Roman culture. It is possible that Cato was more efficient than Xenophōn, but I am not sure of that; he was certainly less lovable.

I have compared Cato with Xenophōn; in his *Parallel lives* Plutarch compared him with the Athenian, Aristeidēs the Just (c. 530–468), much to the latter's credit. Plutarch's portrait of Cato is unforgettable; he helps us to realize his complexity and the extraordinary mixture of greatness and meanness. Cato always spoke of his simplicity and of his dislike of luxury, but he was boastful and miserly. His attitude to the slaves was odious; he loved money above everything. Says Plutarch:

> As he applied himself more strenuously to money getting, he came to regard agriculture as more entertaining than profitable, and invested his capital in business that was safe and sure. He bought ponds, hot springs, districts given over to fullers, pitch factories, land with natural pasture and forest, all of which brought him large profits . . . He used to loan money also in the most disreputable of all ways, namely, on ships . . . He used to lend money

---

room; he was also a kind of foreman. He was assisted in the pressing room by the ladler (*capulator*), who moved the wine or oil away from the presses. The responsibilities of watchman and ladler were heavy.

also to those of his slaves who wished it and they would buy boys with it, and after training and teaching them . . . would sell them again . . . He went so far as to say that a man was to be admired and glorified like a god if the final inventory of his property showed that he had added to it more than he had inherited.[27]

These words illustrate Cato's stinginess and also the versatility of Roman business in his time. He bought ponds (*limnas*) for pisciculture, and hot springs (*hydata therma*) for the exploitation of balneology; [28] lending money on ships was a kind of maritime insurance. Cato wanted to get rich in any way and his love of money was further degraded by sordid avariciousness.

A text as rustic as Cato's was badly transmitted, for editors did not treat it with the same respect as they would have treated a literary text but "corrected" it. It was saved perhaps because of Cassiodorus' [29] interest in it and because of its early association with the *Res rusticae* of Varro. The early manuscripts generally contained both works. That was so for the Codex Marcianus, once in the S. Marco library in Florence and used by early editors, but lost. The five incunabula editions of *Scriptores rei rusticae* contained not only Cato (II–1 B.C.) and Varro (I–2 B.C.) but also Columella (I–2) and R. T. A. Palladius (IV–1). The first was edited by Georgius Merula (Florence: Nicolaus Jenson, 1472) (Figs. 75 and 76); the second was printed by B. Bruschis (Reggio Emilia, 1482); the third was edited by Filippo Beroaldo the elder (Bologna: B. Hectoris, 1494); and the last two were printed by Dion. Bertochus (Reggio Emilia, 1496) and by F. Marzalibus (Reggio Emilia, 1499). This is a good example of the rivalry obtaining between the prototypographers, five Italian incunabula of the *Scriptores rei rusticae*, by five different printers, in three cities; three editions by three different printers in one of them, Reggio Emilia.[30] A far better edition of Cato was prepared by Fra Giovanni del Giocondo (Jucundus) of Verona (Venice: Aldus Manutius, 1514) and still another by Pietro Vettori [31] (Lyons: Gryphe, 1541). After that the tradition of the four *Scriptores* was common.

Modern editions of Cato and Varro by Heinrich Keil (Leipzig: Teubner, 1884–1894), and by Georg Goetz, same publisher, of Cato in 1922 and of Varro in 1929.

English translations of Cato alone

[27] *Lives*, ed. Bernadotte Perrin (Loeb Classical Library; Cambridge: Harvard University Press, 1928), vol. 2, p. 367.

[28] Hot and mineral springs were highly appreciated and exploited by the Romans, as they had been before them by the Greeks, Etruscans, Carthaginians, and Gauls. Balneology began in prehistoric times. *Introduction*, vol. 2, p. 96; vol. 3, pp. 286–288, 1240.

[29] Cassiodorus (VI-1), Ostrogothic statesman, monk, and scholar; founder of a monastery and farm at Squillace, on the southeast coast of Calabria (Scylacium, Bruttium).

[30] Reggio Emilia, or Reggio nell'Emilia, 71 miles north-northwest of Florence.

[31] Pietro Vettori of Florence, better known under his Latin name, Petrus Victorius (1499–1585), was the greatest classical scholar of Italy in the sixteenth century. He was the last to use the Marcianus Codex and was perhaps the cause of its disappearance.

by Ernest Brehaut (New York: Columbia University Press, 1933); of Cato and Varro by a "Virginia farmer," Fairfax Harrison (New York: Macmillan, 1913) and by William Davis Hooper and Harrison Boyd Ash (Loeb Classical Library; Cambridge: Harvard University Press, 1934).

Marcus Terentius Varro (116–27). The time interval between Cato's death in 149 and Varro's in 27 was not more than 122 years, but during that interval enormous changes had taken place. The year 146 was the climax of the Republic, 27 the beginning of the Empire; on the other hand, Cato was the beginning of Latin literature, and by Varro's death the latter had already reached its climax. It was century of literary progress and political regression.

De re rustica is, out of many works of Varro, one of the two extant. It is a treatise on husbandry like Cato's, but the tone of it is very different. Varro well knows that when one is farming for a living one must make it pay and is never unmindful of profits, but he is not a ferocious landowner; he is charitable; he is conservative but without brutality; he has moral scruples. Cato was anti-Greek; Varro had received a Greek education; his erudition and his philosophy were of Greek origin. He shared the enlightened eclecticism of the New Academy. Yet, he is very much of a Roman; we might say that he is a better Roman than Cato. He Romanized Greek learning, even as Cicero Romanized Greek ethics. One of his reasons for writing toward the end of his long life a treatise on husbandry was his consciousness of the agricultural crisis which was jeopardizing his country; that crisis had become far more acute since Cato's day and Varro realized the fundamental need for husbandry for the salvation of the commonwealth – real husbandry, not merely playing with gardens, fishponds, and bird cages.

While Cato in his *De agri cultura* seemed to despise literature, Varro was a genuine writer (*un écrivain de race*). While Cato mentioned no literary sources, Varro mentioned many of them, more than fifty Greek authors. That was the method of a literary man; one must indicate at the beginning one's principles, sources, and methods. From Cato's point of view that was a waste of time, and much of what Varro wrote would have seemed to him utterly irrelevant. But let us consider the work itself.

It is much longer than Cato's (180 pages compared with 78) and is divided into three books of almost equal size (71, 56, 53 pages), dealing with husbandry in general, with domestic cattle, and with smaller stock, such as poultry, game birds, and bees. The literary intention is obvious, because the whole work is dramatized, being written in the form of a dialogue and including many digressions for the reader's solace and pleasure. One of the interlocutors, Gaius Fundanius, was a farmer whose daughter, Fundania, Varro married. The whole work was intended for her; Book I was actually dedicated to her, Book II to a cattleman, Turranius Niger, and Book III to one Pinnius.

MARCI CATONIS PRISCI DE RE RVSTICA LIBER.

ST INTERDVM PRAESTARE MERcaturis rem quærere: ni tam periculoſum ſit: et item fœnerari: ſi tam honeſtum ſit. Maiores .n. noſtri hoc ſic habuere: et ita ĩ legibus poſuere: furẽ dupli cõdemnari: fœneratorẽ q̃drupli. Quãto peiorẽ ciuẽ exiſtimarint fœneratorem q̃ furem: hic licet exiſtimare. Et uirũ bonũ quom laudabant: ita laudabãt: Bonum agricolam bonumq; colonum. Ampliſſime laudari exiſtimabatur: qui ita laudabatᷓ. Mercatorem autem ſtrenuum ſtudioſumq; rei quærendæ exiſtimo: uerum (ut ſupra dixi) periculoſum & calamitoſum. At ex agricolis & uiri fortiſſimi & milites ſtrenuiſſimi gignuntur: maximeq; pius quæſtus ſtabiliſſimuſq; cõſequitur: minimeq; inuidioſus. Minimeq; male cogitantes ſunt: qui in eo ſtudio occupati ſunt. Nunc (ut ad rẽ redeam) quod promiſi inſtitutum primum hoc erit.

Quomodo agrũ emi parariq; oporteat. Caput.i.

P Rædiũ cũ comparare cogitabis: ſic in animo habeto: uti ne cupide emas: neue opera tua parcas uiſere: & ne ſatis habeas circumire. Quotiẽs ibis: totiens magis placebit quod bonũ erit. Vicini quo pacto niteant: id animaduertito. In bona regione bene nitere oportebit: et uti cum introeas et circũſpicias: uti inde exire poſſis: uti bonum cælum habeat: ne calamitoſum ſiet. ſolum bonum ſua uirtute ualeat. Si poteris ſub radice mõtis ſiet: in meridiem ſpectet: loco ſalubri: operariorũ copia ſiet bonum: in quem aquarium oppidum ualidum prope ſiet. Si autẽ mare aut amnis quo naues ambulant: aut uia bona celebris quæ ſiet. In his agris qui non ſæpe dominos mutant: qui in his agris prædia uẽdiderĩt: quos pigeat uẽdidiſſe uti bene ædificatum ſiet. Caueto ne alienam diſciplinam temere contemnas. De domino bono colono bonoq; ædificatore melius emetur. Ad uillam cum uenies: uideto uaſa torcula et dolia multa ne ſient. Vbi non erunt: ſcito pro ratione fructum eſſe. Inſtrumenti magni ne ſiet: bono loco ſiet. Videto q̃minimi inſtrumenti. Sumptuoſus ager ne ſiet. Scito idem agrum quod hominem quãuis quæſtuoſus ſiet: ſi ſumptuoſus erit: relinquere non multum. Prædium quod primum ſiet: ſi me rogabis: ſic dicam de omnibus agris. Optimo loco emito iugera centũ agri. Vinea eſt prima: ſi uino multo eſt: ſecũdo loco hortus irriguus: tertio ſalictum: quarto oletum: quĩto pratum: ſexto campus frumentarius: ſeptimo ſilua cædua: octauo arbuſtum: nono glandaria ſilua.

Fig. 75. First page of Cato's *De re rustica.* The first paragraph is translated in the text. This page occurs in the first edition of the so-called *Scriptores rei rusticae* (folio, 30 cm, 302 leaves; Venice: Nicolas Jenson, 1472), containing the agricultural treatises of Cato the Censor (II–1 B.C.), Varro (I–2 B.C.), Columella (I–2), and Palladius (IV–1). The editor was Franciscus Colucia. [Courtesy of Harvard College Library.]

MARCI TERENTII VARRONIS RERVM RVSTICARVM AD FVNDANIAM VXOREM LIBER.I. PROLOGVS.

Græci & latini qui de re ruſtica ſcripſerunt. Caput primum.

OTIVS ESSEM CONSECVTVS Fũdania & cõmodius: ſi tibi hæc ſcriberẽ: quæ nũc ut potero exponam: cogitãs eſſe ꝓperãdũ: quod (ut dicit) ſi ẽ homo bula: eo magis ſenex. Annus .n. octogeſimus admõet me: ut ſarcinas colligam: ãteq̃ ꝓficiſcar e uita. Quare quoniã emiſti fũdũ: quẽ bene colendo fructuoſum eũ facere uelis: meq; adhibeã curã roges: experiar. Et non ſolum ut ipſe quoad uiuam quid fieri oporteat ut te moneam: ſed etiam poſt mortem. Neq; patiar Sybillam non ſolum cecinisse quæ cum uiueret prodeſſẽt hominibus: ſed etiam quæ cũ periſſet ipſa: & id etiam ignotiſſimis quoque hominibus: ad cuius libros tot annis poſt publice ſolemus redire: cum deſideramus quid faciendũ ſit nobis ex aliquo portẽto: me nedũ uiuo qdẽ neceſſariis meis quod proſit facere: ſed mortuo eqdẽ. Quo circa ſcribã tibi tres libros idices: ad quos reuertare: ſi qua in re quæres: quẽadmodũ qdq; te in colendo oporteat facere. Et quoniam (ut aiunt) dei facientes adiuuant: prius ĩuocabo eos. Nec ut Homerus & Ennius: muſas: ſed duodecim deos conſentis: neq; tamẽ eos urbanos: quorum imagines ad forum auratæ ſtant: ſex mares & fœminæ totidem: ſed illos duodecim deos: qui maxĩe agricolarum duces ſunt. Primũ: qui omnis fructus agriculturæ cælo & terra cõtinẽt: Iouem & Tellurem. Itaque ꝙ hi parentes magni dicuntur Iuppiter pater: Tellus uero mater. Secundo Solem & Lunam: quorum tempora obſeruantur: cum quædam ſerantur & condantur in tempore. Tertio Cererem & Liberum: ꝙ horum fructus maxime neceſſarii ad uictum ſint. Ab his enim cibus & potio uenit e fũdo. Quarto Robigũ & Florã quibus propitiis: neq; robigo frumenta atq; arbores corrumpit: neque non tẽpeſtiue florent. Itaq; Robigo feriæ robigalia: Floræ ludi floralia ĩſtituti. Itẽ aduenerot Mineruã & Venerẽ: quaꝝ unius ꝓcuratio oleti: alterius hortorum: quo nomine ruſtica uinalia inſtituta. Nec non precor Lympham ac Bonum euẽtum: quoniam ſine aqua omnis arida ac miſera agricultura: ſine ſucceſſu ac bono euentu: fruſtratio eſt non cultura. His igitur deis ad uenerationem aduocatis: ego referã ſermões eos quos de agricultura habuimus nup: ex qbus qd te facere oporteat: animaduertere poteris: in quis quæ non inerunt & quæres: indicabo a

Fig. 76. First page of the *De re rustica* of Varro (I–2 B.C.) in the *Scriptores rei rusticae* (folio, 43; Venice: N. Jenson, 1472). [Courtesy of Harvard College Library.]

I trust the reader remembers Cato's blunt exordium; Varro begins:

Had I possessed the leisure, Fundania, I should write in a more serviceable form what now I must set forth as I can, reflecting that I must hasten; for if man is a bubble, as the proverb has it, all the more so is an old man. For my eightieth year admonishes me to gather up my pack before I set forth from life.[32] Wherefore, since you have bought an estate and wish to make it

[32] The same image was used by La Fontaine in "La Mort et le Mourant," *Fables,* VIII, 1:

La Mort avait raison: je voudrais qu'à cet âge
On sortît de la vie ainsi que d'un banquet
remerciant son hôte et qu'on fît son paquet.

profitable by good cultivation, and ask that I concern myself with the matter, I will make the attempt; and in such wise as to advise you with regard to the proper practice not only while I live but even after my death . . . Therefore I shall write for you three handbooks to which you may turn whenever you wish to know, in a given case, how you ought to proceed in farming. And since, as we are told, the gods help those who call upon them, I will first invoke them – not the Muses, as Homer and Ennius do, but the twelve councillor-gods; and I do not mean those urban gods, whose images stand around the forum, bedecked with gold, six males and a like number female, but those twelve gods who are the special patrons of husbandmen. First, then, I invoke Jupiter and Tellus, who, by means of the sky and the earth, embrace all the fruits of agriculture; and hence, as we are told that they are the universal parents, Jupiter is called "the Father," and Tellus is called "Mother Earth." And second, Sol and Luna, whose courses are watched in all matters of planting and harvesting. Third, Ceres and Liber, because their fruits are most necessary for life; for it is by their favour that food and drink come from the farm. Fourth, Robigus and Flora; for when they are propitious the rust will not harm the grain and the trees, and they will not fail to bloom in their season; wherefore, in honour of Robigus has been established the solemn feast of the Robigalia, and in honour of Flora the games called Floralia. Likewise I beseech Minerva and Venus, of whom the one protects the oliveyard and the other the garden; and in her honour the rustic Vinalia has been established. And I shall not fail to pray also to Lympha and Bonus Eventus, since without moisture all tilling of the ground is parched and barren, and without success and "good issue" it is not tillage but vexation. Having now duly invoked these divinities, I shall relate the conversations which we had recently about agriculture, from which you may learn what you ought to do; and if matters in which you are interested are not treated, I shall indicate the writers, both Greek and Roman, from whom you may learn them.

Then follows the long list of his authorities, mostly Greek; curiously enough, it does not include Cato's name, though Cato's opinions are often related in the body of the book. These two beginnings, Cato's and Varro's, are poles apart. Varro's gives us the atmosphere of the book; he is a farmer, but also a religious man and a humanist.

My analysis will be restricted to the third book, which covers many novelties as compared with Cato's. It begins with a contrast between city and country life, going back to legendary times, and the conversation is cleverly enframed in a day reserved for the election of aediles or city magistrates. Among the interlocutors are a senator, the augur (who had to be present to clear up religious difficulties), a member of a consular family. In order to beguile the time needed for the results of the election to be known, they discuss the innumerable problems of minor husbandry, the aviaries, warrens, and fishponds.[33] The aviaries include chickens, wild and domestic, guinea fowl, pigeons and turtle doves, geese and ducks, peafowl, and their care extends to the collection and incubation of eggs and the fattening of birds. The warrens accommodate not only various breeds of

[33] *Ornithones, leporaria, piscinae.*

hare but also boars, stags and roes, and sheep. The fishponds were of two kinds, sweet and salt. Much attention was paid also to dormice [34] and to snails.[35] The two final chapters are especially interesting, as they deal with apiculture and with ponds for the breeding of river fish and of sea fish.

As to the fishponds established near the shore to accommodate salt-water fish, it is noteworthy that the tides of the Tyrrhenian Sea, low as they are (a single foot at the most), were sufficient to refresh the ponds twice a day.

The account of the fishponds, and also to a lesser extent of the bird cages [36] and of the enclosed parks for small animals, illustrates the great complexity of those undertakings. For one thing, it implied the coöperation of many craftsmen, such as fowlers, hunters, and fishermen, with the ordinary plowmen, gardeners, and vinedressers. Some of those establishments required considerable capital but might provide very large profits.

Chapter 16 is the first Latin treatise on beekeeping; a Greek treatise (*melissurgica*) had been published more than two centuries before by Nicandros of Colophōn (III–1 B.C.).[37] Varro's knowledge of bees was still on a low level, the Aristotelian, and the lord of the hive was for him not a queen but a king.[38]

The election and the book end together. These are the last lines:

> Then a noise on the right, and our candidate, as aedile-elect, came into the villa wearing the broad stripe.[39] We approach and congratulate him and

---

[34] *Myoxus glis.* Perhaps it would be better to use the word loir (French and English derivatives from the Latin *glis*), for the European dormouse (the loir) is somewhat larger than the kind known in America.

[35] The Roman vivarium for snails was the prototype of the French *escargotière.*

[36] A. W. Van Buren and R. M. Kennedy, "Varro's aviary at Casinum," *Journal of Roman Studies* 9, 59–66 (1919). Casinum in Latium on the Via Latina near the borders of Campania. Its citadel with a temple of Apollōn occupied the same site as the monastery of Monte Cassino.

[37] The Greeks had used honey from prehistoric times; that was the only form of sugar known west of the semitropical countries where sugar cane grew. However, the bees that produced it were wild bees. The first writer to refer to hives was Hēsiodos, and beekeeping was already so well developed in Solōn's time (he died in 558) that he tried to regulate it; the honey of Mount Hymēttos was then already famous. In his plays, Aristophanēs (d. c. 385) referred to the *melitopōlēs* and to the *cēropōlēs* (the merchants of honey and of wax). Wax was used for casting in metal, modeling, sealing, cosmetics, encaustic painting, lighting, in exceptional cases for preserving dead bodies, covering metal surfaces to prevent oxidation, and, most interesting to us, for the surface of writing tablets. Varro makes many references to wax but not to its collection and sale, in spite of his deep interest in profits. The uses for wax are revealed by the Latin word *cera,* which means not only wax but also a writing tablet, a seal of wax, or an image made of wax (just as we say bronze for a bronze statuette).

[38] The apiculture of Columella (I–2) was further developed than Varro's. After that hardly any progress was made until the seventeenth century. The first anatomical studies of bees were made by the German, Georg Hoefnagel (1592), and with the microscope by the Italian, Francesco Stelluti (1625). The first to recognize that the king was a queen was the Dutchman, Jan Swammerdam, in 1669.

[39] Wearing his official robe, the *toga praetexta,* with a broad border of purple.

escort him to the Capitoline. Thence, he to his home, we to ours, my dear Pinnius, after our conversation on the husbandry of the villa, the substance of which I have given you.

There is an amusing contrast between this ending and the moths and worms at the close of Cato's book. The literary crowd in Rome must have appreciated Varro's book, but I wonder whether the plain farmers did not prefer Cato's manual. With Cato they always knew where they were, while there was much in Varro's book that was to them incomprehensible, irrelevant, and perplexing.

Varro's book was well written, but it must be admitted that it is sometimes confusing. The composition is far from perfect and the humor not always pleasant to our ears. For example, he gives bird names (Blackbird, Peacock, Magpie, Sparrow) to some of the interlocutors; that might be funny if the dialogue was up to it, but it is not. Varro had good intentions; he was trying his best to allure his literary friends, but he was not a great artist, and his book, in spite of its overwhelming literary superiority over Cato's, was not a masterpiece.

Its tradition was mixed up with Cato's,[40] but he was accepted throughout the Middle Ages as one of the great Roman masters, at the side of Cicero and Virgil. It is curious that these three were contemporaries; they shared 27 years (70–43) in common.

C. Julius Hyginus.[41] Gaius Julius Hyginus (I–2 B.C.), who hailed from Alexandria (or Spain) and had been brought to Rome as a prisoner of war by Caesar, was freed by Augustus, who, recognizing his talent and learning, appointed him director of the Palatine Library. He was a teacher of Virgil and Ovid. On the other hand, he wrote a commentary on Virgil.[42] There is no incompatibility between these two facts; he was born six years before Virgil but lived much longer (81 years compared with 51) and survived him by 36 years. As prefect of the Palatine Library he had endless facilities for research, and he wrote many learned books. The most important of these are treatises on agriculture and on beekeeping (the second being perhaps a part of the former). His *De agricultura* and *De apibus* are both lost but frequently referred to by Columella (I–2 B.C.); it was Columella who called him Virgil's teacher.[43]

[40] Hence, for manuscripts and editions, see the end of the section devoted to Cato.

[41] It is well to add his initials when speaking of him, for we have already spoken of another whose initials are not known, Hyginus the astronomer. There is still a third Hyginus (II–1), the surveyor (*agrimensor*, or *gromaticus*), who lived in Trajan's time. The name is of Greek origin, Hygeinos (probably a variant of Hygieinos, healthy), which would confirm the Oriental origin of C. Julius Hyginus (though he may have been taken prisoner in Spain). As this Hyginus lived in Rome and wrote in Latin, it is better to quote his name in its Latin form.

[42] There is no doubt about that, in spite of the fact that his commentary is lost, because it was used by Aulus Gellius (II-2) in his *Noctes Atticae*, and also by the greatest ancient student of Virgil, Servius (end of the fourth century).

[43] Columella of Cadix (I–2) wrote twelve books *De re rustica*, plus one *De*

Virgil. From Cato to Varro was a literary ascension; with Virgil (of whom we shall hear more hereafter) we are taken much higher still, yet without losing touch with reality. Virgil (I–2 B.C.) was not simply a great poet but the leading naturalist of his time.

As far as we can judge from the writings that have come to us, his literary activity did not begin in earnest until he was in his late twenties. His earliest works were the *Bucolica*, composed between 42 and 37 (aet. 28–33), and the *Georgica*, between 37 and 30 (aet. 33–40). We are concerned at present with the *Georgica*, which includes almost the whole of his knowledge on natural history, but let us say a few words first of the *Bucolica*.

The *Bucolica* is a collection of ten short *eclogae* or idyls, ranging from 63 to 111 lines, a total of 829 lines. Theocritos of Syracuse (fl. 285–270) had invented that kind of poetry, and Virgil's *Eclogae* were an unmistakable imitation of Theocritos' *Idyls* (some phrases were translated from Greek into Latin), and yet Virgil's achievement was exceedingly original. Some of the *eclogae* were Theocritan, the frame of all of them was bucolic, but Virgil added deep novelties, whether a prophecy or allusions to the great events of the day. Virgil was the inventor of pastoral poetry in Latin, the inventor of an ideal Arcadia,[44] a land of graceful and lovesick shepherds. The popularity of his poem was due to the combination of pastoral images with contemporary events (the Civil War, Caesar's apotheosis, Octavianus, etc.). The shortest *ecloga*, the fourth, written in 40 B.C., was Sibyllic or messianic; it announces the birth of a boy who will restore the golden age. Some critics professed to see in it a prophecy of Christianity. This mixture of idyls with politics appealed to Roman imagination.

The *Georgica*, written in 37–30, is a much longer work devoted to husbandry. The composition of it may have been suggested by Maecenas, Virgil's friend and patron, to whom it is repeatedly dedicated.[45] The main purpose was a defense of husbandry which was being more and more neglected by the landowners, old and new (the new ones were the veterans to whom pieces of land had been granted). There was an urgent need of good husbandry; farmers were discouraged by the miseries of wars, the attraction of cities, and the huge importation of cereals from Egypt and Africa. And yet the power of Rome was based upon the cultivated land; in order to secure it, it was necessary to restore agriculture, small holdings, religion, and honesty.

The documentation was as thorough as that needed for a scientific treatise. Virgil had studied all the books available in Greek and Latin, too many to be listed here. Among the Greeks he had read Hēsiodos, Aristotle, Theo-

---

arboribus which together form an agricultural collection, larger than Cato, Varro, and Virgil put together. Columella quotes Cato 18 times, Varro 10 times, Hyginus 11 times, and Virgil 29 times; he deeply admired Virgil.

[44] There is a real Arcadia, a mountainous and pastoral region in the middle of the Peloponnēsos, but Virgil's Arcadia was a poetical abstraction like Cockaigne.

[45] Maecenas is addressed at the beginning of each canto.

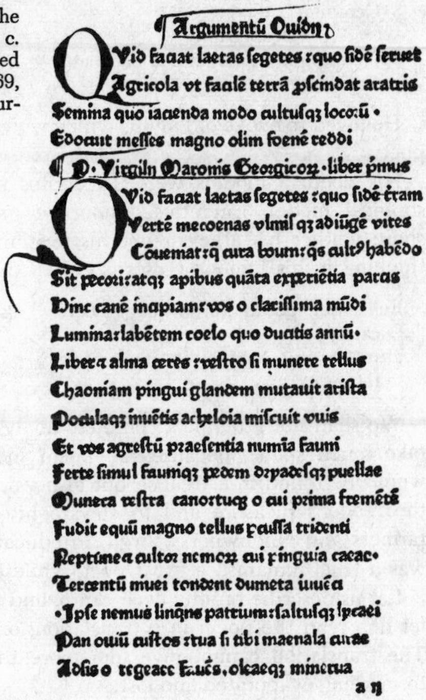
Argumentū Ouidij
Quid faciat laetas segetes: quo sidē seruet
Agricola vt facilē terrā pscindat aratris
Semina quo iacienda modo cultusqz locorū.
Edocuit messes magno olim foenē reddi
P. Virgilij Maronis Georgicorū. liber primus
Quid faciat laetas segetes: quo sidē terram
Vertē mecoenas vlmisqz adiūgē vites
Cōueniat: q̄ cura boum: q̄s cult9 habēdo
Sit pecori: atqz apibus quāta experiētia parcis
Hinc canē incipiam: vos o clarissima mūdi
Lumina: labētem coelo quo ducitis annū.
Liber et alma ceres: vestro si munere tellus
Chaoniam pingui glandem mutauit arista
Poculaqz inuētis acheloia miscuit vuis
Et vos agrestū praesentia numia fauni
Ferte simul faunisqz pedem dryadesqz puellae
Munera vestra cano: tuqz o cui prima frementē
Fudit equū magno tellus pcussa tridenti
Neptune et cultor nemorū cui pinguia ceae.
Tercentū niuei tondent dumeta iuuēci
Ipse nemus linqns patrium: saltusqz lycaei
Pan ouiū custos: tua si tibi maenala curae
Adsis o tegeaee fauēs. oleaeqz minerua
a ij

Fig. 77. First separate edition of the *Georgica* (Deventer: Jacobus de Breda, c. 1486). The *Georgica* had been printed many times before in the *Opera* since 1469, and with the *Bucolica* since 1472. [Courtesy of Library of Congress.]

phrastos, Aratos, Nicandros; among the Latin, Cato, Varro, and perhaps Hyginus. His main source, however, was the experience he had obtained in his father's estate and in the company of other farmers; he was an excellent observer. He knew all that could be known in his time, but his poem was restricted to the essential.

The *Georgica* is divided into four books of almost equal length (about 550 lines each, total 2188): I, agriculture in general; II, trees, especially vines and olives; III, stock farming; IV, beekeeping. The form is perfect; the *Georgica* is the most finished as well as the simplest and the most lovable of Virgil's works. It has the authority of a treatise, yet is not one and is not meant to be read as one would read Cato or Varro. The form was calculated [46] to delight poetical minds and musical ears. Virgil loved to enumerate beautiful names, as in the lines

[46] The word "calculated" is appropriate because prosody as well as music implies a combination of rhythms and a dance of numbers.

Aut Athon, aut Rhodopen, aut alta
  Ceraunia . . .
    O ubi campi
Spercheosque et virginibus bacchata
  Lacaenis
Taygeta . . .
Drymoque Xanthoque Ligeaque
  Phyllodoceque,
caesariem effusae nitidam per candida colla.[47]

He loved to evoke old myths which were for the Romans a kind of national poetry.

His poetical models were Theocritos and Catullus (87–54); his philosophical model, Lucretius, whom he greatly admired though he could share neither his atheism nor his pessimism. It was of Lucretius he was thinking in his famous verses, so often quoted:

Felix qui potuit rerum cognoscere
  causas,
atque metus omnis et inexorabile
  fatum
subiecit pedibus strepitumque Acheruntis avari! [48]

We shall not attempt a complete analysis of the *Georgica*, which would take much space, because the poem includes not only such matters as would be found in a treatise but also various intermezzi meant to increase the reader's pleasure and to elevate his mind. Cato and Varro spoke to farmers and landowners; Virgil, to educated men engaged in farming; he was a true humanist, a great poet, while they were mere technicians.

Let us describe rapidly each canto, and if anyone wishes for more details, let him read the poem in a translation, or preferably in the Latin original. The translation cannot give him more than the substance, the delightful form being evaporated and lost.

The first book or canto begins with a praise of the gods who are the patrons of husbandry and of Octavianus who has brought peace and order to the country and wishes to hearten the discouraged farmers. Then farm work is described, the various methods of culture, the needs of fallow, of manure, ploughing, irrigation, and so on. A great part of the book is devoted to popular astronomy and meteorology. This is derived from Aratos and Eratosthenēs, but also from the farmer lore that Virgil had deeply imbibed in his native country, Cisalpine Gaul.[49]

The invocation of the second canto is addressed to Bacchus, god of vines and trees. The care of trees and their grafting are explained.[50] Different

[47] In the first line (I, 332), he evokes "Athōs or Rhodopē or the high Ceraunia mountains"; in the following (II, 486), he sighs, "Where are the fields, the Spercheios river and the Taygeta hills about which the Spartan virgins are running?" In the last ones (IV, 336) he enumerates the nymphs "Drymō, Xanthō, Ligeia, Phyllodocē, with their golden hair spread over their white necks." The spell is broken in the translation.

[48] "Happy is he who has been able to know the reasons of things and has cast beneath his feet all the fears, inexorable fate, and the noise made about the stingy Acherōn" (II, 490).

[49] It was a part of Gaul and from the Roman point of view Cisalpine and Transpadane, south of the Alps but north of the Po (Padus).

[50] One strange mistake of Virgil's was his belief (II, 80) that any scion could be

trees require different climates and soils. Many climates are reviewed but none is comparable to the glorious climate of Italy.

> Salve, magna parens frugum, Saturnia tellus,
> magna virum: tibi res antiquae laudis et artis
> ingredior, sanctos ausus recludere fontis,
> Ascraeumque cano Romana per oppida carmen.[51]

His love of Rome and of Italy recurs frequently throughout the poem.

Most of this book is devoted to the culture of olives and vines and also of other fruits which do not cause drunkenness. The canto ends with the picture and praise of pastoral life:

> O fortunatos nimium, sua si bona norint
> agricolas! [52]

An invocation to Pales, Italian goddess of flocks and shepherds, announces that the third canto will deal with cattle, horses, and other animals. The poet offers advice concerning their care and their breeding. Every animal that the poet introduces is fully alive and he gives us the feeling of the divinity of life. He sings of the sheep and the goats, explains how to keep their stables in winter or to regulate their pasturage in the good season. Incidentally, he describes the trials of shepherds in Libya and Scythia. He explains how to treat sheep in order to obtain healthy fleeces and creamy milk, the proper care of dogs and hounds, how to protect the animals from snakes by burning cedar and balsam in the stables. The book ends sadly with an account of animal diseases and a description of the plagues that decimated flocks in the Carnic Alps and along the Timavus river.[53] His knowledge of the veterinary art was rudimentary, but he gives us a terrible vision of those epidemics; in spite of the fact that the victims are animals, not men, he warms up our sympathy for them, and his recital is as unforgettable as those of Thucydidēs and Lucretius. Who does not remember his picture of the dying bull and of its sorrowing mate? [54]

The best-known part of the poem is the fourth, dealing with beekeeping; it is perhaps the least scientific part but it is the most poetic, and its practical value — in its own time and for seventeen or eighteen centuries after

---

successfully grafted on any kind of tree. The mistake was not his own but was shared by his contemporaries and successors, such as Columella (I-2) and Pliny (I-2). One does not understand how such a belief could resist experience.

[51] "Hail, great mother of fruits and of men, land of Saturn; for thee I tackle a subject of ancient art and praise, venturing to disclose the sacred wells; I sing the Ascraean poem through the Roman towns" (II, 173). Ascra on Mount Helicōn in Boiōtia was the place wherein Hēsiodos had chosen to reside.

[52] "How fortunate the peasants would be, if they but knew the good reasons they have to be happy" (II, 458).

[53] The Carnic Alps (in Carniola) are north of the top of the Adriatic, and the Timavus (now Timavo) is near the northeast corner of the Adriatic, east of Aquileia.

[54] The account of diseases and plagues covers 125 lines (III, 440–565). The dying bull and the ailing oxen are in lines 515–536.

—was considerable. It was until recent times the best introduction to apiculture. According to Maurice Maeterlinck, it is the only ancient work worth studying; Maeterlinck, it is true, was a poet, who could fully appreciate the humanities as well as the technicalities of Virgil's song. Scientific knowledge of the bees was small, but the folklore was incredibly rich and Virgil was aware of it. He was not alone in believing that the bees participate in the divine spirit. The bees lead him to an account of the lovely gardens that one must provide for them if one wishes to obtain an abundance of good honey. One of the most delightful episodes of the whole poem is his account of the old man, who enjoyed such a beautiful garden near Tarentum, a little garden but full of flowers, vegetables, fruits, and buzzing with bees (IV, 125). Then he explains how to collect the honey and how to care for the bees in health and sickness, for the bees suffer from diseases like other creatures. Toward the end, he tells us the story of Orpheus and Eurydicē, and, after an evocation of Caesar fighting near the Euphratēs to insure Roman order and security, he ends with those sweet lines:

Illo Vergilium me tempore dulcis alebat
Parthenope studiis florentem ignobilis oti,
carmina qui lusi pastorum audaxque juventa
Tityre, te patulae cecini sub tegmine fagi.[55]

This is a very simple and pleasant ending, the more agreeable because "Tityre . . . sub tegmine fagi" is an old friend of ours; he is the graceful shepherd whom we have met many times in the *Bucolica*, the first time in the very first line of the first eclogue. By placing him at the beginning of the *Bucolica* and at the very end of the *Georgica*, the poet united these two works of his youth in a magic circle.[56]

My reflections might be continued almost indefinitely, because every line invites new ones. The main features of the whole poem are Virgil's love of nature, of the beasts, the insects, and the plants, above all, his deep humanity and sensitiveness, his piety and devotion to his country. The *Georgica* is the greatest didactic poem ever written, its greatness being due to a very rare combination of qualities; it is at once serious and sensitive, practical and poetical, simple and stately.

The language is as beautiful as one might wish, except that Virgil was hampered by the lack of scientific terminology; the terminology was necessarily crude because the knowledge was still vague. Some of those deficiencies were perhaps due in part to the uneven development of Latin;

[55] "Sweet Parthenopē was then nourishing me, while I was abandoning myself to the studies of inglorious ease and playfully composing pastoral verses; with the boldness of youth, I sang thee, Tityre, under the open branches of a beech tree" (IV, 563). Parthenopē was the place where the people of Cumae founded the "new town," Neapolis or Naples.

[56] Tityrus is mentioned six times in the *Bucolica*, but only once in the *Georgica*, in the very last line. This is an amusing conceit which would have seemed foolish to Cato and Varro but not to Catullus.

for example, Virgil did not have enough colors to paint with.[57] On the other hand, we must bear in mind that Latin literature was still very young. It was almost as glorious to begin with Virgil as to begin with Homer.

The tradition of the *Georgica* was established by Columella's commentaries, but chiefly by Virgil's own fame, which was warm from the very beginning. He was already considered an immortal before the disappearance of his body, and his name has remained one of the very greatest in all the literature of the West.

[57] The lack of sufficient words to designate the many colors of nature and of the arts is very puzzling, but we English-speaking people have no right to throw stones at others, for our own language is shockingly poor. For example, we say red tape, red blood, red hair, red Indian, and in every one of those phrases "red" means a different color!

XXII

MEDICINE IN THE LAST TWO CENTURIES[1]

There was an abundance of physicians in this sophisticated age, but no great one. It is useful to divide them into two groups, the Greek and the Latin; I do not say the Roman, for the leading practitioners in Rome were of Greek origin, generally spoke in Greek and always wrote in that language.

GREEK MEDICINE

Serapiōn of Alexandria. The work done by the anatomists of Alexandria in the third century was of such a revolutionary nature that it was bound to create a new medical atmosphere. The physicians of the old schools (such as the Hippocratic and the Dogmatic) were not sufficiently aware of anatomic and physiologic realities. A new school capitalizing on the new experience was needed. The foundation of that school, called Empirical (*empeiricos*) meaning practical, factual (as opposed to theoretical dogmatism) has been sometimes ascribed to Philinos of Cōs, who may have been thinking of it, but the real founder was probably Serapion[2] of Alexandria (II-1 B.C.), who flourished about 200 B.C. He rejected every kind of dogmatism and based his practice on three legs: (1) experience and experiment (*tērēsis*),[3] (2) clinical cases (*historia*), (3) analogy (*hē tu homoiu metabasis*). One of his writings, entitled *Dia triōn*, probably explained that trilogy. Its title might be an allusion to a beautiful Hippocratic saying:

[1] For medicine in the third century B.C., see Chapter IX.

[2] The name is typically Greek-Egyptian. There are many other bearers of it, notably one Serapiōn of Antiocheia, a mathematical geographer (*gnomonicus*), contemporary of Cicero, to whom he sent a book in 59 (*Ad Atticum*, II, 4, 1). He claimed that the Sun was 18 times the size of the Earth; Pauly-Wissowa (ser. 2), vol. 4 (1923), 1666. The name Serapiōn spread eastward and we find it in Syriac and Arabic literature (Sarāfyūn).

[3] The word used is *tērēsis*, meaning "watching," "guarding," which is different from "experiment" in the modern sense. When we use the word "experiment," we think of observations made under various conditions determined by the experimenter. *Tērēsis* does not mean more than observation, systematic observation if you please, different alike from vague experience and from deliberate experiment.

"The medical art has three aspects: the disease, the sick man, the physician," but that seems a little farfetched.[4] He wrote two other treatises, one against the medical sects (*Pros tas haireseis*), the other called *Therapeutica*, but only very few fragments remain.[5]

He carried Empiricism so far as to try many popular remedies in spite of their absurdity; we should not judge him, for it all depends upon the experiments that he made and his control of them. It was not unwise to give a good trial to every bit of folklore.

Serapiōn was followed by Glaucias of Tarentum (c. 175 B.C.), Apollōnios of Antiocheia (c. 175), Apollōnios Biblas of the same city (c. 150), Ptolemaios of Cyrēnē (c. 100), Hēracleidēs of Tarentum (c. 75), Zōpyros of Alexandria (c. 80), Apollōnios of Cition (c. 70), one Diōdoros (c. 60), Lycos of Naples (c. 60 B.C.), and so forth. This list shows that the Empirical school had spread from Egypt to Italy, Syria, Cyrēnē, and Cypros. One understands its success because it was a healthy reaction, the reaction of common sense against premature dogmatism. Yet it was itself premature and crude. Empiricism could not help being narrow at a time when diagnostic means were still very poor and when few clinical facts could be interpreted correctly. In spite of their protest against Hippocratic dogmatism, the Empiricists' clinical knowledge was hardly better than that available centuries before in the schools of Cōs and Cnidos. They tended to attach too much importance to popular "empirical" medicines. The use of "analogy" was very risky; we have but to remember all the whimsical conceits of folklore. Analogies and comparisons are the logical tools of primitive and uncritical people. "It was probably Serapiōn who started that wildest of all theories – the wandering of the uterus."[6]

In our judgment of that sect we must not forget that Galen spoke well of Serapiōn and of his followers. There are only three of those (pre-Christian) followers who deserve to detain us: Glaucias of Tarentum, Hēracleidēs of Tarentum, and Apollōnios of Cition.

Glaucias of Tarentum. This Glaucias (I–1 B.C.) wrote many commentaries on Hippocratēs, a treatise on herbs, wherein be paid special attention to the thistle (*acantha*). He compiled a Hippocratic dictionary which was used by Erōtianos (I–2). He is said to have discovered a cure for erysipelas

[4] Hippocratēs, *Epidemics,* I, 5; Emile Littré, *Oeuvres complètes d'Hippocrate* (10 vols.; Paris, 1839–1861), vol. 2, p. 636. The suggestion was made by Karl Deichgräber, *Die griechische Empirikerschule* (Berlin, 1930), p. 256.

[5] The fragments are edited in Deichgräber, pp. 164–168. Deichgräber deals with 19 members of that school, beginning with Philinos of Cōs and Serapiōn (whom he dates c. 225 B.C.), and ending with one Theodosios (after A.D. 200).

[6] According to Allbutt, who does not quote his source. T. Clifford Allbutt, *Greek medicine in Rome* (London, 1921) [*Isis* 4, 355 (1921–22)], p. 170. Sir Clifford makes some witty remarks about the empiricists (pp. 166 ff.); he called them the "Philistines of Medicine"!

(that would have been a remarkable achievement, or rather an impossible one in his time). According to Galen, he invented a kind of bandage for the head which was named after him *tholos Glauciu* (*spica Glaucii*).

Hēracleidēs of Tarentum. Hēracleidēs (I–1 B.C.) is the greatest physician of the ancient Empirical school. He was a disciple of Ptolemaios of Cyrēnē and of the Herophilian Mantias. He was the author of many books of which relatively long fragments exist.[7] He made many experiments, chiefly with opium (*opion*). The earliest treatise on veterinary medicine is ascribed to him (*Pros tas chronius myrmēcias*, "To cure chronic irritations or formications").

Apollōnios of Cition. If empirical medicine could hardly be superior to the Hippocratic, the situation was different in surgery, for the new anatomical experience obtained by Hērophilos, Erasistratos, and their school must have encouraged surgical practice. The greatest surgeon among the Empirics was Apollōnios of Cition (in Cypros; I–1 B.C.), who wrote a commentary on Hippocratēs' treatise on articulations (*Peri arthrōn*). This commentary had a singular fortune, for an early manuscript of it, Codex Laurentianus LXXIV, 7, a Byzantine manuscript of the ninth century, contains surgical illustrations which might well go back to Apollōnios' time, and which are in any case the earliest illustrations of their kind extant (Fig. 78); they explain methods of reposition (to secure the replacement of bones in their normal position). Some of these figures were reproduced during the sixteenth century by il Primaticcio and Guido Guidi, and Guidi's figures were copied by Ambroise Paré and Conrad Gesner (1555). They represent an iconographic tradition of sixteen centuries. Other treatises bear his name, one of them criticizing Hēracleidēs of Tarentum, another on epilepsy, and so on.

Hēgētōr. In one of the fragments of Apollōnios of Cition, there is a reference to an earlier surgeon, Hēgētōr (II–2 B.C.?), who wrote a book on the causes (of disease?), *Peri aitiōn.* The only part of it that is extant deals with the dislocation of the hip and contains the first description of the triangular ligament of the hip joint (*ligamentum teres*).

Attalos III and Mithridatēs VI. An entirely different kind of medicine (if we may call it that) was developed by Oriental despots, who were afraid of being poisoned by their loving subjects.

Thus, Attalos III Philomētōr, the last king of Pergamon (138–133), investigated poisonous plants to find out how they could be used to get rid of a troublesome individual and, equally important, how one could protect oneself against them if one had been inveigled into swallowing their juices.

[7] Deichgräber, pp. 172–202.

Fig. 78. Apollōnios of Citiōn (I-1 B.C.) wrote a commentary on Hippocratēs' treatise on articulations. A Byzantine manuscript of the ninth century contains illustrations of surgical methods which may go back to Apollōnios' own time. Hermann Schöne, *Illustrierter Kommentar zu der hippokratischen Schrift peri arthrōn* (75 pp., 31 pls.; Leipzig, 1896). The plate which is here reproduced is pl. x, serving to illustrate the beginning of Apollōnios' second book.

In the following century another despot, Mithridatēs Eupatōr, king of Pontos,[8] continued those toxicological experiments on a larger scale. It is said that Mithridatēs tried to produce immunity from poison by the administration of gradually increased doses of it and of the blood of ducks, supposed to be immune. He introduced new simples and gave the formula of a universal antidote called after him *Mithridateios antidotos.* Much of that has the earmarks of legend; the naming of an antidote after him was natural enough and does not prove that he invented its formula. In Nero's time a Cretan physician, Andromachos, invented another antidote, the *thēriacē*, which completely superseded the *Mithridaiteiē*. These are futile stories the only meaning of which is that poisons were used for murder in the times of Mithridatēs and of Nero (emperor 54 to 68). That is not astonishing; poisons have always been used for such purposes, and tyrants have always dreaded with good reason being the victims of them.[9]

One final remark: the ascription of botanical knowledge and toxicological investigations to those two kings, Attalos and Mithridatēs, must be taken with a grain of salt. It is as when one says that Augustus built the Pantheon and the Pont du Gard. Those two kings were too busy in other ways to make pharmaco-dynamical experiments, but they probably ordered some of their servants to make them, and their orders were mistaken for actual deeds.

[8] Mithridatēs VI the Great, a dangerous enemy of the Romans, who were obliged to fight three wars against him (88–84, 83–81, 74–64). He was born at Sinōpē (middle south coast of the Black Sea) in 132, and was king of Pontos from 120 to his suicide in Panticapaion in 63, at the age of 69. His name (Mithras-given) shows that his family was Mithraic; it is built upon the same pattern as Theodōros, Isidōros, Dieudonné.

[9] For the study of poisons see Chapter IX above dealing with Nicandros of Colophōn (III-1 B.C.). One of Maimonides' best-known treatises was one on poisons and antidotes which he wrote in 1199; (*Bulletin of the Cleveland Medical Library*, January 1955, p. 16). As to the use of poisons in medieval times, see my remarks in *Introduction*, vol. 3, p. 1241.

Dēmētrios of Apameia. To return to real doctors, Dēmētrios of Apameia [10] (II–2 B.C.), who flourished about the end of the second century, was especially concerned with obstetrics and gynecology; he tried to find the causes of difficult labor. He wrote a treatise on pathology (*Peri pathōn*) which must have been elaborate because it was divided into twelve books, and another on the science of symptoms or diagnosis (*Sēmeiōticon*). He was able to distinguish between pneumonia and pleurisy (?). His writings are known only through quotations by Sōranos of Ephesos (II–1), Galen (II–2), and Caelius Aurelianus (V–1).

Asclēpiadēs of Bithynia. Another Bithynian doctor, Asclēpiadēs (I–1 B.C.), has two claims to distinction, first, to be the first eminent Greek doctor to practice in Rome, and second, to be the founder, or prefounder, of a new medical school, the Methodist.

These claims need qualifications which have been provided above in the form of two words, "eminent" and "prefounder." There had been other Greek doctors in Rome before him, most of them slaves brought by their conquerors and remaining unknown and anonymous. The first of these whose name has emerged was Archagathos.

Asclēpiadēs was born at Prusa,[11] c. 130–124; he was educated in Alexandria in the school of Erasistratos (III–1 B.C.). He practiced medicine in Parion,[12] then in Athens; he was invited by Mithridatēs Eupatōr to come to Pontos, but he preferred to travel westward to Rome, where he opened his office c. 91 B.C.; he died in extreme old age.

He was a disciple of Dēmocritos and Epicuros and introduced atomic ideas into medicine; that is, he based physiologic and therapeutic theories upon those ideas. Disease is a disturbance of the atomic movements or of the atomic equilibrium in the body; healing occurs when that equilibrium is reëstablished. (This had the appearance of a scientific theory but it was unavoidably vague, and therefore as unscientific as the theory of humors.)

The definition of the new doctrine, however, was largely negative. Asclēpiadēs' new ideas were often expressed in the form of criticism of the older ones. For example, he criticized the theory of humors cherished by the Hippocratic and Dogmatic schools and despised the anatomical tendencies of the Empiricists.

He wrote many books but none has been transmitted to us in its integrity. Many innovations have been ascribed to him rightly or wrongly. For ex-

[10] Apameia in Bithynia, not the better-known Apameia on the Orontēs in Syria.

[11] Prusa in Bithynia (Turkish, Bursa). Bithynia, south of the Sea of Marmara and of the southwest coast of the Black Sea, was a country of old culture where Greek, Thracian, Lydian, and Iranian influences had been generously mixed. It gave birth to many illustrious men: Hērophilos of Chalcēdōn, Hipparchos of Nicaia, Dēmētrios of Apameia, Asclēpiadēs of Prusa, Theodosios the mathematician.

[12] Parion in Mysia, on the southwest shore of the Sea of Marmara.

ample, he recommended music for the cure of mentally ill people, but musical means had been used in medicine by his master Dēmocritos (V B.C.), if not before.[13] He is said to have discovered the cause of rabies but Dēmocritos knew something about that and Aristotle recognized its transmission by the bite of a mad dog.[14] He seems to have used massage with discrimination for many purposes: "to dispel and remove stagnant fluids, to open the pores, to induce sleep, to soften and warm the parts." In cases of paralysis, he advised patients to walk "in sandy places to strengthen relaxed parts."

The fragments of Asclēpiadēs and old sayings concerning him were collected by Christian Gottlieb Gumpert, *Asclepiadis Bithyni fragmenta* (204 pp.; Weimar, 1794) (Fig. 79). This work was Englished by the late Robert Montraville Green (1880–1955), who added to it a translation of Asclēpiadēs' biography by Antonio Cocchi (Florence, 1758; Milan, 1824), *Asclepiades, his life and writings* (177 pp.; New Haven, Conn.: Elizabeth Licht, 1955).

Themisōn of Laodiceia. Asclēpiadēs' disciple Themisōn of Laodiceia,[15] who flourished about the middle of the first century B.C.,[16] developed his theories more systematically, and therefore is generally considered the head of the new school, the Methodist school (*methodicē hairesis*). We would rather think that Asclēpiadēs was the founder, but must bow to the judgment of the members of that school, above all Sōranos (II–1), *methodicorum princeps*, and his translator, Caelius Aurelianus (V–1).

The main theory of Asclēpiadēs and Themisōn was called solidism (atomic structure of the body) as opposed to the theories named humoralism and pneumatism. Though those two theories were older than solidism, they continued to emulate it until Galen's time and even later. Solidism made possible a new classification of diseases; either the atoms were too distant and the pores of the body too relaxed (*atonia, rhysis; status laxus*), or the atoms and the pores were too tight (*stegnōsis, sclērotēs; status strictus*); a third, intermediary class was added later (*to memigmenon, status mixtus*). Themisōn's lost works are known only through Sōranos and Caelius Aurelianus. A treatise on acute and chronic diseases (*Peri tōn oxeōn cai chroniōn nosēmatōn*) was formerly ascribed to him, but it has been proved[17] to be a later work by Hērodotos of Rome (I–2).

[13] Dorothy M. Schullian and Max Schoen, *Music and medicine* (New York: Schuman, 1948) [*Isis 40*, 299 (1949)], pp. 53, 74–75, 81–82.

[14] Aristotle, *Historia animalium*, VIII, 22; 604A; Volume 1, pp. 335, 374.

[15] There are many cities thus named after Laodicē, the mother of Seleucos Nicatōr and other Seleucid princesses. This one is Laodiceia hē epi thalassē, a Syrian harbor, now Lādhiqīya (Latakia).

[16] Deichgräber would place him even later, at the end of the first century B.C. or the beginning of the first century after Christ. Pauly-Wissowa (2), vol. 10, (1934), 1632–1638.

[17] By Max Wellmann, *Hermes* 40, 580–604 (1905).

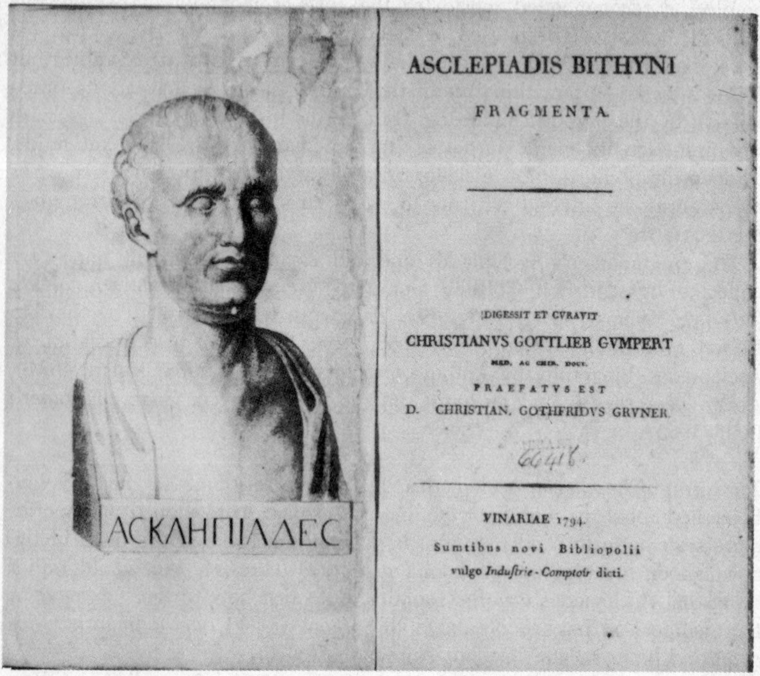

Fig. 79. Earliest edition of the fragments of Asclēpiadēs of Bithynia (I–1 B.C.), by Christian Gottlieb Gumpert (Weimar, 1794). [Courtesy of Armed Forces Medical Library.]

Megēs of Sidōn. The last Methodist to be named now was Megēs (I–2 B.C.), who hailed from Sidōn (in Phoenicia) but flourished in Rome. He was a surgeon and his lost writings are often quoted by later physicians; the most important fragment deals with fistulae (for instance, in the rectum). We have this from Oribasios (IV–2). The Methodist school was not only Roman in the general way; it was located in Rome. Later Methodists, like Thessalos of Tralleis (I–2) and Sōranos of Ephesos (II–1), were also established in the center of the empire. It is well to bear in mind that the time limit of this volume (Christ's birth), however essential it may be in some respects, is arbitrary in others, such as Roman science; but it is impossible to find a limit that applies equally well to every activity.

Ammōnios the Lithotomist and Perigenēs. Among the many other Greek physicians who flourished toward the end of the pre-Christian age, it will suffice to mention two, Ammōnios ho Lithotomos and Perigenēs.

Ammōnios (I–2 B.C.) practiced in Alexandria and received his nickname because he was supposed to have been the first to perform the operation of stone cutting in the bladder; he discovered a special styptic[18] and an eye salve.

Perigenēs (I–2 B.C.) was also a surgeon who invented a kind of bandage for the head and another (cranes-bill) for the luxated humerus. Internal surgery was hardly possible in those days (except perhaps lithotomy) and much of the surgeon's work was necessarily devoted to the reposition of bones and the healing of dislocations, which occurred in the games and the fights, either on the battlefield or in the arena.

LATIN MEDICINE

While Greek physicians remained the leading ones not only in the Roman world but also in the great city, there was a growing body of doctors who were real Romans, and who did not know Greek or knew it only very imperfectly as a foreign language.[19] That growth was remarkably slow. Not only were the old Romans (the 100-percent ones) distrustful of the clever Greeks (that was natural enough), but they tended to distrust medicine itself and treasure their old superstitions. For they had a medical folklore of their own, as every other people is bound to have, however primitive and unscientific.

Cato the Censor. The first stage of distrust is well represented by our old friend, Cato the Censor (II–1 B.C.). He needs no further introduction to our readers. He disliked physicians immensely because they were Greeks and therefore unfit to be trusted with Roman lives. In the (lost) *Praecepta ad filium* he gave advice to his son concerning rules of conduct, country life, hygiene, and precaution against the Greeks. He rejected Greek medicine together with all the Greek arts,[20] but he needed some help for the treatment of diseases that might occur to him or members of his household; it was equally necessary to treat sick or wounded slaves and distempered animals, and many chapters of his *De agri cultura* are devoted to such matters. The reading of them is exceedingly depressing. Cato was a great man in his way, strong and wise, but his scientific outlook was as low as his religion was narrow-minded and as his ethical outlook was mean.

The *De agri cultura* gives a fair idea of his medical knowledge, for his purpose was to help farmers as well as he was able, in sickness as well as in health. It is typical of the disorder of his book that medical chapters occur in many places; there are never more than three hanging together.

[18] *Stypticos*, astringent; producing contraction of blood vessels and stopping bleeding.

[19] Children of the upper classes were given a Greek tutor or were sent to a Greek-speaking country and were thus able to obtain a real and living knowledge of the language. Doctors were more likely to come from modest or poor families and their knowledge of Greek was equally poor.

[20] Toward the end of his life, however, he seems to have relented and began to study Greek literature.

A number of chapters explain how to make laxatives, diuretics, remedies (often called "wines") for gout, indigestion, dyspepsia, strangury, to prevent chafing. Two chapters (156 and 157) deal with the virtues of cabbage (*brassica*); they extend together to 200 lines and constitute the longest section of the book. According to him, "cabbage surpasses all other vegetables."[21] Parts of his book read like old women's recipes. Here is an example (chap. 114):

If you wish to make a laxative wine: After vintage, when the vines are trenched, expose the roots of as many vines as you think you will need for the purpose and mark them; isolate and clear the roots. Pound roots of black hellebore in the mortar, and apply around the vines. Cover the roots with old manure, old ashes, and two parts of earth, and cover the whole with earth. Gather these grapes separately; if you wish to keep the wine for some time as a laxative, do not mix with the other wine. Take a cyathus of this wine, dilute it with water, and drink it before dinner; it will move the bowels with no bad results.[22]

Many remedies are given to cure animal diseases, chiefly oxen and other cattle, to keep scab from sheep (96), to cure snakebite (102).

70. Remedy for oxen: If you have reason to fear sickness, give the oxen before they get sick the following remedy: 3 grains of salt, 3 laurel leaves, 3 leek leaves, 3 spikes of leek, 3 of garlic, 3 grains of incense, 3 plants of Sabine herb, 3 leaves of rue, 3 stalks of bryony, 3 white beans, 3 live coals, and 3 pints of wine. You must gather, macerate, and administer all these while standing, and he who administers the remedy must be fasting. Administer to each ox for three days, and divide it in such a way that when you have administered three doses to each you will have used it all. See that the ox and the one who administers are both standing, and use a wooden vessel.

71. If an ox begins to sicken, administer at once one hen's egg raw, and make him swallow it whole. The next day macerate a head of leek with a hemina[23] of wine, and make him drink it all. Macerate while standing, and administer in a wooden vessel. Both the ox and the one who administers must stand, and both be fasting.

Out of many remarks which such remedies would suggest, this one may suffice. Both the ox who receives the remedy and the man who administers it must be standing, both must be fasting, a wooden vessel must be used. Thus, rational (experimental?) advice is mixed with irrelevant taboos.

Many of the chapters record vows and offerings for the health of cattle or hogs, rituals and sacrifices to purify the land and insure good harvests. There are holidays and working days for animals as well as for men.

[21] In chap. 156: "Brassica est quae omnibus holeribus antistat."

[22] This example and the following are culled with kind permission from the Latin-English edition by W. D. Hooper and H. B. Ash (Loeb Classical Library; Cambridge: Harvard University Press, 1934).

[23] *Hēmina* is a measure of capacity (*demi-setier*, a glassfull). *Cyathos* above is a cup. It is amusing that Cato the Greek-hater could not help using Greek words.

138. Oxen may be yoked on feast days for these purposes: to haul firewood, bean stalks, and grain for storing. There is no holiday for mules, horses, or donkeys, except the family festivals.

One imagines readily that fantastic remedies and various kinds of hocus-pocus might be relied upon to cure internal complaints, which are very mysterious, but it is more surprising to find a charm for use in cases of dislocation. Cato, being a very practical man, must have realized that a dislocation is a mechanical accident to be restored by mechanical means, yet he is stupid enough to tell us the following nonsense:

160. Any kind of dislocation may be cured by the following charm: Take a green reed four or five feet long and split it down the middle, and let two men hold it to your hips. Begin to chant: "motas uaeta daries dardares astataries dissunapiter" and continue until they meet. Brandish a knife over them, and when the reeds meet so that one touches the other, grasp with the hand and cut right and left. If the pieces are applied to the dislocation or the fracture, it will heal. And none the less chant every day, and, in the case of a dislocation, in this manner, if you wish: "haut haut haut istasis tarsis ardannabou dannaustra." [24]

These examples are very depressing, for they leave us with the most dismal impression not only of Roman science but of Roman intelligence. Cato the Censor was not an uneducated man; he was not a silly old fool, and yet his medical recipes were as foolish as could be.

Marcus Terentius Varro. Some 120 years elapsed between Cato and his successor, Varro (I–2 B.C.), and many things happened, the most pregnant being the Hellenization of Rome. In Cato's time it was possible to consider the Greek prisoners and refugees as charlatans whose fancies should not be allowed to debase Roman virtue and Roman knowledge. Such an attitude was no longer acceptable among educated persons of Varro's time. Varro used a good many Greek sources; he did not hide them but was rather boastful in his enumeration of them. He did not repeat silly recipes, as Cato did, but offered rational, or let us say more rational, advice. For example, consider what he had to say concerning the location of a farm; like every intelligent farmer he was aware of the fact that some locations are healthy, while others are not.

Especial care should be taken, in locating the steading, to place it at the foot of a wooded hill, where there are broad pastures, and so as to be exposed to the most healthful winds that blow in the region. A steading facing the east has the best situation, as it has the shade in summer and the sun in winter. If you are forced to build on the bank of a river, be careful not to let the steading face the river, as it will be extremely cold in winter, and unwholesome in summer. Precautions must also be taken in the neighbourhood of swamps, both for the reasons given, and because there are bred cer-

[24] Some words were left untranslated, because they are a meaningless jabber.

tain minute creatures which cannot be seen by the eyes, which float in the air and enter the body through the mouth and nose and there cause serious diseases.[25]

The last sentence is especially remarkable.[26] It suggests the idea of contagion by means of microörganisms, but cannot do more than suggest it. Varro was probably thinking of very small organisms of which one is aware in marshy lands, almost too small to be seen; he could hardly conceive the reality of microörganisms without a microscope. Yet he clearly indicated the possibility of contagion from one organism to another, from exceedingly small ones to those as large as men and animals. In order to measure the full importance of Varro's statement, one has but to realize that it took an enormous time for the idea of contagion to be expressed more clearly.

Columella (I–2 A.D.) simply repeated Varro's idea; he copied that together with all the rest. After that one has to wait a thousand years for the next step. Ibn Sīnā (XI–1) was aware of the contagious nature of phthisis; William of Saliceto (XIII–2) realized the venereal contagion of certain ailments; Bernard of Gordon (XIV–1) made a list of eight contagious diseases (this became a medieval commonplace); Pierre de Damouzy (XIV–1) suggested that the plague might be transmitted by "carriers." The possibility of contagion was well understood by two Spanish Muslims, Ibn Khātimah (XIV–1) and Ibn al-Khaṭīb (XIV–1), but was utterly spoiled by the Egyptian, al-Damīrī (XIV–2), and other Muslims, according to whom diseases are not naturally contagious but God can make them so; the transmission of diseases from one person to another is only a part of fatality.

The scientific idea of contagion was established only in 1546 by Fracastoro in his *De contagione*,[27] and the possibility of contagion by microörganisms was first proved by the Dutchman Antony van Leeuwenhoek in 1675 and 1683, that is, more than seventeen centuries after Varro.

Antonius Musa. The majority of Roman physicians, certainly the most distinguished ones, were Greeks, and that situation continued until the second century after Christ or later. That is not always realized, because some of those Greeks, like Musa and Scribonius Largus, assumed Latin names; after all, they only did what Egyptians and Jews had done before them when they found it convenient to replace their native names with Greek ones. This is a natural practice which should not be misjudged; the purpose may be deception, but is just as likely to be social conformism and admiration.

[25] Varro, *Res rusticae,* I, 12; quoted from the Loeb edition by Hooper and Ash.

[26] The general idea of contagion had been adumbrated by the Babylonians; it was for them a magical rather than a scientific idea, however. The sanitary rules of the early Hebrews suggest that they realized the danger of contagion of some diseases (Volume 1, p. 94).

[27] *Hieronymi Fracastorii de contagione et contagiosis morbis et eorum curatione libri III* (Venice, 1546); Latin text with English translation by Wilmer Cave

We do not know the original name of Antonius Musa[28] (I–2 B.C.); his brother Euphorbos was physician to Juba, king of Numidia (d. 46 B.C.). Antonius was a freedman who was allowed to practice in Rome and was very successful. In 23 B.C., he had the good fortune of saving the life of Augustus with cold-water baths and lettuce. He was richly rewarded and received various privileges such as the permission to wear a gold ring (generally forbidden to freedmen). He became Augustus' ordinary physician and that great honor attracted to him many distinguished patients, such as Virgil, Horace, Maecenas, Agrippa. As always happens in the case of royal archiaters, he was made famous less by his own achievements than by the greatness of his patrons. It is probable that he was a good physician, however, and his failure to save Marcellus does not prove the contrary.[29] On account of his reliance on cold baths, one might be tempted to call him the founder of hydrotherapeutics, but we may be certain that many men had put their faith in cold baths long before him. Here again, his fame is based on ambiguity, not on the use of cold baths but on his saving Augustus' life with one. Writings of his on materia medica (Galen XIII, 463) are lost. Two treatises bearing his name, *De herba betonica*[30] and *De tuenda valetudine ad Maecenatem*, are spurious and late. The first of these was first published in Zurich in 1537, and both in Venice in 1547.[31]

There is no need to speak of other Roman physicians. If Antonius Musa was the most illustrious of them, the others could not amount to much.

Medical knowledge can be inferred from other writings, such as the didactic poems of Aemilius Macer and the *Architecture* of Vitruvius.

Aemilius Macer. Aemilius Macer (I–2 B.C.) of Verona traveled to the East, as so many Romans did, to learn Greek, and he died in Asia c. 16 B.C. He wrote Latin poems, in the manner of Nicandros' Greek ones, dealing with the generation of birds (*Ornithogonia*), venomous creatures and antidotes (*Theriaca*), herbs (*De herbis*). We know nothing of them except the titles.

Vitruvius. As one might expect, there is much of medical interest in the *De architectura*. As Vitruvius put it right at the beginning (I, 1, 10): "The archi-

---

Wright (New York, 1930) [*Isis 16*, 138–141 (1931)].

[28] The Latin word Musa reproduces exactly the Greek one, Musa, one of the goddesses of song, poetry, and the fine arts; there were nine of them. It was an elegant name for a freedman to assume. Compare our word "museum," a temple dedicated to the muses.

[29] Marcellus, born in 41, was Augustus' nephew, adopted son, stepson, and, it would seem, presumptive heir; he died in 23 at the age of 18. He was immortalized by Virgil, *Aeneid*, VI, 860–886: "Tu Marcellus eris. Manibus date lilia plenis . . ."

[30] *Betonica*, betony, a plant of the mint family, which was believed by the author of that treatise to have many medical virtues.

[31] The name Musa was thus resurrected during the Renaissance. Francis I bestowed it upon his own doctor, Antonio Brasàvola, to honor the latter as well as himself. Sarton, *The appreciation of ancient and medieval science during the Renaissance* (Philadelphia: University of Pennsylvania Press, 1955), p. 32.

tect should have a knowledge of medicine on account of the questions of climates, healthiness and unhealthiness of sites and the use of different waters." Those medical concerns are illustrated in various parts of his work, especially in Book VIII dealing with water. For example, he remarks (VIII, 3) that "the tribe of the Medulli in the Alps have a kind of water which causes swellings in the throats of those who drink it" (goiter),[32] and (VIII, 6) that water conducted in a lead pipe is not wholesome; the use of lead affects the health of plumbers, "since in them the natural color of the body is replaced by a deep pallor"; in the digging of wells one has to take special precautions: "let down a lighted lamp and if it keeps on burning, a man may make a descent without danger." Climatic influences to be taken into account when building a house are explained in VI, 1.

Vitruvius was not a medical doctor, but he was intelligent and had enough experience to appreciate the medical requirements of his profession.

[32] For the history of goiter, see Claudius F. Mayer, *Isis* 37, 71–73 (1947).

XXIII

GEOGRAPHY IN THE LAST TWO CENTURIES[1]
CRATĒS AND STRABŌN

While the main treatises, if not the only ones, on architecture and husbandry were written in Latin, almost all the geographic work was in Greek, except at the very end of the period, from Caesar to Augustus, when some found expression in Latin, or was thoroughly Roman, not Roman in Hellenistic garb. The two main heroes were Cratēs of Mallos (II–1 B.C.) and Strabōn of Amaseia (I–2 B.C.).

GREEK GEOGRAPHY

Cratēs of Mallos. Mallos in Cilicia, where Cratēs originated, was a very old Greek settlement, said to have been founded at the time of the Trojan war.[2] He spent his life in Pergamon, where he was chief of the philological school and director of the library. This implied controversies with his Alexandrian colleagues, to which we shall come in Chapter xxvi. The only chronological landmark in his life is the year 168, when he was sent by Eumenēs II as ambassador to Rome to offer his king's congratulations for the victory of Pydna; it has been claimed that his visit affected the development of public libraries in Rome, but it was a little early for that. According to Strabōn (II, 5, 10), he constructed a terrestrial globe; this was the earliest on record (celestial globes had been used before). As the inhabited world (*oicumenē*) is only a small part of the surface, Strabōn remarked that for practical study it would be well to use a large globe of no less than 10 feet in diameter; he does not say that Cratēs' globe was as large as that. It would seem that Cratēs was not interested in geographic details, but rather in the general aspects of the terrestrial sphere. He revived and developed the Pythagorean theory of four land masses: there is not one *oicumenē*; there are four situated on four land masses separated from one

[1] For geography in the third century B.C., see Chapter VI.

[2] The traditional date of the Trojan War is 1192–1183, but the exact date does not matter as far as Mallos is concerned. It suffices to remember that Mallos was founded in hoary antiquity.

another by two oceans, and antipodal, two by two. (Imagine an apple which you cut in four parts by means of two perpendicular planes.) This was of course a gratuitous theory, but it appealed to the imagination and inspired geographic conceptions more than once.[3]

We shall speak a little more briefly of three contemporaries of Cratēs — Polemōn, Agatharchidēs, and Polybios.

Polemōn Periēgētēs. Polemōn (II-1 B.C.) came from Trōas and traveled all over Greece. His nickname, *ho periēgētēs*, meaning the guide, refers to a profession typical of his time; the Greeks had always been fond of wandering and there were now professional wanderers, people who made it their business to know the Greek cities and guide other people, such as Roman visitors, from one to another, pointing out the outstanding monuments. Polemōn's books have reached us only in the form of fragments;[4] he wrote guides and histories of the foundations (*ctiseis*) of many cities. He also discussed archaeological questions and published inscriptions (*peri tōn cata poleis epigrammatōn*). The inscriptions that he collected were chiefly dedications of monuments in Delphoi, Sparta, and Athens. It is not certain whether he acted as a personal guide, but his activities made guiding possible; he was the father of Greek ciceroni.

Agatharchidēs of Cnidos.[5] This Agatharchidēs (II-1 B.C.) was a Peripatetic who flourished in Alexandria in the second quarter of the second century; he was guardian or tutor to one of the kings (Ptolemaios IX Sōtēr II?). He wrote treatises dealing with the geography and history of Asia in ten books (*Ta cata tēn Asian*) and with the geography and history of Europe, in 49 books (*Eurōpiaca*), but his most important was one on the Red Sea (*Peri tēs Erythras thalassēs*)[6] containing geographic and ethnographic information on Ethiopia and Arabia, for example, accounts of Ethiopian gold mines and of the fish eaters (*ichthyophagoi*) of the Arabian coast. The summer flood of the Nile is caused by the waters that accumulate in Ethiopia in winter time.

Polybios. Polybios the Stoic (II-1 B.C.) was primarily a historian, one of the greatest of antiquity, and the importance of his work will be discussed at greater length in the next chapter, but he deserves to arrest our attention right away. Geography to him was ancillary to political history, but he

[3] Hans Joachim Mette, *Sphairopoiia. Untersuchungen zur Kosmologie des Krates von Pergamon* (336 pp., Munich, 1936) [*Isis 30*, 325 (1939)].

[4] Collected by Karl Müller, *Fragmenta historicorum graecorum*, vol. 3 (Paris, 1849), pp. 108-148.

[5] Greek-Latin edition of fragments in Karl Müller, *Geographi graeci minores* (Paris, ed. 1, 1855), vol. 1, pp. 111-195; English translation in E. H. Warmington, *Greek geography* (London, 1934) [*Isis 35*, 250 (1944)], pp. 43-44, 198-207.

[6] This seems to have been a kind of periplus, or manual for the use of navigators, concerning both coasts of the Red Sea. Fragments of it survived in Diodōros of Sicily (I-2 B.C.) and Phōtios (IX-2).

fully realized that a good knowledge of it was one of the fundamental needs of any serious historian. Like other Greeks (he was a genuine one, an Arcadian) he had traveled extensively in the Greek world; unlike most of them he had traveled also in the West (Italy, Gaul, and Spain) and, therefore, he had an unusual familiarity with the western background, which he described to give a good account of Western events. He illustrated the growth of geographic knowledge caused by the Roman conquests; we may say that he was the first to describe the Roman world.

Though Polybios was a *'fin de siècle'* child of the third century, he lived so long (he was 82 at the time of his death, c. 125) that he takes us well into the second half of the second century.

Three other men who were his younger contemporaries deserve the attention of historians of geography – Hipparchos, Artemidōros, and Eudoxos.

Hipparchos of Nicaia. Hipparchos (II–2 B.C.) was primarily an astronomer and as such he helped to establish the mathematical basis of geographic knowledge. One might say that his main merit as a geographer was to insist on the use of rigorous mathematical methods for the determination of places. This was somewhat vitiated by his dislike of Eratosthenēs and his distrust of the new data obtained since the Alexandrian conquests. He wrote a book against Eratosthenēs but had the immense advantage of standing upon the latter's shoulders. He accepted his conclusion on the size of the earth.

He tried to measure the latitudes by means of the ratio of the shortest to the longest day, as opposed to the Babylonian method of measuring the increasing lengths of day (as one goes northward) by means of arithmetic progressions. He was the first to divide the inhabited world into zones of latitude or climates, in evaluating latitudes and longitudes with reference to great circles divided into 360 degrees, and in using those coördinates systematically to establish the position of each locality. For the determination of longitudes he suggested the observation of eclipses from different points; the differences of local times would give the difference of longitudes. The method was excellent, but its systematic application would have involved a degree of political organization that did not exist, and a degree of scientific organization that was almost inconceivable in his day.

There is no evidence that he traveled much. Where and how did he obtain his data? The little we know of his own efforts we owe to Strabōn, and it is probable that Ptolemy's geography, compiled three centuries later than Hipparchos, was partly derived from the data brought together by the latter.

Artemidōros of Ephesos. The geographic data of Agatharchidēs and of Hipparchos were increased by Artemidorōs of Ephesos[7] (II–2 B.C.), who

[7] Not to be confused with another Artemidōros of Ephesos (II–2) of a much

flourished at the very end of the second century (c. 104–100). He traveled extensively as far west as Spain (and Gaul), settled in Alexandria and wrote eleven geographic treatises (*Ta geōgraphumena, Periplus, Geographias biblia*). For eastern geography he depended upon Agatharchidēs and added data concerning the Red Sea and the Gulf of Aden; for India, upon the writers of the Alexandrian age and Megasthenēs. His ambition was to cover the whole inhabited world (*oicumenē*); he twice calculated its length and its breadth without astronomical determinations! Apparently, he objected to Eratosthenēs' and Hipparchos' exclusive interest in longitudes and latitudes and attached more importance to distances. This would mean that his maps were based upon itineraries as well as on astronomical determinations. In judging his method we should bear in mind that the determinations of latitudes were still inaccurate, and those of longitudes much more so. While a map based upon itineraries was theoretically much inferior to one based upon coördinates, it might not be much worse practically. On the other hand, the itineraries were much weakened by the lack of magnetic guidance.[8]

Eudoxos of Cyzicos.[9] Eudoxos' story as told by Strabōn has been doubted because of its strangeness, but I do not think that it lacks plausibility. He had been sent by his native city on a mission to Alexandria and while there he came across an Indian sailor, the only survivor of a ship wrecked on the Red Sea coast (such accidents were not uncommon because the coral reefs of that coast are very dangerous). The Indian told his adventures and offered to lead an expedition back to India if the king (Ptolemaios Evergetēs II, or Physcōn, who ruled until 116) would fit out a vessel. This was done and Eudoxos taken aboard. They sailed to India and back. Their rich cargo was appropriated by the king, but they brought back something important that the king could not steal, namely, the knowledge of the southwest monsoon which made it easier to sail out of the Bab el-Mandeb, from the Red Sea into the Gulf of Aden and the Arabian Sea. We shall come back to this presently, but let us finish Eudoxos' story first.

He made a second voyage to India and this time brought back an ornament, taken from the prow of a ship, which proved that the ship hailed from Gades (Cadiz). Eudoxos concluded that it must have sailed around

later time. This second one, generally called Artemidōros Daldianos, wrote a book on dreams (*Oneirocritica*). The name Artemidōros (Artemis' gift) must have been popular in Ephesos, the city dedicated to Artemis (Diana).

[8] The Greeks had discovered very early the attractive property of a magnet, but its directive property was not recognized until the late Middle Ages (*Introduction*, vol. 2, p. 24). The use of the compass is a late medieval achievement.

[9] Cyzicos is an island of the Propontis (Sea of Marmara); it was one of the earliest Greek settlements in Asia Minor. It is now attached to the south coast of that sea and is named Kapidaği. Our knowledge of this Eudoxos is derived from Poseidōnios through Strabōn.

Africa and decided to do the same. He sailed to Gades, thence down the western African coast and was lost.

It is the first part of the story that is the most interesting, the discovery of the monsoon.[10] It was a discovery the practical importance of which can hardly be overestimated, for sailing from the Red Sea to the Malabar coast and back could be done best with the monsoon and could not be done at all against it. Was it discovered (on the Western side) by Eudoxos? The discovery is generally ascribed to Hippalos, but scholars disagree as to the date. Some believe that Hippalos flourished after the Augustan age, others [11] that he belongs to the late Ptolemaic age. Irrespective of Hippalos, it seems probable that ships of the later Ptolemaioi sailed to India, but that the first voyages direct across the Indian Ocean to South India were not earlier than A.D. 40–50.[12] The later Ptolemaioi had established their control in the straits of Bab el-Mandeb and by 78 B.C., if not before, the general in chief (*epistratēgos*) of Upper Egypt was also admiral of the Red Sea and the Indian Ocean. More Indians appeared in Egypt than before and the products of South India (such as pepper) became more abundant in Egyptian and European markets. The fact that Cleopatra VII could think of abandoning the Mediterranean and ruling the Indian Ocean is the best proof that the Indian trade was already considerable in her time (she died in 30 B.C.), and that trade could not have grown to any size without taking full advantage of the monsoons.

We may now pass to the last century B.C., which is dominated as far as geography is concerned by three great personalities, Poseidōnios, Strabōn, and Isidōros.

Poseidōnios of Apameia.[13] We have already spoken many times of Poseidōnios (I-1 B.C.) and his name will reappear again and again, because he was a man of almost universal curiosity. To compare him with Aristotle, however, or to call him the Hellenistic Aristotle is very misleading. The greatness of Aristotle does not lie as much in the range of his curiosity as in the strength and soundness of his thoughts. It is true that Poseidōnios was the last scholar B.C. to take all knowledge for his province, but he had nothing comparable to Aristotle's genius for synthesis. As far as we can judge

[10] That is, the Western discovery of it. It is probable that Indian or Arabic sailors were aware of its existence, but that cannot be proved. The monsoons are seasonal winds, blowing part of the year in one direction and at another time in the opposite direction.

[11] For example, Michael Ivanovich Rostovtsev (1870–1952). See *Isis 34*, 173 (1942). According to the *Oxford Classical Dictionary*, p. 428, Hippalos flourished in the first century B.C. Hippalos' name was actually given to the southwest monsoon by Pliny, *Natural history*, VI, 104–106.

[12] This and other details in this paragraph are taken from W. W. Tarn and G. T. Griffith, *Hellenistic civilisation* (London: Arnold, 1952), pp. 247–248. I doubt this particular statement, which is hardly compatible with the following facts.

[13] Apameia on the Orontēs; one of the great cities of the Seleucid kingdom and later of the Roman province, Syria Secunda; called Famieh during the First Crusade (1096–1099) when it was ruled by the Norman Tancred.

from the fragments that have come down to us, Poseidōnios was often betrayed by his imagination and his mysticism. It is perhaps more correct to call him "the most intelligent traveller in antiquity," [14] and that is sufficient praise. Many good parts of Strabōn's work were obtained from him.

He wrote a treatise *On the ocean* (*Peri ōceanu*) wherein he repeated the Erastothenian idea that there is but one ocean.[15] He traveled considerably, not only along the Mediterranean coastlands but also deep inland, in such countries as Spain, Gaul, and even England, and made elaborate observations concerning "human" as well as physical geography. He remained a full month at Gades, where he observed the tides, and was one of the first to ascribe that phenomenon to the combined actions of Sun and Moon, calling attention to the spring and neap tides. He observed earthquakes, volcanoes, and the emergence of a new volcanic islet among the Lipari or Aeolian islands (northeast of Sicily). He visited mines in Andalusia and in Galicia and described their galleries and drainage. He witnessed the existence of rock salt, and described the Crau plain, near the mouth of the Rhone, and the abundance of round pebbles which are spread over it. Other details of the same kind might be easily culled from Strabōn's *Geography*, who quoted him repeatedly.

He tried to improve Eratosthenēs' estimate of the size of the Earth, reducing the circumference (wrongly) from 250,000 stadia to 180,000; on the other hand, he overestimated the length of Eurasia and remarked that a man sailing west from the Atlantic for 70,000 stadia would reach India. This error had extraordinary results. It reappeared in one form or another in the writings of Strabōn, Ptolemy, Roger Bacon, Pierre d'Ailly (1410), increased Columbus' optimism, and caused the latter's discovery, not of the eastern limit of Eurasia, however, but of a New World.

Strabōn of Amaseia. Strabōn (I–1 B.C.) is best defined as the author of the *Geography*, and the more so that all that we know of him is derived from this, his main work, the only surviving one. One gathers from it that he was born about the year 64 B.C. in Amaseia,[16] of which he gives a loving description. He belonged to an important family, some of whose members had served the kings of Pontos, Mithridatēs V Evergetēs and Mithridatēs VI Eupatōr, as generals, governors, and priests of Mā (Bellona). His family

[14] Expression used by H. F. Tozer (1829–1916) in his *History of ancient geography* (rev. ed. by M. Cary; Cambridge, 1935), p. 190.

[15] That was an old conceit which can be traced back to Nearchos (IV–2 B.C.), Aristotle, Hecataios (VI B.C.) and even to Homer. For details, see Volume 1, pp. 138, 186, 310, 510, 526. The idea of one ocean is correct, but Homer and Hecataios were wrong in conceiving it as one great river encircling the Earth (Homer's *ōceanos potamos* or *apsorroos*, flowing back into itself). That conception was incompatible with the spherical Earth.

[16] Amaseia (Amasya in Turkish) on the river Iris (Yeşil Irmak) was the capital of the kingdom of Pontos, south of the eastern end of the Black Sea. It was the birthplace also of Mithridatēs the Great. Strabōn, XII, 3, 39; see also XV, 30, 37.

was partly Greek, partly Asiatic, but he was entirely Greek in language and customs; it must have been in good circumstances, for he received an elaborate education. After elementary tuition at home he was sent to Nysa (near Tralleis in Caria), where he studied grammar and literature under one Aristodēmos. In 44 B.C. (aet. 20) he went to Rome for postgraduate studies. His teachers were Tyranniōn of Amisos,[17] grammarian and geographer (he may have received his vocation from him), and Xenarchos of Seleuceia (I–2 B.C.) in Cilicia, a Peripatetic philosopher; he was acquainted with Stoics like Poseidōnios,[18] Boēthos of Sidōn, and Athēnodōros of Tarsos (Cilicia). He himself became a fervent Stoic; he recognized the need of myths, rites, and mysteries for the people, but his own religion was the Stoic one.

He was a great traveler, though not as great as his *Geography* and his own testimony (II, 5, 11)[19] would suggest. He traveled from Armenia in the East to Italy in the West; he was in Greece (at least in Corinth), and in Egypt, where he sailed up the Nile as far as the frontiers of Ethiopia; he was well acquainted with many parts of Asia Minor. Much of his information was derived from books, that is, from Greek books, for there were few others for his purpose.

Some landmarks are given in his *Geography*; he was in Rome in 44 and following years, in 35, 31, 29, 7 B.C.; in Egypt from 25[20] to 20 or later. Much of his information was obtained in the library of Alexandria (for where else could he have obtained all the books he needed?). He flourished throughout the Augustan age and the beginning of Tiberius' rule (A.D. 14–37). It is probable that he spent the end of his life in Amaseia. He died in A.D. 21 or later.

He wrote two great books, one on history, which is lost, and his *Geography*, which has come down to us almost in its integrity and is one of the great monuments of antiquity. It is divided into 17 books, the contents of which are roughly these:

I–II. Prolegomena. This is partly historical; he criticizes Homer and Eratosthenēs, discusses Polybios, Poseidōnios, Eudoxos of Cyzicos. He speaks of mathematical geography, the shape of the Earth, cartography on a sphere and

---

[17] Amisos in Pontos. This Tyranniōn was thus a countryman of Strabōn's, but it was in Rome that they worked together.

[18] Poseidōnios died in 50 B.C., hence Strabōn cannot have met him except in early youth; in 50 Strabōn was only 14, Poseidōnios in his eighties.

[19] Such references are to his *Geography*.

[20] Strabōn in 25 traveled to the Thebaid in the retinue of the prefect of Egypt, Aelius Gallus. Says M. Cary in Tozer's *History of ancient geography*, p. xxviii: "In 25 B.C. Augustus made an ill-advised attempt to break the commercial monopoly of the Himyarite Arabians in the southern Red Sea by directing an overland expedition against one of their towns, Mariaba. His general Aelius Gallus set siege to this town after a laborious march of six months' duration from the Gulf of Akaba across the central Arabian desert, but he failed to reduce it. This was the only serious attempt to open up Arabia in ancient times. The hardships suffered by Gallus' force discouraged the Caesars from further efforts to penetrate Arabia."

on planes. He insists that there is but one ocean, as is proved by the ebb and flow of the tides occurring everywhere. Hence one might sail from Spain to the East Indies (I. 1, 8).

III. Spain. Scillies (Cassiterides).

IV. Gaul, Britain, etc.

V. Northern and central Italy.

VI. Southern Italy and Sicily. The Roman empire.

VII. Central and Eastern Europe (the end is lost).[21]

VIII. Peloponnēsos.

IX. Northern Greece.

X. The Greek islands.

XI. Region of the Black and Caspian seas, Taurus, Armenia.

XII–XIV. Asia Minor.

XV. India, Persia.

XVI. Mesopotamia, Syria, Arabia, Ethiopian coast.

XVII. Egypt.

It is an encyclopedia of geographic knowledge, the books of which were necessarily of unequal value. There is an abundant Strabonian literature, the most valuable parts of which are discussions of each region by scholars who are deeply familiar with it. Such criticism cannot be repeated because it is endless.

Let us consider a few general questions. First, what was Strabōn's purpose? He wanted to write a geographic survey of the world; but, as his training was purely literary, he was not interested in mathematical geography, which he pooh-poohed without sufficient knowledge of it and without a real understanding of its difficulties. On the other hand, he was deeply interested in people and was philosophically minded; his geography is physical but even more so human, historical, archaeological. He wanted to give his readers a general view of the face of the Earth, its physical aspect (rivers, mountains, and so forth), and the differences between its several regions, then to explain how the people lived in each of these, and what kinds of people they were. This implied an account of their vicissitudes and achievements, and enumeration of their cities (when had they been founded?), of the roads and public monuments, and sometimes of their great men.

Being a Stoic, he accepted the general tenets of the astral religion, but his astrology was moderate and there is no evidence that he believed in genethlialogy. He was aware of the "astronomical" pursuits of Egyptians and Chaldean priests.[22] The Phoenicians of Sidōn, he claimed, had transmitted to the Greeks the first rudiments of astronomy and arithmetic.[23]

In politics he was definitely pro-Roman; he realized that the Augustan age had brought peace and unity to the world (VI, 4, 2). For example, by putting an end to the piracy hitherto endemic in the Eastern Mediterranean,

[21] The end of Book VII was still extant in the eleventh century, because there is a summary of it in the Epitome Vaticana, a manuscript written at the end of that century. There are many fragments (some 34 pages) of the last part of VII.

[22] Genethlialogy means the casting of horoscopes or nativities. Everybody in Strabōn's time believed to some extent in astrology; intelligent and well-educated men like Strabōn mitigated their belief with prudence and skepticism. See his remarks on astronomy and astrology in XVI, 1, 6, (Chaldean) and XVII, 1, 46 (Egyptian).

[23] *Geography*, XVI, 2, 24.

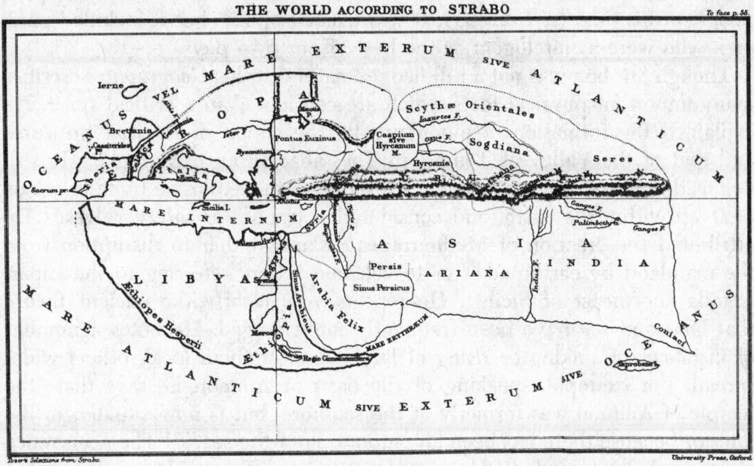

Fig. 80. Map of the world according to Strabōn (I–2 B.C.). [Reproduced from the Rev. H. F. Tozer's *Selections from Strabo* (Oxford, 1893).]

they had established the safety of travel and trade, prosperity. Yet he was proud of being an Oriental and never failed to name the many scholars who were born in the East. Much as he admired Roman government, he had no respect for Roman scholars (one cannot blame him for that).

There has been much discussion as to the date of his work. The bulk of his information was probably obtained before he left Alexandria (c. 20), and the first draft of the *Geography* was already completed by 7 B.C. He made no use of Agrippa's map, which was not yet available in that year. The list of the Roman provinces, on the last page of his work, was written by him not later than 11 B.C. and revised in 7 B.C., far from Rome. He revised the whole work in Amaseia, c. A.D. 18, as is evidenced by his mention of Tiberius (whose rule began in A.D. 14) in some twenty places.

He was fully aware of the size and importance of his work and called it colossal (*colossurgia*); it was indeed and one cannot help wondering how such a large undertaking could be achieved by a single man. An undertaking of comparable scope in our own days would be planned by academies or universities and realized by administrators, directing many scholars and many more secretaries, and using all kinds of machines. We are fortunate in having such an elaborate geographic survey of the Western world in the Augustan age, to which are added an abundance of historical, archaeological, ethnographic information, notes on trades and industries, and other reflections.

The public he had in mind was not one of scientific geographers, which did not yet exist, but the statesmen, men of affairs, and the other educated people of his time (I, 1, 22–23). It was a small public, but it included some men who were as intelligent as the best of our own days.

Though Strabōn was not a full-fledged naturalist, his *Geography* describes many important physical facts which are considered in a critical spirit. He explained the formation of mountains by the action of internal pressures and that of the Valley of Tempē, in Thessalia, by an earthquake. He still believed that volcanic phenomena were due to the explosive force of winds pent up within the earth, and considered volcanoes as safety valves.[24] He attributed the creation of Mediterranean islands either to disruption from the mainland by earthquakes or to volcanic action, referring to the Lipari islands (northeast of Sicily). He repeats very clearly the ancient theory that land and sea have been frequently interchanged. He notes a number of instances of sinking or rising of land, some of them local, others widespread. For example, speaking of the oasis of Ammōn he says that "the temple of Ammōn was formerly at the seashore, but is now situated in the interior because there has been an outpouring of the sea." [25] The occurrence of fossil shells (*conchyliōdēs*) in various places proves, he remarks, that the lands of Lower Egypt where they are found were previously submerged. Such sinkings he ascribes to earthquakes. A similar one might well destroy the isthmus of Suez and open up a communication between the Mediterranean and the Red seas.[26] He relates many observations on the erosive power of water and on the alluvial deposits at the mouths of rivers or along their courses. He gives information on the mining of salt and on its extraction from mineral springs; on the Laurion silver mines, on glass-making in Alexandria; on water wheels; on the slipway (*diolcos*) for moving ships across the isthmus of Corinth; on the ancient canal connecting the Nile with the Red Sea, which it ended at Arsinoē and was closed with a double door as a precaution against the change of current (*euripos*), to permit the passage of ships in both directions.

Strabōn was not a literary artist, but he wrote well in the way a scholar would. He had been well educated and his language was correct and clear without irrelevant ornaments. Men of letters would call it colorless and dull, but he had taken real trouble in its composition and had done his best to introduce diversity and to reward the reader with as many stories as was consistent with his severe purpose. His work is very superior (in style and contents), to the geographic part of Pliny's *Natural history.*

Strabōn said that Aristotle was the first to collect books and that his

[24] The conception of volcanoes as safety valves was still entertained at the end of the eighteenth century by one of the founders of modern geology, James Hutton, *Theory of the Earth* (ed. 2, 2 vols.; Edinburgh, 1795), vol. 1, p. 146.

[25] *Geography,* I, 3, 4. Other examples are given in this section.

[26] Similar geologic observations had been made before by Hērodotos (on Tempē), by Aristotle, by Poseidōnios (Lipari islands); *Introduction,* vol. 3, p. 214.

example was followed by the kings of Egypt.[27] This statement is correct *grosso modo*. Aristotle was perhaps not the first to collect books (what does that mean? how many books must one have to be a "collector"?), but it was certainly because of his influence, transmitted by Dēmētrios of Phalēron and Stratōn, that the early Ptolemaioi decided to found the library of Alexandria.

Strabōn's readings were far more considerable than his travels. He had read the whole of Greek literature available to him, beginning with Homer. He greatly admired the latter (as every Greek did) and overestimated the geographic value of the *Odyssey* (Eratosthenēs had tended to underestimate it). His most abundant source, however, was his older contemporary, Poseidōnios. It was he who transmitted Poseidōnios' wrong estimate of the size of the Earth to posterity.

Considering the immense practical value of his *Geography*, unique of its kind, for statesmen and civil servants of the Roman empire, it is astonishing that relatively little attention was paid to Strabōn in antiquity. Were the earliest copies secreted by their owners for practical use, not scholarship? I can think of no other explanation. Jōsēphos (I–2) knew of it, but no other Greek, not even Ptolemy (I–2), and no Roman scholar, not even, *mirabile dictu*, Pliny (I–2). This ancient neglect may account for the lack of an Arabic translation; Strabōn remained unknown to Muslim geographers and historians.

He was rediscovered in Byzantine times by Stephanos of Byzantion (VI–1) and used by Eustathios of Thessalonicē (XII–2) and Maximos Planudēs (XIII–2). The earliest manuscript is the Parisinus 1397 of the twelfth century, containing only Books I–IX. For Books X–XVII one had to depend upon three later manuscripts, Vaticanus 1329, Epitome Vaticana, Venetianus 640.

The beginning of the printed tradition is due to Guarino of Verona (c. 1370–1460), who brought back a Greek manuscript from Constantinople and translated Books I–X into Latin; Books XI and XII were translated by Gregorio Tifernas and the whole was printed by Sweynheym and Pannartz (Rome, 1469) (Fig. 81), and five times reprinted before 1500: Venice, 1472; Rome, 1473; Treviso, 1480; Venice, 1494, 1495 (Klebs 935.1–6). The Greek princeps was published by Aldus (Venice, 1516) (Fig. 82). A much improved Latin text was prepared by Wilhelm Xylander (Basel: Henricus Petri, 1570); this was the first satisfactory edition.

Isaac Casaubon edited the Greek text, adding Xylander's translation to it (Geneva, 1587). Another remarkable edition was due to the Dutchman Jansson d'Almeloveen (Amsterdam, 1707).

Ademantos Coraēs ("Coray") edited a new Greek text (4 vols.; Paris, 1815–1819) (Fig. 84) and a French translation (5 vols.; Paris, 1805–1819) (Fig. 85). This translation was prepared by Napoleon's order with the collaboration of three French scholars: Laporte du Theil, Letronne, and Gossellin.

The best critical edition is that of Augustus Meineke, first published by Teubner (Leipzig, 1852–53) and frequently reprinted in 3 volumes.

---

[27] *Geography*, XIII, 1, 54.

Greek-English edition in the Loeb Classical Library begun by John Robert Sitlington Sterrett and completed by Horace Leonard Jones (8 vols., 1917–1932).

Marcel Dubois, *Examen de la géographie de Strabon* (416 pp., Paris: Imprimerie Nationale, 1891), reviewing Strabōn literature to 1890. Ernst Honigmann, in Pauly-Wissowa, *Real-Encyclopädie* (2)7, 76–155, 1931.

Henry Fanshawe Tozer, *Selections from Strabo* (388 pp., 6 maps; Oxford: Clarendon Press, 1893); selections in Greek with notes.

Isidōros of Charax.[28] The Greek section of this chapter may be ended with a brief account of this Isidōros (I–2 B.C.), who was a contemporary of Strabōn, though it is impossible to say whether he flourished a little before Christ or a little after. It is simpler to consider him as a geographer of the Augustan age; in fact, his work may have been done by Agrippa's order. Strabōn did not mention him, but fragments of his "description of the world" have been transmitted by Pliny, and one of his *Journey around Parthia* (*Parthias periēgēsis*), dealing with pearl fishing, by Athēnaios of Naucratis (III, 46). We have the complete text of his *Parthian stations* (*Stathmoi Parthicoi*), outlining the caravan road from Antioch to India.[29] This is a good example of a guide or itinerary for travelers, merchants, or government officers of which some were compiled during the Augustan age. We shall come back to them apropos of Agrippa, below.

LATIN GEOGRAPHY

Latin accounts that have survived are far less numerous than Greek, and they begin to appear only at the very end of the pre-Christian age. We shall start with no less a person than Caesar.

Julius Caesar. Caesar's *Commentarii* (written c. 52–50) will be discussed in Chapter XXIV, but we must speak now of their geographic basis. This is difficult, because Caesar's geographic information is meager and the passages wherein there is a little more of it have been suspected of being post-Caesarian interpolations. It has also been argued that some geographic passages, concerning Germany and the Hercynian Forest,[30] were compiled from Greek geographers by a "research assistant" in his employ; we need not be embarrassed about that, however, for "research assistants" do not expect to receive credit. The main point is that much of his information was

[28] Charax in Greek means a stake, hence a palisade made of stakes or a camp surrounded by a palisade. Many camps were called Charax; this place was near the mouth of the Tigris. Isidōros was probably a Chaldean.

[29] Wilfred H. Schoff, *Parthian stations of Isidore of Charax* (47 pp.; Philadelphia, 1914).

[30] Hercynia Sylva, described by Caesar in *De bello Gallico*, VI, 24–25 as stretching across Germany as far as Dacia (60 days long and 9 days wide). This was a vague combination of the Black Forest, Odenwald, Thüringer Wald, Harz, Erzgebirge, Riesengebirge. Note that the words "Harz" and "Erz" are derived from "Hercynia." Of all geographic features, mountains were the most difficult to situate without maps.

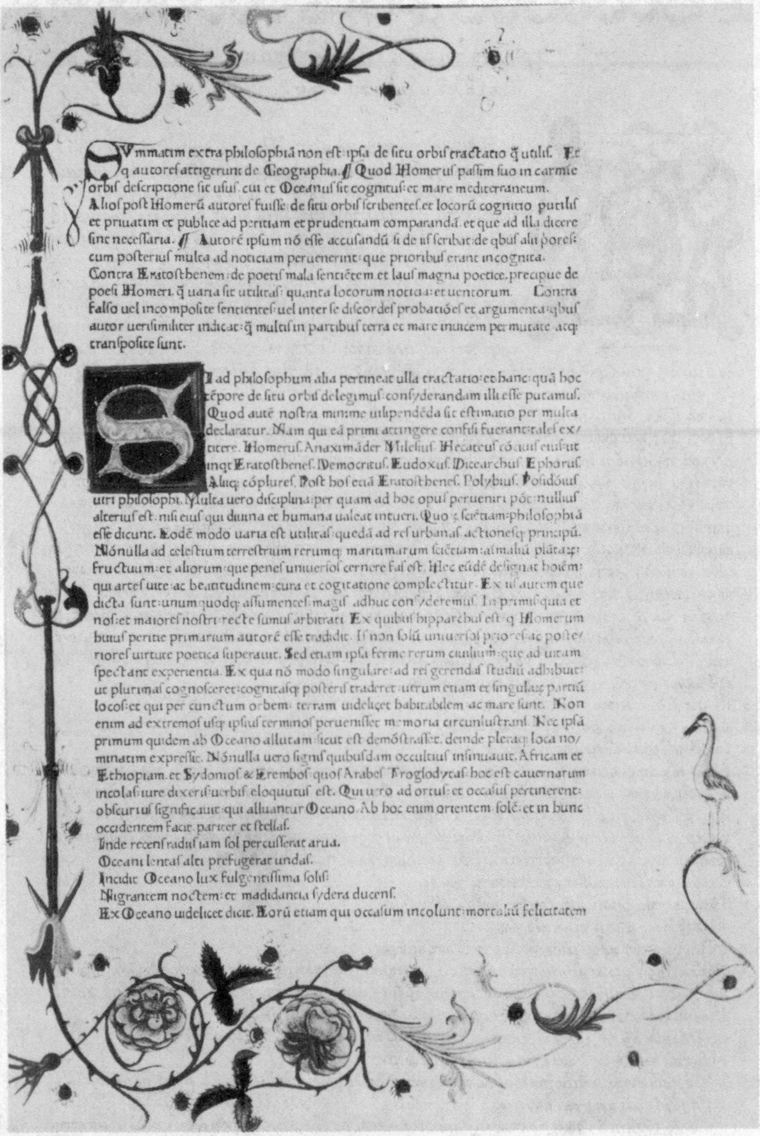
SVmmatim extra philosophiã non est: ipsa de situ orbis tractatio q̃ utilis. Et q autores attigerunt de Geographia. ¶ Quod Homerus passim suo in carmine orbis descriptione sit usus. cui et Oceanus sit cognitus: et mare mediterraneum.
Alios post Homerũ autores fuisse de situ orbis scribentes. et locorũ cognitio utilis et priuatim et publice ad peritiam et prudentiam comparandã. et que ad illa dicere sint necessaria. ¶ Autorẽ ipsum nõ esse accusandũ si de iis scribat de qbus alii poteś: cum posterius multa ad noticiam peruenerint: que prioribus erant incognita.
Contra Eratosthenem: de poetis mala sentiẽtem. et laus magna poetice. precipue de poesi Homeri. q̃ uaria sit utilitas: quanta locorum noticia: et uentorum. Contra falso uel incomposite sentientes: uel inter se discordes probatiões et argumenta: qbus autor uerisimiliter indicat: q̃ multis in partibus terra et mare inuicem permutate atq; transposite sunt.

SI ad philosophum alia pertineat ulla tractatio: et hanc: quã hoc tẽpore de situ orbis delegimus: consyderandam illi esse putamus. Quod autẽ nostra minime uilipendẽda sit estimatio per multa declaratur. Nam qui eã primi attingere confisi fuerant: tales extitere. Homerus. Anaximãder Milesius. Hecateus cõciuis eius: ut inqt Eratosthenes. Democritus. Eudoxus. Dicearchus. Ephorus. Aliiq; cõplures. Post hos etiã Eratosthenes. Polybius. Posidõius uiri philosophi. Multa uero disciplina: per quam ad hoc opus peruenire põt: nullius alterius est: nisi eius qui diuina et humana ualeat intueri. Quo ꝝ scĩetiam: philosophiã esse dicunt. Eodẽ modo uaria est utilitas quedã ad res urbanas actionesq; principũ. Nõnulla ad celestium terrestrium rerumq; maritimarum sciẽtiam: aĩaliũ plãtaꝝ: fructuum: et aliorum: que penes uniuersos cernere fas est. Hec eũdẽ designat hoĩem: qui artes uite ac beatitudinem: cura et cogitatione complectitur. Ex iis autem que dicta sunt: unumquodq; assumentes: magis adhuc consyderemus. In primis quia et nos: et maiores nostri recte sumus arbitrati. Ex quibus hipparchus est: q Homerum huius peritie primarium autorẽ esse tradidit. Is non solũ uniuersos priores: ac posteriores uirtute poetica superauit. Sed etiam ipsa ferme rerum ciuiliũ: que ad uitam spectant experientia. Ex qua nõ modo singulare: ad res gerendas studiũ adhibuit: ut plurimas cognosceret: cognitasq; posteris traderet: uerum etiam et singulaꝝ partiũ locos: et qui per cunctum orbem terram uidelicet habitabilem ac mare sunt. Non enim ad extremos usq; ipsius terminos peruenisset memoria circunlustrans. Nec ipsã primum quidem ab Oceano allutam: sicut est demõstrasset. deinde pleraq; loca nominatim expressit. Nõnulla uero signis quibusdam occultius insinuauit. Africam et Ethiopiam. et Sydonios & Erembos: quos Arabes Troglodytas: hoc est cauernarum incolas: iure dixeris: uerbis eloquutus est. Qui uero ad ortus: et occasus pertinerent: obscurius significauit: qui alluantur Oceano. Ab hoc enim orientem solẽ: et in hunc occidentem facit: pariter et stellas.
Inde recens radiis iam sol percusserat arua.
Oceani lentas alti prefugerat undas.
Incidit Oceano lux fulgentissima solis:
Nigrantem noctem: et madidantia sydera ducens.
Ex Oceano uidelicet dicit. Eorũ etiam qui occasum incolunt: mortaliũ felicitatem

Fig. 81. Latin translation of the *Geography* of Strabōn by Guarino da Verona (Rome: Sweynheym and Pannartz, 1469). This is the beginning of the printed tradition of Strabōn. It is important because Guarino used better Greek manuscripts (now lost) than the editor of the princeps. [Courtesy of the Pierpont Morgan Library.]

ΣΤΡΑΒΩΝΟΣ ΓΕΩΓΡΑΦΙΚΩΝ

ΒΙΒΛΙΟΝ ΠΡΩΤΟΝ.

Τῆς τοῦ φιλοσόφου πραγματείας εἶναι νομίζομεν εἴπερ ἄλλην τινὰ, καὶ τὴν γεωγραφικὴν, ἣν νῦν προῃρήμεθα ἐπισκοπεῖν. ὅτι δ' οὐ φαύλως νομίζομεν, ἐκ πολλῶν δῆλον. οἵ τε γὰρ πρῶτοι θαρρήσαντες αὐτῆς ἅψασθαι, τοιοῦτοί τινες ὑπῆρξαν, ὅμηρός τε, καὶ ἀναξίμανδρος ὁ μιλήσιος, καὶ ἑκαταῖος ὁ πολίτης αὐτοῦ. καθὼς καὶ ἐρατοσθένης φησὶ, καὶ δημόκριτος δὲ, καὶ εὔδοξος, καὶ δικαίαρχος, καὶ ἔφορος, καὶ ἄλλοι πλείους. ἔτι δὲ οἱ μετὰ τούτους, ἐρατοσθένης τε καὶ πολύβιος, καὶ ποσειδώνιος, ἄνδρες φιλόσοφοι. ἥ τε πολυμάθεια, δι' ἧς μόνης ἐφικέσθαι τοῦδε τοῦ ἔργου δυνατὸν, οὐκ ἄλλου τινός ἐστιν, ἢ τοῦ τὰ θεῖα, καὶ τὰ ἀνθρώπινα ἐπιβλέποντος. ὧνπερ τὴν φιλοσοφίαν ἐπιστήμην φασίν. ὡς δ' αὔτως καὶ ἡ ὠφέλεια ποικίλη τις οὖσα, ἡ μὲν, πρὸς τὰ πολιτικὰ, καὶ τὰς ἡγεμονικὰς πράξεις, ἡ δὲ, πρὸς ἐπιστήμην τῶν τε οὐρανίων, καὶ τῶν ἐπὶ γῆς, καὶ θαλάττης ζώων, καὶ φυτῶν, καὶ καρπῶν, καὶ τῶν ἄλλων ὅσα ἰδεῖν παρ' ἑκάστοις ἐστὶ, τὸν αὐτὸν ὑπογράφει ἄνδρα, τὸν φροντίζοντα τῆς περὶ τὸν βίον τέχνης, καὶ εὐδαιμονίας. Ἀναλαβόντες δὲ καθ' ἕκαστον ἐπισκοπῶμεν τῶν εἰρημένων ἔτι μᾶλλον καὶ πρῶτον. ὅτι ὀρθῶς ὑπειλήφαμεν καὶ ἡμεῖς, καὶ οἱ πρὸ ἡμῶν, ὧν ἐστι καὶ ἵππαρχος, ἀρχηγέτην εἶναι τῆς γεωγραφικῆς ἐμπειρίας ὅμηρον, ὃς οὐ μόνον ἐν τῇ κατὰ τὴν ποίησιν ἀρετῇ πάντας ὑπερβέβληται τοὺς πάλαι, καὶ τοὺς ὕστερον, ἀλλὰ σχεδόν τι καὶ τῇ κατὰ τὸν βίον ἐμπειρίᾳ τὸν πολιτικόν. ἀφ' ἧς οὐ μόνον περὶ τὰς πράξεις ἐσπούδασεν ἐκεῖνος ὅπως ὅτι πλείστας γνοίη καὶ παραδώσῃ τοῖς ὕστερον ἐσομένοις, ἀλλὰ καὶ τὰ περὶ τοὺς τόπους τούς τε καθ' ἕκαστα, καὶ τοὺς κατὰ σύμπασαν τὴν οἰκουμένην γῆν τε, καὶ θάλατταν. οὐ γὰρ ἂν μέχρι τῶν ἐσχάτων αὐτῆς περάτων ἀφίκετο τῇ μνήμῃ κύκλῳ περιιὼν, καὶ πρῶτον μὲν τῷ ὠκεανῷ περίκλυστον ὥσπερ ἐστὶν ἀπέφηνεν αὐτήν. ἔπειτα δὲ τῶν χωρίων τὰ μὲν, ὠνόμαζε, τὰ δὲ, ὑπηνίττετο τεκμηρίοις τισὶ, λιβύην, καὶ αἰθιοπίαν, καὶ σιδονίους, καὶ ἐρεμβοὺς, οὓς εἰκὸς λέγειν τοὺς τρωγλοδύτας ἄραβας. ῥητῶς λέγων, τοὺς δὲ πρὸς ταῖς ἀνατολαῖς, καὶ δύσεσιν αἰνιττόμενος ἐκ τοῦ τῷ ὠκεανῷ κλύζεσθαι. ἐνθένδε γὰρ ἀνίσχοντα ποιεῖ τὸν ἥλιον, καὶ δυόμενον εἰς τοῦτον. ὡς δ' αὔτως καὶ τὰ ἄστρα.

» Ἠέλιος μὲν ἔπειτα νέον προσέβαλλεν ἀρούρας
» Ἐξ ἀκαλαρρείταο βαθυρρόου ὠκεανοῖο.
» Ἐν δ' ἔπεσ' ὠκεανῷ λαμπρὸν φάος ἠελίοιο,
» Ἕλκον νύκτα μέλαιναν. — Καὶ τοὺς ἀστέρας λελουμένους ἐξ ὠκεανοῦ λέγει. τῶν δ' ἑσπερίων ἀνδρῶν, καὶ τὴν εὐδαιμονίαν ἐμφανίζει, καὶ τὴν εὐκρασίαν τοῦ περιέχοντος. πεπεισμένος ὡς ἔοικε τὸν ἰβηρικὸν πλοῦτον. ἐφ' ὃν καὶ ἡρακλῆς ἐστράτευσε, καὶ οἱ φοίνικες ὕστερον. οἵπερ καὶ κατέσχον τὴν πλείστην ἀρχήν. μετὰ ταῦτα δὲ ῥωμαῖοι. ἐνταῦθα γὰρ αἱ τοῦ ζεφύρου πνοαί. ἐνταῦθα δὲ καὶ τὸ ἠλύσιον ποιεῖ πεδίον ὁ ποιητής. εἰς ὃ πεμφθήσεσθαι τὸν μενέλαόν φησιν ὑπὸ τῶν θεῶν.
» Ἀλλά σ' ἐς ἠλύσιον πεδίον καὶ πείρατα γαίης
» Ἀθάνατοι πέμψουσιν ὅθι ξανθὸς ῥαδάμανθυς.
» Τῇ περ ῥηΐστη βιοτὴ πέλει ἀνθρώποισιν
» Οὐ νιφετὸς οὔτ' ἂρ χειμὼν πολὺς οὔτέ ποτ' ὄμβρος
» Ἀλλ' αἰεὶ ζεφύροιο λιγὺ πνείοντας ἀήτας — Ὠκεανὸς ἀνίησι. καὶ αἱ τῶν μακάρων δὲ νῆσοι πρὸ τῆς μαυρουσίας εἰσὶ τῆς ἐσχάτης πρὸς δύσιν. καθ' ὃ μέρος συντρέχει καὶ τὸ τῆς ἰβηρίας τὸ ταύτῃ πέρας. ἐκ δὲ τοῦ ὀνόματος δῆλον ὅτι, καὶ ταύτας ἐνόμιζον εὐδαίμονας, διὰ τὸ πλησιάζειν τοιούτοις χωρίοις. ἀλλὰ μὴν ὅτι γε, καὶ οἱ αἰθίοπες ἔσχατοι ἐπὶ τῷ ὠκεανῷ δηλοῖ. ὅτι μὲν ἔσχατοι.
» Αἰθίοπες τοὶ διχθὰ δεδαίαται ἔσχατοι ἀνδρῶν.
Οὐδὲ τοῦ διχθὰ δεδαίαται φαύλως λεγομένου. ὡς δειχθήσεται ὕστερον. ὅτι δ' ἐπὶ τῷ ὠκεανῷ.
» Ζεὺς γὰρ ἐς ὠκεανὸν μετ' ἀμύμονας αἰθιοπῆας — Χθιζὸς ἔβη μετὰ δαῖτα. —
Ὅτι δὲ, καὶ ἡ πρὸς ταῖς ἄρκτοις ἐσχατιὰ παρωκεανῖτίς ἐστιν, οὕτως ᾐνίξατο εἰπὼν περὶ τῆς ἄρκτου.
» Οἴη δ' ἄμμορός ἐστι λοετρῶν ὠκεανοῖο. — Διὰ μὲν γὰρ τῆς ἄρκτου, καὶ τῆς ἁμάξης, τὸν ἀρκτικὸν

Strabo. a α

Fig. 82. Princeps of Strabōn's *Geography* (folio, 31 cm, 366 pp.; Venice: Aldus, 1516). First page of text showing how tightly it was printed; the top ornament, title, and first capital were printed in red. Aldus Manutius (1449–1515) was his own editor of Greek texts, the most active of his age, but he was sometimes helped by the Cretan Hellenist, Marco Musurus (1470–1517) or by others. [Courtesy of Harvard College Library.]

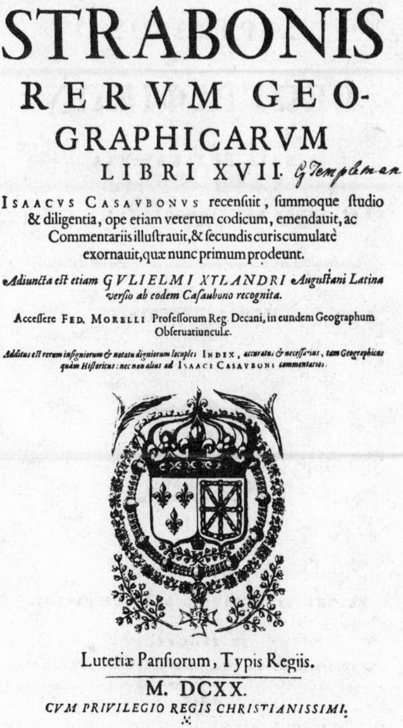
STRABONIS
RERVM GEO-
GRAPHICARVM
LIBRI XVII. G Templeman

ISAACVS CASAVBONVS recensuit, summoque studio & diligentia, ope etiam veterum codicum, emendauit, ac Commentariis illustrauit, & secundis curis cumulatè exornauit, quæ nunc primum prodeunt.

Adiuncta est etiam GVLIELMI XYLANDRI *Augustani Latina versio ab eodem Casaubono recognita.*

Accessere FED. MORELLI Professorum Reg. Decani, in eundem Geographum Obseruatiunculæ.

Additus est rerum insigniorum & notatu digniorum locuples INDEX, *accuratus & necessarius, tam Geographicus quàm Historicus: nec non alius ad* ISAACI CASAVBONI *commentarios.*

Lutetiæ Parisiorum, Typis Regiis.

M. DCXX.

CVM PRIVILEGIO REGIS CHRISTIANISSIMI.

Fig. 83. Greek-Latin edition of Strabōn's *Geography* by Isaac Casaubon (Paris, 1620). Casaubon had published a Greek-Latin edition before (Geneva, 1587), using the Latin translation of Guilielmus Xylander. This is a new edition, however, which is a landmark in Strabonian studies. Its pagination has been reproduced in many later editions; other editors have preferred to cite Strabōn according to the pagination of T. J. Van Almeloveen (Amsterdam, 1707). For example, the beginning of Book II is C 67–A 117. It is a very heavy folio (35 cm high, 8 cm thick without the covers) containing the Greek text with Xylander's Latin version in parallel columns (843 pp.), then a very elaborate index, finally Casaubon's commentary and corrections (282 pp.) with a separate index. [Courtesy of Harvard College Library.]

derived from Greek books, whether by him (he knew enough Greek for that purpose) or by a clerk. His main bookish sources were Eratosthenēs, Polybios, and Poseidōnios, but he was given considerable information on the spot from prisoners or free natives. The names of places and tribes were largely obtained from local informants; so numerous were they that, according to Cicero, every day brought news to him of names previously unknown.[31]

It is difficult for us to imagine Caesar's campaigns and expeditions being carried out without maps. We are so deeply map conscious that we can hardly understand mapless travels. Caesar and his lieutenants had some general information about a country, say Gaul, and got much more from local sources as they proceeded. Some information concerned tribes that even now cannot be placed exactly on any map, because the territory of each tribe was fluid; it might expand or contract according to political circumstances, and it always changed somewhat with the seasons.

Gaul had been visited by Polybios and Poseidōnios, but Caesar's conquest (58–50) increased Roman knowledge of it considerably. It was like the discovery of a new world, full of novelties. A part of it, the Provincia (Provence), was already colonized, but Caesar conquered all the land in-

[31] Cicero, *De provinciis consularibus in Senatu oratio* (chap. 13); dated 56 B.C.

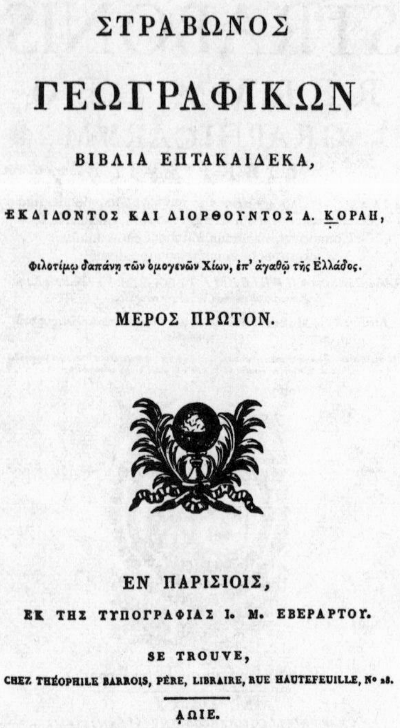

ΣΤΡΑΒΩΝΟΣ

ΓΕΩΓΡΑΦΙΚΩΝ

ΒΙΒΛΙΑ ΕΠΤΑΚΑΙΔΕΚΑ,

ΕΚΔΙΔΟΝΤΟΣ ΚΑΙ ΔΙΟΡΘΟΥΝΤΟΣ Α. ΚΟΡΑΗ,

Φιλοτίμῳ δαπάνῃ τῶν ὁμογενῶν Χίων, ἐπ' ἀγαθῷ τῆς Ἑλλάδος.

ΜΕΡΟΣ ΠΡΩΤΟΝ.

ΕΝ ΠΑΡΙΣΙΟΙΣ,

ΕΚ ΤΗΣ ΤΥΠΟΓΡΑΦΙΑΣ Ι. Μ. ΕΒΕΡΑΡΤΟΥ.

SE TROUVE,

CHEZ THÉOPHILE BARROIS, PÈRE, LIBRAIRE, RUE HAUTEFEUILLE, Nº 28.

ΑΩΙΕ.

Fig. 84. Title page of the first volume of the edition of Strabōn prepared by Adamantios Coraēs and published in four volumes (Paris, 1815–1819). A. Coraēs (Korais, Coray) of Smyrna was a Greek scholar and patriot (1748–1833) who lived in Paris from 1788; he was one of the spiritual founders of modern Greece. See Volume 1, 369. [Courtesy of Harvard College Library.]

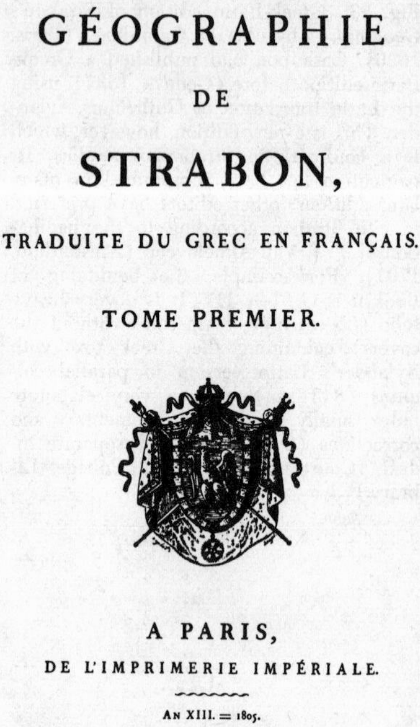

GÉOGRAPHIE

DE

STRABON,

TRADUITE DU GREC EN FRANÇAIS.

TOME PREMIER.

A PARIS,

DE L'IMPRIMERIE IMPÉRIALE.

AN XIII. = 1805.

Fig. 85. Title page of vol. 1 of the French translation of Strabōn made upon Napoleon's order and under his auspices by A. Coraēs, F. J. G. de Laporte du Theil, Jean Antoine Letronne, and P. F. J. Gossellin (5 vols., 29 cm; Paris, 1805–1819), with abundant commentaries and maps. Volumes 1–3 (1805–1812), were issued by the Imprimerie impériale; volumes 4–5 (1814–1819), by the same press, then called Imprimerie royale. [Courtesy of Harvard College Library.]

habited by Galli or Celtae. Under Augustus the whole of Gaul was divided into four provinces, the first of which was the old Provincia, called Gallia Narbonensis (around Narbonne), then the Tres Galliae added by Caesar, to wit, G. Aquitanica from the Pyrénées to the Loire; G. Lugdunensis, between the Loire, the Seine, and the Saône, and around Lyons; and G. Belgica, above the Seine and between the Saône and the Rhine. These "three Gauls" represented the three main nations conquered by Caesar, the Aquitani in the south, the Celts or Gauls in the center, and the Belgae in the north. Caesar was well acquainted with the main rivers which we have already named, plus the Garonne and the Marne, with the mountain chains of the Cévennes in the south and the Jura and the Vosges in the east, and with the forest of the Ardennes in Gallia Belgica. He was acquainted with an abundance of details. Many of the names of places and tribes with which we are familiar in their modern forms first appeared in his *Commentaries*.

Caesar also provided what would be called today ethnographic information, brief accounts of the manners and customs of the "natives."

He invaded Britain twice, in 55 and 54, and made two raids into Germany in 55 and 53. He described the triangular shape of Britain, made a good estimate of its size, and referred to the island of Hibernia or Iernē (Ireland), half of Britain's size and to the west of it; he was the first to notice the Isle of Man.[32] As to Germany his knowledge was much vaguer, as we have already remarked about the Hercynian Forest; he knew short reaches of the Rhine and very little of the Danube.[33]

In brief we find in the *Commentarii* a large number of new geographic and ethnographic names, but we cannot expect to obtain geographic precision, for he was not interested in it and made no effort to obtain it.

Caesar was far less conscious of the need of geographic intelligence than Alexander had been, but the territory that he set out to explore, to conquer, and to colonize was much smaller and less mysterious.

He was the first Roman general to cross the Rhine, and the second was Drusus,[34] whom Augustus in 13 appointed legate of the Gauls; Drusus organized the census of 12 and dedicated an altar to Rome and Augustus in Lyons. In that same year, Augustus ordered him to invade Germany, and this was done from the northern part of Belgica (Batavia, Holland). Drusus' main base on the Rhine was Vetera[35] and later Mainz. His Germanic campaign was continued until 9 B.C., when he reached the Elbe and died; he was buried in the mausoleum of Augustus. In order to facilitate the transportation of supplies, he dug a canal (*fossa Drusiana*) connecting the Rhine with the Zuyder Zee and the ocean; this enabled him to subdue the Frisians but does not seem to have given good results afterwards.[36]

[32] The island which Caesar called Mona and of which he said that it lay halfway between Britannia and Hibernia must be assumed to be Man, not Anglesey. Pliny called it Monapia.

[33] The Greeks were well acquainted with the lower course of the Danube but not with the upper course. During his campaign in Pannonia in 35 B.C., Octavianus (the future Augustus) was first to realize that the Danuvius of south Germany and the Ister of the Balkans were parts of the same river. In 15 B.C., Tiberius visited its sources. It was only then that the whole river was known.

[34] Nero Claudius Drusus (38–9 B.C.), Augustus' stepson, was already quaestor in 18. His older brother, Tiberius (42 B.C.-A.D. 37), succeeded Augustus as emperor (14–37).

[35] Vetera (Castra vetera) on the Lower Rhine, near the modern town of Xanten. It is curious to think that this earliest permanent Roman camp on the Rhine (a legion was stationed there until the end of the empire) is also the place where the Castle of the Niebelungen was located and where Siegfried, the dragon slayer, was born.

[36] Alfred Klotz, *Cäsarstudien nebst*

In 44, when Caesar was consul with Marc Antony, he ordered a general survey of the Roman "empire." His murder, on 15 March 44, prevented him from implementing that project. According to medieval traditions, Caesar had really begun the undertaking. In the *Cosmographia* of Aethicus Ister (VII-2),[37] it is stated that Caesar ordered the survey when he was consul and that Zenodoxus measured the eastern countries in 21½ years, Theodotus the northern ones in 30 years, Polyclitus the southern ones in 32 years. The survey was thus completed in 32 years and submitted to the Roman Senate in 12 B.C. Around the World Map of Richard de Haldingham [38] is a legend stating that the survey was begun by Caesar, who entrusted the work to one Nicodoxus for the east, to Theodoxus for the north and west, and to Policlitus for the south. The three names are very close to those mentioned by Aethicus and must represent the same men. Judging from those names, Caesar's three assistants were Greek.

M. V. Agrippa (63–12 B.C.). The conquest of Germany led us from Caesar to Drusus, and Drusus was simply a lieutenant of Augustus. The same is true of Agrippa, and this section might be entitled "Augustus" as the former was entitled "Caesar." There is an enormous difference between these two men, however. Caesar's expeditions were made by himself and the *Commentarii Caesaris* were his own memoirs. On the other hand, Augustus was simply the lucky man to whom it was given by Fortuna to be the first emperor; he deserved his supreme office and was equal to it, but this obliged him to administer the empire and leave the creative duties and joys to others.

Agrippa's achievements as architect and engineer have already been described. It was his privilege to complete another task begun by Caesar, the survey of the empire. This implied real geographic investigation, such as the measurement of roads. Those roads were built primarily for military purposes but were equally used for trade and travel. The "mapping," or at least the measurement, of those roads served the same necessities of war and peace. This had been begun before Augustus and Agrippa. According to Polybios, the road from the Spanish frontier to the Rhône had already been paced, and the distances along it marked with milestones. From the time of Polybios to that of Augustus many more roads had been built, paced, and marked in this way. The time had come to survey the whole network, and this was the task that the emperor entrusted to Agrippa.

The result of his survey was a map of the world (that is, the Roman empire and some neighboring countries), which was set up by the order of

---

einer Analyse der Strabonischen Beschreibung von Gallien und Britannien (267 pp.; Leipzig, 1910).

[37] Louis Baudet, *Cosmographie d'Ethicus* (Paris, 1843), p. 8.

[38] That is the map preserved in Hereford Cathedral, painted c. 1283; *Introduction*, vol. 2, p. 1050. See the new reproduction of it with memoir by G. R. Crone published by the Royal Geographical Society (London, 1954). This map, or its prototype, was made to illustrate Orosius (V-1).

Augustus on a wall of the Porticus Octaviae in Rome. The map was designed by Agrippa but was not completed at the time of his death; it was explained in a text stating the distances of places and the sizes of regions.

This achievement gave a new impulse to the compilation of itineraries for military and civil purposes. We have given an example in the section above devoted to Isidōros of Charax. It is possible that his *Parthian stations* was a by-product of Agrippa's survey. We may imagine that every governor, conscious of his responsibilities, would order similar itineraries for his own territory, for it would have been difficult, if not impossible, for him to govern otherwise.

Two kinds of itineraries were gradually developed, the *Itineraria adnotata*, describing the roads and countries in terms of words, giving lists of stations and of distances between them, and the *Itineraria picta*, containing maps and other illustrations. Such documents, being necessary tools, were probably compiled before the Augustan age, but they became more numerous from that time on. Nevertheless, very few have survived; their disappearance is an unavoidable consequence of the hard use to which they were put, for they were made for travelers rather than for scholars. The earliest specimen of the first kind is the *Antonine itinerary* dating from the third century. The earliest specimen of an *Itineraria picta* is the *Tabula Peutingeriana* of the same century.[39] In his treatise on the art of war Vegetius (IV–2) sets forth the military need of such itineraries of both kinds, taking their existence for granted; in Vegetius' time they had been available for at least four centuries. There were also directions for sailors, going back to the Alexandrian age (*periploi, stadiasmoi*) and these early exemplars were copied and gradually increased during Byzantine times.[40] The Latin itineraries derived mostly from Agrippa's survey and from independent Greek sources.

King Juba II (d. c. A.D. 20). The influence of Rome, and indirectly of Greece, is splendidly illustrated in the case of a Numidian king, Juba I, who, having taken sides with Pompey, was defeated by Caesar and committed suicide in his capital, Zama,[41] in 46. His son, Juba II (I–2 B.C.), who was a child at that time, graced Caesar's triumph in the same year, was brought up in Rome and educated in the best manner by Greek tutors, and became a very distinguished scholar and a Roman citizen. Augustus, trusting his loyalty, permitted him to return to Numidia and made him king of Mauretania in

[39] For more details, see *Introduction*, vol. 1, p. 323.

[40] For the Byzantine tradition, see Armand Delatte, *Les portulans grecs* (Liége: Faculté de philosophie et lettres, 1947) [*Isis 40*, 71–72 (1949)]. The word *portulan* in Delatte's title refers to the medieval *portolani, Introduction*, vol. 1, p. 167. Similar traditions existed in every civilized country, for example, in China (*Introduction*, vol. 1, pp. 324, 536) and the Muslim countries (*ibid.*, p. 606). The Arabic and Chinese itineraries were independent traditions, caused by the same administrative needs.

[41] Zama Regia, Jama in Tunisia, southwest of Carthage (*Oxford Classical Dictionary*, p. 964).

25 B.C.[42] As a result of his Greek education, he probably wanted to have closer contacts with the Greek world, and he married successively two Greek princesses: first, Cleopatra Selēnē, Mark Antony's daughter by the great Cleopatra; then, Glaphyra, daughter of Archelaos, king of Cappadocia.[43] He tried his best to introduce Greek and Roman culture into his kingdom. He wrote many books in Greek,[44] dealing with Roman history, Libya, Arabia, Assyria; he compared Greek and Latin antiquities, and described the plant *euphorbia* (an African genus), which he thus named in honor of his doctor Euphorbos. His writings are lost but are known to us through Pliny (I–2) and Plutarch (I–2).

This very versatile Numidian-Greek is especially interesting to us because of his geographic curiosity. He made investigations concerning the Fortunate Islands (Canaries), which he believed to be five in number.[45] He knew the river Niger and originated the theory that the Nile had its source in a mountain of western Mauretania, not far from the ocean.[46] Perhaps he had been led astray by Hērodotos (II, 32–33)? At any rate, one cannot blame him for errors that were not corrected until the last century. They were incurable indeed except by long navigations and mathematical cartography.

To speak only of Juba's contemporaries, Lucretius knew that the Nile took its source in southern tropics (*De rerum natura*, VI, 721), and Vitruvius confused the Niger with the Nile. This suggests that other geographic curiosities might be found in Greek and Latin literature, but that would carry us too far, and we have said enough to give an idea of geographic knowledge before the Christian era.

Hyginus (d. c. A.D. 10). This Roman polygraph, who was manumitted by Augustus and put in charge of the Palatine Library, devoted one of his many (lost) publications to the geography of Italy (*De situ urbium Italicarum*). He was in this one of the forerunners of Petrarca and of many Renaissance humanists. That is, he was one of the first, after Polybios and

[42] We might say, roughly, that Numidia corresponded to modern western Tunisia and eastern Algeria, Mauretania to western Algeria and Morocco. Juba I had been king of Numidia; Juba II was made king of Mauretania, another country. That was Roman prudence.

[43] He met both princesses in Rome. Cleopatra Selēnē was taken to Rome after Antony's death in 30 B.C. Archelaos had been made king of Cappadocia by Antony's favor but, in the course of time, was accused of treason, taken to Rome, and obliged to remain there. He died in Rome in A.D. 17.

[44] Hence, it might have been better to speak of Juba together with the Greeks in the first section. His situation was paradoxical because he was definitely a Western man, completely educated in Rome. He illustrates the deep Hellenization of the Latin capital of the world.

[45] Some of the names and other details quoted by Pliny, *Historia naturalis*, VI, 203–205, can be identified with present conditions, for example, the name Canaria. These islands were probably known to the Carthaginians and Juba II was probably inspired in his own research by local traditions.

[46] That theory and others making the Niger a branch of the Nile and considering that fantastic complex Niger-Nile as the African image of the European Danube were very difficult to eradicate; *Introduction*, vol. 3, pp. 1158, 1772.

Strabōn, and probably the first in Latin, to turn geography into the direction of historical geography, the identification of places mentioned by historians and poets with those existing in his own time. For many humanists of antiquity as well as of the Renaissance, a place had no meaning except in its relation to men, not plain men, but statesmen, soldiers, and, above all, philosophers, poets, artists, or mythological heroes.

XXIV

KNOWLEDGE OF THE PAST IN THE LAST TWO CENTURIES[1]

THE GREEK HISTORIANS

Polybios. The greatest historian of the second century was undoubtedly Polybios (II–1 B.C.). Indeed, we may go further than that and say that he was one of the greatest of antiquity, ranking immediately after Hērodotos and Thucydidēs, both of whom had flourished three centuries before him. He is important in himself but also as the symbol of a new age, the first age of Western universalism, the golden age of the Republic. It is paradoxical that the mission and the glory of Rome were first proclaimed by a Greek, and not in Latin but in his own language.

Polybios was born c. 207 in Megalopolis in Arcadia; that is, he was as Greek as one could be; Arcadia is a relatively large region occupying the central part of the Peloponnēsos and separated from the other regions by mountain ranges. Its people considered themselves the oldest and the original Greeks; they were mostly farmers and shepherds. Their main occupation was the breeding and education of cattle; their main sport, hunting; their main gods, Pan and Artemis; their favorite art, music.[2] The Arcadians were able to defend their independence longer than other Greeks, and repeatedly defeated their most dangerous neighbors, the Lacedemonians (Spartans). The hopes of the latter were frustrated by the two Theban heroes, Pelopidas and Epameinōndas, the first of whom drove the Spartans out of Thebes in 379, and the second defeated them at Leuctra in 371. It was upon Epameinōndas' advice that the Arcadians built themselves a new stronghold and capital which they called Megalopolis (the great city). They later joined the Achaian League, shared its vicissitudes, and were finally conquered by Rome.

We must now come back to Polybios, but it was well to see him in his background. The wars with Sparta and with Rome were awful realities to

[1] This continues the story told in Chapter XII, Knowledge of the Past in the Third Century B.C.

[2] In the course of time, the Arcadians were considered the best representatives of the pastoral virtues; that reputation was consecrated by Virgil. Remember *Arcades ambo* (both Arcadians, both skilled in pastoral music) in *Eclogae*, VII, 4.

him. He had deep in his heart remembrances of one of the greatest national heroes, Philopoimēn;[3] his father, Lycortas, was a friend of Philopoimēn and his successor as head of the Achaian League; he defeated the Messenians in 182 and forced the Spartans to enter the League. With such a father as Lycortas we may feel sure that Polybios received the best education that could be obtained as well as the best examples. As to the Romans, the Macedonian wars[4] had made of them familiar enemies. The third war ended with the victory of Pydna,[5] won in 168 by Aemilius Paulus (Macedonicus) over Perseus, king of Macedonia. Perseus adorned Paulus's triumph in Rome and, what was far more important, his Greek library was part of Paulus's spoils, who used it for the education of his elder sons, Q. Fabius and P. Scipio Aemilianus Africanus.[6] A thousand hostages were taken to Rome, Polybios, then forty years of age, being one of them. On account of the distinction of his family and his own merit, he was given hospitality in the victor's family; he was a privileged guest of Scipio Aemilianus, who was the founder and leader of the "Scipionic circle,"[7] a group of highly educated Romans, great admirers of Greek letters and promoters of Latin letters. The two Stoics, our Polybios and Panaitios, were outstanding members of it. Among the Latin members were Gaius Lucilius (180–102) the satirist, Terentius (195–159) the playwright, and Cicero. The importance of that circle in the Hellenization of Rome and the development of Latin philosophy and literature, Roman culture, can hardly be exaggerated. Think of Polybios' privilege in being thus placed in the very center of Roman intelligence. He remained 18 years in Rome (168–150, aet. 40–58) during which he had the opportunity to meet all the leading men of thought, whether Greek or Roman, who resided in the city or visited it. For example, in 155, he had the occasion of meeting the members of the Athenian embassy, Carneadēs of the Academy, Diogenēs the Babylonian, and Critolaos the Peripatetic. In 150, he obtained permission to leave, but by that time

[3] Philopoimēn (253–183) realized the need of strength to defend Arcadian independence, created its means of defense, was a good general, and in 208 became head (*stratēgos*) of the Achaian League. He was made prisoner by the Messenians in 183 and executed by them.

[4] There were four such wars: 215–205, 200–196, 171–168, 149–148. In 148, Macedonia became a Roman province; in 146 the consul, Lucius Mummius (Achaicus), dismembered the Achaian League, destroyed Corinth utterly, and shipped its treasures to Rome.

[5] Pydna is very close to the northwest shore of the Thermaic Gulf (between Macedonia and the three-pronged Chalcidicē).

[6] This was the second Africanus in the illustrious Scipio family. The first was Scipio Africanus major (236–184), who defeated Hannibal at the battle of Zama in 202. Titles like Africanus, Asiaticus, Achaicus, Macedonicus were given to Roman generals in order to celebrate their victories. Compare with the Napoleonic titles of the dukes of Austerlitz or Eckmühl, or English titles such as Nelson of the Nile, Allenby of Megiddo, Montgomery of Alamein.

[7] There were two leaders, Scipio Aemilianus and Gaius Laelius, whose friendship is immortalized in Cicero's *De amicitia*. In spite of their being professional soldiers in wartime, these two men were singularly well educated and richminded.

he had long ceased to be an exile and had become more Roman than most Romans. He did leave Rome, however, and traveled abroad but came back many times to reside with his patron, Scipio Aemilianus, or to accompany him in his campaigns; he was with Scipio in 146 when Carthage was taken and razed. Corinth was destroyed by Mummius in the same year and Polybios was called to help in the reorganization of Greece (*Historiai*, XXXIX, 13 f.). He did his task in Greece (146–145), then completed it in Rome. According to his own statement (XXXIX, 19):

> Having accomplished these objects I returned home[8] from Rome, having put, as it were, the finishing stroke to my whole previous political actions, and obtained a worthy return for my constant loyalty to the Romans. Wherefore I make my prayers to all the gods that the rest of my life may continue in the same course and in the same prosperity; for I see only too well that Fortune is envious of mortals, and is most apt to show her power in those points in which a man fancies that he is most blest and most successful in life.[9]

It is not known where he spent the end of his life. At the age of 82, he had a fall from his horse and died from the effects of it (c. 125 B.C.).

He wrote various books but is immortalized by a single one, which was written during the period 168–140. It was a universal history (*Historiai*) describing the Roman conquest of a great part of the world within a little more than half a century (220–168) and its further Romanization from 168 to 146 when Greece and Carthage were conquered. It extended to 40 books, of which Books I–V are extant, while fragments of the rest (Books VI–XL) are preserved in the writings of Livy (I–2 B.C.), Diodōros (I–2 B.C.), Plutarch (I–2), and Appian (II–2). Books I and II are an introduction (*procatasceuē*) narrating events from the time when Timaios left off, 264, through the First Punic War (264–241) and the Achaian League. Books III to XXX described the Roman conquests to the battle of Pydna, 168 (Polybios had experienced much of this in Macedonia). Books XXXI to XXXIX told the story from 168 to 146; Book XL was probably a recapitulation and chronologic summary by Olympiads[10] of the whole work.

The details do not concern us; it suffices to say that Polybios' history covers the "world" as known to him from 264 to 146 B.C., 118 momentous years. His purpose was severely technical, to teach practical politics to statesmen and civil servants. His experience was as full as could be, for he

[8] Does this mean Megalopolis, Arcadia or Greece?

[9] Polybios, *Historiai*, trans. Evelyn S. Shuckburgh (2 vols.; London, 1889), vol. 2, p. 540.

[10] 264 B.C. = OL. 129.1; 168 B.C. = OL. 153.1; 146 B.C. = OL. 158.3. Polybios used Olympiads because that was the best chronologic scheme in his time. The era *ab urbe condita* could not yet be used. Varro (I–2 B.C.) was the first to establish the era corresponding to A.U.C. 1 = 753 B.C. The Roman practice of giving to each year the name of its two consuls in charge was extremely cumbersome and as unscientific as possible; it made almost impossible the estimation of intervals, say the interval between 264 and 168 B.C. or, in Polybios' style, between OL. 129.1 and OL. 153.1.

had passed the formative years of his life and even more (40 years) in Greece, where he had witnessed the results of political chaos, then the next 40 years in Rome or traveling but always coming back to Rome. He traveled considerably in Greece, Italy, Egypt, Sicily, Mauretania, Spain, Gaul, perhaps Britain; hence, he was well acquainted with the regions and the places. He was well aware of the necessity of describing the physical background of military or administrative endeavors and was well equipped to describe it correctly. He had read every pertinent book available in Greek and Latin, and many public or private documents had passed through his hands. Finally (and best of all), he had been in personal touch with some of the leading men of Greece, and later in the Scipionic circle with the leading men of Rome and of the whole world. He knew the realities of war and peace, the problems of strategy, tactics, diplomacy, the business of political negotiations. He was remarkably impartial; he was a Greek and tried to save his country as long and as much as it could be saved, yet he recognized its weaknesses (as fully as an insider could do it); on the other hand, the merits of Roman discipline and unity were clear to him. He realized that the national religion of Rome was the consecration and best defense of its unity, and that the rulers used religious institutions to coerce the populace (VI, 56). If the Greeks forfeited the right not only to govern the world but also to protect their own independence, there was no way out of political chaos but to trust Roman leadership.

His general views were explained in some parts interrupting the chronologic narrative. For example, in Book VI he discussed the Roman constitution, in XII historical theories, in XXXIV Mediterranean geography.

In the Scipionic circle he had met Panaitios and other Stoics; he had probably come across some of them before leaving Greece; his own philosophy, politics, and religion were Stoic. He tried to explain the vicissitudes of life, to give the causes of events, but recognized that many events, some of them crucially important, are due to chance or fortune.[11] Those events could not be analyzed, but some others could be and it was useful to do so. For example, one could point out the qualities or defects, and chiefly the will power, of definite individuals, the virtues and vices, the constitution and administration of separate nations. He even tried to explain the whole evolution (*anacyclōsis*). In this he was inspired by the Stoic belief in periods, which might be repeated or not.[12]

He was a scientific historian like Thucydidēs, very inferior to him in strength of thought and purity of language, but superior in one respect.

[11] *Eimarmenē* or *peprōmenē*, that which is allotted by Fate, Fate being represented by the goddesses Moirai (Latin Moirae or Parcae), or by the goddess Tychē (Fortuna).

[12] The concepts of the periodic nature of history, eternal return, palingenesis or regeneration were not original with the Stoics. They were Oriental concepts accepted in one form or another by the Pythagoreans, Thucydidēs, Plato, Aristotle, and finally by the Stoics; Volume 1, pp. 321, 475, 515, 602.

He did not introduce speeches in his narrative as Thucydidēs did before him and Livy after him, because he realized the impossibility of doing that exactly.

He was a scholar rather than a man of letters, we might even say a man of science sharing the latter's confidence that truth (if we can but find it) will prevail. His style was criticized early by men like Dionysios of Halicarnassos (I–2 B.C.), who said that Polybios was one of those authors whom nobody can read through. Now, Polybios' work was in his own opinion a treatise or study (*pragmateia*) of practical politics; Dionysios was unable to understand the difficulties and scruples of a scientific mind and the vanity of literary ornaments in a scientific treatise.

Polybios had been very well educated and knew his language as well as any other Greek of his time. That language was no longer the Attic language of the fourth century, however, but the common form (*coinē*) used by gentlemen all over the Greek world from the third century on. He tried to say what he had to say as clearly as possible and managed to do so. He was not trying to amuse his readers or to astonish them with literary conceits, but to teach them.

It is probable that the whole Greek text of the *History* was available to the scholars who worked for Constantinos VII Porphyrogennētos (X–2); a great many Greek manuscripts were lost, however, when the Crusaders sacked Constantinople in 1204. The earliest extant manuscript, Vaticanus 124, dates from the eleventh century and contains only Books I–V. A Latin translation was begun by Leonardo Bruni of Arezzo (1369–1444), but the revival of interest in Polybios was largely due to Nicholas V (pope 1447–1455), founder of the Vatican library, who encouraged Niccolò Perotti of Sassoferrato (1430–1480) to make the new translation of Books I–V printed by Sweynheym and Pannartz (Rome, 1472) (Fig. 86). The Greek princeps of the same books was edited by Vincentius Obsopoeus (Hagenau, 1530) (Fig. 87).

The bibliography of Polybios is very complicated because parts of the lost books, VI to XL, were discovered gradually, hence there are many "first editions" of this or that in Greek and translations. The latest (complete) editions of the Greek text were prepared by Friedrich Dübner, with Latin version and good index (2 vols.; Paris: Firmin Didot, 1839), by Friedrich Hultsch (4 vols.; Berlin: Weidmann, 1866–1872), by Theodor Büttner-Wobst (4 vols.; Leipzig: Teubner, 1867–1889), revised by Ludwig Dindorf (5 vols. in 4; Teubner, 1882–1904), revised a second time by Büttner-Wobst (5 vols.; Teubner, 1889–1904).

French translation of Books I–V by Louis Maigret (Paris, 1552) and of the whole by Pierre Waltz (4 vols.; Paris, 1921).

English translation of Books I–V by Christopher Watson (London, 1568) and (first) of the whole by Evelyn S. Shuckburgh (2 vols.; London: Macmillan, 1889). Greek-English edition by W. R. Paton (Loeb Classical Library, 6 vols.; Cambridge: Harvard University Press, 1922–1927).

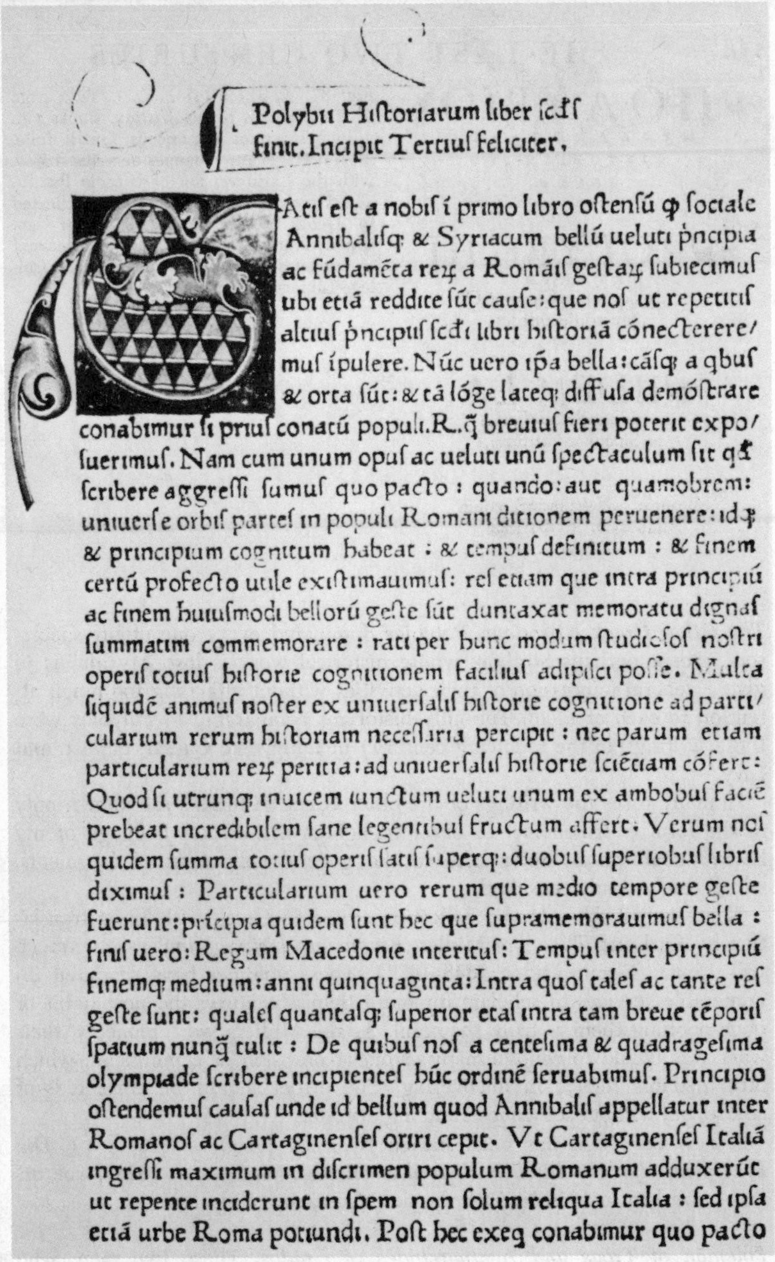
Polybii Historiarum liber scd's
finit. Incipit Tertius feliciter.

Satis est a nobis i primo libro ostensū ꝙ sociale
Annibalisq; & Syriacum bellū ueluti pncipia
ac fūdamēta rerꝫ a Romāis gestarꝫ subiecimus
ubi etiā reddite sūt cause: que nos ut repetitis
altius pncipiis scd'i libri historiā cōnecterere/
mus ipulere. Nūc uero ipa bella: cāsq; a qbus
& orta sūt: & tā lōge lateq; diffusa demōstrare
conabimur si prius conatū populi.R.q̄ breuius fieri poterit expo/
suerimus. Nam cum unum opus ac ueluti unū spectaculum sit qd
scribere aggressi sumus quo pacto: quando: aut quamobrem:
uniuerse orbis partes in populi Romani ditionem peruenere: idq;
& principium cognitum habeat: & tempus definitum: & finem
certū profecto utile existimauimus: res etiam que intra principiū
ac finem huiusmodi bellorū geste sūt duntaxat memoratu dignas
summatim commemorare: rati per hunc modum studiosos nostri
operis totius historie cognitionem facilius adipisci posse. Multa
siquidē animus noster ex uniuersalis historie cognitione ad parti/
cularium rerum historiam necessaria percipit: nec parum etiam
particularium rerꝫ peritia: ad uniuersalis historie sciētiam cōfert:
Quod si utrunq; inuicem iunctum ueluti unum ex ambobus faciē
prebeat incredibilem sane legentibus fructum affert. Verum nos
quidem summa totius operis satis superq;: duobus superiobus libris
diximus: Particularium uero rerum que medio tempore geste
fuerunt: pricipia quidem sunt hec que supramemorauimus bella:
finis uero: Regum Macedonie interitus: Tempus inter principiū
finemq; medium: anni quinquaginta: Intra quos tales ac tante res
geste sunt: quales quantasq; superior etas intra tam breue tēporis
spatium nunq̄ tulit: De quibus nos a centesima & quadragesima
olympiade scribere incipientes hūc ordinē seruabimus. Principio
ostendemus causas unde id bellum quod Annibalis appellatur inter
Romanos ac Cartaginenses oriri cepit. Vt Cartaginenses Italiā
ingressi maximum in discrimen populum Romanum adduxerūt
ut repente inciderunt in spem non solum reliqua Italia: sed ipsa
etiā urbe Roma potiundi. Post hec exeq conabimur quo pacto

Fig. 86. Polybios (II–1 B.C.), *History of Greece and Rome.* The earliest printed edition was the Latin translation of Books I to V by Niccolò Perotti (folio; Rome: Sweynheym and Pannartz, 31 Dec 1473 meaning 1472). This translation was dedicated to Nicholas V (pope 1447–1455) who patronized it. [Courtesy of Pierpont Morgan Library.]

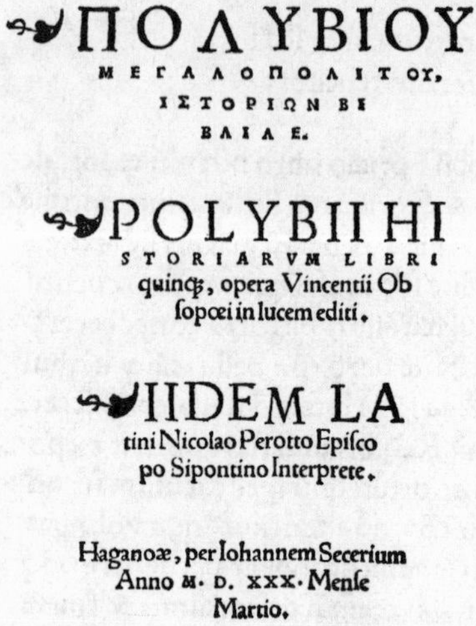
ΠΟΛΥΒΙΟΥ
ΜΕΓΑΛΟΠΟΛΙΤΟΥ,
ΙΣΤΟΡΙΩΝ ΒΙ
ΒΛΙΑ Ε.

POLYBII HI
STORIARVM LIBRI
quinque, opera Vincentii Ob
sopoei in lucem editi.

IIDEM LA
tini Nicolao Perotto Episco
po Sipontino Interprete.

Haganoæ, per Iohannem Secerium
Anno M. D. XXX. Mense
Martio.

Fig. 87. Polybios (II–1 B.C.). Title page of the princeps of his *History*, Books I to V by Vincentius Obsopoeus (small folio, 27 cm; Hagenau: Johannes Secerius, 1530), with the Latin version of Niccolò Perotti. The Greek text (106 leaves) is dedicated to Georg der Fromme, Markgraf von Brandenburg; the Latin text (142 leaves) to Nicholas V. [Courtesy of Harvard College Library.]

The Other Greek Historians. Polybios dominated every one of his successors, except perhaps Strabōn, whose historical work is lost. My aim is to give a general impression of their activities without attaching too much attention to each of them. The only historians comparable to Polybios were those writing (in the following century) in Latin, like Caesar, Sallust, and Livy.

In many cases the writings of the other Greek historians are known only in the form of fragments. In order to lighten the critical baggage of my book I shall simply refer here to the general collections of fragments, which can be easily consulted.

There is first the splendid collection edited in Greek with Latin translation by Karl and Theodore Müller, *Fragmenta historicorum graecorum* (5 vols.; Paris: Firmin Didot, 1848–1872). These volumes have rendered innumerable services to scholars for more than a century; the new habit of pooh-poohing them is truly shameful. As the Müllers were pioneers, their work was bound to contain many errors of omission or commission, which little pedants rejoice in uncovering. The errors should be corrected, of course, but without self-conceit and ingratitude.

A new collection has been started by Felix Jacoby (1876–), *Die Fragmente der griechischen Historiker* (Berlin: Weidmann, 1923); vol. 3B was published in 1950 by Brill, Leiden; Greek text only.

Polemōn of Trōas and Agatharchidēs of Cnidos. These two men, who flourished in the first half of the second century B.C., were primarily geographers, but, as they were interested in archaeology, they might be called historians. This is especially true of Polemōn Periēgētēs, who copied Greek inscriptions, and was perhaps the first epigraphist.[13] See my account of them in Chapter XXIII.

Apollodōros of Athens. Apollodōros (II–2 B.C.) spent part of his life in Alexandria and another in Pergamon. He was probably the disciple in Alexandria of the famous philologist, Aristarchos of Samothracē (II–1 B.C.); about the middle of the century he moved to Pergamon and dedicated to Attalos II Philadelphos (king 159 to 138) a Greek chronology in verse (*Chronica*), from the fall of Troy to 144 (later extended to 119). This work was partly derived from Eratosthenēs. He was a philologist and a mythographer as well as a historian and wrote commentaries on early poets, Epicharmos of Cōs (540–450) and Sophrōn of Syracuse (fl. 460–420), inventor of a form of comedy (*mimos*, mime), but chiefly on Homer, for example, on the latter's catalogue of ships. His most ambitious work was an elaborate history of the gods (*Peri theōn*) in 24 books, a kind of encyclopedia of Greek mythology. Such a work was becoming more necessary than ever before, because the literate people did not know the stories of the gods as well as their fathers did, and, what was worse, found it harder and harder to believe them. Apollodōros was a Stoic and he tried to interpret myths in rational terms.

This work should not be confused with another, written in a very different spirit, less rational, more mythological in the conventional manner, by another Apollodōros, who was also an Athenian or at any rate was called Apollodōros of Athens.[14] This second work, entitled *Apollodōru bibliothēcē* (*Apollodoros' library*) is a later production, almost certainly post-Christian. It dates from the first three centuries after Christ, perhaps from the time of Hadrian (emperor 117–138); it may be even later, of the time of Alexander Severus (emperor 222–235). It is impossible to date it from context, because the most recent events referred to are the death of Odysseus and the return of the Hēracleidai (prehistoric undatable events). *Apollodōros' library* should not be dealt with at all in this volume, except to stop confusion with the treatise on the gods of the older Apollodōros. It was practically unknown in antiquity and the first scholar to refer to it was Phōtios (IX–2) in his own *Library*. It was first edited (in Greek and Latin) by Benedict Aegius (Rome: Ant. Bladus, 1555); it appealed to Renaissance imaginations and was often reprinted. It can be easily consulted in the Greek-English edition prepared for the Loeb Classical Library by James George Frazer (2 vols., 1921).[15]

Poseidōnios. Poseidōnios (II–1 B.C.) began in 74 the redaction of a universal history which was to be a continuation of Polybios; it covers the period extending from 144 to 82. His account contained many details, but,

[13] As far as Romans were concerned, the attraction of Trōas was nothing then, as compared with what it would become after the publication of Virgil's *Aeneid*.

[14] The name of Apollodōros must have been fairly common in Athens.

[15] Sir James's edition is an excellent tool for the student of ancient mythology,

as far as it is fair to judge from the fragments, it was superficial rather than deep. Some of the details were picturesque and unexpected. For example, he originated the tradition that the Celtic hierarchy was divided into three orders, Bards, Prophets, and Druids. He tried to explain the contemplated alliance between Athens and Mithridatēs against Rome. His most durable work for good and evil was done in the field of geography.

He was a popular lecturer and a successful teacher (Pompey as well as Cicero sat at his feet); his fame as a man of science and as the leader of the Stoics in Rhodos gave him a prestige and authority that he did not really deserve. His admirers considered him the most profound philosopher of his time and went so far as to call him a new Aristotle.[16] He was apparently as unable as the majority of his contemporaries to distinguish facts from marvels. One cannot help having the impression that he was one of those overrated individuals such as exist in any community, but too small a part of his writings has reached us to change that impression into certainty.

All the historical fragments have been edited by Felix Jacoby, *Fragmente der griechischen Historiker*, vol. 2A (1926), pp. 222–317.

Castōr of Rhodos. Castōr (I–1 B.C.) was a contemporary of Poseidōnios, and he flourished for a time in Rhodos, but we do not know whence he came there. He married a girl in the family of Dēiotaros, the pro-Roman tetrarch of Galatia, and rendered services to Pompey; later, he bore witness in Caesar's court against Dēiotaros and died a victim of the latter's vengeance. He wrote a history (*Chronica*) in six books followed by chrolonogic tables from the legendary founders of Babylon and Nineveh, Bēlos and Ninos, down to 61 B.C. We conclude that he died only after that date. His tables are important as a part of the chronologic tradition leading to the Christian chronologists such as Eusebios (IV–1), the medieval ones, and our own.

Castōr was the last Greek historian of the second century. In the first century five more deserve to be remembered, who came from five different parts of the world, Diodōros of Sicily, Nicolaos of Damascus, Dionysios of Halicarnassos, Strabōn of Amaseia, and Juba of Numidia.

Diodōros of Sicily. Diodōros (I–2 B.C.) is called the Sicilian (Siceliōtēs, Siculus), because he was born in Agyrion[17] c. 85, but he lived mostly in Rome, flourishing under Caesar and Augustus until 21 B.C. or later. In 30, he

because he has added ethnographic comparisons taken from everywhere (apropos of the origin of fire, the renewal of youth, the chariot of the Sun, and so forth) in the appendix, vol. 2, pp. 309–455.

[16] This reminds us of the Arabic title "Arisṭū-al-zamān," "the Aristotle of his time," many times awarded to individuals unworthy of it, say university presidents.

[17] Agyrion (modern Agira) was one of the oldest Greek colonies in central Sicily.

completed, after 30 years of travel and study, a Greek compilation of historical excerpts, which he called a *Historical library* (*Historiōn bibliothēcē*).[18] It was supposed to survey the whole past from the beginnings to his own day. It was divided into three sections: (1) before the Trojan War (6 books); (2) from that war to Alexander's death (11 books); (3) from 323 to the beginning of Caesar's conquest of Gaul in 58 (23 books). It thus extended to 40 books, of which 15 are extant, plus fragments of the others. We have Books I to V of the first part, seven books of the second part covering the years 480–323, and three of the third part covering the years 323–302. The project was very ambitious, for Diodōros wanted to describe the part of every nation, but it was uncritical and low-minded. Diodōros had no general views and his style was as mediocre as his thought; yet a number of facts have been preserved because of his diligence.

It is noteworthy that he tried to understand the whole past; as a Sicilian a certain international disinterestedness was easier to him than it would have been to an Athenian, an Alexandrian, or a Roman. His language was Greek, but he had learned Latin in his youth. It is significant that for him the main "cuts" in the past were the war of Troy and Alexander's death; that was not a bad choice.

Nicolaos of Damascus. Nicolaos (I–2 B.C.), son of Antipatros, takes us not only from Sicily and Italy to Syria but also from the pagan world to the Roman-Jewish court of Herod the Great (king of Judaea 40–4 B.C.). He was born in Damascus in 64; his father was a man of substance, who valued education and made sure that Nicolaos would receive the very best available. He was probably taught by Greek tutors, and he distinguished himself sufficiently to attract the king's attention. Herod had become king of the Jews in 40, thanks to Antony's favor; he promoted the Hellenization and Romanization of Judaea, and he needed Greek assistants. Nicolaos was the most notable of these; he spent his life in Herod's service and accompanied him twice to Rome during the last ten years (14–4) of his rule.

Nicolaos was the king's secretary, dealing with political and diplomatic matters, but even more with philosophy, history, and general education. It was his business to explain Herod's anti-Arabic (anti-Nabataean) politics to the Roman senate, but also to explain the past to Herod himself. After the king's death (in 4 B.C.), Nicolaos tried to retire but was obliged to continue in the service of Herod's son Archelaos and went to Rome for his defense. In spite of that, Archelaos was banished by Augustus to Vienne (on the Rhone) and died there; we do not know what happened to Nicolaos himself. Did he spend his final years in Jerusalem or in Rome?

[18] This is the earliest use (or one of the earliest) of the word *bibliothēcē* to mean not a book case or a book house but simply a collection of writings issued together in a single series. The word "library" has experienced the same semantic evolution. Compare "Harvard College Library" and "Loeb Classical Library."

ΔΙΟΔΩΡΟΥ ΤΟΥ ΣΙΚΕΛΙΩΤΟΥ ΒΙΒΛΙΟΘΗΚΗΣ ΙΣΤΟΡΙΚΗΣ

βίβλοι πεντεκαίδεκα ἐκ τῶν τεσσαράκοντα.

DIODORI SICVLI

Bibliothecæ hiſtoricæ libri quindecim de quadraginta.

Decem ex his quindecim nunquam prius fuerunt editi.

ANNO M. D. LIX

EXCVDEBAT HENRICVS STEPHANVS

illuſtris viri HVLDRICI FVGGERI typographus.

Fig. 88. Diodōros of Sicily (I–2 B.C.). Title page of the princeps of his *Histories*, edited by Henri Estienne (folio, 35 cm, 848 pp.; Geneva, 1559). It was dedicated by him to his patron, Huldric Fugger. It contains only the Greek text. This is the princeps of the fifteen books extant (out of forty); but the Greek text of Books XVI–XX had been printed before (Basel, 1539). Diodōros' complete work must have been truly enormous. [Courtesy of Harvard College Library.]

His main literary work was the writing of a universal history like that of Diodōros, but on a larger scale. It was meant to cover the history of mankind from the beginnings to Herod's death, in 144 books. We do not know exactly how it was divided, but it became naturally more elaborate as it came closer to the author's present. Book 96 told of the wars of Mithridatēs the Great and his ally Tigranēs, king of Armenia; [19] this means that some fifty books, one third of the whole work, dealt with the first century B.C. A large part was given to Herodian and Jewish history and this was a primary source of Josēphos (I–2).

Nicolaos also wrote a biography of Augustus, an autobiography (*Peri idiu biu cai tēs eautu agōgēs, Concerning his own life and education*), and

[19] To repeat, the Mithridatic wars with Rome occurred in 88–84, 83–81, 74–64. Mithridatēs was finally defeated by Pompey and fled to the Crimea, where he committed suicide in 63. Tigranēs ruled Armenia from 96 to 56. He married Mithridatēs' daughter, Cleopatra. By 83 he was master not only of Armenia but of the Seleucid kingdom from the Euphratēs to the sea.

ΔΙΟΝΥΣΙΟΥ ΤΟΥ ΑΛΙΚΑΡΝΑΣΣΕΩΣ

ΡΩΜΑΙΚΗΣ ΑΡΧΑΙΟΛΟΓΙΑΣ

ΒΙΒΛΙΑ ΔΕΚΑ.

Dionysii Halicarnassei antiquitatum Romanarum Lib. X.

EX BIBLIOTHECA REGIA.

Βασιλεῖ τ' ἀγαθῷ κρατερῷ τ' αἰχμητῇ.

LVTETIAE.

Ex officina Rob. Stephani, Typographi Regii, typis Regiis.

M. D. XLVI.

Ex priuilegio Regis.

Fig. 89. Dionysios of Halicarnassos (I–2 B.C.). Princeps of the *Roman antiquities* (Paris: Robert Estienne, 1546–1547). It is a folio, 35 cm, 2 vols. often bound in one, 540 + 500 pp. [Courtesy of Harvard College Library.]

a curious collection of the manners and customs of some fifty nations (*ethōn synagōgē*). It is a pity that all his historical works are known only in the form of fragments. His "ethnological" collection might have been very instructive. He was a Peripatetic and wrote commentaries on Aristotle, the loss of which is less regrettable. His treatise on plants, which has remained a part of the Aristotelian corpus, has been briefly described in Chapter XXI.

Dionysios of Halicarnassos. Dionysios came to Rome at the end of the Civil Wars and flourished there from c. 30 to c. 8 B.C. He was primarily a teacher of Greek and a literary critic; his profession was that of schoolmaster or private tutor; that was a good profession at that time in Rome, for many young Romans could not afford to reside in Greece and yet were anxious to know Greek as well as possible. Most of his writings dealt with literary and grammatical subjects, but we must speak of him now because of his treatise on the beginnings of Roman history (*Rhōmaicē archaiologia*), completed in 8 B.C. He was completely acclimatized in Rome and his purpose was to explain the origins of Roman destiny and the causes of her greatness. His book, which was probably rhetorical, covered Roman history from the foundation of the city to the First Punic War (264–241), but is lost.

Strabōn of Amaseia. The last but one of those pre-Christian historians, Strabōn (I–2 B.C.), was the greatest next to Polybios, and he is generally known because of his treatise on geography, which is one of the main legacies of antiquity. We have no sufficient means of appreciating the historian in him, because his historical studies (*Historica hypomnēmata*) are lost. They were written at the beginning of the Augustan age and filled 47 books. After an introduction covering early history (Books I–IV), it continued the history of Polybios,[20] that is, the bulk of the work (Books V–XLVII) dealt with a relatively short period, from the destruction of Carthage in 146 to the beginning of the principate in 27.

He wrote the *Geography* later in life and in it refers to his historical work in the following characteristic words:

> In short, this book of mine should be generally useful — useful alike to the statesman and to the public at large — as was my work on *History*. In this work, as in that, I mean by "statesman," not the man who is wholly uneducated, but the man who has taken the round of courses usual in the case of freemen or of students of philosophy. For the man who has given no thought to virtue and to practical wisdom, and to what has been written about them, would not be able even to form a valid opinion either in censure or in praise; nor yet to pass judgment upon the matters of historical fact that are worthy of being recorded in this treatise.[21]

It is clear that the two works were written for the same public, educated people in general but chiefly officers of state and leading men (*tus en tais hyperochais*). Judging from the *Geography*, the disappearance of his *History* is a very great loss. He was not a rhetor, like Diodōros and Dionysios, nor a royal counselor, like Nicolaos, but a man of Polybios' stature and genius, earnest and independent.

Juba II. This Greek section may be closed with a brief reference to a Greek historian who came from Numidia but was educated in Rome. When his father, Juba I, king of Numidia, was beaten by the Romans in 46, he was a small boy about four years old and was taken to Rome for Caesar's triumph. He received the Greek-Latin education of a Roman aristocrat, became a Roman citizen, and served in Octavianus' army. He was eventually permitted to return to Numidia and in 25 B.C. the Romans created him king, not of his native country but of Mauretania to the west of it.[22]

[20] He called it *ta meta Polybion* (the sequel to Polybios).

[21] *Geography,* I, 1, 22, translated by Horace Leonard Jones (Loeb edition), vol. 1, p. 47.

[22] Numidia corresponded more or less

All his writings were in Greek and all are lost. He had formed a collection of works of art, chiefly statues, remains of which have been found at Julia Caesarea (Cherchel, an Algerian seaport, west of Algiers).

THE LATIN HISTORIANS

Note that this section is not entitled the Roman historians, for all the men dealt with in the first section were as Roman as those of this section and most of them were students of Roman history. They wrote in Greek, however, while the following wrote in Latin and were really the creators of Latin historiography. The Greek authors might be born anywhere, East or West, though most of them lived in Rome or visited the great city at one time or another. The Latin authors, on the contrary, were all of them children of Italy. There are six of them, divided into three groups: the pioneers, Ennius and Cato the Censor; then Caesar and Varro; finally, Sallust and Livy.

Ennius. Ennius (II–1 B.C.) was called the father of Roman poetry; he might be called as well the father of Roman history. Two other historians, Q. Fabius Pictor and L. Cincius Alimentus, had written annals of Rome before Ennius, but they had written them in Greek. Both stopped their accounts at the time of the Second Punic War (218–201).

Ennius was a son of Calabria, where he received a Greek education, but he learned Latin in the Roman army (if not before); he was a centurion in Sardinia in 204 and was brought from there to Rome by Cato the Censor. He wrote his *Annals* (*Annalium libri XVIII*) in Latin verse; the poem began with Aeneas and extended to c. 181, within a dozen years of his own death. This was an epic rather than a scientific history; his verse were generally rugged and commonplace but sometimes magnificent. He had seen service in the Second Punic War under Scipio Africanus, and Books I and XV ended there. They obtained so much success, however, that he was moved to add the following three books in the form of yearly supplements. This destroyed the unity of the whole work but may have satisfied the patriotic readers and held their attention.

Ennius' *Annals* created a great subject and prepared an appreciative public for Virgil's *Aeneid.*

Cato the Censor. The first Roman historian in Latin prose was Cato (II–1 B.C.), and his main historical work (lost) was entitled *Origines.* It was divided into three books, of which the first dealt with the Trojan origins and Aeneas, the foundation of Rome (753) and the kings (to 510), and the second and third books with the origins of other Italian communities and the foundations of Italian cities.[23] Later, he added four books bringing

---

to western Tunisia and eastern Algeria; Mauretania to western Algeria and Morocco.

[23] Insisting on "origins" was a Hellen-

the story down to the year of his death (149); or it may be that these four books written by him in his old age were added to the *Origines* in a posthumous edition. The *Origines* is generally said to be divided into seven books, but the title does not apply very well to Books IV to VII. The contents of those books are as follows:

IV. First Punic War and Second Punic War to 216.[24]

V. Macedonian wars and Rhodian affairs. Rhodos and Pergamon inveigled Rome into Eastern politics in 201; Rhodos was an ally of Rome but let her down in the Third Macedonian War (171–167). This caused a violent crisis in 167, ending with Rhodos' political downfall.

VI. War against Antiochos III the Great, king of Syria (223–187).

VII. Spanish wars, with special emphasis on the prosecution of Servicius Sulpicius Galba, praetor in Hispania Ulterior (151–150), accused of having caused the massacre of Lusitanians suing for peace. Cato supported the prosecution in 149, but it failed.

It is clear that Books IV to VII were very different from Books I to III. Cato had started his work with the purpose of explaining the origins of Roman power and greatness; he conceived it as a kind of introduction to universal history. He was not concerned only with the nations of Italy but also with the Ligurians,[25] the Celts, and the Spaniards, and not only with the past and the present but also with the destinies of Rome as yet unfolded. As far as we can judge from the fragments, he was not interested only in wars and politics but also in geography, climate, agriculture, mining, economic matters of various kinds.

The main point of view, however, was political, namely, to explain how Rome had accomplished and was accomplishing her imperial duties. He was exceedingly well prepared for this because of his long experience as a soldier and public servant. He fought as a youth in the Second Punic War (218–201) and did his best to cause the Third, which began in the year of his death. He was quaestor in Sicily in 204 and returned home via Sardinia bringing Ennius with him; [26] he was plebeian aedile in 199, praetor in Sardinia in 198, consul in 195, senator, and so forth. His public services ended only with his life at 85. Therefore, he had personal knowledge of every aspect of Roman politics and administration. He was a democrat,

---

istic trait. The Hellenistic historians liked to speak of the foundations (*ctiseis*) of cities.

[24] To the battle of Cannae (in Apulia, southeast Italy), where the Romans were utterly beaten by Hannibal in 216. The Romans never suffered a worse military defeat.

[25] The Ligurians were established around the Gulf of Genoa as far as the Maritime Alps in the west and Cispadane Gaul (Aemilia) in the east. Liguria and Cispadane Gaul were both just south of the Po river, the first under the Upper Po in the west, the second under the Middle and Lower Po in the east.

[26] Ennius, born in 239, was only a few years older than Cato, born in 234, and it was the latter who brought him to Rome from Sardinia in 204. He died in 169, twenty years before Cato (149) and preceded him as historian. Ennius' *Annals* ended in 181. Cato began his *Origins* about the time of Ennius' death and ended them at the time of his own in 149.

despising the luxury and vanity of the great lords, and was more anxious to praise the common people, the rank and file, and even the elephant Surus, than the generals and magistrates.

In Books v to vii he exploited his own experience of the events described, and he sometimes inserted his own orations; those orations were genuine enough but perhaps a little irrelevant. Cato was prejudiced but honest; he was not rhetorical but matter-of-fact. His history, lopsided as it was, was derived from good sources; its loss (but for fragments) is truly irreparable.

Caesar. A century after Cato's death there appeared another historian, a far greater one, greater as a man, as a writer, and in every respect, one of the outstanding heroes in the whole past. Caesar (I–1 B.C.) was primarily a statesman and a politician; he became a general and proved his military genius relatively late in life. When he began his Gallic campaigns, he was already older than Alexander had been at the time of his death and almost as old as Napoleon at the time of his defeat.[27] In a sense, his literary career began even later, in spite of the fact that he was a born man of letters.

The only writings of his that have survived are his *Commentaries*, which are the reminiscences of his military campaigns. They inaugurated a new literary genre and have remained models of their kind.[28] Few men have the opportunity of accomplishing great military deeds, and few of these have the literary power of describing them.[29]

The *Commentaries* include two separate works, the *Gallic war* (*De bello Gallico*) in seven books, each of which covers one of the years 58–52,[30] and the *Civil war* (*De bello civili*) 49–45, in three books.

They are our main sources for the events described, and they describe

[27] He was 43. Alexander was 33 when he died and Napoleon 44 at the time of the battle of Leipzig and only 46 when he reached St. Helena. Caesar began his military deeds at an age when they were already out.

[28] Caesar was not the very first general to write down military reminiscences. This was done before him, in Greek, by Ptolemaios Sōtēr (d. 283), but Ptolemaios' book is lost.

[29] Caesar's military deeds have been summed up in a single sentence by Plutarch, *Caesar,* xv: "Although it was not full ten years that he waged war he took by storm more than 800 cities, subdued 300 nations and fought pitched battles at different times with 3,000,000 men of whom he slew one million in hand-to-hand fighting and took as many more prisoners." I have not tried to check Plutarch's "statistics."

[30] An eighth book was added, carrying the story to 50 B.C., by Aulius Hirtius, one of Caesar's officers. The same Hirtius was perhaps the author of the *Bellum Alexandrinum,* continuing the *De bello civili.* This book deals not only with Caesar's battles in Alexandria but with other events down to his victory over Pharnacēs, king of Pontos, in 47, at Zēla (in south Pontos). That was the victory which was so easy that Caesar informed the Senate of it with the famous words "Veni, vidi, vici" (I came, saw, and conquered). There has been much discussion concerning the veracity of Caesar and Hirtius. Michel Rambaud, "L'art de la déformation historique dans les Commentaires de César," *Annales de l'Université de Lyon, Lettres* (vol. 23, 410 pp.; Paris: Belles Lettres, 1953).

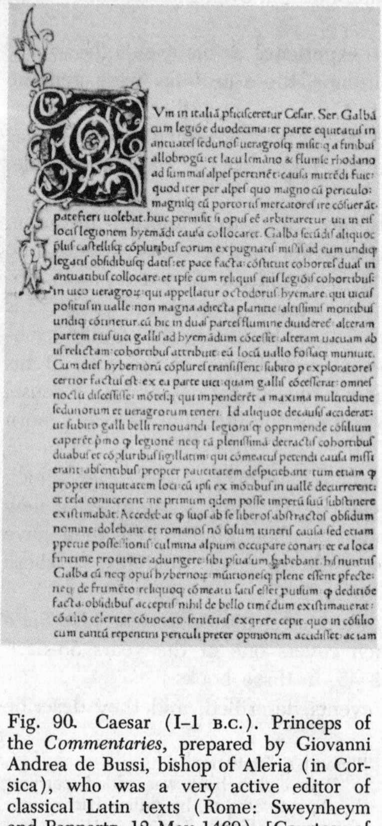

Vm in italiã pficisceretur Cesar. Ser. Galbã cum legiõe duodecima: et parte equitatus in antuates sedunos ueragrosq; misit: q a finibus allobrogũ: et lacu lemãno & flumĩe rhodano ad summas alpes pertinẽt: causa mittẽdi fuit: quod iter per alpes quo magno cũ periculo: magnisq; cũ portoris mercatores ire cõsuerãt: patefieri uolebat. huic permisit si opus eẽ arbitraretur uti in eis locis legionem hyemãdi causa collocaret. Galba secũdis aliquot pliis castellisq; cõpluribus eorum expugnatis missis ad eum undiq; legatis obsidibusq; datis: et pace facta: cõstituit cohortes duas in antuatibus collocare: et ipse cum reliquis eius legiõis cohortibus in uico ueragrorũ: qui appellatur octodorus: hyemare. qui uicus positus in ualle non magna adiecta planitie: altissimis montibus undiq; cõtinetur. cũ hic in duas partes flumine diuideretur: alteram partem eius uici gallis ad hyemãdum cõcessit: alteram uacuam ab iis relictam: cohortibus attribuit. eũ locũ uallo fossaq; munuit. Cum dies hybernorũ cõplures transissent: subito p exploratores certior factus est: ex ea parte uici: quam gallis cõcesserat: omnes noctu discessisse: mõtesq; qui impenderẽt a maxima multitudine sedunorum et ueragrorum teneri. Id aliquot de causis acciderat: ut subito galli belli renouandi legionisq; opprimende cõsilium caperẽt: pmo q legionẽ neq; tã plenissimã: detractis cohortibus duabus et cõpluribus sigillatim: qui cõmeatus petendi causa missi erant: absentibus: propter paucitatem despiciebant: tum etiam q propter iniquitatem loci: cũ ipsi ex mõtibus in uallẽ decurrerent: et tela conicerent: ne primum qdem posse impetũ suũ sustinere existimabãt. Accedebat q suos ab se liberos abstractos obsidum nomine dolebant: et romanos nõ solum itineris causa sed etiam ppetue possessionis culmina alpium occupare conari: et ea loca finitime prouintie adiungere sibi psuasum habebant. his nuntiis Galba cũ neq; opus hybernorũ munitionesq; plene essent pfecte: neq; de frumẽto reliquoq; cõmeatu satis esset puisum: q deditiõe facta: obsidibus acceptis nihil de bello timẽdum existimauerat: cõsilio celeriter cõuocato: sententias exqrere cepit. quo in cõsilio cum tantũ repentini periculi preter opinionem accidisset: ac iam

Fig. 90. Caesar (I–1 B.C.). Princeps of the *Commentaries*, prepared by Giovanni Andrea de Bussi, bishop of Aleria (in Corsica), who was a very active editor of classical Latin texts (Rome: Sweynheym and Pannartz, 12 May 1469). [Courtesy of Pierpont Morgan Library.]

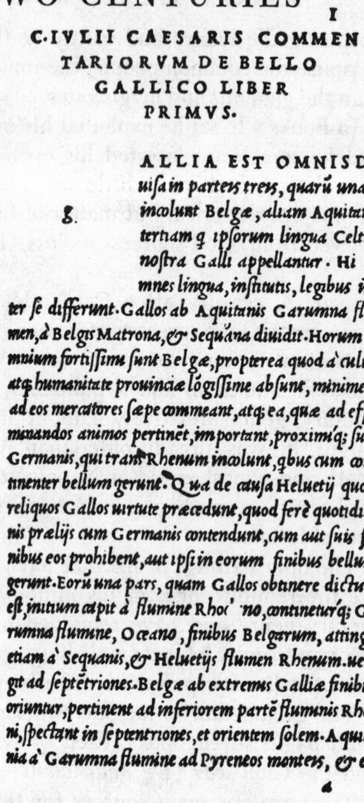

I

C. IVLII CAESARIS COMMEN TARIORVM DE BELLO GALLICO LIBER PRIMVS.

g. ALLIA EST OMNIS DI *uisa in parteis treis, quarũ unam incolunt* Belgæ, *aliam* Aquitani, *tertiam q ipsorum lingua* Celtæ, *nostra* Galli *appellantur.* Hi *omnes lingua, institutis, legibus inter se differunt.* Gallos *ab* Aquitanis Garumna *flumen, à* Belgis Matrona, *&* Sequana *diuidit.* Horum *omnium fortissimi sunt* Belgæ, *propterea quod à cultu, atq; humanitate prouinciæ lõgissime absunt, minimé q; ad eos mercatores sæpe commeant, atq; ea, quæ ad effœminandos animos pertinẽt, important, proximi q; sunt* Germanis, *qui trans* Rhenum *incolunt, qbus cum continenter bellum gerunt.* Qua *de causa* Heluetij *quoq; reliquos* Gallos *uirtute præcedunt, quod ferè quotidianis prælijs cum* Germanis *contendunt, cum aut suis finibus eos prohibent, aut ipsi in eorum finibus bellum gerunt.* Eorũ *una pars, quam* Gallos *obtinere dictum est, initium capit à flumine* Rhodano, *continetur q;* Garumna *flumine,* Oceano, *finibus* Belgarum, *attingit etiam à* Sequanis, *&* Heluetijs *flumen* Rhenum. *uergit ad septẽtriones.* Belgæ *ab extremis* Galliæ *finibus oriuntur, pertinent ad inferiorem partẽ fluminis* Rheni, *spectant in septentriones, et orientem solem.* Aquitania *à* Garumna *flumine ad* Pyreneos *monteis, & eã*

a

Fig. 91. Caesar. Very remarkable edition of his *Commentaries* by Fra Giocondo (16 cm; Aldus: Venice, 1513), with woodcuts. We reproduce the first page of the *De bello gallico* and elsewhere (Fig. 73.) a part of the bridge built by Caesar across the Rhine. [Courtesy of Harvard College Library.]

them excellently. Caesar explains his campaigns with perfect simplicity and clearness; as he was a born writer as well as a born general, his *Commentaries* are one of the masterpieces of historical literature.

Varro. Caesar was murdered at the age of 56, while Varro (I–2 B.C.) was permitted to live 89 years; thus, while Varro was the older by sixteen years, he survived the younger man by seventeen and seemed to belong to a later generation. While Caesar became an author by the force of circumstances (he had to justify his tremendous deeds), Varro was an irrepressible bookmaker.

The eyght bookes of Caius Iulius Cæsar conteyning his martiall exploytes in the Realme of Gallia and the Countries bordering vppon the same translated oute of latin into English by Arthur Goldinge G.

Imprinted at London by Willyam Seres

Anno. 1565

THE FIRST BOOKE of Caius Iulius Cæsars Comentaries of the warres in Gallia. fol. 1.

ALL GALLIA IS deuided into thre partes: Of the whiche, one is inhabited by ye Belgies. An other by the Aquitanes, and the. iii. by them who in their toung are called Celtes, and in our Galles. All theis dyffer eache from other in Language, Customes, and Lawes. The ryuer of Geronde deuideth the Celtes from the Aquitanes: and the riuers of Seane and Marne do deuide them from the Belgies. Of all theis, the Belgies be moste puissante, as they whiche are furthest distant from the delycatenes and ciuilitie of the Prouince, and vnto whom is little or no resort of marchauntes, to bringe in thynges that might effeminate their mindes. Besides that, they border vpon the Germanes

The whole countrey of Fraunce

A manes

Fig. 92. First English edition of Caesar. Translation by Arthur Golding (small thick volume, 13.5 cm; London: Willyam Seres, 1565). Dedicated to Sir Willyam Cecill, principal secretary to Queen Elizabeth. We reproduce the title page and the beginning of the first book of the *Gallic war* (the Latin text is shown in Fig. 91.) [Courtesy of Harvard College Library.]

Every one of Varro's books, except the one on agriculture, was historical in intention; he wanted to set forth the origin and growth of institutions, the lives of great men. He was less of a historian, however, than a man of letters fond of historical subjects. On the other hand, Caesar was more than a conventional historian; he was the main actor and the best witness of the events that he described. His books are not so much history as first-rate documents to be used by historians, as the French put it, "mémoires pour servir à l'histoire." The contrast between the two writers could not be greater; Caesar was ahead of the other historians; Varro was a long way behind them.

The most ambitious of Varro's historical works was his treatise on *Profane and sacred antiquities in 41 books* (*Antiquitatum rerum humanarum et divinarum libri XLI*), which he wrote in 47. Many fragments enable us to reconstruct its ordonnance, which is very original and symmetric. It was divided into two main parts, profane antiquities (25 books) and sacred antiquities (16 books); the first was subdivided into $(6 \times 4) + 1$ books, the second into $(3 \times 5) + 1$ books. Let us examine this a little more closely.

His account of "profane antiquities" was divided into four sections, which might be entitled men, places, times, things (answering to the queries who? where? when? what?) and each section was subdivided into six books. Book I was a general introduction to the whole work. Books II–VII dealt with men, from Aeneas down, who were the actors of Roman history. Books VIII–XIII dealt with places; this was a kind of historical geography of Italy; Books XIV–XIX with Roman chronology; Books XX–XXVI with things, institutions (there are very few fragments of that final section).

The second part, the "sacred antiquities," is equally symmetric though the pattern is different. There is again an introductory book, followed by four sections dealing with (sacred) men, places, times, and things. Thus, Books II–IV concern three classes of sacred men, pontiffs, augurs, and the *quindecimviri*;[31] Books V–VII, three classes of sacred places, private altars, temples, other sanctuaries; Books VIII–X three kinds of holy times, the *feriae* (holidays), circus days, and theater days; Books XI–XIII three kinds of sacred things, consecrations, private and public sacrifices. A fifth section (Books XIV–XVI) was devoted to three classes of gods, certain gods, uncertain gods (foreign ones), and the principal or selected gods.

Varro revealed the same fondness for classification or symmetric grouping in other books, chiefly in the *Hebdomades*; such a fondness could be traced back to Pythagorean origins and beyond those to Oriental ones,[32] and his *Antiquities* was built almost as regularly as a Greek temple, but I believe that this literary architecture was an invention of his own. At any rate, I do not know of any Greek treatise that was built in the same fashion.

It is equally clear that such a work as Varro's *Antiquities* was full of historical data and yet was far removed from a conventional historical treatise.

Two other historical works of his were entitled *De gente populi Romani*, a history of the Roman family or the master race, and *De vita populi Romani*, which was presumably a social history of the Roman people and implied a philosophy of history. While it was written before the consecration of Augustus as the first emperor, Varro realized that the evolution of the Roman people was like that of a single living being, passing from childhood to youth, maturity and senility. It was a cyclical conception similar in a humbler way to the more elaborate ones of Oswald Spengler (1880–1936) and Arnold Toynbee (1889–).

[31] The *quindecimviri* (the fifteen men) were an extension of an earlier college, the *decemviri sacris faciundis*, or *sacrorum* (the ten men in charge of sacred things). They were sometimes considered priests of Apollōn, and were in charge of the Sibylline books, of the celebration of the games of Apollōn, and of the secular games. As against the pontiffs and the augurs responsible for the good execution of Roman rites, the rites which they were supervising were of Greek origin.

[32] For grouping, see my *Introduction*, indexes of vols. 1, 2, 3, under the words Number 1, Number 2, . . . , Number 40, and Volume 1, index, *s.v.* Number. No people has carried that mania to the same extent as the Chinese — see, for example, *Isis* 22, 270 (1934) — but it is almost universal.

Instead of calling Varro a historian, it would probably be more correct to call him a learned man, and he was indeed the greatest scholar of his nation. During the whole duration of the Roman empire, including its real decadence, his books were used as we today use classical dictionaries or encyclopedias; our tools are infinitely better, but we should remember that the tools provided by Varro, however primitive and imperfect, were the first of their kind. Whenever I consult an encyclopedia like "*PW*," I am very grateful to its authors, but do not forget their own predecessors as far back as Varro and his own predecessors, Greek and Latin; my gratitude extends to all of them. Glory to the pioneers!

One more word in Varro's praise. It would be very unfair to consider him only a compiler who put together all the learning of earlier scholars. He was, to some extent, a philosopher or at least a thinker, who tried to understand and to explain the origin and evolution of social phenomena. For example, he tried to justify the rites of the Roman religion, irrespective of its mythology, which had become incredible. He distinguished three kinds of religion, that of the poets, that of the state, and that of the philosophers, and he himself preferred the last one. Though the bulk of his information was necessarily of Greek provenience, he tried to include as many Roman data as possible, and to explain Greek matters in Roman terms and vice versa. His main purpose was to improve Roman institutions or to justify them, and he was convinced that religion was the main cause of purity, strength, and unity. It is for that reason that he wrote the *Antiquities*, and Cicero recognized its value in noble words:

> We were wandering and straying about like visitors in our own city, and your books led us, so to speak, right home, and enabled us at last to realize who and where we were. You have revealed the age of our native city, the chronology of its history, the laws of its religion and its priesthood, its civil and its military institutions, the topography of its districts and its sites, the terminology, classification and moral and rational basis of all our religious and secular institutions, and you have likewise shed a flood of light upon our poets and generally on Latin literature and the Latin language, and you have yourself composed graceful poetry of various styles in almost every metre, and have sketched an outline of philosophy in many departments that is enough to stimulate the student though not enough to complete his instruction.[33]

Sallust. The youngest of the historians of the Republic, Gaius Sallustius Crispus (I–2 B.C.), was born thirty years later than Varro, in 86, at Amiternum.[34] He was of plebeian origin, became a senator, but was expelled from the Senate in 50 for immorality (?); he was appointed a quaestor by Caesar in 49, and obtained sufficient wealth to buy a fine estate and lay out

[33] Cicero, *Academica*, I, 3, written in 45 B.C. Cicero wrote a letter to Varro (*Ad familiares*, IX, 8) dedicating the second edition of *Academica* to him.

[34] Amiternum is about 60 miles northeast of Rome, in the Sabine country, and was claimed to be the cradle of the Sabine people.

beautiful gardens (*horti Sallustiani*). His main works were written c. 43–39, and he died in 34 B.C.

There is much mystery in his life; he was a politician, a defender of the popular party, and subject to accusations which may have been calumnies. He was disillusioned and pessimistic. His models were Thucydidēs and Cato the Censor.

He did not attempt, like Cato and Varro, to cover too wide a field, but, on the contrary, wrote what might be called "monographs" of limited scope. His largest work, the *Historiae*, in five books, covers a period of a dozen years (78–66). The two others are even more restricted. The *De bello Catilinae* is an account of Catiline's conspiracy during Cicero's consulship, in 63; it might be called a political pamphlet. The *De bello Jugurthino* describes the Roman wars (112–105) against Jugurtha, king of Numidia.

He tried to imitate Thucydidēs' impartiality, but was too much involved in politics to succeed in that; he was more successful in his imitation of Thucydidēs' style. His writings are brilliant analyses of political events, the first models of their kind in world literature.

Livy. There was but one Latin historian during the Augustan age, but he was the most famous of all Latin historians. Titus Livius [35] (I–2 B.C.) was born in 59 in Patavium (Padova), then the most important city of North Italy.[36] He belonged to a patrician family of his province, obtained some distinction as a rhetorician and as the author of philosophical dialogues, and found a place in the Augustan court, where the need of a historian was felt and his ability was readily appreciated. He probably traveled, but we do not know where and when; he spent most of his life in Rome and in his native city where he died in A.D. 17.[37]

He is the man of one work, but a gigantic one to which the whole of his mature life was devoted. It was a complete history of Rome from the very beginnings, *Ab urbe condita libri*, to his own days. The first book was ready in 28, when he was 31, and the work was continued to the end of his life at the age of 75.

The whole work contained no fewer than 142 books [38] and it apparently survived in its integrity until the end of the fourth century after Christ. During the dark ages preceding the medieval rebirth, much of it was lost,

[35] Curiously enough, in French he is always called Tite-Live and indexed under *T*, in spite of the fact that Titus is one of the eighteen Roman first names (abbreviated T.).

[36] Patavium was in northeast Italy, that is, not in Transpadane Gaul like Mediolanum (Milano) but in the territory of the Veneti.

[37] It may be that he returned to Patavium after Augustus' death (A.D. 14), because the favorites of one emperor are seldom welcome to the next. In 14, Livy was 72 and perhaps longing for quiet.

[38] He possibly planned to write 150 books and to carry his history to Augustus' death in A.D. 14. That would have been a fine ending, but his life was not long enough. His work actually ended with the death of Nero Claudius Drusus in 9 B.C.

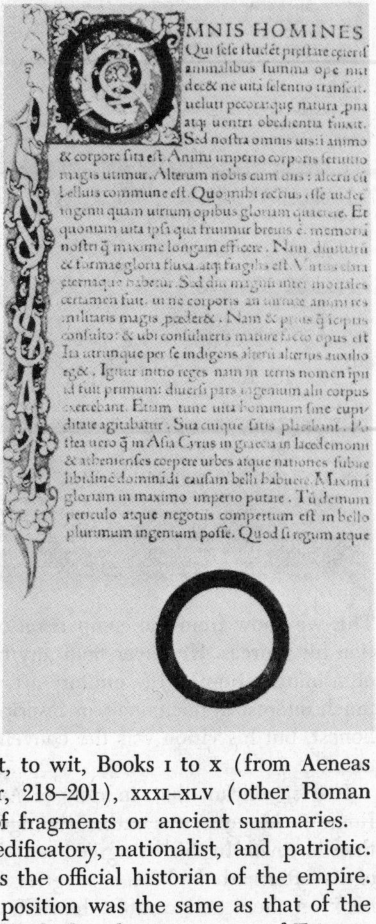

OMNIS HOMINES
Qui sese studēt prestare ceteris
animalibus summa ope niti
decet ne uita silentio transeant.
ueluti pecora: quę natura prona
atque uentri obedientia finxit.
Sed nostra omnis uis in animo
& corpore sita est. Animi imperio corporis seruitio
magis utimur. Alterum nobis cum diis: alterū cū
beluis commune est. Quo mihi rectius esse uidetur
ingenii quam uirium opibus gloriam quaerere. Et
quoniam uita ipsa qua fruimur breuis ē. memoriā
nostri q̄ maxime longam efficere. Nam diuitiarū
& formae gloria fluxa. atque fragilis est. Virtus clara
aeternaque habetur. Sed diu magnū inter mortales
certamen fuit. ut ne corporis an uirtute animi res
militaris magis procederet. Nam & prius q̄ incipias
consulto: & ubi consulueris mature facto opus est
Ita utrunque per se indigens alterū alterius auxilio
eget. Igitur initio reges nam in terris nomen imperii
id fuit primum: diuersi pars ingenium alii corpus
exercebant. Etiam tunc uita hominum sine cupi-
ditate agitabatur. Sua cuique satis placebant. Po-
stea uero q̄ in Asia Cyrus in graecia lacedemonii
& athenienses coepere urbes atque nationes subigere
libidinē dominandi causam belli habere. Maximā
gloriam in maximo imperio putare. Tū demum
periculo atque negotiis compertum est in bello
plurimum ingenium posse. Quod si regum atque

Fig. 93. Sallustius (I–2 B.C.). Princeps of *De bello Catilinae* and *De bello Jugurthino* (Venice: Vindelinus de Spira, 1470). [Courtesy of Pierpont Morgan Library.]

and no more than 35 books are extant, to wit, Books I to X (from Aeneas to 293), XXI–XXX (Second Punic War, 218–201), XXXI–XLV (other Roman conquests to 167),[39] plus a number of fragments or ancient summaries.

The purpose of Livy's labor was edificatory, nationalist, and patriotic. Being patronized by Augustus, he was the official historian of the empire. He did not carry such a title, but his position was the same as that of the historiographers who were eventually attached to the main courts of Europe. He had full access to official papers, including Augustus' memoirs, and was thus as well informed as one could be from the government's point of view; of course, he could use also and did use the books already published, not only in Latin[40] but also in Greek (chiefly Polybios and Poseidōnios).

[39] In early days (say in the fourth century) the work was divided into decades. We have the first, third, and fourth, and half of the fifth. During the Renaissance and later it was the lost decades of Livy that were spoken of rather than the lost books. Attempts at restoration were made, the most remarkable being that of Johannes Freinsheim of Uppsala, who tried to restore six lost decades (Strassburg, 1654).

[40] My readers are already familiar with the leading ones (Cato, Caesar, Cicero, Varro), but there were many more (now

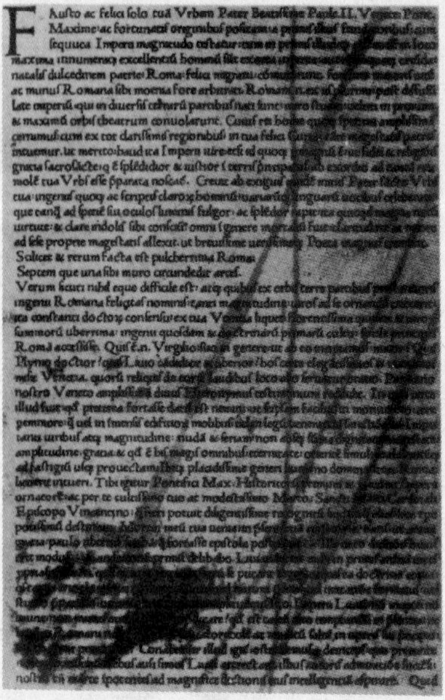
Scilicet & rerum facta est pulcherrima Roma:
Septem que una sibi muro circumdedit arces.

Fig. 94. Titus Livius (I–2 B.C.). Princeps of the *Historiae Romanae decades* (Rome: Sweynheym and Pannartz, 1469). The text was edited by Giovanni Andrea de Bussi, bishop of Aleria (in Corsica). Dedication to Paul II (pope 1464–1471). This Venetian pope was a patron of scholars and probably was responsible for the introduction of printing in Rome. [Courtesy of Library of Congress.]

This we know from the comparison of texts, for he did not generally mention his sources. He never held any office and had no technical knowledge of administration, of the military art, or even of historiography; he had not much interest in documents or inscriptions. He was a man of good will and honest, but his vision was the conventional vision of his class and environment.

It is significant that, in spite of Augustan prejudices, the golden age of Rome in his eyes was rather the age of Cato and of the Scipionic Circle than his own. In this he agreed with Varro. The evils of the Civil War and of its aftermath had been so grievous that Livy was driven to turn his back on them and comfort himself with the vision of the brave days of old (he tells us so in his preface).

His task was like Virgil's, though in a different way — the vindication of Roman honor and greatness.[41] It was very largely a literary task; his duty was not only to tell the official version of events but also to tell it in the best language, the ceremonial and rhetorical language of the best people.

---

lost), so many, indeed, that Livy apologized in his preface for adding one more to the long list.

[41] Many national histories were composed in the same spirit for the glory of France, or England, or Switzerland. This kind of infatuation is even more obvious in books written for the glory of Christianity, Islām, or any other religion. National or religious successes are interpreted as being not accidental but rather the result of a providential plan. The nation (or religion) was magnified above all others because of God's will.

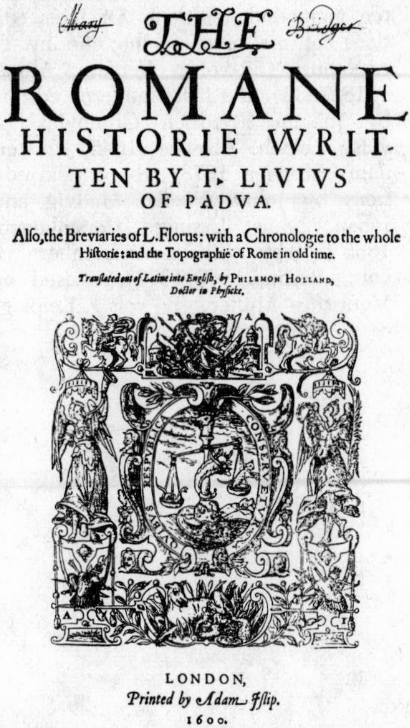
THE ROMANE HISTORIE WRITTEN BY T. LIVIVS OF PADVA.

Also, the Breviaries of L. Florus: with a Chronologie to the whole Historie: and the Topographie of Rome in old time.

Translated out of Latine into English, by PHILEMON HOLLAND, Doctor in Physicke.

LONDON, Printed by Adam Islip. 1600.

Fig. 95. First English translation of Livy, by Philemon Holland (London: printed by Adam Islip, 1600). This includes the translation of an *Epitome* of Livy and other Roman historians made by L. Florus in the first half of the second century; that *Epitome* was a popular schoolbook in the seventeenth century. It is a large and thick folio (33 cm, 1403 pp. + index); dedicated "to the most high and mightie monarch Elizabeth (my dread soveraine) . . ." The text of Livy covers 1233 closely printed pages, followed by that of Florus (pp. 1234–1264), an elaborate chronology from Romulus to 9 B.C. (pp. 1265–1345), the topography of anicent Rome (pp. 1346–1403), an elaborate index, and a glossary. The book was meant to be a kind of encyclopaedia of pre-Christian Roman history. [Courtesy of Harvard College Library.]

This conception of history was remote from that of Hērodotos, Thucydidēs, or even Polybios, yet, as long as the Roman prestige lasted and the ideal of Renaissance humanists dominated, Livy's history was considered the outstanding work of its kind.

The Livian tradition is as continuous as the Virgilian, for both authors walked together on the road to immortality; Livy's work was more vulnerable, however, because of its immense size. A complete manuscript was available in the fourth century, and the Verona palimpsest of Books III–VI dates of that time. A papyrus of the third century, discovered at Oxyrhynchos in 1903, contains an abstract of Books XLVIII–LV. A great part of the text was lost during the period of chaos between Antiquity and the Middle Ages.

The work was so large that it was early divided into decades, and each decade had its own tradition. This makes the study of the manuscript more complicated; in addition to the Verona manuscript of the fourth century, there are many of the ninth to the thirteenth centuries.

The princeps was edited by Giovanni Andrea, Bishop of Aleria,[42] and printed by Sweynheym and Pannartz (Rome, 1469) (Fig. 94). There are at least

[42] Aleria on the east coast of Corsica.

ten incunabula editions. Of later editions let me mention the one by F. Asulanus (5 vols.; Venice: Aldus, 1518–1533), the first "modern" edition by Johann Friedrich Gronovius (3 vols.; Leiden: Elzevier, 1645; Amsterdam: Elzevier, 1678), the critical editions by Johan Nikolai Madvig and Johan Louis Ussing (Copenhagen, 1865 ff.), by Wilhelm Weissenborn (9 vols.; Berlin, 1867–1879), revised by Mauritius Müller (6 vols.; Leipzig: Teubner, 1910–1911). These critical editions have been very often reprinted. There are many other editions of all extant works and innumerable ones of separate decades or books or selections.

The earliest English translation was made by Philemon Holland (London, 1600) (Fig. 95).

Very convenient Latin-English edition by Benjamin Oliver Foster (13 vols., Loeb Classical Library, 1919–1951).

XXV

LITERATURE[1]

In the Greek and Latin worlds, even as in our own, there was a distinction between literature and the more technical books the purpose of which was to teach rather than to entertain. In Greek one did not even have to speak of "belles-lettres" but simply of "letters" (*ta grammata*) and the man of letters was a *philologos* and if his profession was to teach letters, he was called a *logodidascalos.* In Latin the belles-lettres were called *litterae*, and their study, *humanitas*, *artes ingenuae*, *optimae*, *honestae*, *liberales*, *studia litterarum*, and so forth. We recognize some of those expressions in our own language, as when we speak of humanities or liberal arts. In both worlds, Greek and Latin, the distinction was sometimes blunted by the existence of didactic poetry, such as the poems of Aratos and Nicandros; that kind of poetry was likely to be poor, but the *De rerum natura* of Lucretius and the *Georgica* of Virgil are glorious exceptions.

Treatises on the history of Greek or Latin literature are naturally centered upon the poets and the writers of beautiful prose, and the men of science, such as Hipparchos or Vitruvius, are left out or dealt with in an offhand manner. In this book we are obliged to do just the opposite. Our main heroes are the great men of science, but it would be shocking to leave the artists out. Therefore, in this chapter and in the 27th, I shall introduce some of them, the greatest, in order that my readers may be reminded of the artistic and literary glories of this age. As language is the vehicle of literature, it is more necessary than ever to divide our account into two main parts, the Greek and the Latin. This affords a very striking contrast, because the Greek letters are going down, while the Latin are coming to birth and growing with the energy of youth.

GREEK LETTERS

As compared with the poets of the third century, and especially with Theocritos, who was a great master, those of the last two centuries B.C. cut a sorry figure. I cannot think of any one flourishing before the end of the second century and do not find it possible to mention more than a very

[1] This continues the story begun in Chapter XIII.

few – Meleagros, Philodēmos, Archias, and Parthenios – all characteristically Alexandrian in the bad sense.

Meleagros of Gadara. The greatest of these was Meleagros of Gadara [2] (c. 140–70), the son of a Greek father. Gadara was a little center of Greek culture, the birthplace of Menippos. Meleagros was educated in Gadara and submitted to Menippos' influence, then went to Tyros which was the nearest metropolis. He wrote many erotic poems, some of which are gracious, and a treatise on the Graces (*Charites*), somewhat in the Menippean style, a mixture of prose and verse, cynicism and wisdom. He had the idea of collecting his poems together with those of other poets, some forty, of all ages. That collection was entitled the *Crown* (*Stephanos*), and it was literally an *anthologia,* because in the preface he compared individual poems to single flowers; the whole was then a bouquet. This was not the first collection of its kind, but it was richer than previous ones, attracted considerable attention, and was the model of later ones, especially the famous ones put together by Constantinos Cephalas (fl. 917) and by Maximos Planudēs (in 1301).[3] That is really a remarkable achievement.

Philodēmos of Gadara was an Epicurean poet, a contemporary of Cicero. His poems (some thirty) were eventually included in the *Crown,* not in the first edition, but in the second, prepared by Philippos of Thessalonicē (c. A.D. 40).

Archias of Antioch wrote a poem on the Mithridatic Wars. His main title to fame is the fact that he was one of Cicero's clients. He was also included in the second edition of the *Crown.*

Parthenios of Nicaia was taken prisoner during the Mithridatic Wars and brought to Rome, but he was soon manumitted because of his learning. He was well received in literary circles and was befriended by the poet Cornelius Gallus (c. 66–26) and by Virgil; it is said that he was Virgil's teacher of Greek. All of his poetry (elegiac and mythologic) is lost, but a collection of love stories in prose (*Peri erōticōn pathēmaton*) has been preserved. It was written for Cornelius Gallus's education and dedicated to him.

It has been claimed that Parthenios lived until the end of the Augustan age; that seems impossible unless he became a centenarian, because the Mithridatic Wars did not end until 64 B.C., and Augustus lived until A.D. 14.

[2] Gadara, a town in Palestine, southeast of the lake of Tiberias. Readers of the New Testament are acquainted with its inhabitants, the Gadarēnoi (Mark 5:1, Luke 8:26, 37).

[3] For more details on those two Byzantine anthologies, the *Anthologia Palatina* and the *Anthologia Planudea,* see *Introduction,* vol 2, p. 974.

Parthenios' memory was still fresh enough, however, in Tiberius' time for the latter to wish to imitate his poems.

Except for Meleagros, immortalized by the *Greek anthology* all the others are remembered only because of their Roman connections, Philodēmos and Archias with Cicero, Parthenios with Cornelius Gallus and Virgil.

Lesser prose writers. Greek prose was hardly more distinguished than Greek verse, that is, if one leaves out the authors who were primarily philosophers or men of science, such as Panaitios, Hipparchos, Polybios, Poseidōnios, or Strabōn. We have paid sufficient tribute to these, and we shall now deal only with lesser men, who might be attached to this or that philosophic school but were rather grammarians and rhetoricians. Let us introduce briefly a few of them.

There are, first, two men bearing the name Apollōnios of Alabanda,[4] and both teaching rhetoric in Rhodos. The earlier one was nicknamed Malacos (gentle); among his pupils he counted Quintus Mucius Scaevola the Augur (c. 121) and M. Antonius the Orator (98). The younger one was nicknamed Molōn,[5] and distinguished himself as a pleader in the law courts as well as the head of a school of rhetoric. In 81, when Sulla was dictator, Apollōnios Molōn was sent to Rome as ambassador of the Rhodians. Cicero heard him then and again in Rhodos (c. 78). Caesar was another of his auditors. Molōn wrote discourses, rhetorical treatises, and perhaps historical ones. The school of Rhodos obtained some reputation because its teachings were a compromise between Asiatic luxuriance and the austerity of Roman Atticists; it was inspired by the vigorous grace of Hypereidēs.[6]

Two Epicureans must now be named, Phaidros (140–70),[7] head of the Epicurean school in Rome, who counted Cicero as one of his pupils, and Philodēmos, already listed above among the poets. The fame of Philodēmos of Gadara was much increased when some of his treatises were found in papyrus rolls excavated in Herculaneum. One of Phaidros' books helped to inspire the *De natura deorum* of Cicero.

Another of Cicero's teachers was Philōn of Larissa, a member of the Academy. The city of Athens, having sided with Mithridatēs against Rome, was besieged and taken by Sulla (87–86). At that time, if not a little

[4] Alabanda in Caria. Could it be that these two men are but one, named Malacos in his youth and Molōn in his old age?

[5] I do not understand the nickname Molōn (*molōn* is the aor. 2 part. of *blōskō,* to come or go). Cicero called him Molon or Molo. He was also called Apollōnios of Rhodos, but it is better not to do so, to avoid confusion with a greater Apollōnios of Rhodos (third century B.C.), author of the *Argonautica.*

[6] Hypereidēs (c. 400–322) was one of the "ten Attic orators" listed in the Alexandrian canon (Volume 1, p. 258).

[7] Phaedrus the fabulist (c. 15 B.C.–c. A.D. 50) bore the same name, which in his case must be written Latin-wise. He flourished somewhat later than Phaidros, but also in Rome. He came from Macedonia and was one of Augustus' freedmen. Indeed, his collection of fables was entitled *Phaedri Aug. Liberti Fabulae Aesopiae.*

before, Philōn moved to Rome where he opened a school of philosophy and rhetoric. He is mentioned many times in Cicero's *Academica* and his *De natura deorum*.

Two representatives of the Peripatetic school have obtained fame in another way. They are Apellicōn of Teōs and Andronicos of Rhodos. Apellicōn was a rich collector of books who had managed to obtain Aristotle's manuscripts; when Sulla sacked Athens, he bought or took those inestimable treasures and brought them to Rome. Those manuscripts were put in order by Tyranniōn and the first edition was prepared by Andronicos.[8] Apellicōn died shortly before Sulla's capture of the manuscripts; Andronicos was still living in 58 B.C.

The Skeptic school was represented in Rome in Cicero's days by Ainēsidēmos, who wrote Pyrrhonic discourses (*Pyrrōneioi logoi*) in eight books. Ainēsidēmos came from Cnossos in Crete. He seems to have been a philosopher of some independence, who tried to combine Skepticism with Academism. His works are lost, but Sextos Empeiricos (II–2) was deeply indebted to him.

Apollodōros of Pergamon established himself in Rome a little later and was chosen by Caesar as tutor (teacher of rhetoric) for the young Octavius. Apollodōrus was primarily a teacher, not a writer, and he influenced the Romans chiefly by his explanation of the best Attic prose. A similar task was accomplished by Caecilius Calactinus[9] and by Dionysios of Halicarnassos, teachers of Atticism in the Augustan age.

All these men were rhetoricians and philosophers; they were necessarily philosophers because every rhetorician had a philosophic color and belonged to a definite school. Every main school was represented in Rome: the Academy, the Lyceum, the Porch, and the Garden, and a Skeptical voice could be heard as a correction to the others. These Greek writers were all living in Rome, or they were in contact with Roman leaders abroad. They were not patronized by Hellenistic princes but by Romans, such as Scipio Aemilianus, Cicero, Caesar, Maecenas, Augustus. Their outstanding merit was the transmission of the Greek language and Greek thought to the Roman aristocracy.

Instead of entitling this section "Greek letters," it would have been more explicit to entitle it "the progress of Greek literature in Rome."

LATIN LETTERS

When one remembers that the first year of the first Olympiad equals 776 B.C., and the first year of Rome, 753 B.C. (these dates are conventional and arbitrary but may serve as a first approximation), one cannot help being astonished by the lateness of Latin literature; when one remembers

[8] More details in Volume 1, pp. 477, 494, 495.

[9] He was called Calactinus, because he came from Calē Actē in Sicily. (Actē was an ancient, poetical name of Attica; *calē* means "beautiful.") Caecilius is Latin, therefore I leave the whole name in its Latin form.

that Greek literature began triumphantly with Homer (in the ninth century if not before), it is even more astonishing to be told that Quintus Ennius (II–1 B.C.), "the founder of Roman poetry," died as late as 169, almost seven centuries later than Homer. Such a shift in the case of two cultures that were, to some extent, concurrent is truly immense. The truth is simply that for the Greeks poetry came first, while the Romans had no time to think of it except after their political fortune had been made. They were a little like those businessmen who think that there will be time enough to educate themselves after the first million, but it is then almost always too late.

Livius Andronicus and Naevius. We must not exaggerate, however. Ennius was the first great poet, as great at his best as Virgil, but there had been Latin poets a generation before him. The very first worth mentioning was a Greek, Andronicos, who had been made prisoner at Tarentum in 272 and brought to Rome. His master, Livius, appointed him the tutor of his children and manumitted him, giving him his name as was the custom. From that time on, Andronicos was named Livius Andronicus, and it is under that Latin name only that posterity remembered him. Lucius Livius Andronicus opened a school, explained the Greek poets, and translated the *Odyssey* into Latin verse. He also translated Greek tragedies and comedies and his example was followed by many.

Another poet, a little ahead of Ennius, was Naevius (c. 270–201), who created the *fabula praetexta*, a new kind of tragedy based upon Roman subjects (Romulus' childhood, the defeat of the Gauls in 222, the first Punic war, 264–241). He was a true Roman, but he ventured to criticize the authorities, was thrown into prison, and died in exile, c. 201, at Utica, not very far from Carthage. It is impossible to judge his plays, for fragments only have survived.

Ennius. The first great poet was undoubtedly Ennius (239–169). Like Livius Andronicus he was of Greek origin, being born at Rudiae, Calabria in 239, but he soon learned to speak Latin as well as Greek. He was a centurion in the Roman army and was brought to Rome by Cato (he was the latter's Greek tutor). He flourished in Rome, was befriended by Scipio Aemilianus and others, and died at the age of seventy. He translated Greek plays, chiefly those of Euripidēs, into Latin, and wrote *Annals of Rome*, from the time of Aeneas to his own, in Latin verse (this was the first history of Rome in Latin). He wrote also two philosophical poems, *Epicharmos*, summarizing Pythagorean doctrine, and *Euhēmeros*, a rationalistic interpretation of religious tradition.[10] He was the forerunner of Lucilius, Lucretius, and Virgil.

[10] The heroes of these two poems were real philosophers, Epicharmos of Cōs (V B.C.) and Euhēmeros of Messina (IV–2 B.C.). Lucretius imitated Ennius' example

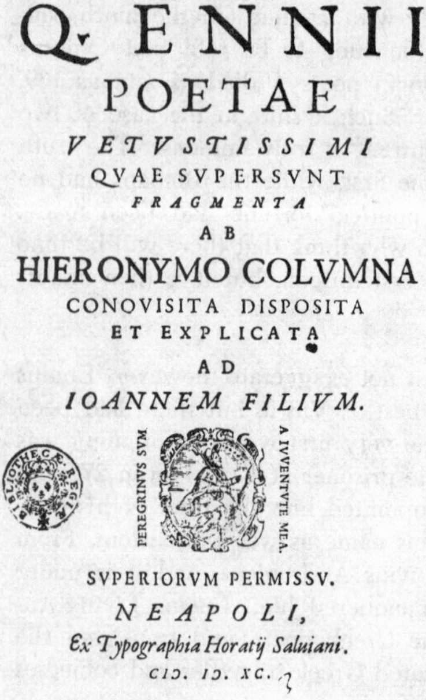
Q. ENNII
POETAE
VETVSTISSIMI
QVAE SVPERSVNT
FRAGMENTA
AB
HIERONYMO COLVMNA
CONQVISITA DISPOSITA
ET EXPLICATA
AD
IOANNEM FILIVM.
PEREGRINVS SVM A IVVENTVTE MEA.
SVPERIORVM PERMISSV.
NEAPOLI,
Ex Typographia Horatij Saluiani.
CIↃ. IↃ. XC.

Fig. 96. Ennius (II–1 B.C.). First separate edition of his relics by Girolamo Colonna (Naples, 1590). Fragments of Ennius had been printed before in the *Fragmenta veterum poetarum latinorum,* collected by Robert and Henri Estienne (Geneva: Henri Estienne, 1564). [Courtesy of Bibliothèque Nationale, Paris.]

Plautus and Terentius. Ennius' plays introduce the two leading Latin playwrights of his own time, Plautus and Terentius. Plautus was born at Sarsina in Umbria, c. 254, and died in 184. He wrote comedies derived from the Greek "New Comedy," mainly from Menandros. He copied the plays but took many liberties with them and his style is highly original and racy. He knew how to adapt an old story to the needs of a Roman audience, and he obtained considerable popularity.

Terentius (c. 195–159), born half a century later, was far more sophisticated than Plautus but less amusing. As Caesar put it, he lacked the "*vis comica.*" His plays were mainly derived, like those of Plautus, from the New Comedy, chiefly Menandros, but with infinitely more freedom. For one thing, he would not translate a single play but took inspiration in one play of his own from many Greek ones. He was not of Italian birth like Plautus but was born in Carthage [11] of Libyan stock and brought to Rome as a slave. His master gave him an excellent education, and, as soon as his genius became obvious, he received all the needed encouragements. He lacked Plautus's energy, but he exceeded him in urbanity; he was kind and

a century later, when he dedicated his *De natura rerum* to the glory of Epicuros.

[11] Greek culture was strong in North Africa, especially in Carthage, and therefore it is probable that Terentius had already obtained a living knowledge of Greek in his childhood.

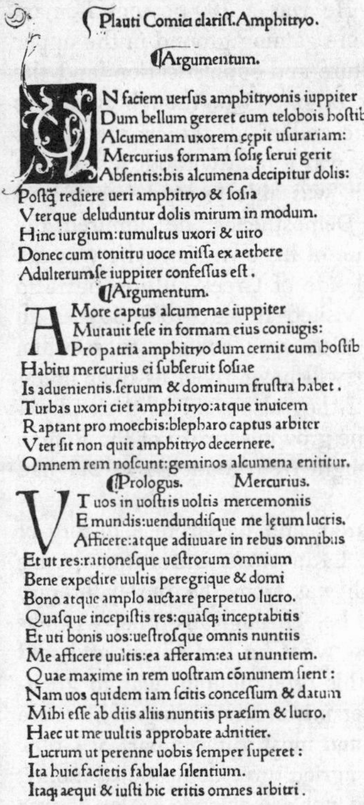
Plauti Comici clariss. Amphitryo.
Argumentum.
IN faciem uersus amphitryonis iuppiter
Dum bellum gereret cum telobois hostibus:
Alcumenam uxorem cępit usurariam:
Mercurius formam sosię serui gerit
Absentis: bis alcumena decipitur dolis:
Postq̃ rediere ueri amphitryo & sosia
Vterque deluduntur dolis mirum in modum.
Hinc iurgium: tumultus uxori & uiro
Donec cum tonitru uoce missa ex aethere
Adulterum se iuppiter confessus est.
Argumentum.
AMore captus alcumenae iuppiter
Mutauit sese in formam eius coniugis:
Pro patria ampbitryo dum cernit cum hostib
Habitu mercurius ei subseruit sosiae
Is aduenientis. seruum & dominum frustra habet.
Turbas uxori ciet amphitryo: atque inuicem
Raptant pro moechis: blepharo captus arbiter
Vter sit non quit amphitryo decernere.
Omnem rem noscunt: geminos alcumena enititur.
Prologus. Mercurius.
VT uos in uostris uoltis mercemoniis
Emundis: uendundisque me lętum lucris.
Afficere: atque adiuuare in rebus omnibus
Et ut res: rationesque uestrorum omnium
Bene expedire uultis peregrique & domi
Bono atque amplo auctare perpetuo lucro.
Quasque incepistis res: quasq; inceptabitis
Et uti bonis uos: uestrosque omnis nuntiis
Me afficere uultis: ea afferam: ea ut nuntiem.
Quae maxime in rem uostram cõmunem sient:
Nam uos quidem iam scitis concessum & datum
Mihi esse ab diis aliis nuntiis praesim & lucro.
Haec ut me uultis approbare adnitier.
Lucrum ut perenne uobis semper superet:
Ita huic facietis fabulae silentium.
Itaq; aequi & iusti hic eritis omnes arbitri.

Fig. 97. Plautus (c. 254–184). Princeps of his *Comedies,* edited by Georgius Merula, alias Giorgio Merlani (folio; Venice: Vindelin de Spira, 1472). [Courtesy of Pierpont Morgan Library.]

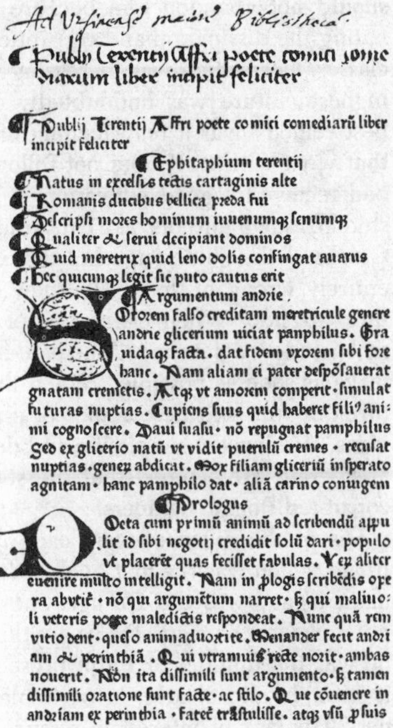
Publij Terentij Affri poete comici comediarum liber incipit feliciter
Publij Terentij Affri poete comici comediarũ liber incipit feliciter
Ephitaphium terentij
Natus in excelsis tectis cartaginis alte
Romanis ducibus bellica preda fui
Descripsi mores hominum iuuenumq; senumq;
Qualiter & serui decipiant dominos
Quid meretrix quid leno dolis confingat auarus
Hec quicunq; legit sic puto cautus erit
Argumentum andrie
Sororem falso creditam meretricule genere andrie glicerium uiciat panphilus. Grauidaq; facta. dat fidem vxorem sibi fore hanc. Nam aliam ei pater despõsauerat gnatam cremetis. Atq; vt amorem comperit. simulat futuras nuptias. Cupiens suus quid haberet filiꝰ animi cognoscere. Daui suasu. nõ repugnat pamphilus Sed ex glicerio natũ vt vidit puerulũ cremes. recusat nuptias. generũ abdicat. Mox filiam glicerium insperato agnitam. hanc pamphilo dat. aliã carino coniugem
Prologus
Poeta cum primũ animũ ad scribendũ appulit id sibi negotij credidit solũ dari. populo vt placerẽt quas fecisset fabulas. Verũ aliter euenire multo intelligit. Nam in plogis scribẽdis opera abutit. nõ qui argumẽtum narret. sed qui maliuoli veteris poete maledictis respondeat. Nunc quã rem vitio dent. queso animaduortite. Menander fecit andriam & perinthiã. Qui vtramuis recte norit. ambas nouerit. Non ita dissimili sunt argumento. sed tamen dissimili oratione sunt facte. ac stilo. Que cõuenere in andriam ex perinthia. fateẽ trãstulisse. atq; vsũ psuis

Fig. 98. Terentius (c. 190–c. 159). Princeps of his *Comedies* (folio; Strassburg: Johann Mentelin, not after 1470). [Courtesy of Huntington Library, San Marino, California.]

humane. Everybody remembers at least one of his lines:

Homo sum, humani nil a me alienum puto.

His comedies were less popular with the average person but more agreeable to people of education and refinement.[12] The gentle Terentian spirit was revived in the English plays of William Congreve (1670–1729) and in the Italian ones of Carlo Goldoni (1707–1793).

Cato the Censor. We gave in Chapter XXIV a long account of Cato's agricultural notes, put together c. 160. That is certainly not literature, but we

[12] René Pichon in his *Histoire de la littérature latine* (Paris, 1898), p. 8, went so far as to compare him with Marivaux (1688–1763). That was very high praise indeed.

should not abandon him because of that. He was a 100-percent Roman, hating the dissipation and corruption that were gaining ground in the upper classes. Corruption was increasing with culture and sophistication, and the highest culture was undoubtedly of Greek origin. Cato thought that the best remedy was a reawakening of rural life and of all the simple virtues that went with it. It does not follow that he was a man without letters. He had received a good education in his youth, was able to read Greek, and studied such authors as Thucydidēs and Dēmosthenēs; he admired the Greeks of the golden age but distrusted those of his own time (he was not entirely wrong in that). He knew the good side of Greek culture but also the bad side. When Carneadēs of Cyrēnē visited Rome (156–155) as an ambassador defending Athenian interests, Cato was anxious to see him leave as soon as possible; "When the Greeks will have given us their literature our ruin will be completed, especially if they also send us their physicians." He loved a simple life and despised the growing luxury of the Roman aristocrats. He claimed that the statues brought from Syracuse in 211 had corrupted Roman manners.

His public speeches were carefully prepared, and he wrote a history of Rome which was the first book of its kind in Latin prose. Unfortunately, his main legacy was the *De agri cultura* which was from the purely literary point of view as mediocre as anything could be. Through his other writings he was the founder of Latin prose. He knew what he had to say and said it strongly and clearly; in particular cases this brought him near to greatness. His lack of interest in science and utter misunderstanding of it made it impossible for him to appreciate the best and most durable part of Greek culture. For him science was futile, except agriculture, domestic economy, and jurisprudence. In other words, he could not see science where it was most brilliant, and saw it only where it was of necessity rudimentary.

His popular fame was artificially increased in modern times by two confusions. In the first place, he was confused with his great grandson, Cato of Utica (95–46), who committed suicide at Utica, when he was defeated by Caesar and preferred death to submission; Cato of Utica is one of the most heroic figures of the Roman Republic, and to many people the name Cato meant him. In the second place, he was supposed to be the author of the *Moral distichs*, which were as popular in medieval schools as were Aesop, Avianus, and Romulus.[13] When Chaucer (in the "Miller's tale") remarked, "He knew no Catoun for his wit was rude," this "Catoun" was the author of the *Distichs*. The confusion had begun in very early

[13] Aesop (Aisōpos) is the legendary author of the Greek fables bearing his name. According to Hērodotos (II, 134), he was a slave in Samos during the rule of Amasis (king of Egypt, 569–525) (Volume 1, p. 376). Avianus was a medieval fabulist writing in Latin verse. "Romulus" was not a man but the title of a version of Phaedrus in Latin prose. To the common people, Aesop, Avianus, Romulus were words of the same kind, the titles of schoolbooks.

days[14] and it continued at least until the eighteenth century. The *De moribus ad filium* (or *Disticha de moribus, Dicta Catonis*) was immensely popular in Latin and in many vernaculars. For most people, if not for everyone, the author was the old Cato and his fame was based on it. It was probably the first (and last) Latin book studied by Franklin and he himself printed an edition of it in English form.[15] This pseudo-Cato was one of the teachers of "Poor Richard."

We have now reviewed the first century (250–150) of Latin literature. It produced no Homer, no Hēsiodos, but half a dozen writers of merit: Livius Andronicus, Naevius, Ennius, Plautus, Terentius, Cato the Censor. That was not a bad beginning, even if it was a late one.

Scipio Aemilianus and C. Lucilius. One of the chief characteristics of that literature is that it was imitative. The best of it was translated from the Greek. Out of six writers, three were of Greek birth or had received a Greek education in childhood, Livius Andronicus, Ennius, Terentius. Even Cato, who denounced the Greek danger, could not help using Greek words.

That first century which witnessed the birth of Latin poetry and Latin prose was followed by a period of transition, which might be called that of Scipio Aemilianus (185–129) and of the Scipionic Circle, when the seeds of the golden age were sown. It was a period of intense Hellenization. Scipio's friends were Greeks like Panaitios and Polybios, but also Latin writers, like Terentius and Lucilius. It was during Scipio's leadership that Perseus's[16] library was brought to Rome (168) and that event stimulated a new and deeper interest in Greek letters.

Gaius Lucilius (c. 180–102) was born at Suessa Aurunca[17] in Latium and came to Rome after 160. He was a man of substance and a poet, the author of some thirty books, of which 1300 lines have survived. He was somewhat of a dilettante, composing *saturae* (which means medleys) about many subjects of the day, sometimes with a satirical intention but in a gentle vein. He was thus the forerunner of Horace, Persius (34–62), and

[14] Some at least of the *Distichs* were already in circulation in the second century after Christ, and the collection was known to the African Vindicianus (IV-2). Wayland Johnson Chase, *The distichs of Cato, a famous mediaeval textbook* (43 pp.; Madison, Wisconsin, 1922), Latin text with English version. This particular text contains 144 distichs, with a medieval addition of 56 very short lines.

The text was ascribed to Cato the Censor in the fourth century and a little later to one "Dionysius Cato," which compounded the error.

[15] *Cato's Moral distichs Englished in Couplets* (Philadelphia, printed and sold by B. Franklin 1735). The translator was James Logan (1674–1751). This humble booklet was the first Latin classic translated and printed in the British Colonies of North America. Franklin was aware that Cato the Censor was not the author. The booklet was reproduced in facsimile with a foreword by Carl Van Doren (Los Angeles: Book Club of California, 1939).

[16] Perseus was the last king of Macedonia (179–168). He was defeated at Pydna by Aemilius Paulus (Scipio, Aemilianus' father), captured in Samothracē, and brought to Rome to adorn Paulus' triumph. He died in 166.

[17] Now called Sessa Aurunca, 33 miles north-northwest of Naples.

Juvenal (fl. 100–130). Toward the end of his life, he retired to Naples where he died c. 102.

The last century B.C. is truly the golden age of Latin literature. Some of the great men of letters, who were far more than men of letters, have already been introduced to our readers, men like Lucretius, Caesar, Cicero, Varro, and Virgil, but we shall come back to them, as much as may be necessary to complete our picture.

Catullus. About Lucretius we need not add anything, however, for his only book, the *De rerum natura*, was fully discussed in Chapter XVII. His contemporary, Catullus, was a great contrast to him. They were almost perfect contemporaries, Lucretius dying in 55 at the age of 44, Catullus in 54, at the age of 30. Lucretius was inspired by Greek models, chiefly Epicuros; Catullus, by Hellenistic ones, the Greek-Oriental literature that grew up in Egypt and in the Asiatic kingdoms of the post-Alexandrian breakdown. Naevius and Ennius had exploited Greek letters for their country's education and advantage; Catullus had no such ideal; he was interested in poetry, whether Alexandrian or his own, only for the sake of literary elegance. His main subject of interest was himself, and the main events of his life were the sudden death of his brother in 59 and the treason of his mistress, Lesbia, a few years later. He wrote a great many poems, lyrical, elegiac, epigrammatic, of which 113 have come down to us; the artificiality of his art is redeemed by relative sincerity and by a few glimpses of deep feeling.

Gaius Valerius Catullus (c. 84–54) was born in Verona, being thus an Italian of the North, above the Po, like his friend Cornelius Nepos, like Virgil, Titus Livius, and the two Plinys. He loved his native country, especially Lake Garda (*lacus Benacus*). He came to Rome c. 62 and spent the rest of his life there except for short journeys.

He was a man of means who could indulge himself, and did. His art is "l'art pour l'art," without political or social convictions of any kind. In this he was very much like the Alexandrian poets whom he imitated; he shared their sophistication and wrote, like them, for the happy few; this was terrible because those happy few were not necessarily the best people, and sometimes they were very mean; he was better than his Alexandrian models in that he was simpler, less abstruse, and less allusive. His Roman public was on the whole more virile and less sophisticated than the Alexandrian or the Asiatic one. Catullus was not alone of his kind in mid-century Rome; there were many others who considered themselves the new writers or, let us say, the new clique (*neōteroi*), and we have many worse specimens of Roman Alexandrianism than those given by Catullus, for example, the poems that were formerly ascribed to Virgil's youth, but are most probably apocryphal.[18]

[18] For a list and bibliography, see the *Oxford Classical Dictionary* under the

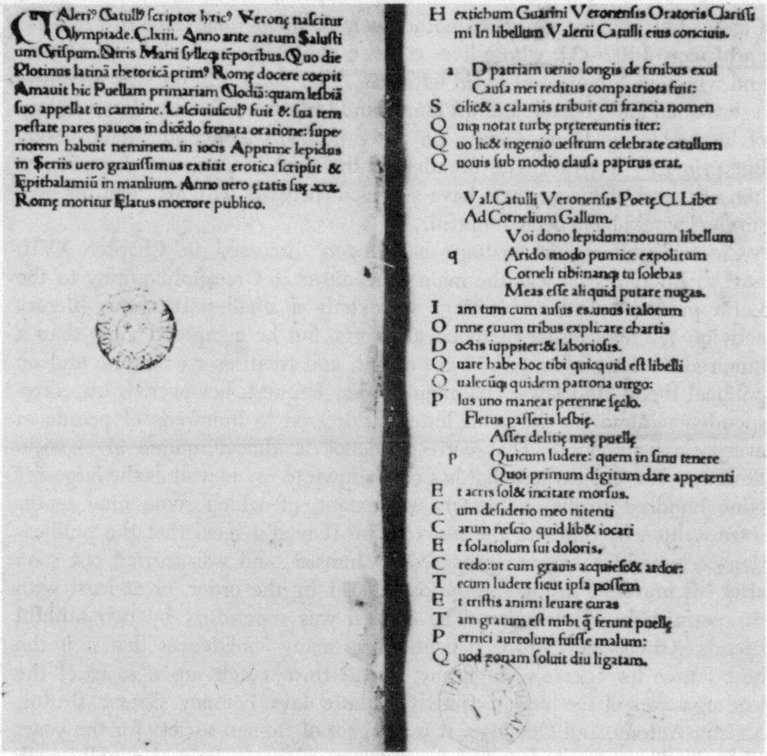

VAleriꝰ Catullꝰ scriptor lyricꝰ Veronę nascitur
Olympiade.C.lxiii. Anno ante natum Salusti
um Crispum. Diris Marii sylleqū tēporibus. Quo die
Plotinus latinā rhetoricā primꝰ Romę docere coepit
Amauit hic Puellam primariam Clodiā: quam lesbiā
suo appellat in carmine. Lasciuiusculꝰ fuit & sua tem
pestate pares paucos in dicēdo frenata oratione: supe
riorem habuit neminem. in iocis Apprime lepidus
in Seriis uero grauissimus extitit erotica scripsit &
Epithalamiū in manlium. Anno uero ętatis suę .xxx.
Romę moritur Elatus moerore publico.

H exticbum Guarini Veronensis Oratoris Clarissi
mi In libellum Valerii Catulli eius conciuis.

a D patriam uenio longis de finibus exul
Causa mei reditus compatriota fuit:
S cilic& a calamis tribuit cui francia nomen
Q uiq; notat turbę pętereuntis iter:
Q uo lic& ingenio uestrum celebrate catullum
Q uouis sub modio clausa papirus erat.

Val.Catulli Veronensis Poetę.Cl. Liber
Ad Cornelium Gallum.
Voi dono lepidum: nouum libellum
q Arido modo pumice expolitum
Corneli tibi: nanq; tu solebas
Meas esse aliquid putare nugas.
I am tum cum ausus es unus italorum
O mne ęuum tribus explicare chartis
D octis iuppiter: & laboriosis.
Q uare habe hoc tibi quicquid est libelli
O ualecūq; quidem patrona uirgo:
P lus uno maneat perenne sęclo.
Fletus passeris lesbię.
Asser delitię meę puellę
p Quicum ludere: quem in sinu tenere
Quoi primum digitum dare appetenti
E t acris sol& incitare morsus.
C um desiderio meo nitenti
C arum nescio quid lib& iocari
E t solatiolum sui doloris.
C redo: ut cum grauis acquiesc& ardor:
T ecum ludere sicut ipsa possem
E t tristis animi leuare curas
T am gratum est mihi q̄ ferunt puellę
P ernici aureolum fuisse malum:
Q uod zonam soluit diu ligatam.

Fig. 99. Princeps edition of Catullus, Tibullus, Propertius, and Statius, *Carmina Catulli quibus accessere Tibulli et Propertii carmina et Statii Sylvae* (Venice: Vindelinus de Spira, 1472). Catullus, Tibullus, and Propertius flourished in the first century B.C., but Statius (45–96) belongs to the following century and is not dealt with in this volume. [Courtesy of the Bibliothèque Nationale, Paris.]

It was necessary to speak of Catullus, because he was (before Ovid) the best representative of Alexandrianism, a dangerous phase in Roman culture, which justified Cato's fear and contempt, but must be taken into account. In the second place, he is noteworthy because of his influence; every Latin poet coming after him, even the greatest, like Virgil and Horace, was a little different because of him. Not to mention prosodic novelties, Catullus introduced into Latin poetry a new ingredient, a combination of elegy and preciosity, which could no longer be eradicated, and continued not only in Latin but in the Italian of Petrarca and the French of Ronsard.

heading *Appendix Vergiliana*. One of those poems, *Culex* (meaning gnat or midge) is the subject of a large volume by Charles Plésent, *Le Culex. Etude sur l'Alexandrinisme Latin* (514 pp.; Paris, 1910).

Cicero. Let us now consider another pair of great men, Caesar (102–44) and Cicero (106–43), whose lives covered almost the same interval of time and who have dominated Latin letters in other ways. Caesar was primarily a statesman and general, and his fame would be immense even if no writings of his had come to us; Cicero, on the contrary, was a political busybody but primarily a writer, the founder of the best Latin prose; happily for himself, most of his writings have survived, because without them his own survival would have been doubtful.

Cicero's philosophic writings have been discussed in Chapter XVII; next to Lucretius he was the main transmitter of Greek philosophy to the Latin public, but those writings were only a small part of his literary activity. We may forget his poetical efforts, but he composed more than a hundred orations, of which 58 are extant, and treatises on rhetoric and on political theory and laws. His main literary bequest, however, is his correspondence, almost a thousand letters addressed to hundreds of people of every class and kind. That correspondence is almost unique in classical literature; it is the earliest that has come down to us, as well as the largest.[19] Nine hundred thirty one letters are extant, of which some nine tenths were written by him and the rest to him. It would seem that the publication of some of them was prepared by himself, and was carried out soon after his murder (say in the period 43–31) by the order, or at least with the permission, of Octavian. The edition was supervised by two faithful friends, Atticus and Tiro.[20] It contains so many confidences that it is the best source for Cicero's biography, and it throws light upon some of the leading actors of the human comedy in those days, Pompey, Caesar, Brutus, Atticus, Antony, and Octavian. It is a mirror of Roman society for the years 68 to 43. There is no period of ancient history that is lighted equally well, for those letters take us behind the scenes. There has been much discussion among scholars as to Cicero's sincerity.[21] It would not do to claim that he always said what he thought or that he never veiled his opinions, but I am more tempted to believe him than not. After all, that rich correspondence

[19] The publication of Cicero's correspondence was imitated by Seneca (I-2) in his *Epistulae morales ad Lucilium*, by Pliny the Younger (c. 61–114), and by Marcus Cornelius Fronto (c. 100–66), the friend of Marcus Aurelius, but Cicero's correspondence is considerably larger.

[20] Titus Pomponius Atticus (109–32) was so called because he had lived long enough in Athens to be a true Atticist. He was a wealthy aristocrat and businessman, debonair, cautious, and tolerant, an easy Epicurean, and Cicero's most intimate correspondent until the end. Marcus Tullius Tiro (I-1 B.C.) was Cicero's freedman and secretary, the inventor of a kind of shorthand (*notae Tironianae*). He wrote a biography of Cicero and a few other books, but his main merit was to have helped to preserve and to publish his master's books and letters.

[21] For Cicero's correspondence, see Gaston Boissier, *Cicéron et ses amis* (Paris, 1865), translated into English in 1897 and often reprinted. Jérome Carcopino, *Les secrets de la correspondance de Cicéron* (2 vols.; Paris: Artisan du livre, 1947; English trans., London: Routledge, 1951). Boissier defended Cicero and believed in his sincerity; Carcopino distrusts him completely.

iussit:&cum eo.L.Cornelium Balbum Sextumq; peduceum.Hos ut uenisse uidit in cubitum innixus:quan'am inquit curã diligentiaq; in ualitudine mea tuẽda hoc tẽpore adhibuerim:cum uos testes habeam nihil opus est pluribus uerbis cõmemorare:quibus quoniam ut spero satisfeci me nihil reliqui fecisse quod ad sanandũ me ptiner&:reliquũ est ut egomet mihi consulã:id uos ignorare nolui.Nam mihi stat alere morbũ desinere.Nãq; his diebus quicquid cibi sũpsi ita produxi uitam: ut auxerim dolores sine spe salutis.Quare a uobis peto primũ ut cõsilium pbetis meũ:deide ne frustra dehortando conemini.hac oratione habita tanta constantia uocis atq; uultus:ut non ex uita:sed ex domo in domũ uideret' migrare.Cum quidẽ Agrippa eum flens atq; osculans oraret atq; obsecraret:ne ad id quod natura cogeret ipse quoq; sibi accelerar& : Et quoniam tum quoque poss& temporibus superesse se sibi suisq; reseruaret.preces eius taciturna sua obstinatione depressit. Sic cum biduũ cibo se abstinuisset subito febris decessit:leuiorq; morbus esse cœpit. Tamen propositum nihilo secius peregit.Itaq; die quinto postq̃ id cõsilium inierat pridie kalẽdas aprilis Cn.Domicio.C.Sosio cõsulibus decessit.Elatus est in lecticula ut ipse præscripserat sine ulla põpa funeris comitantibus omnibus bonis maxima uulgi frequentia. Sepultus est iuxta uiam Appiam ad quintum lapidem in monumẽto Quinti Cecilii auunculi sui.

.FINIS.

Attice nunc totus ueneta diffunderis urbe:
 Cum quondam fuerit copia rara tui.
Gallicus hoc Ienson Nicolaus muneris orbi
 Attulit:ingenio dædaliq; manu.
Christophorus Mauro plenus bonitate fideq;
 Dux erat.auctorem lector opusq; tenes.

MARCI.T.C.EPISTOLAE AD ATTICVM BRVTVM:
Et Quintum Fratrem cum ipsius Attici uita feliciter Expliciunt.

.M.CCCC.LXX.

Fig. 100. Cicero (I–1 B.C.), *Epistolae ad Atticum Brutum Quintum fratrem cum Attici vita* (folio, 30 cm, 182 leaves; Venice: Nicolas Jenson, 1470). These letters were discovered by Petrarch (XIV–1) at Verona in 1345 and copied by Coluccio Salutati (XIV–2). This is the first (or second) printed edition of them. Cicero addressed these letters to his old friend Atticus, to Brutus the tyrannicide, and to his own brother Quintus. The Frenchman Nicolas Jenson issued them during his first year of printing in Venice. We reproduce the last page with colophon. Another edition of the same letters was published by Sweynheym and Pannartz in Rome, same year, 1470, before 30 August. Did it appear before Jenson's edition the date of which is simply M.CCCC.LXX? [Courtesy of Harvard College Library.]

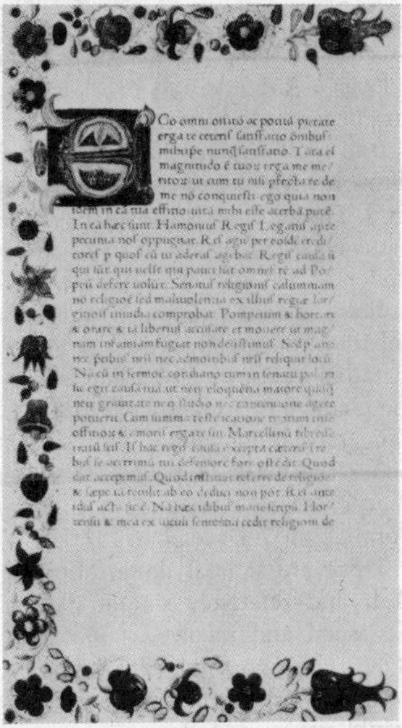

Fig. 101. Cicero (I–1 B.C.). Letters to his friends, *Ad familiares* (Rome: Sweynheym and Pannartz, 1467). [Courtesy of the John Rylands Library, Manchester.] When Coluccio Salutati (XIV–2) learned in 1389 that the Verona and the Vercelli manuscripts of Ciceronian letters were in Milan, he caused a copy to be made of the Vercelli manuscript and found that it contained the *Epistolae ad familiares*. In 1392, he received a copy of the Verona manuscript (discovered by Petrarch). The Vercelli original and the two copies made for Salutati are now in the Laurentian Library in Florence. Salutati was the first modern man who knew the bulk of Cicero's correspondence.

reveals a man very similar to the political man, statesman, senator, and advocate whom we know from his other writings or from contemporary opinion. He was very intelligent and, therefore, a liberal (left of center) rather than a fanatical partisan; being at the hub of affairs he was very well informed and understood the passions of men without being able to share them; he was a Stoic moralist rather than a politician. We know that he was full of vanity and his vanity shines repeatedly in his letters. Because his intelligence was combined with generosity his emotions changed and also his decisions; was that insincerity or greater sincerity to himself? He was a lover of the arts and belles-lettres, a true humanist, but not a complete one because of his ignorance of science.

It is unfortunate that many students become acquainted with Cicero through his orations, which are considered to be the very models of their kind. Those orations cannot be understood without a deep familiarity with the events leading up to them and teachers of Latin (not to mention their students) are seldom sufficiently well equipped to bring those orations back to life. His best political discourses are perhaps the *Philippica*,[22] which he delivered during the two last years of his life, against Mark Antony. The dictators used peace as a bait to buy the people's servitude, but Cicero had the courage to protest "Why do I not want peace? because it is shameful, dangerous, impossible . . . it is not that I reject peace, but I fear war under the mask of peace," [23] and he asked again and again, "Is slavery peace?" Mark Antony had his revenge and caused Cicero's murder at Formiae on 7 December 43.[24]

There are at least three different styles in Cicero's writings. First is the clear and relatively simple style of the treatises; second, the more impassioned and involved one of the political orations and the pleadings. It is strange to think that such long sentences, which drive students into despair, were necessary to persuade senators, judges, and jurymen; that would not work today, at any rate not in the United States, or rather it would work against the orator. Third is the style of the letters, which is the simplest and the best, especially in the letters that had to be written in haste. That style shows the man with all his weaknesses but also with his humanity and other virtues. It is spontaneous, elegant, full of images. One can imagine the enthusiasm of scholars when those letters, which had been lost, were rediscovered in the fourteenth century; innumerable editions of them have appeared since 1467.[25]

Cicero completed the creation of the Latin language. It was thought for a time that it was impossible to improve on the Ciceronian style or to add good words to his vocabulary. Such excessive claims caused a reaction best illustrated by Erasmus' *Ciceronianus* (Basel: Froben, 1528).

[22] The title is derived from that of the orations pronounced by Dēmosthenēs against Philip of Macedonia in defense of Greek freedom. The analogy was correct; Cicero was defending Roman freedom against Mark Antony. The word *philippica* ("philippic" in English) is often used in that general sense, a defense of freedom against dictators or potential dictators.

[23] "Cur igitur pacem nolo? quia turpis est, quia periculosa, quia esse non potest . . . nec ego pacem nolo, sed pacis nomine bellum involutum reformido" (*Philippica* 7, III, 9; VI, 19).

[24] Some horrible details are given by me in "The death and burial of Vesalius and, incidentally, of Cicero," *Isis* 45, 131–137 (1954).

[25] The *Epistolae ad familiares* were printed by Sweynheym and Pannartz (Rome, 1467). These were followed by

It is clear that no writer, however great, can ever stabilize the language; if he could stabilize it, not only its growth but its very life would be stopped.

Caesar. Caesar was not a professional author as Cicero was, but he had no difficulty in writing because he had been very well educated and was in the fullest sense a two-language man.[26] He was a humanist, an Atticist, a lover of belles-lettres.

His style was simple and direct; as his books were accounts of his own military deeds, they were autobiographical; he was primarily a man of action, a general who must seize fleeting opportunities and improve them to the limit, and this gave his expressions spontaneity and strength. It is for that reason that we observe similar qualities in the writings of other men of the same kidney, such as Frederick the Great and Napoleon.

Caesar is not only one of the great writers of his age, but he is unique in Latin literature and his *Commentaries* created a new literary genre.

M. T. Varro. The life of Marcus Terentius Varro was told when we spoke of his treatise on agriculture, but he holds a very high place in literature because of other writings which are lost. He was born before Lucretius, Caesar, and Cicero, but lived to be almost ninety and survived them by many years; he thus appeared to be of a later generation and lived long enough to greet the beginning of the Augustan age. On 16 January 27 B.C. the Senate gave Octavianus the title of "Augustus," and Varro died in the same year.

He was not a literary master like Cicero, nor even like Caesar, but he was a polymath of incredible fertility. According to Aulus Gellius (II–2),[27] who may have exaggerated, by the beginning of his eighty-fourth year Varro had already composed 490 books [28] and he continued to wield his

---

other letters and by various collections of them: *Epistolae ad Brutum* (Rome, 1470; Venice, 1470); *Epistolae ad Atticum* (Venice: Aldus, 1513); *Opera epistolica* (Paris, 1531); *Epistolae ad Octavium* (Paris, 1539); *Epistolae ad Quintum fratrem* (Lyons, 1543); and others. The immense vogue of Ciceronian writings during the Renaissance can be judged from the fact that 185 editions were listed by Margaret Bingham Stillwell, *Incunabula in American libraries* (New York: Bibliographical Society, 1940).

[26] To know a second language is always a spiritual asset, but at certain times it was a social necessity. A Roman of this time must know Latin and Greek, just as a Frenchman of the Renaissance had to know French and Latin, a German of the eighteenth century, German and French, a Canadian of the twentieth, French and English.

[27] *Noctes Atticae*, III, 10. The whole chapter is devoted to the virtues of seven and to Varro's *De hebdomadibus*. Varro remarked in that book that he composed it when he began his twelfth hebdomad of years (84) and that he had written seven times seventy books (490). See notes 28 and 33.

[28] Ancient and medieval scholars counted books, not works as we do. They would say that Galen wrote 262 books, while we would rather say that he wrote 122 treatises. The seven works of Varro considered by me, out of many more, amount to 319 books. The number 490 is thus not as terrible as it seems.

stylus or to dictate for half a dozen years. Whatever the outcome of that may have been, there are only seven works of his that have escaped oblivion. Only two of the seven are extant, of which one is the agricultural treatise discussed in Chapter XXI and the other a fragment of his treatise on the Latin language which will be dealt with in the next chapter.

The five others, to be mentioned presently in chronologic order, are significant if one considers them in the Greco-Roman background of the second century.

(1) *Satyrarum Menippearum libri CL* (*Menippean satires*). These are not satires in our sense but essays in prose mixed with verse, according to the model left by the Cynic philosopher Menippos of Gadara.[29] These essays are humorous rather than harshly satirical, though one of their purposes seems to have been the denunciation of luxury and other social weaknesses. They were composed between 81 and 67.

(2) *Antiquitatum rerum humanarum et divinarum libri XLI*. This was a history of profane antiquities (25 books) and sacred ones (16 books), written in 47. It has been discussed in Chapter XXIV.

(3) *Logistoricon libri LXXVI*,[30] dialogues written after 45 on many subjects. Judging from what remains, each dialogue had a double title, such as "Catus on the education of boys," "Marius on fortune," "Atticus on numbers," "Pius on peace."[31]

(4) *Hebdomades vel de imaginibus libri XV* (*Hebdomads or images*), written in 32. The second word of the title, *images*, revealed the main purpose; it is a collection of 700 biographies of famous Greeks and Romans, presumably short sketches because there were so many of them. According to Pliny,[32] the text was illustrated with 700 portraits; this is not impossible but extraordinary. It may be that one manuscript was illustrated in that way (?). The word *hebdomades*, occurring first in the title, recalls the importance attached to the number seven or to hebdomadic cycles.[33]

[29] See the account of Menippos and of the *Satire Ménippée* in Chapter XIII. That form (mixture of prose and verse) was recognized by Quintilian (I-2) in his *Institutiones oratoriae* (10, 1, 95); it was imitated not only by Varro but also in Petronius' *Satyricon* (time of Nero, emperor 54–68), and by Seneca (I-2) and Martianus Capella (V-2).

[30] This title is Greek; *logistoricōn* is the genitive plural of *logistoricos*, meaning "skilled in computation, rational." A Greek would not have given that title to the same book.

[31] I have used the fragments edited in Latin and Italian by Ettore Bolisani, *I logistorici varroniani* (123 pp.; Padua, 1937).

[32] *Historia naturalis*, xxxv, 2. Pliny refers to Varro's innovation as a *benignissimum inventum* and it has been argued that the portraits were probably transferred from one manuscript to another by means of stencils. Of course, that might have been done; it may be that the Egyptians already used stencils to reproduce large hieroglyphics on monuments. At any rate, Varro realized the value of graphic portraits to complete the literary ones. That is very remarkable.

[33] Hebdomadic fancies, probably of Oriental origin, were cultivated by the Pythagoreans (Volume 1, pp. 214, 215, 491). Examples occur in the Old and New Testaments, and the Christian religion is full of them: seven sacraments, seven capital sins, seven choirs of angels, seven sorrows, seven joys of Mary, seven

(5) *Disciplinarum libri IX.* We have already referred to Varro's *Disciplinae*, which was a kind of encyclopedia, or program covering the whole field of studies of a gentleman, all the "liberal arts" as distinguished from technical knowledge, such as agriculture, medicine, or business methods.

Other books of his dealt with history, law, geography, music, medicine, and what not. He was a learned man anxious to make Greek knowledge available to his Roman brothers who needed it; he was equally anxious to illustrate the Roman past, sacred and profane. For example, he continued the studies of his teacher, Stilo,[34] on Plautus's comedies. He did not investigate anything deeply, because the field that he had started to till was too large, but he filled a real need: he explained Greek and Roman antiquities to the growing mass of citizens who were not able to drink from the very sources. He did on a lower level what Cicero was doing on a higher one, and the efforts of the two men were equally useful.

Varro's merit was recognized not only by his contemporaries, Cicero to begin with, but even more by such men as St. Augustine (V–1), who was witnessing the twilight of antiquity. Later still, in Dante's time, he was regarded as one of the great masters, in the company of Cicero and Virgil. This may shock us, but we should not forget that they knew of him much that is lost to us.

Varro was not discussed in Chapter XVII (philosophy), because he was not a philosopher in the technical sense, less so even than Cicero and Lucretius, yet he was a serious thinker, deeply concerned with the main problems of life. For example, we know from St. Augustine[35] that he had been cogitating on the supreme good (*summum bonum*)[36] and calculated that there must be 288 different opinions about it; then he analyzed those opinions and found that the differences between them were often specious; he reduced the number of sects from 288 to 12, then to 6, finally (like Cicero) to 3. It must be the supreme good of the body, or of the soul, or of both. The last-named possibility was his final choice (it was the Academic choice, not the Stoic).

Sallust. Both Sallust and Livy have been discussed in Chapter XXIV, but it is necessary to insist here on their literary qualities which were perhaps the main causes of their popularity and of their influence. Both were masters of Latin prose and created some of the best examples of its golden age. They were contemporaries in the way fathers and sons are; when

canonical hours. More examples could be found in other fields. See note 27.

[34] More about Stilo in the next chapter.

[35] I cannot put my finger on St. Augustine's statement but quote it from Gaston Boissier, *Etude sur . . . Varron* (Paris, 1861), p. 117.

[36] This was truly a crucial problem, for, as Cicero put it, "Qui autem de summo bono dissentit, de tota philosophiae ratione dissentit" (*De finibus bonorum et malorum,* v, 5): "He who disagrees on the supreme good disagrees on the whole philosophy."

Sallust died in 34, he was 52 and Livy 25; but the difference in time was really much greater than the years suggest. Sallust was closer to Caesar and Cicero, while Livy did most of his work during the Augustan age and died only in A.D. 17 under Tiberius.

The style and outlook of the older man were modeled upon those of Thucydidēs. He tried to imitate the impartiality of the Greek historian; his style was concise, clear, and dramatic. His chief merit lay in the drawing of portraits, either directly or indirectly in the form of discourses which each man was supposed to have delivered and which revealed his own passions and weaknesses, his personality. So great was his desire to express himself briefly and strongly that some of his phrases are truly epigrammatic. For example, he will say, "The wasting of other people's wealth is (now) called liberality, and criminal daring, fortitude," or "What is friendship between good people becomes conspiracy between evil ones."[37] These two sayings illustrate also his love of contrast of words and ideas, and his psychological tendencies. He was a disgruntled man of action who found his revenge and consolation in the form of literary candor and bitterness.

Livy. Livy's chief models were Polybios and Cicero. He was considerably more of an artist than the former but less erudite; his conception of history was rhetorical. His general aim was to justify the glory of Rome; his *Decades* are as patriotic as the *Aeneid*, but, while Virgil used poetry, Livy wrote in prose, the most eloquent Ciceronian prose he could achieve. He was as honest as a good patriot can be, but that is not enough for a true scholar. The Romans as described by him were a good deal better than nature had made them; he wanted to educate and to improve his own readers and made use of history as a mirror which would show them as they had been at their best and could be again, if they were worthy of their ancestors. Such methods are terribly out of fashion today, and Livy has lost most of his prestige; in fact, modern readers can hardly bear him, but he gave the Romans of the Augustan age just what they needed. His history was as welcome to them as the *Aeneid*, and when the people of later times tried to evoke the grandeur of Rome and the dignity of the old Romans they appealed to Livy, whom they admired almost as much as they did Cicero and Virgil. Dante praised him[38] and scholars of the Renaissance were just in the right mood to appreciate his oratorical conception of history and his literary mannerisms. We are not in that mood any more and can hardly recapture it.

[37] The Latin original is terser and more effective. "Bona aliena largiri liberalitas, malarum rerum audacia fortitudo vocatur" (*Catilina*, 52, 11). "Inter bonos amicitia, inter malos factio est" (*Jugurtha* 31, 15).

[38] "Come Livio scrive che non erra" (*Inferno*, XXVIII, 12).

THE LATIN POETS OF THE AUGUSTAN AGE

Maecenas. While Livy has left us the best image of the Augustan age available in Latin prose, the true literary splendor of that age is represented by its poetry. Before examining its poetic offerings, it is well to give a moment's attention to a man who was not a poet but a friend of poets, not a creative writer but the patron of Augustan literature. He was so magnificent in his devotion to arts and letters that he has become the eponym of his successors. When we wish to give the highest reward to a patron of the humanities, we call him a Maecenas.[39]

Who was the original Maecenas? Our first surprise is to learn that Gaius Maecenas was a scion not of the Roman aristocracy but of the Etruscan, and this may help us to remember that the vigorous children of the Roman republic were educated by the Etruscans before the Greek cornucopia was opened to them. Maecenas' father and grandfather were Roman citizens and knights. We know the very day of his birth, 13 April,[40] but not the exact year (c. 68). He received the best training in Greek as well as in Latin and wrote compositions of his own in verse and in prose. He became acquainted with Octavius at Apollōnia (Illyria), before Caesar's death, but was first mentioned as Octavianus'[41] friend in 40. Their friendship prospered from that time on; he was employed by Octavianus as a counselor and diplomatic agent, and by the emperor as a confidential servant. Even as Agrippa was Augustus' right hand for military matters and public works, Maecenas was his first adviser for public letters and humanities. This was not an official position but one of immense importance which he filled in the best manner. Maecenas was the friend of Horace, Virgil, and Propertius and, in the emperor's name as well as in his own, their patron. He died in 8 B.C., bequeathing his vast estates to Augustus.

Maecenas was probably an Epicurean; he was gentle and generous. His enlightened patronage of letters was essentially a form of imperial patronage, but no other could have lasted during the principate. We may be grateful that Augustus found the right man to promote the literary glory of his reign.

Virgil (70–19). The two greatest poets of ancient Rome, Virgil and Horace, and their promoters, Augustus and Maecenas, all came to light within a very few years (70–63). Augustus was the youngest, and Virgil possibly the oldest.

We have already spoken of him apropos of the *Georgica*, but we must

[39] The homage is truly international. A great patron of letters is called Mécène in French, Mecenate in Italian, Mecenas in Spanish, Maicēnas in Greek.

[40] Horace, *Ode* IV, 11. Maecenas was born on the Ides of April.

[41] Octavius was Caesar's heir, in 44, and recognized as such under the name of Gaius Julius Caesar Octavianus. In 27, he was hailed Augustus by the senate. I feel obliged to repeat this because my readers may forget that Octavius, Octavianus, Augustus are the same person at different dates (63, 44, 27).

come back to him and celebrate him as elaborately as the frame of this book allows, because his is one of the greatest personalities in the history of the whole West. He belongs to a very small group of universal poets; behold him standing between Homer and Dante; no others are equal to these, though the Portuguese might mention the name of Camões and English Protestants that of Milton.

The relation between Virgil and Homer is particularly close, because the former imitated the latter. This is a new example, and the most magnificent, of the dependence of the Roman genius upon the Greek. Even as Lucretius and Cicero explained Greek philosophy in Latin, so did Virgil create a Latin epic that was modeled upon Greek patterns, the *Iliad* and the *Odyssey*.

The relation between Virgil and Homer was even deeper. The ancients admired Homer so much and knew him so well that learned Romans were led into the fantastic belief that Rome had been founded by descendants of the Trojan kings. The *Aeneid* was the mature development of that legend. From its point of view the *Iliad* was a kind of introduction not only to Greek history, but to Roman!

Virgil (Publius Virgilius Maro) was born on the Ides of October (15 October) 70, in a village near Mantua, in Venetia, north of the Po. His father was a small farmer who earned his living by beekeeping and was sufficiently prosperous to send his promising son at twelve to a good school in Cremona. We find Virgil there celebrating his fifteenth birthday (15 October 55) with the assumption of the *toga virilis*; that is, he was considered a man at 15, a little earlier than was usual. He went the same year to Milan for a short time and to Rome, to complete his education. This meant the study of rhetoric and perhaps of astronomy and medicine; later still, he became the disciple of Siro the Epicurean. His interest in poetry began very early, under the influence of the Alexandrian poets, of their imitator Catullus, and, above all, of Lucretius. He was probably in Rome before, during, and after the period 53–46, and one can easily imagine the perturbations and confusion of a sensitive boy, poor in health, lost in the great city, suffering the anxieties of the Civil War and the indignities of political corruption; one can imagine also his nostalgia for the sweet land of his birth; he returned to Mantua about 44 or 43. Unfortunately, soon afterward (42), that part of the country (including his father's estate) was seized to be distributed to veterans of the Civil War. Virgil returned to Rome to obtain some compensation.

When Octavian had put an end to chaos, Virgil was ready to worship him, soon got into Maecenas' good graces, and enjoyed Horace's friendship. His patrimony was not returned to him, but he was given instead a villa in Nola.[42] This was a decisive event, for he came to love Campania and

[42] Nola was one of the most ancient towns of Campania, inland but not very far from Naples. It was there that Augustus died in A.D. 14.

the Bay of Naples even more than the country of his birth; there is no appearance that he ever returned to Mantua. The *Bucolica* were probably written by him in 42–37, that is, in Rome, but the *Georgica* in 36–29, in Nola, and the *Aeneid* in Nola and Cumae; his main creations were the fruits of his life in Campania. What a country to live in for a poet, a country full of natural beauty and of glorious remembrances. The reader was given some idea of this in the chapter on the Phlegraean Fields. If one wants to visualize Virgil, it is there that one must seek for him, not in the land of his birth but rather in the one that was the nursery of his genius.

The *Aeneid*[43] is the story of Aeneas, a royal prince of Troy, and of his wanderings after the capture of his ancestral city. For seven years he and his companions wandered from the Troad, to Thrace, Crete, Ēpeiros, Tarentum, Sicily, Carthage, then back to Sicily, to Cumae, where Aeneas consulted the Sibyl, and to Latium, where he married the king's daughter, Lavinia. This is a legend of the distant origins of Rome, such as was already known to Naevius and Ennius, but which Virgil amplified with great learning and enthusiasm. His purpose was to write the national epic of his own country; it was also to emulate the Greeks. He imitated in it both the *Iliad* and *Odyssey* and borrowed not a little from other Greek poets as well as from the Latin poets who have just been named.

The *Aeneid* is divided into twelve books of almost equal length.[44] A summary of it would be vain and tedious. Considered as a narrative it is disappointing, for the story is disjointed and perplexing; the reader is taken here and there and gets lost. Instead of following the poet in the labyrinth of innumerable episodes, it is better to indicate the characteristics of the whole work.

After many vicissitudes the Roman state had reached its climax and the time was ripe to explain it and to justify it. The greatness of Rome is not an accident but the necessary triumph of a providential evolution. Virgil's extraordinary conception is to explain the development of Roman glory as understood prophetically by his hero Aeneas, who lived many centuries before. The glory of Rome and that of the emperor, Augustus, are combined in the same prefiguration.

Much of it seems very artificial to the modern reader, but he must try to put himself in the place of the first readers, Augustus and his friends. Our knowledge of the old history, the mythology, and the legends is so weak that we cannot follow the poet except with the help of a multitude of footnotes, and reading with so many halts is not pleasant. The learned

[43] *Aeneid* is the English title, which has been used so long that it has become a part of our language. The original Latin form is *Aeneis*, and the hero's name is Aeneas (Aineias in Greek).

[44] It contains 9893 lines, being much shorter than the *Iliad* and even than the *Odyssey*. For the length of those epics and others, see Volume 1, p. 134. The average length of the books is 824 lines; the shortest (IV) has 705 lines, while the last book is the longest, 950 lines.

Romans could enjoy a double pleasure, first that of recognizing Greek reminiscences, and second that of attending the unfoldment of Roman destinies, the gratification of Roman wishes and ambitions.

When all that is said, it must be admitted that Virgil's mythology is dry, the gods and goddesses very conventional and lifeless, or rather they behave and talk like Roman gentlemen. Few persons are really alive and memorable; I can think only of Dido,[45] who is truly moving, and, to a much smaller degree, Aeneas himself, "pious Aeneas."

We expect an epic poem to be simple and naïve, but the *Aeneid* is very sophisticated. It is almost impossible to read it from end to end, but it contains many touching episodes and a great many admirable lines. As long as Latin was a living language, educated people knew many of those lines by heart and could quote them without need of reference. Everybody knew them just as English people know lines of Shakespeare, without being able in many cases to replace them in their context. There was no need of that: those lines were beautiful in themselves, and it was a great pleasure to quote them oneself or to acknowledge them immediately in the sayings or writings of one's friends.

Much of the *Aeneid* was meant to be studied, just as one must study Lucretius. For example, Book VI begins with Aeneas' arrival at Cumae, his consultation with the Sibyl, and his request to be permitted to visit the underworld; this turns into a treatise on eschatology, explaining the theory of rewards and punishments after death, the Pythagorean idea of metempsychosis, and the Stoic idea of a world soul. This book also contains the best presentation of Rome's actual and future greatness. Every citizen must have recited with pride and joy the lines (VI, 851–853):

> Tu regere imperio populos, Romane, memento;
> hae tibi erunt artes, pacique imponere morem,
> parcere subjectis, et debellare superbos.[46]

It was such lines that raised Virgil above every other poet of the Roman age; they survived the empire but immortalized him.

It has been remarked that the *Aeneid* is essentially different from the Greek epics in that it is impregnated with religious piety and with moral earnestness. It is concerned with Aeneas' wanderings and with civil wars, but more deeply with a kind of pilgrimage and holy war. For Virgil, Roman religion was an essential part of the Roman empire; the latter could not exist without the former. The good Roman must be pious like Aeneas and strong like Augustus.[47]

[45] Dido was the legendary daughter of a king of Tyre, the founder and queen of Carthage. She fell in love with Aeneas, who after a while was obliged to leave her in obedience to divine command; she threw herself on a pyre.

[46] "Remember, O Roman, thy destiny to rule the peoples imperially; these shall be thine arts, to impose peaceful customs, to spare thy subjects and subdue the proud."

[47] This was from the truly imperial

Some of the most agreeable features of the *Aeneid* are the poet's love of nature (some of its lines are comparable to lines in the *Georgica* and the *Bucolica*) and his love of humanity, his tenderness. He was not simply pious, which is good, but genuinely kind, which is rarer. Such qualities were the more valuable in very hard times. The age when Virgil flourished was not simply a golden age, it was an age of blood and tears, of cruelty and bestiality.[48] One cannot praise Virgil too much for having preached better ideals; he was the highest civilizing influence of his time and place, and he has continued to be one throughout the centuries.

The composition of his masterpiece took Virgil eleven years (30–19); he worked at it in Campania, and perhaps during occasional visits in Sicily. Much remained undone and, like a good craftsman, he was displeased with many lines and wanted to replace them with better ones. In particular, Book III, describing Aeneas' journey, did not satisfy him; he wanted to visit Greece and Asia, which would enable him to add details and put more color in the background. He left in 19, planning to devote three years to his journey, but he fell ill in Megara and managed painfully to reach Athens, where Augustus, ending a two years' vacation in Greece, was planning to return home. Augustus realized that the poet was in no condition to continue his own journey and persuaded him to return with him. After a difficult crossing, they landed in Brindisi; Virgil was now exceedingly ill and so depressed that he ordered the destruction of his work; but before this could be done, he died (at Brindisi) on 21 September 19 B.C.

He was buried, according to his own wish, in a tomb two miles out of Naples on the road to Pozzuoli; we have already taken the reader to it in a previous chapter.

His wish to have the *Aeneid* destroyed was wisely put aside. Instead, two friends of his, L. Varius Rufus and Plotius Tucca, were ordered by the emperor to revise the poem and publish it; they added nothing to it and their revision was restricted to minor corrections.

As Virgil was already illustrious at the time of his death and hailed by the emperor himself as the Poet of Rome, high above all others, his tradition was secure from the beginning.

Public lectures on his achievements were offered soon after his death by Quintus Caecilius Epirota,[49] headmaster of a Roman school (since

---

point of view a natural combination. The empire must be a holy empire or cease to be. Similar ideas were developed by Muslim writers during the golden age of Islam, by Russian writers (like Dostoevski) apropos of the Orthodox empire, and by English ones associating imperialism with the Church of England.

[48] Think only of the inhumanity of the circus games and of the sadistic joys of their spectators.

[49] The name Epirota (= *ēpeirōtēs*) shows that this Caecilius was of Greek origin; he was freed by Atticus, Cicero's friend, before 32. Atticus had vast estates in Epeiros and had inherited additional

26 B.C.). Such lectures had been devoted to Homer and possibly other Greek authors, but Epirota's were the first helping to celebrate a Latin one.

There were two reasons for celebrating Virgil, the first and the best his being a great poet and the national one; the second was his erudition. The learning displayed in the *Aeneid* was of the very kind to attract the critical attention of grammarians and exegetes. Explanations were written by Aelius Donatus (IV–1), a grammarian of such fame that his name, Donat or Donet, came to mean a grammar, and by others; all those commentaries were put together by Servius, who flourished in the fourth and fifth century. Servius' work was designed for school purposes. This illustrates another aspect of Virgil's fame. He was established very early as a classic, that is, an author to be used in the classrooms, for the edification of some children and the torture of many more.

On account of his gentleness and piety and of his messianic eclogue (No. IV; 40 B.C.), Virgil was taken to be a herald of Christ, and his poetry was popular in Christian circles. While pagan authors were occasionally denounced by the clergy and the reading of their works was discouraged, this was never the case for Virgil; hence, his tradition was never interrupted in the Latin West. One startling example of the veneration he was held in is Dante's use of him as his guide through Hell and Purgatory.

His fame was so great in medieval times that it degenerated into superstition. He became a legendary figure, a man of superhuman wisdom, a magician or necromancer.[50] Idiotic admirers of Virgil's poetry used it for the making of centos, poems made up entirely of Virgilian lines or hemistichs arranged in such manner as to make sense in any unvirgilian way. Finally, many people used Virgil's poems, especially the *Aeneid*, to take auguries, say, from the first word or the tenth line of a page taken at random. These auguries were called *sortes Virgilianae*. The same practice was resorted to with the Bible (*sortes sanctorum*) and by Muslims with the Qur'ān or with the poems of Ḥāfiẓ (XIV–2).[51]

The manuscripts of Virgil are the best witnesses of an early tradition which continued throughout the ages. There is no other Latin writer of whom we have so many ancient manuscripts; no less than seven date from the second to the sixth century, all written in capitals without separation between the words, on sheets of vellum forming books (codices). Many more manuscripts were written in minuscules during the Carolingian age (ninth century), and by that time the text was well established.

---

wealth from his uncle, Q. Caecilius. Does this account for Epirota's name?

[50] This has been explained in great detail by Domenico Comparetti, *Virgilio nel Medio Evo* (2 vols.; Leghorn, 1872; rev. ed., Florence, 1896); English trans. by E. F. M. Benecke (London, 1908). John Webster Spargo, *Virgil the necromancer* (Cambridge, 1934) [*Isis* 22, 265–267 (1934–35)].

[51] *Introduction,* vol. 3, p. 1457.

We must next consider the early editions, not only because the *Georgica* is a scientific book, one of the most important of its time, but also because Virgil is one of the outstanding personalities of our culture. The early editions established the tradition on an imperishable basis.[52]

The princeps of the *Opera* was printed by Sweynheym and Pannartz (Rome, 1469) (Fig. 102). It was followed by two other editions within a year (Strassburg, 1469–70; Venice, 1470). There were altogether 91 editions in the fifteenth century, 184 in the sixteenth, 82 in the seventeenth, and so on, crescendo. The editions prepared by Niklaas Heinsius (Amsterdam, 1664, 1671, 1676) marked the beginning of a more critical tradition.

The first edition of the *Bucolica* and *Georgica* was printed in Paris in 1472, and of the *Georgica* alone in Deventer c. 1486.

The earliest printed translations are: Italian, *Aeneid* (Vicenza, 1476), *Georgica* (c. 1490); French, *Aeneid* (Lyons, 1483); English, *Book of Eneydos compyled by Vyrgyle* (London: William Caxton, 1490) (Fig. 103); German, *Aeneid* (Strassburg, 1515), *Georgica* (Görlitz, 1571–72); Spanish, *Aeneid* (Antwerp, 1557), *Georgica* (Salamanca, 1586); Polish, *Aeneid* (Cracow, 1590).

These short notes are sufficient to prove that by the year 1600 it was easy to obtain printed editions of Virgil not only in Latin (275 editions!) but also in six vernaculars.

Horace (65–8). Horace's fame as a Latin poet was second only to Virgil's; he was never as popular as Virgil, and he was never believed to be a magician, but he was duly admired and revered by every man who knew Latin well enough, and until the last century that meant every educated man. He had been wondering about the meaning of fame and at the end of one of his epistles he addressed it thus: "You may be thumbed by the hands of the vulgar or made food for moths, or be exiled to Africa or Spain. . . Ah, something more dreadful yet may be in store; you may become a textbook for beginners in the outskirts of Rome." [53] That kind of fame has been realized. Horace is one of the classics [54] and his books have become textbooks, but what would you have? Is not that the common punishment of literary genius everywhere?

Quintus Flaccus Horatius was born at Venusia [55] on 8 December 65, his father being a former slave who had gained freedom and wealth. Young Horatius was sent to the best school in Rome and for graduate studies to

[52] Giuliano Mambelli, *Gli annali delle edizione virgilianae* (392 pp.; Florence: Olschki, 1954); *Gli studi virgiliani nel secolo XX* (2 vols.; Florence: Sansoni, 1940).

[53] *Epistles*, I, 20, written c. 20 B.C.

[54] He was already a school classic a century after his death, if not before. Augustus' praise was the beginning of his glorification.

[55] Venusia (modern Venosa) in Apulia; near Mount Vultur. It is a place in southern Italy as deep inland as possible, a little town too far away to be visited by tourists.

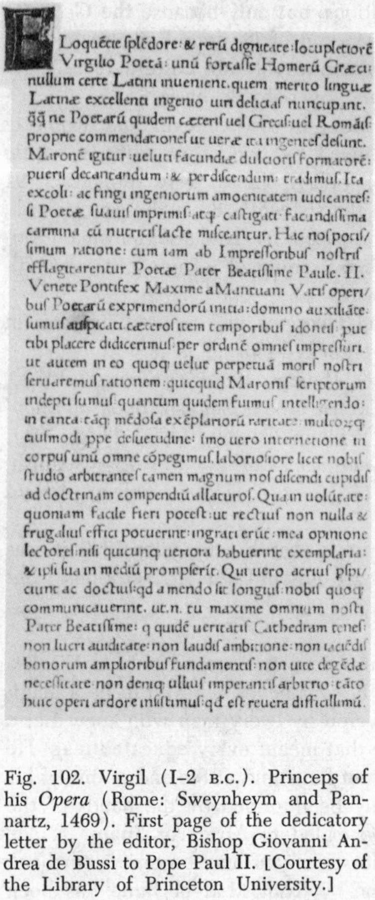

Eloquẽtie splẽdore: & rerũ dignitate: locupletiorẽ Virgilio Poetã: unũ fortasse Homerũ Græci: nullum certe Latini inuenient. quem merito linguæ Latinæ excellenti ingenio uiri delitias nuncupant. q̃q̃ ne Poetarũ quidem cæterisuel Græcis: uel Romãis: proprie commendationes ut ueræ ita ingentes desunt. Maronẽ igitur: ueluti facundiæ dulciorisformatorẽ: pueris decantandum: & perdiscendum: tradimus. Ita excoli: ac fingi ingeniorum amoenitatem iudicantes: si Poetæ suauis imprimis: atq̧ castigati facundissima carmina cũ nutricis lacte misceantur. Hac nos potissimum ratione: cum iam ab Impressoribus nostris efflagitarentur Poetæ Pater Beatissime Paule. II. Venete Pontifex Maxime a Mantuani Vatis operibus Poetarũ exprimendorũ initia: domino auxiliãte: sumus auspicati cæteros item temporibus idoneis: put tibi placere didicerimus: per ordinẽ omnes impressuri. ut autem in eo quoq; uelut perpetuã moris nostri seruaremus rationem: quicquid Maronis scriptorum indepti sumus: quantum quidem fuimus intelligendo: in tanta: tãq; mẽdosa exẽplariorũ raritate: multorũq; eiusmodi ppe desuetudine: imo uero internetione in corpus unũ omne cõpegimus. laboriosiore licet nobis studio arbitrantes tamen magnum nos discendi cupidis ad doctrinam compendiũ allaturos. Qua in uolũtate: quoniam facile fieri potest: ut rectius non nulla & frugalius effici potuerint: ingrati erũt: mea opinione lectores: nisi quicunq; ueriora habuerint exemplaria: & ipsi sua in mediũ prompserit. Qui uero acrius pspiciunt ac doctius: qd a mendo sit longius: nobis quoq; communicauerint. ut. n. tu maxime omnium nosti Pater Beatissime: q quidẽ ueritatis Cathedram tenes: non lucri auiditate: non laudis ambitione: non uel cũdis bonorum amplioribus fundamentis: non uite degẽdæ necessitate: non deniq; ullius imperantis arbitrio: tãto huic operi ardore insistimus: qđ est reuera difficillimũ.

Fig. 102. Virgil (I–2 B.C.). Princeps of his *Opera* (Rome: Sweynheym and Pannartz, 1469). First page of the dedicatory letter by the editor, Bishop Giovanni Andrea de Bussi to Pope Paul II. [Courtesy of the Library of Princeton University.]

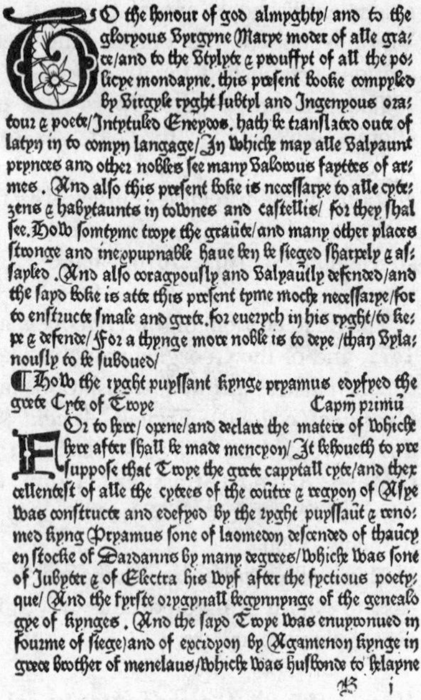

To the honour of god almyghty/ and to the gloryous Vyrgyne Marye moder of alle gra: ce/and to the Vtylyte & prouffyt of all the po: licye mondayne. this present booke compyled by Virgyle ryght subtyl and Ingenyous ora: tour & poete/Intytuled Eneydos. hath be translated oute of latyn in to comyn langage/ In whiche may alle valyaunt prynces and other nobles see many valorous faytes of ar: mes. And also this present boke is necessarye to alle cyte: zens & habytaunts in townes and castellis/ for they shal see. How somtyme troye the graūde/and many other places stronge and inexpupnable haue ben be sieged sharpely & as: sayled. And also coragyously and valyaūtly defended/and the sayd boke is atte this present tyme moche necessarye/for to enstructe smale and grete. for eueryche in his ryght/to ke: pe & defende/ For a thynge more noble is to kepe /than vyla: nously to be subdued/

How the ryght puyssant kynge pryamus edyfyed the grete Cyte of Troye Capm̄ primū

For to here/ opene/and declare the matere of whiche here after shall be made mencyon/ It behoueth to pre suppose that Troye the grete cappytall cyte/and the ex cellentest of alle the cytees of the coūtre & regyon of Asye was constructe and edefyed by the ryght puyssaūt & reno: med kyng Pryamus sone of laomedon descended of thaūcy en stocke of Dardanus by many degrees/whiche was sone of Jubyter & of Electra his wyf after the fyctious poety: que/ And the fyrste orygynall begynnynge of the genealo gye of kynges. And the sayd Troye was enuyronued in fourme of siege)and of excidyon by Agamenon kynge in grece brother of menelaus/whiche was husbonde to helayne

B i

Fig. 103. Virgil (I–2 B.C.). First English edition of the *Aeneid* (folio; London: William Caxton, 1490). This is a very rare book of which only nineteen copies have been traced by Seymour De Ricci, *A census of Caxtons* (Oxford, 1909), pp. 98–100. [Courtesy of Pierpont Morgan Library.]

Athens, where he found himself in 44 soon after Caesar's murder; he joined Brutus's army as *tribunus militum* (say major); when Brutus and Cassius were defeated by Octavian and Antony at Philippoi in 42, young Horace returned home "with clipped wings"[56] and had the same misfortune as Virgil; his father had died in the meanwhile, and his estate had been confiscated.[57] He obtained a post of clerk (*scriba quaestorius*) in

[56] "decisis humilem pennis."

[57] Horace's and Virgil's misfortune happened in the same year, 42, and the cause was the same: their estates were taken to be distributed to the veterans of the Civil War. Note that Virgil's estate was far north, near Mantua, while Horace's was far south.

the government, wrote poetry, won the friendship of Virgil and Varius,[58] and through them that of Maecenas. He was well launched, devoted more and more time to poetry, and received various gifts including his beloved farm at Tibur (Tivoli) in the upper Anio valley not far from Rome. After Virgil's death he was the leading national poet. He died on 27 November 8 B.C., a few months after his patron Maecenas.

His work cannot be analyzed for it is a mass of poems composed for diverse occasions on many subjects. Many of those poems would deserve a separate description, but it is impossible to describe the whole. His inspiration was derived from the Greek poets, from Lucretius and Virgil, and above all from the events and feelings of each day.

The poems were published in a number of books or collections: the *Epodes*, 17 poems written between 41 and 31; the *Satires* in two books, 18 poems c. 35–30; four books of *Odes* (*Carmina*), some 103 lyrical poems ranging from 8 to 80 lines. It was in one of these (III, 30) that he claimed (rightly) to have built a monument more durable than bronze (*monumentum aere perennius*). Then followed the *Epistles*, letters in verse; Book I, 19 epistles, 20 B.C.; Book II, two long letters, the first addressed to Augustus c. 13, in defense of poetry, the second on style and education, c. 18. Two single poems must still be named, the *Carmen saeculare* written by Augustus' order for the secular games in 17, to be recited by a chorus of boys and girls, and the *Ars poetica* composed toward the end of his life.

Some of the poems are purely lyrical, others are didactic, discussing education, public and private morality, style. His outlook was Epicurean to begin with, but was more and more impregnated with Stoicism; it was the outlook of the gentle people of his time and afterward, as long as it was not corrected or displaced by Christianity. He was a defender of good morals and good manners, without heroism and without more than polite enthusiasm for anything. Virgil was, to some extent, a scientific poet, one of the Roman authorities on agriculture; there was no science at all in Horace's poems, but he was one of the great educators of antiquity. At his best, his language and prosody were close to perfection and many of his poems are little jewels, which have remained unsurpassed not only in Latin literature but also in any other.

Tibullus and Propertius. Let us speak more briefly of three more poets of the Augustan age, all somewhat younger than Virgil and Horace: Tibullus, Propertius, and Ovid, born c. 54, c. 51, and in 43; the first two died before Horace in 18 and 16,[59] the third, Ovid, survived Augustus and lived until A.D. 17. All three are distinguished; none is great in comparison with Horace or Virgil.

[58] Varius Rufus, elegiac poet. He was one of the editors of the *Aeneid* soon after Virgil's death.

[59] More exactly, we know that Propertius was still alive in 16 B.C., but he may have lived a little longer; he may even have survived Horace in obscurity.

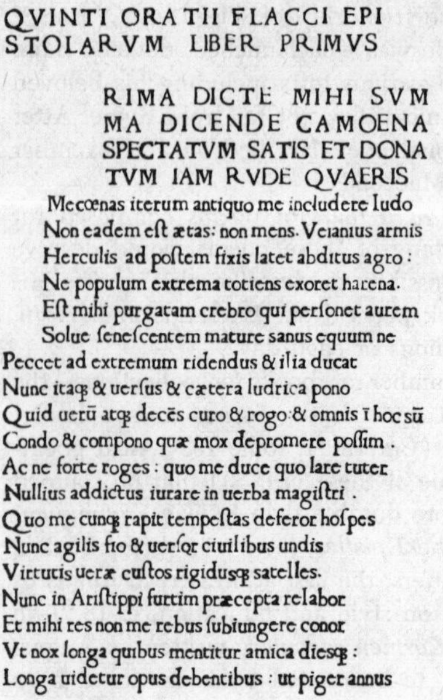
QVINTI ORATII FLACCI EPI
STOLARVM LIBER PRIMVS

RIMA DICTE MIHI SVM
MA DICENDE CAMOENA
SPECTATVM SATIS ET DONA
TVM IAM RVDE QVAERIS
Mecœnas iterum antiquo me includere ludo
Non eadem eſt ætas: non mens. Veianius armis
Herculis ad poſtem fixis latet abditus agro:
Ne populum extrema totiens exoret harena.
Eſt mihi purgatam crebro qui perſonet aurem
Solue ſeneſcentem mature sanus equum ne
Peccet ad extremum ridendus & ilia ducat
Nunc itaqꝫ & uerſus & cætera ludrica pono
Quid uerū atqꝫ decēs curo & rogo: & omnis ī hoc sū
Condo & compono quæ mox depromere poſſim.
Ac ne forte roges: quo me duce quo lare tuter
Nullius addictus iurare in uerba magiſtri.
Quo me cunqꝫ rapit tempeſtas deferor hoſpes
Nunc agilis fio & uerſor ciuilibus undis
Virtutis ueræ cuſtos rigidusqꝫ satelles.
Nunc in Ariſtippi furtim præcepta relabor
Et mihi res non me rebus ſubiungere conor
Vt nox longa quibus mentitur amica diesqꝫ:
Longa uidetur opus debentibus: ut piget annus

Fig. 104. Horace (65–8 B.C.). Princeps of the *Opera* (Venice, c. 1471–1472). [Courtesy of Pierpont Morgan Library.]

Albius Tibullus wrote a number of idylls and erotic elegies; his verses are clear and elegant, often melodious. Renaissance editors divided them into four books, of which the first two only are certainly genuine; those of the first book were published c. 26 B.C. His death occurred in the year following Virgil's.

We do not know whence Tibullus came, but Sextus Propertius was born in Umbria, probably at Assisi. His four books of elegies were written between 35 and 16. They deal chiefly with love, sometimes with Roman myths. There was a growing public of well-educated people, men and women, who loved such gracious and light compositions as those of Tibullus and Propertius, because their souls had been thwarted and disillusioned, first, by the miseries of the civil wars, and next by Augustan totalitarianism.

Ovid (43 B.C.–A.D. 17). From Virgil to Propertius one can witness a decadence of which the last of these poets, Ovid, marks the end. Alexandrianism, which had been repressed by the healthy genius of Virgil and Horace, reappeared in him in the most aggravated form. On the other hand, he wrote considerably more than his friends, Tibullus and Propertius, and was more famous; indeed, his fame approached that of Horace and in bad periods exceeded it.

We know his life much better than those of Tibullus and Propertius. He was born of equestrian rank in a lovely mountain town, Sulmo,[60] in 43, was educated in Rome and Athens, and traveled in Asia Minor and Sicily (this was a kind of Roman grand tour). Virgil, Horace, and Tibullus never married; whether Propertius did or not, we do not know; but Ovid was married thrice. He was a man of independent means and devoted his time to society and poetry. His first book on love (*Amores*) met with success and each of his many publications increased his popularity among the people of fashion. About the year A.D. 8, Ovid was (at the age of fifty) the recognized leader of poets, the "poet laureate," but he incurred Augustus' anger because of his immorality and more certainly because of some political indiscretion. He was in the island of Elba when he received the news of his disgrace and of his exile to a very distant and barbarous place, Tomis,[61] on the western shore of the Black Sea. This would have been a severe punishment for anybody; for such a fashionable poet, such a worldling as Ovid it was terrible. The inhabitants of Tomis were Getae (a Thracian-Danubian tribe) and a minority of Greeks; the languages spoken were Greek but chiefly Getic [62] and Sarmatian. Just imagine the famous poet exiled to a place where nobody understood Latin! The weather was hard (very hot in summer, very cold in winter) and life insecure. Yet Ovid managed to win the friendship of some inhabitants and to continue his work. He spent nine or ten years in exile and died there in 17 or 18.

A brief enumeration of his main works must suffice. Each is a collection of poems: (1) *Amores*, love poems divided into five books (16 B.C.); (2) *Heroides*, fictitious letters written by ladies (such as Sapphō) to their lovers; (3) *Ars amatoria* or *Ars amandi*, the art of love, in three books (c. 1 B.C.); this might be called the art of loving without love; (4) *Metamorphoses*, fifteen books of mythology; (5) *Fasti*, a poetical calendar of the first six months of the Roman year, completed c. A.D. 8 and revised in exile; (6) *Tristia*, letters to friends in defense of himself and to plead for the mitigation of his sentence; (7) *Epistulae ex Ponto*, letters from the Black Sea, somewhat like the *Tristia*; one was written as late as A.D. 16; (8) *Halieutica*, on fishes of the Black Sea.

Next to mythology (a kind of sophisticated folklore), his main source of inspiration was Alexandrine poetry; he was well acquainted, of course, with every Latin poet (many of whom were his personal friends); his own compositions were learned and brilliant, irreverent, frivolous, witty; his poetry was fluent and facile; it probably gave pleasure to many people

[60] Sulmo (modern Sulmona) in the Abruzzi e Molize, some 90 miles east of Rome.

[61] Tomis in lower Moesia, just south of the Danubian delta; now Constanza in southeast Romania; it is now the chief Romanian sea harbor. Ovid's exile began in November of the year A.D. 8 and he reached Tomis in the spring or summer of 9.

[62] Sarmatian was a Slavonic language, and Getic, a kind of Gothic or Teutonic language. Ovid is said to have learned Getic well enough to write a poem in that language. Would that it had been preserved!

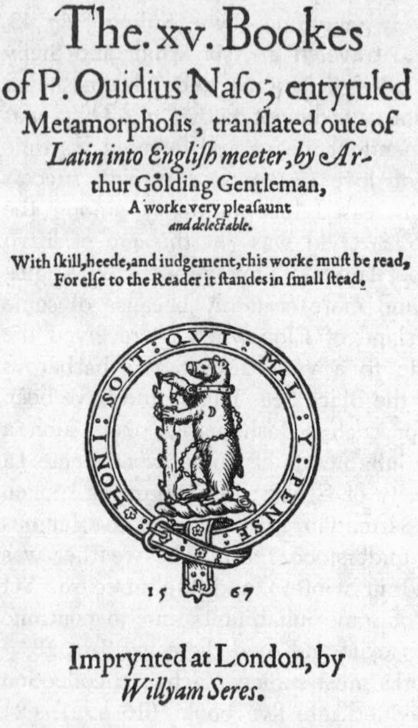
The. xv. Bookes
of P. Ouidius Naso; entytuled
Metamorphosis, translated oute of
Latin into English meeter, by Arthur Golding Gentleman,
A worke very pleasaunt
and delectable.
With skill, heede, and iudgement, this worke must be read,
For else to the Reader it standes in small stead.
HONI SOIT QUI MAL Y PENSE
15 67
Imprynted at London, by
Willyam Seres.

Fig. 105. Ovid (43 B.C.–A.D. 18). First edition of the *Metamorphoses* in English, as translated by Arthur Golding (London: Willyam Seres, 1567). [Courtesy of Harvard College Library.]

who were as sophisticated and superficial as himself; it was typical of the Augustan Age, or, more exactly, of the seamy side of the higher classes, who indulged their desires and combined superstition with luxury. It was convenient to be rich, but even the rich had no freedom, except to satisfy material appetites, sexual indulgences, or at best poetical fancies. For Virgil and Horace, the writing of poetry was a sacred mission; for Ovid, it was hardly more than relaxation, a delicate amusement.

The most popular of his poems was the *Ars amatoria* and others of the same kind; the most harmful was the *Metamorphoses*. This was also his most ambitious one; it is a vast collection of mythological adventures involving metamorphoses.[63] This book had a tremendous popularity, especially during the Renaissance,[64] when many scholars were exceptionally hungry for mythological romances and nonsense.

There was mythology in all the Latin writings, of course, but Ovid offered something new, a kind of mythological encyclopedia the influence of which

[63] The fifteen books describe in verse more than a hundred examples of metamorphoses. The very last one (xv, 8) is the story of Julius Caesar being changed into a star; this is followed by the praise of Augustus. The work was substantially finished before A.D. 8 but revised in exile.

[64] First edition of Ovid's *Opera* in 3 vols., folio (Rome: Sweynheym and Pannartz, 1471). First edition of the *Metamorphoses* (Milan, 1475); this may have been preceded by another edition without place and date. Another edition appeared in Louvain, c. 1475.

can be traced in many medieval writings, such as the *Roman de la Rose* of Jean de Meung (XIII-2) and Chaucer's translation thereof. The *Metamorphoses* was translated into Greek by Maximos Planudēs (XIII-2).

This means that the *Metamorphoses* survived paganism, or, more correctly, kept pagan conceits alive in Christian times. One of the most curious fruits of medieval literature was the *Ovide moralisé*, an enormously long poem written by Chrétien Legouais of Sainte More,[65] in order to interpret the *Metamorphoses* in Christian terms; this was the greatest and last effort to Christianize pagan literature.

The *Metamorphoses* was one of the favorite books of Renaissance poets and scholars. It inspired many books, such as the *Orlando innamorato* (1487) of Matteo Maria Boiardo and the *Orlando furioso* (1516) of Lodovico Ariosto. For the artists as well as the poets, it was a treasure chest of divine stories which they opened whenever they needed to refresh their memories. Their Christian heads were still chockfull of pagan mythology, and they frequently mixed pagan with Christian symbols. The *Metamorphoses* favored irrationalism and, therefore, helped to delay the progress of science throughout the Renaissance.

It is sad to think of the golden age of Latin poetry ending with those mythological fireworks.[66]

[65] It extends to 62,000 lines. It was formerly ascribed to the musician, Philippe de Vitry (XIV-2) (*Introduction*, vol. 3, pp. 48, 743).

[66] There is still a curious (and, to my mind, perverse) interest in the *Metamorphoses*, as is proved by many recent translations. It will suffice to mention three English versions, by F. J. Miller (Loeb Classical Library, 2 vols., Cambridge: Harvard University Press, 1951), by A. E. Watts with etchings by Pablo Picasso (Berkeley: University of California Press, 1954), by Rolfe Humphries (Bloomington: Indiana University Press, 1955).

XXVI

PHILOLOGY IN THE LAST TWO CENTURIES[1]

GREEK PHILOLOGY

The Greeks of the Golden Age had the privilege of using one of the most beautiful languages with a minimum of grammatical conscientiousness. In this respect they were at the opposite pole from the Indians, who were acutely conscious of grammatical niceties (especially such as concerned morphology and phonology) in very early times, centuries ahead of other peoples.[2] The linguistic innocence of the happy Greeks was disturbed by logicians, like Prōtagoras of Abdēra (V B.C.), and by philosophers, like Aristotle and Zēnōn of Cition (IV–2 B.C.), but grammar did not begin to shape itself until the third century. The efforts of Alexandrian grammarians in the third century have been explained in Chapter XIII. It is clear that it was impossible to investigate ancient texts, as was done by Zēnodotos, Callimachos, Eratosthenēs, without having to solve difficulties that were lexicographic or grammatical. Literary criticism implies grammatical problems. On the other hand, the logical analysis of sentences begun by Zēnōn was developed by other Stoics, like Chrysippos of Soloi (III–2 B.C.) and Diogenēs the Babylonian (II–1 B.C.), and Stoic logic led to grammar. Diogenēs was sent to Rome in 156 as a member of an Athenian embassy and he brought with him Stoicism, logic, Greek grammar, and the seeds of the Latin one. The analysis of any language leads not only to its grammar but to grammatical consciousness in general.

In the first half of the second century B.C. philological studies were much

[1] This is a continuation of Chapter XIII.

[2] See my notes on Yāska (V B.C.) and Pāṇini (IV–1 B.C.) in *Introduction*, vol. 1, 110, 123. Sanskrit grammar was not known outside of India before the end of the eighteenth century and hence could not exert any influence on European grammar until then. Its influence upon the development of comparative grammar was considerable in the nineteenth century, but that is another story.

The Sanskrit grammarians were the world pioneers in phonetics, but they did not discover the alphabet. That was a Semitic invention (Volume 1, pp. 109–111). Oral traditions were (and still are) exceptionally strong in India.

stimulated by the increasing rivalry between the old school of Alexandria and the new one established in Pergamon. In both cases the center of such studies was the library. The leading grammarians of Alexandria were Aristophanēs of Byzantion (d. c. 180) and Aristarchos of Samothracē (c. 220–c. 145).

Aristophanēs prepared better editions of Homer and of Hesiod's *Theogony*, the first collected edition of Pindar's poems, and recensions of Euripidean and Aristophanic plays; he compiled a Greek dictionary (*lexeis*) and made a study of grammatical analogies or regularities.

His main title to fame is to have invented (or improved) means of indicating accentuation and punctuation. It would be misleading to say that he invented accentuation and punctuation themselves, for these are as old as the spoken language. Indeed, it is impossible to speak correctly and to be understood clearly without accenting and grouping the words and without dividing the sentences. When the language was reduced to its written form, it was found necessary, or at least convenient, to indicate the accentuation of the words and the subdivisions of the sentences by means of symbols. Was Aristophanēs the very first to do that? Probably not, but he did it better and more thoroughly than his predecessors.

These were innovations of the first importance, as will be readily appreciated by scholars who have been obliged to read texts without punctuation and capitals (for example, in Arabic).[3] It is noteworthy that they did not obtain any popularity for a considerable time. The oldest Greek and Latin manuscripts were not only innocent of punctuation, but there were no separations between the words. There are manuscripts as late as the thirteenth century that are not punctuated. The printers of incunabula, imitating the manuscripts, pared punctuation to the minimum and reduced blank spaces between words and sentences as much as possible. Punctuation was not generally adopted until the sixteenth century – seventeen centuries after Aristophanēs – when it was sufficiently advertised and stabilized by the printer's art.

This immense delay was not caused simply by inertia but rather by the supremacy of the vocal tradition over the written one. Writing (and even early printing) was a method of representing the real language (the spoken one), which was meant to be suggestive rather than complete and detailed. This is obviously true of the writing of Semitic languages, wherein short vowels are not marked, and it is even more true of languages like

[3] The reader will be helped to imagine the difficulties caused by the absence of punctuation and word separation if he tries to puzzle out the following passage: "thetheoryofrelativityisintimatelyconnected withthetheoryofspaceandtimeishalltherefore beginwithabriefinvestigationoftheoriginof ourideasofspaceandtimealthoughindoingso iknowthatiintroduceacontroversialsubjectthe objectofallsciencewhethernaturalscienceor psychologyistocoordinateourexperiencesand tobringthemintoalogicalsystemhowareour customaryideasofspaceandtimerelatedtothe characterofourexperiences" Albert Einstein, *The meaning of relativity* (Princeton: Princeton University Press, 1945), opening paragraph.

the Chinese, wherein the pronunciation and accentuation are not indicated at all.[4] The fundamental inadequacy of the written language, as compared with the spoken one, is well illustrated by English spelling in spite of the fact that the English language is very highly developed in other ways. There are many English words that a foreigner cannot read aloud, if he has not been told beforehand how to pronounce them.

Aristarchos of Samothracē was Aristophanēs' successor as head of the Library, c. 180; in his old age, he moved to Cypros, where he died c. 145. He was primarily a literary critic, who composed commentaries (*hypomnēmata, syngrammata*) on the classics, and compared the Homeric language with the Attic one. Such comparison concerns the words themselves (lexicography) or their morphology and accidence, or the structure of sentences (syntax). Prōtagoras had distinguished the genders of nouns and certain tenses and modes; Aristotle recognized three parts of speech, noun, verb, and the rest; Aristarchos was aware of eight parts, noun (and adjective), verb, participle, pronoun, article, adverb, preposition, and conjunction.

In the meanwhile, Cratēs of Mallos was carrying on similar investigations in Pergamon and was bound to reach the same conclusions. His grammatical consciousness was increased by his comparison of Greek with Latin; indeed, it is impossible to preserve one's grammatical innocence when one uses two languages. He is said to have composed the first Greek grammar. This statement should be taken with caution. The analysis of a language is like the analysis of the human body; it is impossible to say who began either, and it is difficult to say when it is completed. Grammar, like anatomy, was invented not at one time with a single effort, but again and again in many instalments. Cratēs' achievement was certainly a great one, but we cannot measure it exactly because his grammar is lost. The earliest grammar that has come down to us was composed by one of Aristarchos' disciples, Dionysios Thrax[5] (II-2 B.C.), who was born c. 166 and flourished in Alexandria and in Rhodos. His grammar (*technē grammaticē, ars grammatica*)[6] was the prototype of all later grammars, not only Greek but also Latin and Armenian[7] and indirectly of the other Indo-European languages. Said Gilbert Murray, "It was one of the most successful textbooks in the world, which remained the basis of Greek grammar till well into the nineteenth century, and was actually used in Merchant Taylor's School when a great uncle of mine was a boy there . . . Dionysios did for grammar what

[4] The pronunciation of a Chinese character may be suggested by the phonetic element included in it (Volume 1, p. 22).

[5] Thrax or Thracos means Thracian, but it does not follow that he himself was born in Thrace; he may have inherited the name from his father or from his ancestors.

[6] Edited by Gustav Uhlig, *Ars grammatica* (224 pp.; Leipzig, 1883). Alfred Hilgard, *Scholia* (703 pp.; Leipzig, 1901). The *Ars* was Englished by Thomas Davidson, *Journal of Speculative Philosophy* (St. Louis, 1874), 16 pp.

[7] Jacques Chahan Cirbied (1772-1834), *Grammaire de Denis de Thrace tirée de deux MSS arméniens de la Bibliothèque du Roi* (125 pp.; Paris, 1830), text in Armenian, Greek, and French.

Euclid did for geometry, and his textbook has lasted almost, though not quite, as long as Euclid's." [8]

Its publication in the second half of the second century B.C. marks definitely the end of the age of grammatical innocence. Happy the little Athenians of old who spoke the most beautiful language of the world without having to learn it painfully as we do. It takes us great efforts to master it, and, if we do not use it enough, we risk forgetting it and must learn it all over again. The Athenian boys of the fourth century did not have to learn their mother tongue and could not possibly forget it.

The account of early Greek grammar has caused me to overlook two philologists who are remembered for other achievements, Dēmētrios of Scēpsis and Apollodōros of Athens. We would call them archaeologists rather than grammarians.

Dēmētrios (c. 200–130) flourished at Scēpsis in Trōas, wrote a commentary on the Homeric catalogue of ships (*Iliad*, II, 816–877), which was called *Trojan arrangement* (*Trōicos diacosmos*),[9] filled 30 books, and was a mine of useless knowledge. The Athenian Apollodōros was a pupil of Aristarchos in Alexandria, which he left in 146. He probably settled in Pergamon, for his main work (*Chronica*) was dedicated to Attalos II Philadelphos (king of Pergamon 159–139). He wrote commentaries on Homer and studies on etymology, geography, mythology. His book on mythology (*Peri theōn, On the gods*) was confused with another mythological treatise, the so-called *Library of Apollodōros* (*Apollodōru bibliothēcē*), which is at least two centuries younger.

Linguistic consciousness was much developed in the Greco-Roman world by polyglottism. Educated men of the West were obliged to know two languages, Greek and Latin; in the East, they did not always know Latin but were familiar with Oriental languages. Ennius said that he had three souls because he could speak three languages, Greek, Oscan, and Latin.[10] The second of these, Oscan, was the most popular of Italian dialects, for it was used all over South Italy (the Latin dialect on the contrary, which became the official language of Rome, was at first the most restricted geographically). It continued to be used even after the Osci or Opici had been defeated and driven into obscurity; it was kept alive in Rome in the *fabulae Atellanae*, rustic comedies, farcical and ribald, spoken in Oscan and very popular.

If Ennius had three souls, how many had Mithridatēs the Great? Perhaps

[8] Gilbert Murray, *Greek studies* (Oxford: Clarendon Press, 1946), p. 181.

[9] Dēmocritos of Abdēra wrote two treatises entitled *Megas* and *Micros diacosmos* (The large and the small arrangement), *diacosmos* meaning arrangement or explanation.

[10] "Quintus Ennius tria corda habere sese dicebat quod loqui Graece et Osce et Latine sciret." Aulus Gellius (II–2), *Noctes Atticae*, XVII, 17. This is the reference also for the following statement concerning Mithridatēs' polyglottism.

as many as twenty-five. Indeed, Gellius tells us that Mithridatēs spoke the languages of the twenty-five nations which he had subjugated or with which he had dealings. This may seem unbelievable to monoglot Americans but is simply a reflection of the multiplicity of languages having currency in the Near East. Listen to Pliny's testimony apropos of Dioscurias in Colchis,

> the Colchian city of Dioscurias on the river Anthemus, now deserted, but once so famous that according to Timosthenēs [11] 300 tribes speaking different languages used to resort to it; and subsequently business was carried on there by Roman traders with the help of a staff of 130 interpreters.[12]

It is thus not surprising that Mithridatēs had mastered twenty-five languages; circumstances had made of him a collector of languages, even as they had obliged him to collect plants and poisons, electuaries and *thēriaca.* A curious afterglow of this was Mithridatēs' reputation as a polyglot during the Renaissance. When the great naturalist, Conrad Gesner, published his study of languages he entitled it *Mithridates.*[13]

There were two distinguished Greek grammarians in the Augustan age, Dionysios of Halicarnassos and Didymos.[14]

Dionysios hailed from Halicarnassos but flourished in Rome; we have already spoken of his history of Rome (*Rhōmaicē archaiologia*), but he was primarily a man of letters, a rhetorician and grammarian, who took pains to insure the purity of the Greek language. He was perhaps the outstanding literary critic of his age; he wrote books explaining the literary merits of the Ancient Orators, of Thucydidēs, Plato, and others (he disliked Plato's style), other books on the need of imitating the good authors, on the choice of words and their best arrangement. For him it was not enough to know the Greek language, one must know it well, one could not know it too well. He was one of the best defenders of that language at a time when it was jeopardized by its very popularity among the Romans and other barbarians.

Didymos, nicknamed Chalcenteros ("with brazen bowels") because of his prodigious industry, was fighting for the same cause in Alexandria, where

[11] Timosthenēs was an admiral of Ptolemaios Philadelphos (285–247); his geographic writings were used by Eratosthenēs and by Strabōn.

[12] *Naturalis historia,* VI, 5 (Loeb edition, vol. 2, p. 349). Pliny wrote some 130 years after Mithridatēs' death but referred to Timosthenēs who flourished in the third century B.C. Dioscurias was at the eastern end of the Black Sea, just east of Pontos, Mithridatēs' kingdom.

[13] *Mithridates* (Zürich, 1555). See my *Appreciation of ancient and medieval science during the Renaissance* (Philadelphia: University of Pennsylvania Press, 1955), p. 111. Gesner dealt with 130 languages, as many languages as there were Roman interpreters in Dioscurias! A more ambitious study of languages was published by Johann Christoph Adelung under the same title *Mithridates* (4 vols. in 6; Berlin, 1806–1817). Hence, Mithridates had for many people the meaning of general linguistics, even as Euclid meant geometry.

[14] Dionysios flourished from 30 to 8 B.C.; Didymos was born c. 65 B.C. and died c. A.D. 10.

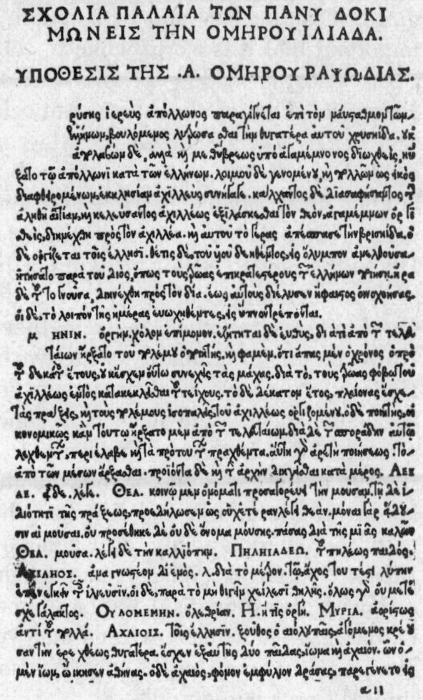
ΣΧΟΛΙΑ ΠΑΛΑΙΑ ΤΩΝ ΠΑΝΥ ΔΟΚΙ
ΜΩΝ ΕΙΣ ΤΗΝ ΟΜΗΡΟΥ ΙΛΙΑΔΑ.

ΥΠΟΘΕΣΙΣ ΤΗΣ .Α. ΟΜΗΡΟΥ ΡΑΨΩΔΙΑΣ.

Fig. 106. Didymos (I–2 B.C.). Ancient annotations of the best kind to the *Iliad* of Homer. *Hypothesis of the Homeric rhapsody*, edited by Janus Lascaris and printed by Aggelos Collōtios (folio, 30 cm; Rome, 1517). A Latin letter of recommendation written by Leo X on 7 Sept. 1517, in the fifth year of his pontificate, is printed at the end just above the Greek colophon. Janus Lascaris (1445–1535) was, next to Aldus Manutius, the leading Greek editor of the early Renaissance. Emile Legrand, *Bibliographie hellénique* (Paris, 1885), vol. 1, pp. 159–162. [Courtesy of Harvard College Library.]

the Greek language was even more corrupted than in Rome, because it was abused by ignorant people. He wrote studies on Greek literature and edited Homer, Thucydidēs, and the Ancient Orators.

The task that men like Dionysios and Didymos were struggling to accomplish was comparable to that of Englishmen (or Frenchmen) trying to maintain as high standards as possible of English (or French) letters in remote countries. It is a task of great difficulty but of great merit. Good literature is the main vehicle of civilization.

The Greek scholars of whom we have spoken above devoted themselves to the defense and illustration of the Greek language and Greek culture either among the Greek-speaking peoples of Egypt and Asia whose language and manners had been corrupted by a barbaric [15] environment, or among Romans for whom Greek was a foreign language. There is still a third body of people to be considered, the Jews, who were diffused all over the Greco-Roman world, especially in the large cities of Syria and Egypt, in Rome, and in other cities of the West. What was their reaction to Greek? This problem has already been discussed when we dealt with the Septua-

[15] The word "barbaric" is used here in the Greek way to designate anything non-Greek or foreign. Greek-speaking people were a very small minority in Egypt and Asia; but many more people were able to jabber in pidgin Greek.

gint in Chapter XIV and with Jewish religion in Chapter XVI, but it is so important that it is worth while to consider it again.

In Syria the leading people were Greeks who were anxious to defend Hellenism. One of the rulers, Antiochos IV Epiphanēs, carried Hellenization so far that he caused the Maccabean revolt (in 168). The Jews succeeded in defending their religion, but they were not so successful in defending their language. Hellenization was continued not only by Greek princes but also by some of the Maccabees (or Hasmoneans), and by Herod the Great (40–4 B.C.), king of Judaea under Roman patronage.

A large number of Egyptian and Asiatic Jews were Greek-speaking. Greek was their mother language; they did not have to learn it; they knew it. However, their familiarity with two very different languages, Greek and Hebrew, may have awakened their grammatical consciousness, and if that happened they would have been among the first to know of the *Technē grammaticē* of Dionysios Thrax and to avail themselves of it; indeed, Dionysios (b. c. 166) was one of their neighbors and his activities would be best known in Alexandria and Rhodos where he flourished.[16]

In the Herodian city of Caesarea[17] there were Jews who read the *Shema*[18] in Greek! What is more, there was an academy of Greek wisdom in Jewish Palestine. Said Rabban Simeon, the son of Rabban Gamaliel the Patriarch, "There were a thousand young men in my father's house, five hundred of whom studied the [Jewish] Law, while the other five hundred studied Greek wisdom."[19]

The rabbis knew Greek fairly well; as to the Jewish masses, their knowledge of Greek was not more debased than that of other Greek-speaking Orientals. The Hellenization of the Eastern Jews is proved by a great many Jewish antiquities and artistic objects of that time[20] and by the existence of many Greek words in Hebrew writings.[21]

[16] In spite of Dionysios' example, no Greek-speaking Jew thought of deducing Hebrew grammar from the realities of Hebrew speech. Hebrew grammar was not created until much later, under Arabic influence, by Saadia Gaon (X–1). The cause of the delay was the double fact that Hebrew grammar is very different from Greek grammar but very close to Arabic grammar.

[17] This particular Caesarea was on the coast of Samaria, 55 miles northwest of Jerusalem. It was rebuilt on a grandiose scale by Herod the Great. Under the imperial procurators and legates it became the capital of Roman Judaea.

[18] The *Shema* is a selection of short passages taken from Deuteronomy, 6:4–9, 11:13–21 and Numbers 15:37–41 which express the main articles of the Jewish faith. The name is the first word of the first passage, *shema'* meaning "hear!"

[19] Saul Lieberman, *Greek in Jewish Palestine* (New York: Jewish Theological Seminary, 1942), p. 1. See also his *Hellenism in Jewish Palestine* (New York: Jewish Theological Seminary, 1950). Both books deal rather with the post-Christian period. Rabban Gamaliel the Elder died c. A.D. 50; he was St. Paul's teacher (Acts, 22:3).

[20] Ernst Cohn-Wiener, *Die jüdische Kunst* (Berlin, 1926). Franz Landsberger, *History of Jewish Art* (Cincinnati: Union of American Congregations, 1946). Erwin Ramsdell Goodenough, *Jewish symbols in the Greco-Roman period* (6 vols., quarto; New York: Pantheon, 1953–1956).

[21] See the index of Greek words by Immanuel Loew in Samuel Krauss,

Many Jews wrote Greek prose; some of them even wrote Greek poetry. For example, Philōn the Elder wrote an epic poem on Jerusalem (*Peri ta Hierosolyma*); Theodōtos wrote one on the history of Shechem; [22] Ezechiel, a tragedy on the Exodus (*exagōgē*).[23] These three poets flourished probably in the second century B.C. Note that the first two had Greek names, as was the case for many other Jews of that time.

The Jews of the Western diaspora, even those of Rome, often had a better knowledge of Greek than of Latin.

LATIN PHILOLOGY

The Latin-speaking people were very slow in adapting the "grammatical art" of Dionysios Thrax to the realities of their own language, but when they finally did so, their borrowings were obvious enough. The very grammatical terms with which we are familiar (genitive, accusative, infinitive, and so on) are mistranslations of the Greek terms. Nevertheless, the Latin authors never enjoyed the linguistic innocence of the early Greeks. That innocence was spoiled during the Hellenistic age not only for the Greek-speaking people but for everybody else around them. That paradise was lost forever.

We all have to learn the grammar not only of foreign languages but also of the one that fell from our mother's lips. We have to know it so thoroughly that it becomes a part of our substance, and then we may forget it (or think that we have forgotten it); after a climax of grammatical consciousness, our grammar becomes subconscious and then we really have it. Is not the same true of every kind of knowledge?

To return to Latin: it is clear that every Roman who tried to master the Greek language, as well as every Greek who studied Latin, was bound to make grammatical comparisons and to ask himself grammatical questions. It must be assumed that the Greek tutors, explaining the fine points of Atticism to their students, were giving them, wittingly or not, grammatical lessons. Hence, the relative inertia or slowness of Latin grammarians is surprising.

The great majority of Romans did not study Greek however, but if they were intelligent, or, more exactly, if their intelligence was properly oriented, their philological consciousness was awakened by the comparison of Latin with various Italian dialects. One is apt to forget that Latin was at the beginning the language of a relatively small territory, the city of Rome and of Latium. That territory was gradually expanded by the Roman conquests

---

Griechische und Lateinische Lehnwörter im Talmud (2 vols.; Berlin, 1898–99).

[22] Shechem or Sichem, an ancient Biblical town; chief city of Samaria; home of Jacob; Jacob's well and Joseph's tomb are there. Later called Nāblus and Neapolis.

[23] Emil Schürer, *Geschichte des jüdischen Volkes im Zeitalter Jesu Christi* (vol. 3, Leipzig, ed. 3, 1909); English trans. of vol. 3 by Sophia Taylor and Peter Christie (New York, 1891), pp. 156–320, chiefly pp. 222–223.

of Italy and other countries, but it does not follow that Latin immediately replaced the local dialects. That could not be. The replacement of each dialect by Latin was a very slow process which began at different dates in different places according to the dates of conquest, and proceeded more slowly in regions where dialectical resistance was greater than in others. Latin was obliged to oust Italic dialects like Oscan and Umbrian, and non-Italic ones like Etruscan and Ligurian. Outside of Italy it had to compete with Celtic, Iberic, Lybic, Punic, and many more.

As Latin was the language of administration, every Roman citizen who wanted to be employed by the government or the municipalities had to learn it. The best school of Latin, however, was the Roman army, which recruited soldiers in every province. Moreover, Roman officers, officials, traders established themselves in every foreign country as soon as it was conquered and brought their language, manners, and customs with them.

By the time of Christ, many dialects were still alive and yet Latin had attained the status not only of a national language – the language of Rome – but even of an international one.

All considered, the internationalization of Latin was more rapid than that of Greek, but it was not as durable, for Greek is still a living language today, spoken in many cities of the world, while Latin speech is obsolete, except in some ecclesiastical or monastic circles.

The first Roman grammarian or philologist was Lucius Aelius Stilo Praeconinus, who flourished in the second half of the second century B.C. This Stilo was a native of Lanuvium (in Latium) and had thus spoken the Latin dialect from his childhood. He was an archaeologist, a literary critic, and wrote linguistic commentaries on the Twelve Tables [24] and other ancient writings, and critical editions of Ennius and Lucilius. In short, he was the first to do in Rome what Greek grammarians had begun to do in Alexandria more than a century before. He counted Varro and Cicero among his pupils. His inspiration was derived from Greek models, chiefly Stoic logicians and grammarians like Chrysippos of Soloi (III–2 B.C.), because he was himself a Stoic.

It would be an exaggeration to call Marcus Tullius Tiro, Cicero's freedman and secretary, a philologist, and yet he was the inventor of Latin stenography. His abbreviated notes, later called *notae Tironianae*, enabled him to preserve Cicero's orations and other writings. Of course, every intelligent secretary, having to record endless dictations, would invent sooner or later some means of facilitating his task. He wrote letters (*Epistulae*) of his own and a treatise on the Latin language, of which nothing is known

[24] The Twelve Tables are a collection of the earliest rules of Roman law, derived from ancient customs (*mores maiorum*); it is impossible to date them except by saying that they belong to the earliest Roman culture. They mark the beginning of an evolution the climax of which was to be the *Corpus iuris* of Justinian (VI–1).

except the alluring title *De usu atque ratione linguae Latinae* (*On the use and philosophy of Latin*).

We are on firmer ground with Varro, because of his great treatise on the Latin language, *De lingua Latina libri XXV*, one of the two treatises of his that are extant (the other being the one on agriculture). Unfortunately, we have only a part of the former one, Books v to x (v and vi complete). The general plan can be reconstructed as follows. Book i was introductory, a general view of the whole subject; ii–vii explained the origin of words and their applications to concrete objects and to abstractions; viii–xiii dealt with declension and conjugation; xiv–xxv, with syntax. Books v–xxv were dedicated to Cicero and the whole work was completed shortly before Cicero's murder in 43. Varro's etymologies were more often than not popular and fanciful, as were all those of the ancients (Hebrew or Greek), because none of them had sufficient linguistic knowledge to understand the principles of the subject.[25] He was philosophically minded, however, in the Stoic way, and was perhaps the first to grasp a fundamental idea of modern grammar: linguistic standards (*le bon usage*) are not static and never final; "consuetudo dicendi est in motu" (habits of speech are always in flux). His book has preserved various quotations and ancient terms and forms that would otherwise have been lost; it is a thousand pities that we have only a fifth of it.

The last grammarian to be named is perhaps already outside our present frame. Varrius Flaccus belonged to the Augustan age, but rather to the end of it. He was a freedman who proved to be an excellent teacher and was put in charge of the education of Augustus' grandsons. He composed various grammatical and educational books, *Libri rerum Etruscarum* (*On Etruscans*), *Libri rerum memoria dignarum* (*Things worth remembering*), *De orthographia*, and others. His main work was a kind of encyclopedic dictionary, the earliest one in Latin, *De verborum significatu*, which is partly preserved in the abridgments of Pompeius Festus (second century) and Paulus Diaconus (VIII–2). Some of his data were copied and transmitted to us by Pliny the Elder (I–2).

The earliest Latin grammar was composed only later (c. 67–77) by Quintus Remmius Palaemon (I–2). He was probably a freedman or a Greek because his name (Palaimōn) is obviously Greek. Thus, in spite of the great example given by another Greek, Dionysios Thrax, Latin grammar was not established in the pre-Christian age.

At the very time when the Romans had conquered the world, the Latin language was already immortalized by literary masterpieces, yet its analysis had not yet been completed and its vocabulary was still very insufficient. Latin writers were still dependent upon the Greek, as was frequently realized by philosophers like Cicero or technicians like Vitruvius. It was im-

[25] For Greek and Latin etymology in general, see Peter Barr Reid Forbes, *Oxford Classical Dictionary*, p. 341.

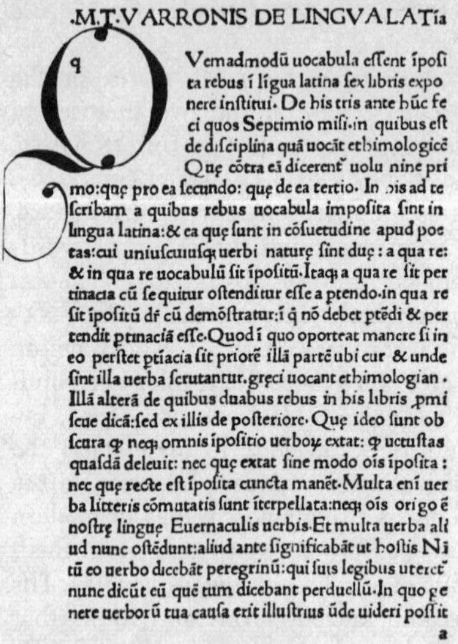
.M.T.VARRONIS DE LINGVA LATina

Q q Vemadmodũ uocabula essent iposi ta rebus ĩ lĩgua latina sex libris expo nere institui. De his tris ante hũc fe ci quos Septimio misi: in quibus est de disciplina quã uocãt ethimologicẽ Quę cõtra eã dicerent uolu nine pri mo: quę pro ea secundo: quę de ea tertio. In his ad te scribam a quibus rebus uocabula imposita sint in lingua latina: & ea quę sunt in cõsuetudine apud poe tas: cui uniuscuiusq; uerbi naturę sint duę: a qua re: & in qua re uocabulũ sit ĩpositũ. Itaq; a qua re sit per tinacia cũ sequitur ostenditur esse a ptendo: in qua re sit ĩpositũ dr cũ demõstratur: ĩ q nõ debet ptẽdi & per tendit ptinaciã esse. Quod ĩ quo oporteat manere si in eo perstet ptiacia sit priorẽ illã partẽ ubi cur & unde sint illa uerba scrutantur. greci uocant ethimologian. Illã alterã de quibus duabus rebus in his libris ꝑmi scue dicã: sed ex illis de posteriore. Quę ideo sunt ob scura ꝙ neq; omnis ĩpositio uerboꝝ extat: ꝙ uetustas quasdã deleuit: nec quę extat sine modo oĩs ĩposita: nec quę recte est ĩposita cuncta manẽt. Multa enĩ uer ba litteris cõmutatis sunt ĩterpellata: neq; oĩs origo ẽ nostrę linguę e uernaculis uerbis. Et multa uerba ali ud nunc ostẽdunt: aliud ante significabãt ut hostis Nã tũ eo uerbo dicebãt peregrinũ: qui suis legibus uterẽt nunc dicũt eũ quẽ tum dicebant perduellũ. In quo ge nere uerborũ tua causa erit illustrius ũde uideri possit

a

Fig. 107. Varro (I–2 B.C.), *De lingua latina*. Edited by Pomponius Laetus (1425–1498), who was the founder of the Roman Academy and one of the leading Latin humanists of his time. The Roman Academy enjoyed its golden age under Pomponius and during the pontificate of Leo X (1513–1521). It was one of the victims of the sack of Rome by Charles Quint in 1527. Quarto, 22 cm, 84 leaves (Rome: Georgius Lauer, undated). The date is probably 1471. First page of Varro's text. The capital Q was written in red by a limner. [Courtesy of Harvard College Library.]

possible to discuss philosophy or the arts without borrowing Greek words and the greatest Latin poets were still plundering Greek models.

This illustrates the difficulty of the grammatical task as well as Roman slowness in approaching it. The Greeks themselves would not have accomplished it if they had not been driven by increasing cosmopolitanism and polyglottism. As the centuries passed, even Greek-speaking people became less and less able to speak correctly without continuous efforts; as to the "barbarians," they had to learn Greek in an artificial and painful way; they needed grammars, dictionaries, and other tools. It is thus not surprising that this age saw the birth of grammar.

Men of science should not think that this was a mean achievement. Of course, nowadays to compile a grammar of a known language can hardly be considered a scientific work. But the creators of grammar, like Dionysios Thrax or even his predecessors, Diogenēs the Babylonian and Cratēs of Mallos, who were the first to systematize the result of a long evolution, had been accomplishing a scientific task of considerable importance and merit. The discovery of the logical structure of language was as much a scientific achievement as the discovery of the anatomical structure of the body, but, as linguistic self-consciousness appeared very gradually, the discovery was very slow and largely anonymous.

The construction of the first grammar of any language may still be considered a scientific achievement, but of much smaller importance. Indeed, the philologist undertaking such a work is aware that every language has a grammar and knows very clearly what to look for. He might be compared to the zoölogist who dissects for the first time a recently discovered animal; all the tissues and organs are already familiar to him and the anatomy of this animal is but a variant of the anatomy of many others. In short, to describe a new grammar is not at all comparable to the discovery of grammar itself, as was done by the earliest Sanskrit and Greek grammarians.

It should be noted also that scientific endeavors in other directions were bound to induce some sort of grammatical elaboration. Every scientific investigation entails sooner or later the use of special words and phrases; it introduces new thoughts which must be adequately expressed. It is not even sufficient for a scientist to use a correct language unconsciously, for he must know exactly the peculiarities and limitations of his tools, and language is one of them. He must be sure of his ability to express his thoughts with precision and without ambiguity. Scientific progress necessarily involved a sufficient analysis and determination of the language. Grammatical work was thus an essential step in the development of science.

Euclid, Hērophilos, Cratēs, Hipparchos, Dionysios Thrax were children of the same environment. Their curiosities were differently oriented, but they performed similar tasks.

XXVII

ART IN THE LAST TWO CENTURIES[1]

HELLENISTIC SCULPTURE IN GREECE, EGYPT, AND ASIA

The division between the art before Alexander and that after him is somewhat arbitrary, and that between the third century and the two following ones even more so. It is impossible to draw lines that apply everywhere because style did not change with the same speed in different places (it would be extraordinary and almost unbelievable if it did). There were many art centers during the last two centuries B.C., the most important in Pergamon and Rhodos, others in Athens, Alexandria, Sicyōn, and elsewhere. Sculptors who traveled to distant places to execute commissions entrusted to them gathered local assistants and pupils around them and began new schools. Moreover, some artists were conservative and behind the times, while others were reckless and aggressive, ahead of them. In any case, I do not intend to write a history of Hellenistic art but to give the reader a general idea of what the artists thought and wrought.[2]

The exact dating of Hellenistic sculpture is out of the question except when works of art were ordered by kings whose reign is known. The main patrons of art in the second century B.C. were the Attalid, Eumenēs II, king of Pergamon from 197 to 159, and the Seleucid, Antiochos IV Epiphanēs, king of Syria from 175 to 164.

The first of these was the more munificent. He continued on a grander scale the work so well begun by his immediate predecessor, Attalos I (ruled 241–197). Both kings were anxious to raise Pergamon to the same cultural level as Alexandria or higher still.[3] It was Attalos who decided to

[1] This continues the story begun in the third part of Chapter XIII.

[2] All the pertinent illustrations will be readily found in the latest book of Margarete Bieber, *The sculpture of the Hellenistic age* (quarto, 244 pp., 712 ills.; New York: Columbia University Press, 1955). Many are available also in José Pijoán, *Summa artis* (Madrid, vol. 4, 1932; vol. 5, 1934) and in other books.

[3] It is often as difficult to divide the credit between Attalos Sōtēr and Eumenēs II as it is to divide it between Ptolemaios Sōtēr and Ptolemaios Philadelphos. It is simpler to ascribe the Pergamene renaissance (and the Alexandrian renaissance) to both kings.

express his gratitude to Zeus for the victory over the Galatians by the erection of a grandiose altar upon the terraces of the upper town. The altar was 40 feet high and decorated with immense friezes in very high relief representing the battles between gods (the Pergamenians) and giants (the defeated Galatians). The slabs bearing the friezes were 7½ feet high and their combined length about 350 feet, of which nearly three quarters were preserved (at least until World War II). The sculptors had treated their subject, the episodes of a gigantic war, with imagination and astounding verve. In spite of its enormous size, it is probable that the whole altar was completed by the end of Eumenēs' rule. It was very familiar to the modern world because it was taken to Germany and admirably exhibited in the Berlin museum [4] (Fig. 108).

While the altar was being built, a large number of sculptors and assistants were kept busy. Even if there had been but few of them in Pergamon when Attalos began, there would have been many (summoned from many places) by the end of Eumenēs' rule. The Pergamene school was created in the best manner by the steady coöperation of many artists and enthusiastic patrons in a great undertaking. Many other monuments were produced by them while the altar was being constructed or later when the great work was done and their hands were idle. Some of their works have found their way to Western museums, for example, the "Dying Gaul" in the Capitoline Museum in Rome, and the "Wounded Gaul" (Paris). "The Gaul killing himself with his sword after having killed his wife" in the Museo Nazionale (Rome) is a replica of a Pergamene work of that golden age.[5]

It was probably in Pergamon (if not in Alexandria) that the critical study of Greek sculpture was begun. This would have been very natural because the Pergamene renaissance of sculpture took place in an atmosphere of learning. The erudite scholars who worked in the library of Pergamon would wish to know the lives of the great sculptors of the past. There was a Pergamene canon of ten sculptors which might be called a companion piece to the Alexandrian canon of the ten Attic orators.[6]

That canon is very interesting because it represents the critical judgment

[4] That is, the sculptural part was taken to Berlin and a gigantic hall was built in the Muesum to house a reproduction of the altar together with the original friezes. This was one of the glories of the Berlin museum. These monuments have been taken away by the Russians and their present location cannot be ascertained. (Letter from Dr. Gerda Bruns dated Berlin 31 January 1952.)

[5] In these two cases and in others the word Gaul often misleads visitors, because they think of a Gaul of the territory now called France, while these Gauls were really Asiatic Gauls, Galatians. Piotr Bieńkowski (d. 1925), *Die Darstellungen der Gallier in der Hellenistischen Kunst* (184 ills.; Vienna, 1908); *Les Celtes dans les arts mineurs gréco-romains* (336 figs.; Cracow, 1928).

[6] The ten Attic orators are listed in Volume 1, p. 258; they all date from the fifth and fourth centuries. Hence, that canon was established early in the third century. The Pergamene canon of sculptors covers roughly the same period of time.

Fig. 108. General view of the great altar of Zeus in Pergamon (middle of second century B.C.) as it was reconstructed in the Berlin Museum before World War II. The sculpture was genuine. More photos of it can be seen in Gerda Bruns, *Das grosse Altar von Pergamon* (74 pp.; Berlin: Mann, 1949) or in Margarete Bieber, *The sculpture of the Hellenistic age* (New York: Columbia University Press, 1955), Figs. 458–470.

of the Pergamene sculptors on their predecessors whom they were trying to emulate. Here they are in chronologic order, and when no city is mentioned after their name, it means that they were Athenians (six were out of ten): Callōn of Aigina (fl. 520), Hēgias (beginning of fifth century), Calamis (fl. 470), Myrōn of Boiōtia (born 480), Polycleitos of Argos and Sicyōn (fl. 452–412), Pheidias (500–432) and his most famous disciple Alcamenēs (fl. 444–400), Praxitelēs (c. 370–330), Lysippos of Sicyōn (Alexander's contemporary), and one Dēmētrios(?).

In spite of the vicissitudes of his rule, Antiochos IV Epiphanēs was as anxious to adorn his capital, Antiocheia on the Orontēs, as his rival Eumenēs was to embellish Pergamon. Among other things he ordered copies to be made of the heroic-sized statues of Zeus and Athens by Pheidias. Boēthos of Chalcēdōn received a commission for a bust or statue of Antiochos IV for the city of Dēlos. This Boēthos flourished c. 180 in Lindos (Rhodos). He is well known for his "Boy who almost strangles a goose while embracing it," a type often copied. One of his works has been found in a ship sunk in the first century B.C. near Mahdia (Tunisia) and is now in the Bardo Museum (Tunis). It is a statue of an adolescent, Agōn, divine patron of athletic games, together with a bronze herma, a monument of singular originality. Did Boēthos flourish at Antiochos' court? Unfortunately, Antiochos was an impulsive and unsteady sovereign who caused considerable trouble to himself and to others. It was he who tried to uproot the Jewish religion and to replace Adonai with Greek gods; he did not succeed but precipitated the Maccabean revolt. He was accused of sacrilegious crimes by the Greeks as well as by the Jews and died in 163 in a state of madness.[7]

Another definite personality of the same age is Damophōn (or Dēmophōn) of Messēnē,[8] who was charged with repairing Pheidias' Zeus at Olympia, probably after the earthquake of 183, and thus became familiar with sculpture on a heroic scale. He made various statues of gods and

[7] His enemies changed his title Epiphanēs (coming to light, illustrious) to Epimanēs (furious).

[8] Messēnē, capital of Messēnia (southwest Peloponnēsos); not to be confused with Messēnē on the northeast coast of Sicily (modern Messina).

goddesses for sanctuaries of the Peloponnēsos: his native Messēnē, another in Achaia, others in Megalopolis and Lycosura in Arcadia. The gigantic group created by him for the sanctuary of Dēmētēr and Despoina near Lycosura was seen by Pausanias (II–2) who described it (*Description of Greece*, VIII, 37). Many fragments of it have been found (for example, heads in the National Museum, Athens). There were four gods in a row, Despoina and Dēmētēr seated at the ends, Artemis and the Titan Anytos standing in the middle.[9]

This kind of work, inspired by Pheidias' and comparable to his, at least with regard to the heroic size, was done also in the same century by Eubulidēs in Athens and by Eucleidēs in Aigeira.

The ancient glory of Rhodos was held up during the second century not only by Boēthos, but also by Philiscos and by the two brothers, Apollōnios and Tauriscos of Tralleis. Philiscos of Rhodos was probably the author of a group of nine Muses; that famous group and his statues of Apollōn, Lētō, and Artemis[10] found place eventually in the temple of Apollōn near the Porticus of Octavia, in Rome. Apollōnios and Tauriscos of Tralleis, sons of Artemidōros, were adopted by Menecratēs of Rhodos;[11] they were said to be the authors of the "Farnese Bull,"[12] a baroque composition of heroes and animals that originated a new tradition of which the Laocoōn was to be the climax.

It is strange that we do not know the names of other sculptors of the second century, of whom there must have been a good many, not only in Rhodos but elsewhere, in every Hellenistic city that enjoyed a modicum of prosperity. This was for them a matter of civic emulation and pride. It is possible that various Hellenistic reliefs are of Rhodian origin or inspiration, for example, the "Apotheosis of Homer" ascribed to Archelaos of Priēnē[13] (British Museum), the "Votive Relief" (Glyptothek, Münich), the "Couple on horseback" from Capri (Museo Nazionale, Naples), the

[9] Despoina, meaning "the mistress," represented other goddesses, chiefly Persephonē (or Corē). Persephonē was the Proserpina of the Romans. The Titans (12 or 13 in number) were the gigantic offspring of Uranos and Gē (Heaven and Earth).

[10] Lētō was the daughter of a Titan and the mother of Apollōn and Artemis by Zeus. The Latin-speaking Romans called Lētō and Artemis Latona and Diana.

[11] This Menecratēs was perhaps the leading artist in the building of the great altar in Pergamon. There were active connections, artistic as well as political, between Pergamon and Rhodos. Menecratēs' adoption of Apollōnios and Tauriscos places them and the "Farnese Bull" in the second century B.C.

[12] "Farnese Bull" is the conventional and easy name of an immense group now in the Museo Nazionale in Naples. It represents the brothers Amphiōn and Zēthos binding the girl Dircē to the horns of a bull (a complicated mythological story for which we have no space). The bull dominates the group, the general shape of which is pyramidal. It is called the "Farnese Bull" because it was a part of the collection of antiquities that belonged to the illustrious Farnese family, dukes of Parma, eventually bequeathed to the Museum of Naples.

[13] Priēnē, one of the twelve Ionian cities, on the coast of Asia Minor, was in northwest Caria. Communications with Rhodos and other islands were easy.

Fig. 109. Helenē and Aphroditē with the boy Alexandros (or Paris) and a male angel. [Museo Nazionale, Naples.]

"Scene outside a house" (British Museum), "Dionysios' visit to a mortal" (Louvre), "Youth with courtesans" (Naples), and especially that exquisite one, "Helenē and Aphroditē with the boy Alexandros (or Paris) and a male angel" (also in Naples). We reproduce it (Fig. 109) because it is not as well known as it deserves to be, and will help the reader to imagine the others.

One understands readily how the creation of such reliefs appealed to the sophisticated minds of those artists, whether in Rhodos or other cities. Sculpture was very limited in scope; one could reproduce one person or a few but hardly more, and it was impossible to suggest their environment.[14] On the contrary, any relief could easily evoke not only people and animals but all kinds of objects, even buildings and trees; it was even possible to hint at a landscape. In short, reliefs were the sculptural equivalents of paintings. As the Hellenistic paintings are lost, we do not know how the painters managed to suggest the feelings of a group of people or the mood of the country around them. Fortunately, some of the Hellenistic reliefs give us that kind of information.

Another characteristic of Hellenistic sculpture is the love of portraits. Most of the portraits in our museums are either Hellenistic or Hellenistic copies of Greek originals (I am referring only to the oldest available; the later ones are Roman copies of Hellenistic originals or of Hellenistic copies). The earliest form of portrait was the herma.[15] It is probable that portraits

[14] Unless one adopted the charming method of the Egyptians as in the ceramic hippopotamus of the Metropolitan Museum (Twelfth Dynasty, c. 1950 B.C.), with lotus flowers, birds, and leaves painted upon its body. That is all right for a hippopotamus, but how could one apply it to an Apollōn or an Aphroditē?

[15] From Hermēs, probably because some of the earliest specimens represented that very god, a bearded head placed at the top of a stone pillar. The word "herma" is used to designate a portrait restricted to the head and upper part of the chest; the

were duly made of the Hellenistic kings and queens, princes and princesses, but how can one be sure that a given portrait represents Seleucos Nicatōr or any one of the Ptolemaioi? In a few cases, comparisons may be made with coins, but none of those comparisons is convincing to me. As to the busts or statues of Homer, Dēmosthenēs, Aischylos, Sophoclēs, Euripidēs, Hippocratēs, Aristotle, Plato, they are nothing more than symbols. Art critics have been so eager to give names to anonymous busts or statues that they have helped to create innumerable ghosts.[16] As soon as one bust had been baptized "Aristotle" by a reputed scholar, all the busts resembling it were Aristotles *ipso facto*. Our museums are full of ancient busts (most of them Hellenistic or Roman) to which the names of illustrious men have been arbitrarily attached. Hellenistic sculptors produced many such portraits because there was a ready market for them, and that market was much enlarged when Western buyers began to compete with those of Eastern Europe and Asia.

The best trade of all concerned the gods, goddesses, and heroes, for there was an inexhaustible need of their effigies in the civic buildings, the temples, and the private palaces. Without having made any count, I have the impression that effigies of Aphroditē were in greater demand than any others. We have many types of her: Aphroditē rising from the sea, arranging her hair, preparing to bathe, unfastening her sandal, kneeling, and so on. Her body is almost always nude, rarely semidraped. Whatever her gestures might be, one could not imagine a simpler composition – the nude body of a woman – and yet such was the artistry of those sculptors that they have created unforgettable types.

The popularity of those statues of Aphroditē in the Hellenistic age suggests a comparison with the popularity of Madonnas during the Renaissance and later, but, as far as artistic matters are concerned, there is an immense difference. The best-known Madonnas are paintings; while we remember in many cases the type of woman who served as model, our memory is always helped by the persons and objects that surround her.[17] Sculpture does not lend itself to that, and, moreover, Aphroditē was almost always naked and without any accessory, and yet many of the statues of

---

full bust was a later (Roman) development. The earliest specimens (extant) are the helmeted heads of Attic generals; the best-known of them is Periclēs.

[16] G. Sarton, "Portraits of ancient men of science," *Lychnos* (Uppsala, 1945), pp. 249–256, 1 fig.; shorter note in *Horus* (Waltham: Chronica Botanica, 1952), pp. 42–43.

[17] Not only are the Madonnas distinguished from one another by details indicating a special phase of her life (Purification, Annunciation, Assumption, and so on), but many are named for an outside addition to the painting which helps us to remember her; maybe definite saints or angels, picturesque rocks, or a burning bush, a protective mantle or a rosary, or a bunch of grapes, roses, violets, a pear or an apple, or a monkey, or various kinds of birds, plover, golden oriole, goldfinch. Madonnas with goldfinches are so numerous that they must be separated from one another in other ways. Herbert Friedmann, *The symbolic goldfinch* (New York: Pantheon, 1946) [*Isis* 37, 262 (1947)].

her emerge from the crowd and are as definite in our remembrance as any person can be.

Two Aphrodita out of many had already become popular in ancient times and their fame must have reached its climax in Hellenistic days; one (Fig. 110) was the "Aphroditē of Cnidos," sculptured by Praxitelēs (370–330) and the other the "Aphroditē of Cōs," painted by Apellēs (c. 332). It will tickle medical historians to realize that each one of these two goddesses was connected with one of the rival medical schools of Greece; it is said that the statue and painting were both made from the same model, Phrynē. She was an exceedingly beautiful woman and may have served as a model to either artist or to both, or she may have inspired that beautiful story. Whether the story is true or not, it helps us to remember that these two masterpieces of ancient art were created in almost the same environment and at about the same time, the age of Alexander the Great. Apellēs' painting is irretrievably lost, and the "Aphroditē of Cnidos" is known only from a very early copy in the Vatican. The Medici "Aphroditē" (Uffizi, Florence) dates probably from the same time as the Vatican copy just mentioned (say from the end of the third century B.C.). An admirable ancient replica of the Medici "Aphroditē" has been recently added to the treasures of the Metropolitan Museum.[18]

The most popular statues of Aphroditē today are the one of Cyrēnē in Rome and her sister of Mēlos in Paris. Both are so beautiful and so original that early critics ascribed them to the fourth century; the critics of today are now agreed that both are Hellenistic, though it is impossible to be more precise. They were both found in relatively recent times in out-of-the-way places. The one that the Italians found at Cyrēnē in North Africa may be a Rhodian work of the end of the second century B.C. Its sensuality is truly disturbing. The other "Aphroditē," just as beautiful but purer, was discovered by French naval officers in the remote island of Mēlos [19] and brought to the Louvre in 1820. She is as enigmatic as she is serene and seems to discourage every attempt to determine her age. It is difficult to describe these two masterpieces but impossible to forget them.

In the last half century B.C., two distinguished sculptors were active probably in Athens. Apollōnios of Athens, son of Nestōr, created the Belvedere torso in marble and a boxer in bronze. Glycōn of Athens made a copy of the "Farnese Hēraclēs" [20] which was later placed in the baths of Caracalla in Rome. These two men might be called Greco-Roman, for they represent the end of Hellenistic art.

That end is even better symbolized by two colossal monuments, the

[18] Christine Alexander, *Bulletin of the Metropolitan Museum of Art* (New York, May 1953), pp. 241–251, 14 figs.

[19] Remote as compared with the other islands clustered along the Asiatic coast. Mēlos is the most westerly of the Cyclades, almost as far south as the southeast of the Peloponnēsos.

[20] The "Farnese Hēraclēs" (in Naples) is a type created by Lysippos, the favorite sculptor of Alexander the Great. An enormous number of works (some 1500) were

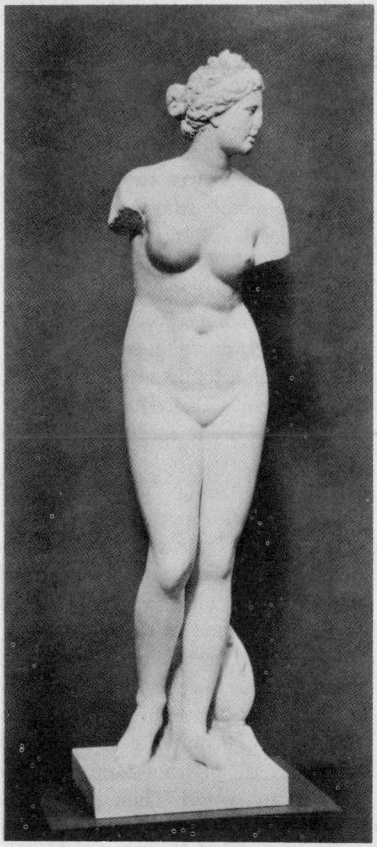

Fig. 110. Plaster reconstruction of an ancient imitation of the Aphroditē of Cnidos by Praxitělēs. The Metropolitan Museum acquired the original in 1952. The original plinth, one foot, and the dolphin were preserved separately and are also in the Metropolitan. This ancient replica of the Medici Aphroditē (Florence) had been hidden away in a Silesian castle since the time of Winckelmann (1711–1768) or before. We reproduce it because it is not yet as well known as the "Venus de Milo" and her Cyrenaic sister in Rome. [Courtesy of Metropolitan Museum.]

"Nile" and "Laocoōn." The "Personification of the Nile" in the Vatican is a copy of an earlier Greco-Egyptian group; the copy was made for the sanctuary of Isis and Osiris in Rome.[21] Father Nile is a giant surrounded by sixteen children, and various motives recall animals of Egypt.[22]

ascribed to him; helped by Alexander's munificence he must have employed many other artists. The adjective "Farnese" was explained in note 12.

[21] The temple of Isis and Osiris (or Serapis?) was dedicated in Rome by Mark Antony in 43 B.C. It was destroyed by order of Tiberius in A.D. 17 because of the scandals which were alleged to have taken place in it.

[22] The sculptural evocation of the Nile (or call it the genius of the Nile) was an old artistic conceit, represented in Egyptian monuments, for example, in the Pyramid of King Sehurē, Abuṣīr (Fifth Dynasty, c. 2550 B.C.) and in a bas-relief in the British Museum of the Twenty-First Dynasty (c. 1000 B.C.). The gigantic monument preserved in the Vatican is something very different, however; it is the Greco-Roman interpretation of an Egyptian idea. There is also a representation of the

The "Laocoōn"[23] (also in the Vatican) is the very climax of Hellenistic baroque (Fig. 111). It was the work of three artists, Agēsandros, Polydōros, and Athēnodōros of Rhodos, who completed it, c. 50 B.C. According to Pliny (*Natural history*, xxxv, 37), it was set up in the palace of Titus (emperor 79–81) on the Esquiline hill in Rome; it was discovered there in 1506 and that discovery was one of the most sensational events of the Renaissance. The "Laocoōn" was duly admired by great artists, such as Michelangelo and El Greco,[24] sung by poets, was considered to be one of the greatest masterpieces of antiquity by such enlightened men as Johann Joachim Winckelmann (1755),[25] Lessing (1766), and Goethe (1798). A century later, when knowledge of ancient sculpture was richer and purer, the "Laocoōn" found fewer admirers and more contemners.

It was gradually recognized that "la difficulté vaincue" is a very poor criterion of artistic merit. The technique of the creators of the "Farnese Bull" and of "Laocoōn" was incomparable, but their artistic vision was poor. Great art is as different from technical virtuosity as wisdom is from learning.

The story of the "Laocoōn" tradition is very instructive for the understanding of the change of taste throughout the ages.[26] The monument itself was the ripest fruit of the Hellenistic climate. It was admired at first because of its unrestrained pathos and also, no doubt, because of the stupendous technical difficulties which the artists overcame. It offered a large scope to the commentaries of poets and of the old-fashioned art critics.

Such grandiose monuments as the "Farnese Bull," the "Nile," and "Laocoōn" give us a very poor impression of Hellenistic art, but we should beware of that and remember other monuments, such as the "Nicē of Samothracē," the "Aphroditē of Mēlos," and even the "Aphroditē of Cyrēnē,"[27] which are the favorite examples of Greek art for millions of educated people all over the world. Their tradition is, unfortunately, too short to be as in-

---

sources of the Nile in Hadrian's gateway, Philae's temple (Philae island, Upper Egypt). Other rivers were evoked in a similar way, for instance, the Tiberis (Tevere) in the Louvre.

[23] Laocoōn was a Trojan prince and priest of Apollōn, who had violated the sanctity of the temple. While he was sacrificing at the altar, assisted by his two sons, two large pythons coming from the right and the left coiled themselves around the bodies of the three men. The monument represents their death agony. It is dramatic and pathetic to an unbearable degree; the technical difficulties that the artists overcame were stupendous.

[24] See El Greco's astonishing painting of Laocoōn with Toledo in the background (National Gallery, Washington, D. C., formerly in the collection of Prince Paul of Serbia, Belgrade).

[25] Winckelmann (1717–1768) is often called the father of classical archaeology and the first expounder of classical art. In justice to him (as well as to Lessing and Goethe), it must be remembered that the best examples of Greek art were still unknown.

[26] The story has been told and the relevant texts quoted by Margarete Bieber, *Laocoon. The influence of the group since the rediscovery* (22 pp., 29 ills.; New York: Columbia University Press, 1942).

[27] The first two, being in the Louvre, are known to most people under their French names, "Victoire de Samothrace" and "Vénus de Milo." Their popularity was partly increased by the Louvre; it would

structive as that of the "Laocoōn." The "Aphroditē of Mēlos" was discovered in 1820, the "Nicē" in 1863,[28] and the "Aphroditē of Cyrēnē" only on 28 December 1913.[29] Their great popularity is mysterious; it may seem extravagant and somewhat impure, but it must be taken into account.[30]

It is paradoxical that those three monuments, which the majority of people would raise above all others, do not belong to the golden age of Greek sculpture but to the silver age, to an age of decadence when Greek ideals were steadily corrupted by the admixture of Egyptian and Asiatic influences.

HELLENISTIC SCULPTURE IN ROME

The introduction of Hellenistic art into the city of Rome was a fruit of the Roman conquest of Greek lands. It is a history of war and loot which makes one wonder what the artistic feelings of the Romans really were. For the stealing of objects of art implies a certain love of them, at least a modicum of admiration and appreciation, and yet was the character of the looters improved by the beauty of their loot? Certainly not, but human nature is very complex, and it is perhaps better not to judge those Roman "connoisseurs" and "art lovers" too harshly.

The tragedy of war has always implied sacking and looting; that is terrible, but is it worse than killing men and raping women? What the Romans did during their conquests was certainly not worse than what the good Christians of the Fourth Crusade did when they looted Constantinople in 1204, or the Christian soldiers of Charles V, when they sacked the Eternal City in 1527, or Bonaparte, when he extorted Italian treasures in 1796–97, or the European nations, when they looted Peking in 1860 and again in 1900–1901. That damnable list is not up to date and only very few examples have been mentioned, yet enough, not to justify Roman greed, but to place it in its context. The Romans were not worse than other conquerors and the worst atrocities in the whole past were perpetrated by "civilized"

---

have grown more slowly if they had been in a smaller museum; on the other hand, remember that a great many works of art have been shown in the Louvre for centuries without becoming popular.

[28] The ship's prow serving as pedestal to the "Nicē" and exalting her was brought to the Louvre in 1883. The popular fame of the "Nicē" began only some time after 1883. As the "Nicē" may be a work of the third century, we spoke of it in Chapter XIII. Margarete Bieber would ascribe it to the beginning of the second century (200–190) and to one Pythocritos of Rhodos.

[29] Gilbert Bagnani, "Hellenistic sculpture from Cyrene," *Journal of Hellenic Studies 41*, 232–246 (1921). No wonder that the Aphroditē of Cyrēnē is not as well known as she of Mēlos, for the latter has been haunting men's imaginations a century longer.

[30] Popularity is always impure, for it expresses the opinion of incompetent people and their opinion is often based upon irrelevancies. It is mysterious because it is generally impossible to know how it begins, grows, and establishes itself. How and why did Queen Nefertete and the "Venus of Milo" become "best sellers"? We may call them best sellers because reproductions of them have been sold (and are selling) by the millions.

Fig. 111. The Laocoōn as it appeared soon after its discovery in Rome in 1506. Engraving by Marco Dente of Ravenna (d. 1527). Comparison with photographs of the Laocoōn at present exhibited in the Vatican will show many differences due to wise and unwise restorations. [Courtesy of Metropolitan Museum.]

people, who flourished more than two millennia later, within our own time. Therefore we could not condemn the Romans as looters without condemning ourselves as hypocrites.

The main steps in the history of Roman looting of works of art are the following. The first important date is 212, when Syracuse was plundered by Claudius Marcellus. The rich city was full of Greek statues which were shipped away to grace the Roman temples. Historians of science will have no trouble in remembering that date, because Archimēdēs was killed during the sack of his native city. Marcellus whetted Roman appetite for Greek art and gave the Roman generals and proconsuls an example that they did not forget.

Tarentum, in Calabria, was taken and plundered in 209 by Fabius Cunctator.[31] In 198, Eretria[32] was sacked by Quintius Flamininus, who brought to Rome the first examples of Lysippos' art; it was this same Flamininus, curiously enough, who two years later (196), at the Isthmian games in Corinth, proclaimed in the name of the Roman senate the freedom and independence of Greece (conquerors always consider themselves liberators). In 187, Gnaeus Manlius Vulso returned from a long expedition in Syria and Anatolia with a very large booty; though much of it was lost while crossing Thrace, he brought enough objects of art and Asiatic ornaments to "demoralize" the Romans; it was Manlius Vulso's veterans who introduced the taste for foreign luxuries in the capital. After Perseus' defeat at Pydna in 168 by Aemilius Paulus Macedonicus, his library and art collection were taken to Rome. In 146, Corinth was utterly sacked by L. Mummius, who sold many of the art objects to the king of Pergamon, yet brought a great many of them to Rome; according to Polybios, most of the Greek sculpture available in Rome at the turn of the second century came from Corinth. When Felix Sulla, the dictator, took Athens by storm in 86, the glorious city was given up to rapine and much Athenian treasure found its way to Rome. Mummius and Sulla were emulated by C. Verres; during the latter's propraetorship in Sicily (73–71), his extortions and exactions knew no bounds; Verres was interested primarily in wealth, but by that time Greek sculpture was worth a lot of money in the Roman market and Verres took statues as readily as jewels or coins. That ugly story is well documented because Verres' plundering activities were so outrageous that he was prosecuted, and the prosecution was entrusted to Cicero, who had been quaestor in Sicily in 75 and loved the Sicilians. Cicero wrote no fewer than seven orations or documents against Verres and in spite of enormous difficulties (Verres was supported by the whole aristocracy) he succeeded in obtaining the condemnation of that criminal *in absentia*.[33] Verres had taken refuge in Marseilles and had kept so many treasures that, in 43, he

[31] Quintus Maximus Fabius was nicknamed Cunctator (the delayer) because of his delaying and evasive tactics in the war with Hannibal (Second Punic War). It was his name Fabius which inspired the title of the Fabian Society organized in England in 1884 to spread socialism without violence.

[32] Eretria is in Euboia, the largest island of the Aegean Sea. That island is so close to the Greek mainland that it might almost be considered a part of it. At Chalcis, the strait Euripos, between the island and Boiotia, is so narrow that a bridge was built across it.

[33] In George Long's edition of the *Orationes*, the whole of vol. 1 (London, 1851) is devoted to *Verrinarum libri VII*. See par-

was proscribed by Mark Antony, who, it is said, coveted them for himself. Perhaps he needed them to adorn the temple that he was dedicating to Isis and Osiris? Indeed, many of the plunders were inspired by religious piety; the plunderers wanted to embellish the temples that appealed most to their devotion.[34] The sack of Rhodos by Cassius Longinus [35] in 43 did not enrich the Roman temples so much, but it was a death blow to the glorious artistic school of that island.

The discriminating lovers of Greek sculpture in Rome were anxious to favor the creation of new masterpieces, and, in the meanwhile, the artists who still managed to flourish in Athens and other Greek cities had realized that the Romans might become their best patrons. Much work done in Athens during the last two centuries B.C. was probably inspired or encouraged by Roman commissions. For example, Polyclēs the Athenian and his two sons, Timoclēs and Timarchidēs, had obtained some fame in Greece; one of Polyclēs' statues was in Olympia and his two sons were the authors of a statue of Asclēpios in Elateia.[36] They established themselves in Rome, possibly upon the advice of Caecilius Metellus Macedonicus. After his conquest of Macedonia in 146 and before his death in 115, Metellus built the Porticus of Octavia in Rome and included in it some of their works, chiefly Polyclēs' "Apollōn holding a cithara." Other Greek sculptors followed their example, for Rome was now a better market for Greek art than Athens. Thus, Arcesilaos, for example, executed commissions for the rich aesthete Lucullus (c. 117–56), for Asinius Pollio, founder of the first public library, for Varro, and for Caesar. His "Venus genetrix" was in a temple dedicated by Caesar in 46.[37] Another good example is that of Pasitelēs, who flourished in Rome c. 60–30. Pasitelēs did not come from Greece, however, but from Magna Graecia, and as such was one of the many "Italians" who benefited from the Lex Plautia Papiria [38] granting the Roman franchise to all the people of Italy south of the Alps. He was not simply a sculptor but an exponent of Greek art, whose task was similar to that of many other Greeks who ex-

---

ticularly the beginning of Book IV, *De signis*. Verres was the most thoroughgoing and unscrupulous "collector" of Greek art in Sicily. He used spies and informers such as Tlēpolēmos and Hierōn. In Messēnē (modern Messina), he "collected" the Hēraclēs of Myrōn (fl. 480–455), the Canēphorai (Basket-bearers) of Polycleitos (fl. 452–405), and the Enōs (Cupido) of Praxitelēs (fl. 370–330).

[34] Compare the stealing of relics by Christian fanatics who did not hesitate to commit crimes in order to increase the sanctity of their favorite churches (*Introduction*, vol. 3, pp. 1044; 266, 291).

[35] Often called Cassius the Tyrannicide, because he and Brutus were the leaders of the conspiracy against Caesar, and the main authors of Caesar's murder on the Ides of March 44 B.C.

[36] Elateia in Phōcis, the most important city in that country next to Delphoi.

[37] Pliny, *Natural history*, XXXV, 156.

[38] This law was enacted in 89, thanks to M. Silvanus Plautius, who was tribune in that year, and to C. Papirius Carbo, consul in 85–84 and 82, executed by Pompey in 82. The Roman franchise was not granted to every Italian but only to those who satisfied certain requirements.

plained Greek letters. He wrote a treatise on Greek art (*Quinque volumina nobilium operum in toto orbe*); it is a great pity that that work is lost, because it was the last of antiquity to be written by a professional artist. He was a connoisseur and probably helped collectors with his criticism. He formed a school and his best pupils were Stephanos and Menelaos.[39]

ROMAN SCULPTURE

This introduces the subject of Roman sculpture, or, more correctly, Greco-Roman sculpture. The line is difficult to draw between the work done by Greek sculptors in Athens to satisfy Roman taste, the work done by Greek sculptors in Rome, and the work done by their Roman disciples. There never was a sharp discontinuity; Roman characteristics became more common but were never sufficient (at least before the Augustan age) to obliterate or even to dominate the Greek style. The Greco-Roman sculptors of the Republic were obviously more deeply under Greek influence than were such writers as Lucretius, Cicero, and Virgil.

Indeed, the influence of Greek sculpture in Rome was ubiquitous and much more tangible than that of Greek letters. The latter had no power at all upon people who did not know the Greek language or did not know it well enough. On the other hand, almost every statue in the Roman temples and palaces was Greek, and their message could be understood immediately by any person having artistic feelings.

Rome had become the greatest market of Greek art; there were regular merchants and brokers, notably, Avianus Evandros, who was Cicero's friend.[40] Any rich man who wished to embellish his favorite temple, or his own house, could easily satisfy his needs in Roman shops.

The fashion for portraits (busts and statues) was increasing. It was in this field that Roman qualities had the best chance of asserting themselves – realism, for good or evil. It is possible that Etruscan portraits helped to divert Roman sculptors from the Greek fascination. The most significant Roman portraits, however, did not appear until the end of the Augustan age and later.

As there was a greater concentration of Greek art in Rome than in any Greek city it is not surprising that our knowledge of it is derived not so much from Greek sources like Pausanias (II–2) but rather from a Latin source, the *Natural history* of Pliny the Elder (I–2). The main result of that concentration, however, was to retard the development of a genuine Roman art. Moralists would be tempted to remark that this was a condign punishment for the wholesale confiscation and importation of Greek art. Never was that done on so large a scale and with the same thoroughness,

[39] Pliny, *Natural history*, xxxv, 156.

[40] C. Avianus Evandros, a freedman of M. Aemilius Avianus, was in the antique business in Athens when Cicero got to know him. He was taken to Rome as a prisoner in 30 B.C. C. Robert, Pauly-Wissowa, vol. 11 (1907), 843.

but in the course of time the Roman holdings were scattered all over Europe and America.[41]

The Romans introduced two new architectural forms at about the same time, the beginning of the second century B.C.; these were the basilica and the triumphal arch.

The basilica [42] was not a simple portico but a closed building of oblong shape which served as a court of law or an exchange and place of meeting for businessmen or politicians. The first in Rome was the Basilica Porcia erected by Cato the Censor in 184. Many more, some twenty, were gradually built in Rome. Some were open to the sky and were thus like a portico built all around a courtyard. In the course of time these basilicas were transformed into Christian churches, and the very name evokes at present a Christian church built upon the same model.[43]

The *Arcus triumphalis* was a Roman development of a simpler structure, the *Porta triumphalis* (the triumphal door), through which a victorious general might enter a city. The earliest triumphal arch was built by one L. Stertinius in Rome c. 196, and the next by P. Scipio Africanus in 190. Some thirty-eight triumphal arches were eventually erected in Rome and many more in the Roman world; of those in Rome only five remain and none of them is pre-Christian.

The best example of Roman sculpture was perhaps the *Ara pacis* (peace altar) solemnly dedicated by the Senate in 9 B.C. as a memorial to the peace which Augustus had given to the Roman world (Fig. 112). The altar was surrounded by a marble wall about 3 m high, upon which were represented in low relief a solemn procession of the imperial family and of high magistrates. As far as one can judge from the remains, that national monument was a masterpiece; it was obviously Roman in intention, yet reminiscent of Greek art, a perfect symbol of the highest Roman culture of that age, Greek loveliness grafted upon the vigorous Roman tree.

The *Ara pacis* has long been dismantled but fragments of it, discovered at various times, can be seen in the Uffizi (Florence), the Louvre, the Vatican, and, chiefly, in the Museo Nazionale in Rome, where casts of the

[41] The only comparable cases of wholesale importation of foreign art are the importation of Chinese art into Japan, and that of European and Asiatic art into the United States. American connoisseurs did not steal, however, but paid so much that they raised considerably the world prices of art objects.

[42] This is the Latin name, preserved in English, derived from the Greek feminine adjective *basilicē* (royal). The Greek spoke of a *stoa basilicē* (royal portico).

[43] For example, St. Clemente in Rome and St. Ambrogio in Milan. At present, the term basilica has an ecclesiastical meaning independent of architecture. Some churches are called basilicas because of their eminence, and they enjoy certain privileges. There are seven basilicas in Rome (St. Clemente is not one of them). In Paris, Ste. Clotilde, Ste. Jeanne d'Arc, and the Sacré Coeur are basilicas; the oldest of them, Ste. Clotilde, was begun as late as 1846, the Sacré Coeur in 1876, and Ste. Jeanne d'Arc only in 1932.

Fig. 112. *Ara pacis Augustae* (Augustus' altar of peace) completed in Rome in 13 B.C. and dedicated by the Senate in 9 B.C. Only fragments remain, but attempts have been made to reconstruct the whole. This is one of the friezes; it represents members of the imperial family; Agrippa (d. 12 B.C.) stands in the middle wearing the head scarf of a pontiff. For explanations and more photos see José Pijoán, *Summa artis* 5, 271–79 (Madrid, 1934). [Uffizi, Florence.]

other fragments are exhibited as well as a tentative reconstruction of the whole.

The best "Tanagra" figurines were made in the third century B.C. (see Chapter XIII); they were made in many places and it is possible that some were baked in Italy by Greek artists. The Romans used terra cotta for larger statues or for the decoration of buildings (decorative sculpture); they probably derived the idea from Etruscan examples (cinerary urns, sepulchral masks, recumbent groups on sarcophagi). That Roman art was fairly ancient but its architectural applications continued until the end of the Republic. In 195, Cato the Censor complained that terra-cotta antefixes of the Roman temples seemed mean and ridiculous as compared with the Greek ones made of marble.

The same material was also used for mural decoration or for hiding beams and cornices. Terra-cotta reliefs were molded, and Cicero wrote to Atticus to ask for Athenian forms. Their use diminished during the Augustan age because increasing luxury favored marble instead of baked clay.

HELLENISTIC AND ROMAN PAINTING [44]

Our knowledge of Hellenistic and early Roman painting is curiously deficient. We are far better informed concerning the earlier and later periods. As to the first, the evolution of vase painting is very instructive, we realize all the qualities of Greek drawing; as to the later period, we have the wall paintings of Pompeii and Herculaneum, which reflect Hellenistic models.[45]

The names of a few "Roman" painters have come down to us. One of

[44] For illustrations see Ernst Pfuhl, *Meisterwerke griechischer Zeichnung und Malerei* (160 ills.; Munich, 1924); English trans. by J. D. Beazley (152 pp., 126 pls.; London: Chatto and Windus, 1955).

[45] Pompeii and Herculaneum were destroyed by an eruption of Vesuvius in A.D. 79, but both were ancient cities. The wall paintings cover the period 300 B.C. to A.D. 79; they have been divided into three groups: (1) the earliest, first style; (2) after Sulla (138–78), second style; (3) after Augustus (d. A.D. 14), Egyptianizing third style. Most of the paintings, all the significant ones, are of the second or third style.

the earliest was a woman, Iaia[46] of Cyzicos, who flourished in Rome at the time of Varro's youth (say c. 100 B.C.). She painted portraits, chiefly of women, including herself, and received higher fees than the best of her male rivals, Sopolis and Dionysios; she remained unmarried. Two other painters deserve mention. Timomachos of Byzantion, who flourished in Caesar's time, treated mythological subjects and made portraits. Ludius (or Tadius?), who belonged to the Augustan age, introduced "a very pleasant style of wall-painting" representing "villas, porticoes, landscape gardens, sacred groves, woods, hills, fishponds, straits, rivers, shores,"[47] with various personages engaged in all kinds of activities. No example of their handicraft is known but one can imagine it on the basis of some of the Pompeii paintings.

ENGRAVED GEMS

The most significant of the decorative arts was the engraving of gems and cameos,[48] which came to Rome from Greece. The story is the same as for sculpture and painting; at first, the art objects were imported, later the artists, and in a final stage those artists educated Roman apprentices. This final stage was not yet reached in Christ's time, and the best Roman gems were made by Greeks.

Mithridatēs the Great was a great collector of gems[49] and after his death, in 63, his treasures were given by Pompey to the temple of Jupiter Capitolinus. The first Roman gem collector was M. Aemilius Scaurus, Pompey's quaestor in the Mithridatic war and later (c. 61) conqueror of the Nabatean king Aretas. Julius Caesar was also an eager collector; he presented many gems to the temple dedicated by him to Venus Genitrix. One must always remember that the gems were believed to have occult qualities, and giving them to a temple was somewhat like giving to a church not simply precious objects (such as a frontal or a ciborium) but relics.

Following Oriental and Greek examples Roman generals and rulers used signets to authenticate their orders. Julius Caesar was perhaps the first to have a special guardian of his seal (*custos anuli*), who was the ancestor of similar officers in later governments (*Custos sigilli*, Keeper of the Great Seal, the Privy Seal, Garde des sceaux, and so on). Augustus had three

[46] Iaia or Lala, Laia, Maia? Pliny, *Natural history*, xxxv, 147; Pauly-Wissowa, vol. 17 (1914), 612.

[47] Pliny, *Natural history*, xxxv, 116. He probably meant that Ludius introduced that new style in Rome; it had been practiced already by Hellenistic painters (Vitruvius, VII, 5).

[48] A cameo is a gem carved in relief, especially in a stone such as onyx or sardonyx, that has layers of different colors. The engraver tries to have the figure in one color and the background in another.

[49] Mithridatēs had accumulated such immense collections that it took the Romans thirty days to make a catalogue of those kept in a single warehouse of his (at Talaura; I do not know where this was). For more information on him as a patron of arts and a collector, see Théodore Reinach, *Mithridate Eupator* (Paris, 1890), pp. 286, 399.

signets, the first bearing a sphinx, the second a head of Alexander by Pyrgotelēs, and the third his own head, by Dioscoridēs. The first seal was probably Egyptian, the second Greek, the third Greco-Roman. Dioscoridēs flourished in Rome, was the greatest engraver of the Augustan age, and was succeeded by his three sons, Eutychēs, Hērophilos, and Hyllos.

A number of ancient gems and cameos can be examined in the Cabinet des Médailles attached to the Bibliothèque Nationale, Paris, and in similar collections. A description of them would be tedious and, without illustrations, futile.[50]

[50] Let us make a single exception for the "Grand camée de la Sainte Chapelle," which is one of the outstanding treasures of the Cabinet des Médailles. It is the most famous ancient cameo, as well as the largest (30 × 26 cm), and is ascribed to Dioscoridēs, Augustus' engraver. It represents the glorification of Germanicus. Julius Caesar Germanicus (15 B.C.–A.D. 19) was adopted by Tiberius in A.D. 4 and celebrated a triumph in Rome in 17. Hence, this cameo is a little post-Christian. Reproduction, description, and history in Ernest Babelon, *Catalogue des camées antiques de la Bibliothèque Nationale* (2 vols.; Paris, 1897), no. 264, vol. 1, pp. 120–137; vol. 2, pl. XXVIII.

XXVIII

ORIENTALISM IN THE LAST TWO CENTURIES[1]

The story of orientalism is less startling in the last two centuries of Hellenism than it was in the first, but we should bear in mind that some activities begun in the third century B.C. were continued in later centuries. This is true, for example, of the Septuagint.

BORDERLANDS: THE PARTHIAN EMPIRE AND THE RED SEA

The Hellenistic world was half Oriental; Greek or Macedonian princes ruled in the islands, in Egypt, and in many countries of Eastern Asia. There were strong Greek or Hellenizing colonies in all those countries and outposts as far as the Thebaid in the south and the Oxus and Indus rivers in the east. On the other hand, those colonies were permeated with Oriental influences, not only the local ones, but also the more distant ones which originated in Babylonia, Iran, or India.

From the middle of the third century on, the Parthian empire was the main borderland between East and West. It started as an offshoot of the disintegrating Seleucid empire when the Scythian brothers, Arsacēs and Tēridatēs, satraps of Bactria, revolted against their overlord, Antiochos II Theos (261–246); c. 250, Arsacēs became the first independent king of Parthia, with his capital at Hecatompylos.[2] He was the founder of the Parthian empire, which was gradually increased by his successors, and of the Arsacid dynasty, which lasted almost five centuries (476 years), thirty kings ruling from 250 B.C. to A.D. 226.[3]

The Parthians overran the neighboring satrapies until their empire extended from the Euphratēs[4] to the Indus, and from the Oxus in the north

[1] This continues the story told in Chapter XIV. Oriental religions, Essenian activities, Hebrew writings are discussed in Chapter XVI.

[2] Hecatompylos (hundred gates) was founded by the Seleucids south of the southeast corner of the Caspian Sea; modern Damghan in northeastern Iran.

[3] The last Arsacid, Artabanos IV, was defeated in 226 by Artaxerxēs (Ardashir), founder of the Sassanid dynasty which governed until the Muslim conquest in 651. Note that the Arsacid dynasty ruled Parthia at about the same time as the Han dynasties ruled China (206 B.C.–A.D. 221).

[4] Their empire was so far extended to

to the Indian Sea in the south. This empire did not imperil Rome in the way that the Achaemenidian empire (which lasted until 330 B.C.) had jeopardized Greece, but it was a solid barrier in the Roman path to the East. The Parthian victories were partly due to their cavalry tactics; they combined excellent horsemanship with archery; in this they were the forerunners of the Mongol invaders.[5] They were temporarily checked on their northwest boundary by Tigranēs,[6] c. 88 and following years, but inflicted a terrible defeat on the Romans at Carrai[7] in 53 B.C., when Crassus[8] lost his army and his life. Further advance westward was stopped in 39–38 by two victories of Ventidius, Antony's legate, and by internal disunion, thanks to which Augustus was able, c. 20, to reëstablish peace on the Parthian frontier. There was still great rivalry between the Roman and Parthian empires, however, especially about the control of the Armenian kingdom, which they were equally anxious to "protect."

The essential difference between the Parthian empire and the Seleucid one which it partly replaced lies in the fact that the Seleucid rulers were of Greek origin and the main champions of Hellenism in Asia, while the Arsacids were Scythians or Asiatics, who were not at all hypnotized by Greek culture. As to international business, it is difficult to know whether the Parthians managed, or not, to improve it to their own advantage, for we know far too little about Hellenistic trade. As Tarn remarked, "Hellenistic trade is largely a palimpsest buried under that of the Roman Empire as the Hellenistic road system beneath the Roman; and one cannot merely argue backward from the better known Roman phenomena."[9] The main

---

the west that they had to establish new capitals at Ectabana and at Ctēsiphōn, on the Tigris (just south of modern Baghdād). Ectabana (now Hamadan) had been the capital of Median kings and later of the Achaemenids from whom the Arsacids claimed descent.

[5] See *Introduction*, vol. 3, p. 1865, about bowmen on horseback. Parthian skill in that kind of warfare is immortalized by the phrases "Parthian shot" and "Parthian arrow" (already in Virgil and Horace). The Parthian horsemen continued hoary traditions of Anatolia. The early Hittites used light battle chariots, and a Hittite treatise on horse training of the fourteenth century B.C. has come down to us (Volume 1, pp. 64, 85, 125). On the other hand, cavalry tactics were hardly developed by the Greeks and Romans, and few generals distinguished themselves as cavalry leaders; I can think only of Xenophōn (IV–2 B.C.) and Mark Antony, the triumvir (c. 83–30).

[6] Tigranēs I, the Great, was king of Armenia from 96 to 56. He increased its territory so much that he could call himself king of kings. His capital was Tigranocerta (Siirt, southeastern Turkey). He owed his first opportunities to the Parthians, yet fought them off later. He became as powerful a king as Mithridatēs the Great, whose daughter Cleopatra he married. He was Mithridatēs' ally and later his enemy.

[7] Carrai (or Carrae) in Osroēnē, northwest Mesopotamia, just south of Edessa. One might say that the battles of Cannae in 216 and Carrae in 53 were the worst disasters suffered by Roman armies (B.C.), respectively, west and east. Edessa and Carrae are now called Urfa and Haran.

[8] M. Crassus, called the triumvir because he was in 60 a member of the first triumvirate with Pompey and Caesar. (The second triumvirate was that of Antonius, Octavianus, and Lepidus in 43.)

[9] W. W. Tarn and G. T. Griffith, *Hellenistic civilisation* (London: Arnold, ed. 3, 1952), p. 249.

center of Oriental trade in the Mediterranean was still Alexandria, but were the Alexandrian warehouses replenished through Parthian roads or otherwise? The Arabian trade is out of the question, because this always came through the Red Sea, but were the Indian and Chinese caravans encouraged to travel across Parthian territories? A part of the Indian trade came across the Arabian desert or along the Red Sea and its importance may be judged from the wonderful growth of the Nabataean city of Petra.[10]

The main source of iron was the region of the Chalybes (south of the eastern Black Sea); the easiest path for that iron to reach the West was across the Black Sea and the Bosporos and the main warehouse was at Cyzicos, in the Sea of Marmara. A better kind of iron came from China across Soghdiana and the rest of the Parthian empire. Many things were imported from India, for example, cotton goods (muslins). The Chinese statesman, Chang Ch'ien (II–2 B.C.), traveled west as far as Soghdiana and Bactria, and by 115 "he had established regular intercourse between China and the West." [11] It is probable that the "Chinese silk road" did not function before that time and that the importation of Chinese silk remained small until much later; [12] indeed, lovers of silk in the Mediterranean world were using rather the wild silk of Cōs and Syria.[13]

It is hardly possible to give more precise information concerning the Eastern-Western trade which came across Parthia rather than across other roads south of that empire. Our doubts extend to cultural exchanges. Iranian influences, such as Mithraism, expanded south of the Caucasus, across Armenia and the Black Sea; most of them, however, had already reached the West and begun a new life there before the Parthian empire was constituted. The Chaldean astronomers did much of their work after the substitution of Arsacid for Seleucid rule, but it remained unknown in the West until our own times.[14] On the other hand, a modicum of Greek art traveled

[10] Petra in the northwest Arabian desert, halfway between the Dead Sea and the Gulf of 'Aqaba. It was my great privilege to spend a few days in the ruins of Petra in 1932. The presence of such vast and beautiful ruins in the middle of the desert is astounding. For details, see Michael Rostovtsev, *Caravan cities* (246 pp.; Oxford, 1932); the cities dealt with are Petra, Jerash, Palmyra, and Dura. On Rostovtsev's map (p. 2), the trade road feeding Petra comes from Ctēsiphōn; it might have come more directly from the Persian Gulf or from the Gulf of 'Aqaba. Beautiful color photographs of Petra in Julian Huxley, *From an antique land* (New York: Crown, 1954).

[11] Quoted from *Introduction*, vol. 1, p. 197, where many references will be found. "The West" in this context means the Parthian empire, but Chinese goods reaching that empire could find their way to Milētos, Petra, or Alexandria, and thence easily to Rome. For Chang Ch'ien, see also W. W. Tarn, *The Greeks in Bactria and India* (Cambridge, 1938).

[12] For details see Florence E. Day, *Ars Orientalis 1*, 232–245 (1954), an elaborate review of Adele Coulin Weibel, *Two thousand years of textiles* (New York: Pantheon, 1952).

[13] Similar to the tussah silk of India (see Florence Day, p. 236); it is produced by another kind of moth than the Chinese silk. On Coan silk, see Volume 1, p. 336.

[14] G. Sarton, "Chaldaean astronomy of the last three centuries B.C.," *Journal of the American Oriental Society* 75, 166–173 (1955).

eastward,[15] but the main drive of Greek art, all the way to Gandhara and further, did not occur until later (after Christ). The best monuments of Parthian art are coins; their use was a Greek idea, but the coins became more and more Oriental. All considered, it would seem that the Parthian empire was (at least in pre-Christian times) a barrier to the Hellenization of the East and the Orientalization of the West, rather than a channel for them. It was not a solid barrier, however, but a kind of grille or trellis permitting a little silk, as well as peaches and apricots, to move westward, and pomegranates to go east.

TRADE WITH INDIA AND CHINA

We have thus far considered only the eastern boundaries, but Oriental influences never ceased to pour in from Egypt. The Red Sea was a link between Egypt, on the one side, and Arabia and all the Indies on the other; the Upper Nile was a link with the Sudan, Ethiopia, and West Africa.[16] The monsoons never ceased to bring ships from the Malabar coast to Arabia or to Somaliland, whence Indian men, goods, and ideas moved northward to the Mediterranean world.

Most of our knowledge of the East-West exchanges of ideas and goods concerns later times, however. For example, a great many Roman coins have been found in India but almost all are post-Christian.[17]

Polybios. Much information on the Eastern countries came through the Greek historians, chiefly Polybios (II–1 B.C.). For example, in his account of the war between Antiochos the Great and Arsacēs (212–205), there is a good account of the very remarkable system of subterranean channels (*qanāt*) built in Iranian lands; he also described the wonderful palace of Ectabana.[18] The Greek and Roman readers of his *History* were given, if not by any means a full knowledge of the East, at least some vivid and unforgettable images.

Ptolemaios V Epiphanēs; the Rosetta Stone. To the young king Ptolemaios V Epiphanēs (210–180) we owe a contribution to modern orientalism that was as important and extraordinary as it was unconscious. A general council of Egyptian priests, assembled in Memphis in 196, passed a decree in his

[15] Sir Aurel Stein found in Parthia (more exactly near Fasa, in Persis or Fārs) a small marble head of a woman dating probably from the third or second century B.C. See his "Archaeological tour in the ancient Persis," *Iraq* 3, 111–225 (1936), p. 140.

[16] Anthony John Arkell, "Meroe and India," in *Aspects of archaeology presented to O. G. S. Crawford* (London: Edwards, 1951), pp. 32–38.

[17] See, in the same Crawford Festschrift, R. E. M. Wheeler, "Roman contact with India, Pakistan and Afghanistan," pp. 345–381, map of Roman coins in India, p. 374.

[18] Polybios, x, 27–28. His description of the *qanāt* can be read also in A. V. Williams Jackson's delightful book, *From Constantinople to the home of Omar Khayyam* (New York, 1911), p. 159. Some of the *qanāt* are functioning to this very day.

honor which was engraved on a stone (45 × 28 in.) in the demotic character, together with translations in the ancient hieroglyphics and in Greek. That inscription was lost to mankind for almost two thousand years, discovered by the French conquerors of Egypt, in 1799, at Rashīd,[19] surrendered in 1801 to the English, and taken to the British Museum. Its immense importance was realized at once by the French, especially by General Bonaparte, who ordered rubbings of it to be taken and distributed among the scholars of Europe; as soon as it was in England (in 1802), the English distributed casts and copies. Thus, the trilingual text could be investigated by many scholars and it revealed to them gradually the secret of the hieroglyphics; their interpretation was completed by the Frenchman, Jean François Champollion, in 1822.[20] As there is no bilingual inscription of equal importance, the science of Egyptology could not have been constituted without it.

The Rosetta stone was the very key to the understanding of one of the greatest civilizations of the past.

Mithridatēs VI the Great. The name of Mithridatēs the Great (I–1 B.C.) has frequently occurred in these pages, and I trust that it will engrave itself upon my reader's memory. In ancient times he was very famous and some of his admirers went so far as to compare him with Alexander; he did not deserve so much praise, but neither does he deserve present-day oblivion. He is one of the outstanding rulers of the past, one of the few "barbarians" who instilled fear into Roman souls.[21] As his name suggests, Mithridatēs was of Persian origin, but he had received a Greek education and knew many Oriental languages. He was a true Orientalist and perhaps the first whose name has come down to us. Not the very first, of course, because the diversity of languages was so great in the Oriental countries that an intelligent man could not help picking up many of them if he had to deal with various categories of people or was obliged to travel often enough from his home. Mithridatēs' international contacts were not restricted to the many nations of Eastern Asia; they extended to the Greek and Roman world in the West and, if we assume, as we may, that Chang Ch'ien's efforts had converged with his, as far as China in the East.

[19] Rashīd or Rosetta. The stone bearing the inscription is called the Rosetta Stone. Rashīd is in the Delta near Abuqīr, in whose offing the Battle of the Nile was fought in 1798, Nelson destroying the French fleet. It was also in Abuqīr that Bonaparte defeated the Turkish army in 1799 and that Sir Ralph Abercromby routed what remained of the French army, in 1801, and precipitated the evacuation of Egypt.

[20] E. A. Wallis Budge, *The Rosetta Stone* (8 pp., quarto; London, 1913). Champollion le Jeune, *Lettre à M. Dacier relative à l'alphabet des hiéroglyphes phonétiques* (52 pp., 4 pls.; Paris, 1822). Facsimile reprint with introduction by Henri Sottas (84 pp.; Paris, 1922).

[21] Two others have already been mentioned, Hannibal and Cleopatra. All three were finally subdued by Rome and driven to suicide, Hannibal in 183, Mithridatēs VI in 68, Cleopatra VII in 30 B.C.

THE END OF THE FIRST CENTURY

Mithridatēs died in 63 B.C. By the time of his death many more Greeks and Romans were becoming interested in Oriental matters.

Among the many books composed by Alexander of Milētos, nicknamed Polyhistor, were monographs on the Jews (*Peri Iudaiōn*), on Egypt, Syria, Babylonia, India. This Alexander had been brought to Rome as a prisoner of war in Sulla's time; he flourished in Rome and in Laurentum,[22] where he died in old age during the conflagration of his house. He had probably some knowledge of Eastern matters before his abduction, but he could and did obtain much more in the Roman libraries, public and private.

The *Library of history* (*Historiōn bibliothēcē*) that Diodōros of Sicily completed, c. 30 B.C., devoted as much attention to the East as to the West. For example, the first part, ending with the Trojan War, dealt with Egypt, Assyria, Media, Arabia, and the islands, including Panchaia,[23] in the Indian Sea.

Juba II, king of Mauretania, composed in Greek histories of Assyria and Arabia.

Nicholas of Damascus dedicated to his patron, Herod the Great, his ethnographic collection (*Ethōn synagōgē*) describing the manners and customs of many nations. His universal history treated the Achaemenidian empire, the Mithridatic Wars, the Jewish wars, and so forth.

The second half of Strabōn's geography concerned Egypt and Asia, and was richer in knowledge than the first. His lost history was more Asiatic than European.

It is clear that until the Augustan age (and for a few centuries afterward) humanism was still in large part Oriental, because the learned men were still as conscious of their Asiatic heritage as they were of their Greek or Western one. Egypt and Babylonia were as real to them as Crete, Greece, or Etruria; the Romans looked for the origin of their national traditions not in Rome but in Troy.

[22] Laurentum was one of the most ancient towns of Latium; it was near the sea and close to Lavinium, a religious center founded by Aeneas (?). The two places were later united into a single town.

[23] Panchaia is the island where Evhēmeros (IV-2 B.C.) found the "sacred inscriptions" (*Introduction*, vol. 1, p. 136).

XXIX

CONCLUSION

Let us ask ourselves what was accomplished during those three centuries of the Hellenistic age. We can easily measure that period of time; it is the same as that which elapsed between the landing of the Pilgrim Fathers on the shores of Massachusetts in 1620 and our own days. In this brief review we shall consider only scientific activities.

In the first place, scientific research was organized as it had never been before in the Museum of Alexandria, while the accumulation and transmission of knowledge was given splendid instruments in the libraries of Alexandria, of Pergamon, and later of Rome.

The main philosophic school was the Stoic, illustrated by Cleanthēs of Assos, Chrysippos of Soloi, Diogenēs the Babylonian, Panaitios and Poseidōnios of Rhodos. The best representatives of the New Academy were Carneadēs of Cyrēnē and Cicero. The leading defender of the Garden of Epicuros was another Roman, Lucretius. The tradition of the Lyceum was continued by Strabōn of Lampsacos, and Andronicos of Rhodos prepared the first scientific edition of Aristotle and Theophrastos.

This was a golden age of mathematics, the like of which did not occur again until the seventeenth century. Think of such a galaxy as Euclid of Alexandria, Archimēdēs of Syracuse, Eratosthenēs of Cyrēnē, Apollōnios of Pergē, Conōn of Samos, Hypsiclēs of Alexandria, Hipparchos of Nicaia, Theodosios of Bithynia, Geminos of Rhodos.

Much astronomical work was done not only by Greeks but also by Chaldeans. The outstanding men were Aristarchos of Samos and Seleucos the Babylonian, Hipparchos, Cleomēdēs, and Geminos. The greatest of all, one of the greatest of all times, was Hipparchos.

Physical investigations were carried out by Stratōn, Euclid, Aristarchos of Samos, Archimēdēs, Ctēsibios of Alexandria, Philōn of Byzantion. Sostratos of Cnidos built the Pharos, the lighthouse, which was one of the seven marvels of the ancient world. Greek and Roman engineers and architects built roads, aqueducts, harbors, and innumerable monuments. Vitruvius wrote the main architectural treatise of antiquity.

The methods of agriculture were explained by Cato the Censor, Mago of Carthage, Varro of Reate, Virgil of Mantova. Botanical studies were pursued by Cratevas and by Nicholas of Damascus.

Hērophilos of Chalcēdōn and Erasistratos of Ceōs were the creators of anatomy and physiology. The medical record is not so good, yet there were a number of distinguished physicians – Archagathos of Rome, Serapiōn of Alexandria, Asclēpiadēs of Bithynia, Themisōn of Laodiceia, Hēraclēdēs of Tarentum, Apollōnios of Cition, and Antonius Musa.

Geographic studies were cultivated by Eratosthenēs, Cratēs of Mallos, Hipparchos, Poseidōnios, Isidōros of Charax. Strabōn of Amaseia composed the most elaborate description of the world; Caesar and Agrippa ordered a survey of it which was completed in 12 B.C.

The main Greek historians were the Arcadian Polybios and Poseidōnios; the main Latin ones, Caesar, Sallust and Livy. The legendary background of Roman history was recreated in Virgil's *Aeneid*.

Greek grammar was invented and the foundations of Greek philology were laid by Zēnodotos of Ephesos, Aristophanēs of Byzantion, Aristarchos of Samothracē, Cratēs of Mallos, Dionysios Thrax, Dionysios of Halicarnassos. Latin philology was developed by Varro and Verrius Flaccus.

The main achievement in the field of international literature and religion was the *Septuaginta*, the translation of the Old Testament from Hebrew into Greek.

Surely this is an astonishing record, equally astonishing in its wealth and in its scope. Would that we had done as well in the three centuries from the "Mayflower" until now. The record is even more remarkable than it seems to be, if we remember the catastrophies, wars, revolutions that jeopardized it almost without interruption.

The political conflicts and wars remained essentially the same during this period and later, but the religious conflict was deeply modified.

During the whole Hellenistic age there flourished in close rivalry three kinds of popular religion – first, the old Greek paganism; second, Judaism; and third, various Oriental mystery cults, such as the cults of Mithras, Cybelē and Attis, Isis and Osiris. The appearance of a new incomprehensible mystery, that of Jesus Christ, and its gradual triumph characterized an entirely new period.

GENERAL BIBLIOGRAPHY

Heath, Sir Thomas Little (1861–1940), *History of Greek mathematics* (2 vols; Oxford, 1921) [*Isis 4*, 532–535 (1921–22)].

—— *Manual of Greek mathematics* (568 pp.; Oxford, 1931) [*Isis 16*, 450–451 (1931)].

—— *Greek astronomy* (250 pp.; London, 1932) [*Isis 22*, 585 (1934–35)].

Isis. International review devoted to the history of science and civilization. Official journal of the History of Science Society. Founded and edited by George Sarton (43 vols., 1913–1952); vols. 44–48 (1953–1957), edited by I. Bernard Cohen.

The many references to *Isis* in this volume are generally made to complete in the briefest manner the information given about this or that book or memoir. From them the reader may obtain quickly, if he wishes, either a critical review of the book or some other additional information the development of which would take too much space.

Klebs, Arnold C. (1870–1943), "Incunabula scientifica et medica," *Osiris 4*, 1–359 (1938).

Osiris. Commentationes de scientiarum et eruditionis historia rationeque. Edidit Georgius Sarton (12 vols. Bruges, 1936–56); vol. 12, edited by Canon A. Rome and the Abbé J. Mogenet.

Oxford classical dictionary (998 pp.; Oxford: Clarendon Press, 1949).

Pauly-Wissowa, *Real-Encyclopädie der klassischen Altertumswissenschaft* (Stuttgart, 1894 ff.).

Sarton, George, *Introduction to the history of science* (3 vols. in 5; Baltimore: Williams and Wilkins, 1927–1948). Referred to as *Introduction.*

—— *The appreciation of ancient and medieval science during the Renaissance, 1450–1600* (243 p.; Philadelphia: University of Pennsylvania Press, 1955). Abbreviated *Appreciation.*

—— *A history of science. Ancient science through the Golden Age of Greece* (xxvi + 646 pp., 103 ills.; Cambridge: Harvard University Press, 1952). Referred to as Volume 1.

—— *Horus: A guide to the history of science* (Waltham, Mass.: Chronica Botanica, 1952).

Tannery, Paul (1843–1904), *Mémoires scientifiques* (17 vols.; Paris, 1912–1950); see *Introduction*, vol. 3, p. 1906.

Tarn, W. W., *Hellenistic civilisation*; third edition, revised with G. T. Griffith (xi + 372 pp.; London: Edward Arnold, 1952).

INDEX